AF595057

Crops and Vegetation Monitoring with Remote/Proximal Sensing

Crops and Vegetation Monitoring with Remote/Proximal Sensing

Editors

Kenji Omasa
Shan Lu
Jie Wang

Basel • Beijing • Wuhan • Barcelona • Belgrade • Novi Sad • Cluj • Manchester

Editors
Kenji Omasa
Takasaki University of Health and Welfare
Takasaki
Japan

Shan Lu
Northeast Normal University
Changchun
China

Jie Wang
China Agricultural University
Beijing
China

Editorial Office
MDPI
St. Alban-Anlage 66
4052 Basel, Switzerland

This is a reprint of articles from the Special Issue published online in the open access journal *Remote Sensing* (ISSN 2072-4292) (available at: https://www.mdpi.com/journal/remotesensing/special_issues/761R5154T9).

For citation purposes, cite each article independently as indicated on the article page online and as indicated below:

Lastname, A.A.; Lastname, B.B. Article Title. *Journal Name* **Year**, *Volume Number*, Page Range.

ISBN 978-3-0365-9446-0 (Hbk)
ISBN 978-3-0365-9447-7 (PDF)
doi.org/10.3390/books978-3-0365-9447-7

Contents

About the Editors

Kenji Omasa

Kenji Omasa received a Ph.D. in Engineering from the University of Tokyo, Japan. Kenji was a Department of Agricultural and Life Sciences professor at the University of Tokyo from 1999 to 2016 after working as the head of Biotron and Environment Plant Science at the National Institute for Environmental Studies, Japan, from 1976 to 1999. Kenji was also president of the Agricultural Academy of Japan from 2018 to 2022 and an executive board member of the Science Council of Japan from 2014 to 2017. Kenji is currently a professor emeritus at the University of Tokyo, an emeritus researcher at the National Institute for Environmental Studies, the dean of the Faculty of Agriculture at Takasaki University of Health and Welfare, the vice president of the Association of Japanese Agricultural Scientific Societies, and a board member of Japan Academy of Life Sciences. Kenji is interested in imaging cell level to plants, remote sensing, smart agriculture, modeling of ecosystems, and analysis of global change effects on plants and ecosystems. To date, Kenji has contributed to 650 scientific publications, 250 of which are in journals with references, and 400 are in books, reviews, and other publications. Kenji has received 38 academic awards, including a Medal with Purple Ribbon from the Government of Japan in 2013.

Shan Lu

Shan Lu graduated with an M.Sc. in Physical Geoscience from the Northeast Normal University, China, in 2002 and a Ph.D. in Biological Environmental Engineering from the University of Tokyo in 2007. She is now a Professor at the School of Geographical Sciences at Northeast Normal University, China. Shan Lu has participated in twenty scientific projects, coordinating nine projects. Her current research interests include hyperspectral remote sensing, multiangular remote sensing, polarized remote sensing, and their applications in biophysical parameter assessment of vegetation and soils. She has contributed 45 papers, and 33 are in international journals with references. She is also a reviewer for different papers in international conferences and journals.

Jie Wang

Jie Wang graduated with a Ph.D. in Ecology from the University of Oklahoma, USA, in 2017 and received her M.Sc. and B.S. degrees in Geography Information Systems (GIS) from the Chinese Academy of Sciences and Nanjing University, China, respectively. Now, she is a professor at the College of Grassland Science and Technology at the Chinese Agricultural University, China. Her research interests include land use/cover change and carbon cycle, multi-source remote sensing, and their applications in biophysical parameter assessment of vegetation and soils. She has participated in more than 10 scientific projects as PI or Co-PI. She has published over 40 papers and contributed to 35 international journals.

Article

Potential of Satellite Spectral Resolution Vegetation Indices for Estimation of Canopy Chlorophyll Content of Field Crops: Mitigating Effects of Leaf Angle Distribution

Xiaochen Zou [1,*], Jun Jin [1] and Matti Mõttus [2]

1 Technology Innovation Center for Integration Applications in Remote Sensing and Navigation, Ministry of Natural Resources, School of Remote Sensing and Geomatics, Nanjing University of Information Science & Technology, Nanjing 210044, China
2 VTT Technical Research Centre of Finland Ltd., P.O. Box 1000, FI-02044 Espoo, Finland
* Correspondence: xiaochen.zou@nuist.edu.cn

Abstract: Accurate estimation of canopy chlorophyll content (CCC) is critically important for agricultural production management. However, vegetation indices derived from canopy reflectance are influenced by canopy structure, which limits their application across species and seasonality. For horizontally homogenous canopies such as field crops, LAI and leaf inclination angle distribution or leaf mean tilt angle (MTA) are two biophysical characteristics determining canopy structure. Since CCC is relevant to LAI, MTA is the only structural parameter affecting the correlation between CCC and vegetation indices. To date, there are few vegetation indices designed to minimize MTA effects for CCC estimation. Herein, in this study, CCC-sensitive and MTA-insensitive satellite broadband vegetation indices are developed for crop canopy chlorophyll content estimation. The most efficient broadband vegetation indices for four satellite sensors (Sentinel-2, RapidEye, WorldView-2 and GaoFen-6) with red edge channels were identified (in the context of various vegetation index types) using simulated satellite broadband reflectance based on field measurements and validated with PROSAIL model simulations. The results indicate that developed vegetation indices present strong correlations with CCC and weak correlations with MTA, with overall R^2 of 0.76–0.80 and 0.84–0.95 for CCC and R^2 of 0.00 and 0.00–0.04 in the field measured data and model simulations, respectively. The best vegetation indices identified in this study are the soil-adjusted index type index SAI (B6, B7) for Sentinel-2, Verrelts's three-band spectral index type index BSI-V (NIR1, Red, Red Edge) for WorldView-2, Tian's three-band spectral index type index BSI-T (Red Edge, Green, NIR) for RapidEye and difference index type index DI (B6, B4) for GaoFen-6. The identified indices can potentially be used for crop CCC estimation across species and seasonality. However, real satellite datasets and more crop species need to be tested in further studies.

Keywords: broadband vegetation indices; chlorophyll content; leaf angle distribution; Sentinel-2; WorldView-2; RapidEye; GaoFen-6

Citation: Zou, X.; Jin, J.; Mõttus, M. Potential of Satellite Spectral Resolution Vegetation Indices for Estimation of Canopy Chlorophyll Content of Field Crops: Mitigating Effects of Leaf Angle Distribution. *Remote Sens.* **2023**, *15*, 1234. https://doi.org/10.3390/rs15051234

Academic Editors: Kenji Omasa, Shan Lu and Jie Wang

Received: 15 January 2023
Revised: 18 February 2023
Accepted: 22 February 2023
Published: 23 February 2023

1. Introduction

Foliar chlorophyll content is a very important photosynthetic pigment that governs light absorption and conversion to chemical energy [1,2]. Canopy chlorophyll content (CCC), defined as the total amount of chlorophyll in plant leaves per unit ground area [3,4], is related to plant photosynthetic productivity and light use efficiency [5], and contributes to the vegetation response to the environment [6,7]. It is usually calculated as the product of leaf chlorophyll content (C_{ab}) and leaf area index (LAI) [8,9], defined as the total of the single-sided leaf area per area unit of horizontal ground [10]. From the perspective of agricultural applications, the instantaneous value and dynamics of CCC indicate the crop growth potential and actual development [11–13]. CCC is also strongly correlated with plant nutritional status and crop yield [8,14–17], so it needs to be accurately determined for precision agriculture.

CCC drives visible light absorption and transmission within a canopy and hence it can be detected by optical remote sensing technology [8]. Instead of laborious time-consuming regional scale in situ measurements, spatially and temporally resolved CCC can be determined from remote sensing data. The numerous approaches developed for this [18,19] can be categorized into two general types, physically- and empirically-based methods. Physically-based CCC estimation approaches mainly rely on canopy radiative transfer models to determine the relationship between CCC and radiometric signals [20,21]. The empirical approach is to establish a statistical relationship between the measured CCC and observed spectral features [4,22]. One of the commonly used empirical approaches is via the use of spectral vegetation indices, mathematical combinations of remote sensing instrument band readings designed to enhance the sensitivity of the outcome to variables of interest and to minimize the impact of other factors [23–25].

Due to its simplicity, adaptability and computational efficiency, many vegetation indices have been designed to estimate CCC [26], such as the MERIS terrestrial chlorophyll index (MTCI) [27], normalized difference red edge index (NDRE) [28] and red edge chlorophyll index ($CI_{red\text{-}edge}$) [3]. CCC is related to specific spectral features making it easier to detect using narrow-band indices [2,11,29,30]. Specifically, chlorophyll is visible in the reflectance spectrum between 680 and 760 nm (known as the red edge) [31,32], which can be efficiently utilized for estimating CCC [33]. For large-scale practical applications, the use of low-cost (or in many cases, free for the end user) spatially and temporally continuous multispectral satellite data simplify the design of the vegetation index and makes estimation of CCC feasible regionally or globally [9]. Fortunately, modern multispectral satellite sensors are equipped with red edge bands, such as Sentinel-2, RapidEye, WorldView-2 and GaoFen-6. Sentinel-2-based vegetation indices have been assessed for CCC estimation for several crop species, including potato, soybean, maize and winter wheat [33–35], but RapidEye, WorldView-2 and GaoFen-6 have received little attention in the estimation of crop CCC.

In addition to leaf optical properties, affected strongly by chlorophyll absorption in the visible part of the spectrum, remotely sensed canopy reflectance is affected by ground (soil) and canopy structure [36–40]. The canopy of field crops is usually assumed to be horizontally uniform, which means that its architecture can be simply characterized by the amount of leaves and their orientations within a canopy. These can be characterized using two physical parameters—LAI and leaf inclination angle distribution or leaf mean tilt angle (MTA), the leaf area-weighted average of all the leaf inclination angles in a canopy. To a large extent, MTA is a species-specific characteristic, and it has been reported to have more variation among species than within species [41–44]. In addition, MTA is affected by biome, genotype and growth conditions. As LAI is included in the computation of CCC, MTA is the only independent canopy structure parameter affecting the relationship between CCC and canopy reflectance in horizontally homogeneous canopies.

There are only a few studies on the removal or minimization of the influence of MTA on CCC estimation from satellite remote sensing data [45], mainly because of a lack of measured MTA and corresponding spectral observations, either true satellite measurements or the equivalent hyperspectral data resampled to simulate satellite spectral bands. To address this shortcoming, the objectives of this study are to (1) evaluate the performance of four multispectral satellites with red edge channels for CCC estimation of field crops with diverse canopy architectures using vegetation indices and (2) develop CCC-sensitive and MTA-insensitive vegetation indices for CCC estimation.

2. Materials and Methods

2.1. Study Area and Field Measurements

The empirical datasets acquired in this study include airborne imaging spectroscopy data acquisitions and field measurements at Viikki Experimental Farm (60.224°N, 25.021°E),

Helsinki, Finland (Figure 1). The experimental area is located in southern Finland with a mean annual temperature of 6 °C. The study site area is approximately 4 km × 4 km with an altitude no more than 10 m above sea level. The study site encompasses six crop species, faba bean, narrow-leafed lupin, turnip rape, wheat, barley and oat. Three crop biophysical and biochemical parameters were collected including LAI, C_{ab} and MTA from 162 plots. The maximum plot size is 50 m × 12 m and the minimum is 2 m × 10 m. A detailed description of the field plots is given in [46].

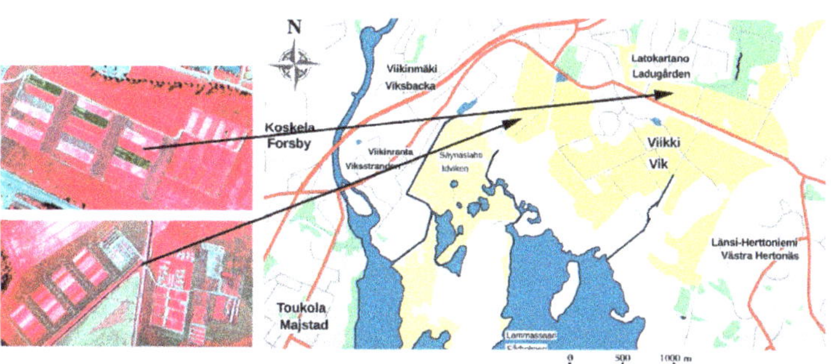

Figure 1. A map of the field site and aerial imagery of field plots.

Canopy MTA was measured using the photographic method developed by [47] and validated and extended to field crops [46,48]. Leaf inclination angle measurements were taken on 6th July 2012. The photographs of leaves were acquired outside of the field plot approximately one meter away from the plot edge with a Nikon D1X digital camera. The photograph of the canopy was acquired using the camera attached and leveled on a tripod during acquisitions under windless conditions. The camera height was adjusted depending on crop height, ranging from 30 cm to 50 cm to cover the whole plant vertically. With the help of ImageJ software, leaf angles were visually measured from photographs for each species. Leaf inclination is defined so that increasing MTA indicates more vertical leaves. As suggested in [49], 75–100 leaves are sufficient to represent the leaf inclination angle distribution. This method keeps the MTA measurement error within 4° [48]. Full details of the method are given by [46].

The leaf area index of field crops was indirectly measured using a SunScan SS1 probe (Delta-T Devices). The 1 m long SunScan probe with 64 radiation microsensors was inserted below the crop canopy from the plot edge orthogonally to plant rows to minimize the row effects. An additional beam fraction sensor recorded the incident direct and diffuse downwelling irradiances simultaneously outside of field plots. The leaf area index was calculated through a canopy radiative transfer (RT) model implemented in the SunScan device. A one-parameter ellipsoidal leaf angle distribution model was assumed in this RT model, and the leaf clumping effect was not considered for this instrument. The ellipsoidal LAD model input parameter χ can be derived using Equation (16) in [50] as:

$$\chi = -3 + \left(\frac{\text{MTA}}{553}\right)^{-0.6061} \tag{1}$$

MTA was assumed to be a species-specific characteristic. The details of the LAI calculation algorithm are fully described in SunScan user manual version 2.0.

The C_{ab} of leaves was measured with a portable SPAD-502 device in the field. Based on the size of the field plot, 15–30 leaves were randomly sampled. This device acquired dimensionless readings that were converted into absolute C_{ab} values using the formula [51,52]:

$$C_{ab}\left(\mu\text{g cm}^{-2}\right) = 0.0893\left(10^{\text{SPAD}^{0.625}}\right) \tag{2}$$

which has achieved a strong correlation between laboratory-determined C_{ab} and SPAD-502 readings for field crops (soybean, maize and barley). After the LAI and C_{ab} were acquired, the canopy CCC was calculated as:

$$\text{CCC } (\mu\text{g cm}^{-2}) = C_{ab} \times \text{LAI} \quad (3)$$

Airborne imaging spectroscopy data of the study plots were acquired using an AISA Eagle II spectrometer on 25 July 2011 under cloudless conditions between 09:36 and 10:00 local time. The instrument provided data in 64 spectral bands covering the spectral range between 400 and 1000 nm, and the resolution of the spectra was between 9 and 10 nm. The average flight altitude was 600 m and achieved a ground spatial resolution of approximately 0.4 m. Radiometric correction of the raw image was completed using Specim CaliGeo software. The radiometrically calibrated imagery was georectified using Parge (ReSe Applications Schläpfer) by means of ground control points and the navigation data acquired during the flight. Atmospheric correction was carried out with ATCOR-4 (ReSe Applications Schläpfer). The plot scale spectra were visually extracted from each plot and averaged. A detailed description of imaging spectroscopy data acquisition is given in [46].

2.2. Validation Datasets from the PROSAIL Model Simulation

Canopy reflectance was simulated with the widely used PROSAIL model, which is a coupled model of the leaf reflectance model PROSPECT-5 [53] and canopy reflectance model SAILH [54,55]. In the PROSAIL model, homogeneous randomly distributed leaves are presumed to form a one-dimensional turbid medium [54], which is suitable for simulating the canopy reflectance of field crops. PROSPECT-5 simulates leaf reflectance and transmittance from 400 nm to 2500 nm as a function of six input parameters: C_{ab}, the mesophyll structure parameter (N), carotenoid content (C_{car}), brown pigment content (C_{brown}), equivalent water thickness (C_w), and dry matter content (C_m). In addition to leaf optical properties, eight canopy structural parameters were used as inputs for PROSAIL: LAI, MTA (assuming an ellipsoidal distribution), solar zenith angle (t_s), observer zenith angle (t_o), relative azimuth angle (φ), soil reflectance, fraction of diffuse radiation (skyl) and hot spot size parameter. The PROSAIL model inputs, summarized in Table 1, were set in accordance with in-situ measurement conditions and scientific literature: C_{ab} was set between 20 and 90 μg cm^{-2}, in steps of 5 μg cm^{-2}, C_{car} was set to 20% of the C_{ab} value based on the LOPEX93 dataset [56], C_w was fixed to 0.001, N was fixed to 1.55—a mean value for various crops [57], C_m was set to 0.005 g cm^{-2}—the mean value of the six crop species [58–61], C_{brown} was fixed to 0 assuming no senescent leaves during the measurements. LAI was set between 1 and 5 with a 0.1 interval, and MTA ranged from 20 to 70 with a 2-degree interval. Based on the conditions of airborne imaging spectroscopy data acquisition, the three illumination and view geometry parameters t_s, t_o and φ were set to 49.4°, 9.0° and 90.0°, respectively. The 6S atmosphere radiative transfer model was used to calculate the parameter skyl [62]. The hot spot parameter was fixed to 0.01 and the soil reference was measured using a handheld Analytical Spectral Devices spectroradiometer (ASD). In total, 15,990 canopy spectra between 400 nm and 1000 nm were simulated and resampled to satellite broadband reflectance.

Table 1. The variable settings of the PROSAIL model.

Model	Variable	Value or Range
PROSPECT	Leaf structure parameter (N)	1.55
	Leaf chlorophyll content (C_{ab})	20:5:90 μg cm^{-2}
	Equivalent water thickness (C_w)	0.001 cm
	Dry matter content (C_m)	0.005 g cm^{-2}
	Brown pigment content (Cbp)	0 μg cm^{-2}
	Carotenoid content (C_{car})	Linked to C_{ab} (0.2 × C_{ab}) μg cm^{-2}
SAIL	Leaf area index (LAI)	1, 1.1, ... , 5.0
	Leaf mean tilt angle (MTA)	20, 22, ... , 70°
	Hot spot size	0.01
	Solar zenith angle (t_s)	49.4°
	Observer zenith angle (t_o)	9°
	Azimuth angle (φ)	90°
	Fraction of diffuse radiation (skyl)	6S model (Wm^{-2} nm^{-1})
	Soil reflectance	ASD measurement

2.3. Satellite Broadband Reflectance Simulations

The airborne imaging spectroscopy data and PROSAIL model-simulated canopy reflectance in Visible to NIR spectral region (VNIR) were resampled to the broadband resolution of selected satellite sensors that had red edge channels: Sentinel-2, RapidEye, WorldView 2 and GaoFen-6. The MultiSpectral Instrument (MSI) of Sentinel-2 has 10 bands with three different spatial resolutions (10–60 m) in VNIR, including two red edge channels. RapidEye is a commercial Earth observation mission that offers high spatial resolution (6.5 m) imagery in five bands. The WorldView-2 satellite acquires very high spatial resolution (1.84 m) imagery in eight bands. The GaoFen-6 satellite, launched in 2018, has a multispectral sensor with 16 m spatial resolution in eight bands. The spectral response functions (SRFs, Figure A1 and Table A1) of the four multispectral instruments were used to convolve the modeled and measured narrow-band reflectance. The resampled four satellite broadband reflectance from the mean spectra of six crop species are presented in Figure 2.

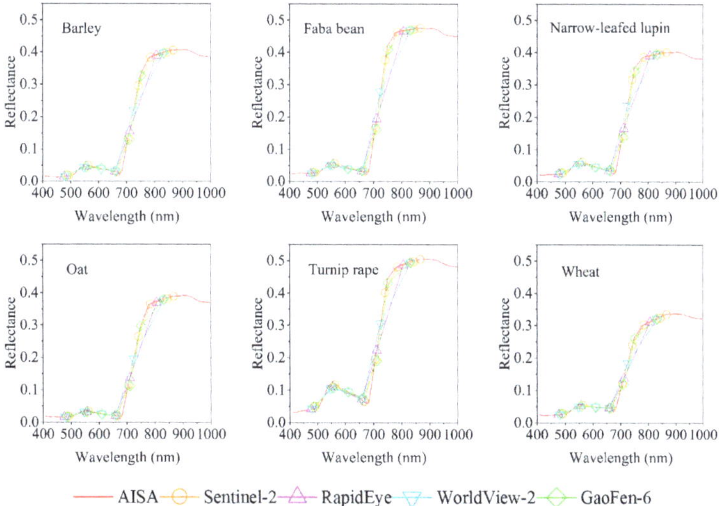

Figure 2. Mean reflectance spectra of the six crop species used in the study: the four simulated satellite broadband spectra and AISA spectra.

2.4. Tested Vegetation Indices

A wide range of vegetation indices has been used to estimate vegetation canopy chlorophyll content, a product of LAI and C_{ab}. In this study, twelve widely used vegetation indices that have been used to estimate chlorophyll content or LAI were evaluated (Table 2). Some of these use reflectance in VNIR: the normalized difference vegetation index (NDVI), enhanced vegetation index (EVI) and its two-band version (EVI2), optimized soil-adjusted vegetation index (OSAVI), renormalized difference vegetation index (RDVI), pigment-specific normalized difference index (PSND) and transformed chlorophyll absorption reflectance index/OSAVI (TCARI/OSAVI). These indices are used to extract one or more vegetation parameters, such as LAI, canopy cover fraction, biomass and pigment content. Other indices have been formulated with the red edge bands: the red-edge transformed chlorophyll absorption reflectance index/OSAVI (TCARI/OSAVI$_{\text{red edge}}$), which has a red edge band instead of the NIR band, the MERIS terrestrial chlorophyll index (MTCI), two versions of normalized difference red-edge vegetation indices (NDRE1 and NDRE2, see Table 2 for details) and the red-edge chlorophyll index (CI$_{\text{red edge}}$) (rows 1–12 in Table 2). These indices were used to extract chlorophyll content in previous studies. To identify the CCC-sensitive and MTA-insensitive band combinations, eleven general index types were selected from the literature next, including six two-band and five three-band formulations (Table 2): ratio index (RI), normalized difference index (NDI), difference index (DI), soil adjusted index (SAI), modified simple ratio (MSR) and modified soil adjusted index (MSAI), triangular index (TI), Gitelson three-band index (Git), Tian's three-band index (BSI-T), Verrelts's three-band index (BSI-V) and Wang's three-band index (BSI-W) (rows 13–23 in Table 2). When calculating TI, the central wavelength of the broadband was used to calculate the wavelength difference.

Table 2. The vegetation indices used in this study: indices 1–12 are existing indices with fixed wavelengths; 13–23 are general indices with wavelengths found by optimization.

No	Index	Abbreviation	Formulation	Reference
1	Normalized difference vegetation index	NDVI	$\frac{R_{NIR}-R_{Red}}{R_{NIR}+R_{Red}}$	[63]
2	Enhanced vegetation index	EVI	$\frac{2.5(R_{NIR}-R_{Red})}{R_{NIR}+6R_{Red}-7.5R_{Blue}+1}$	[64]
3	Two-band enhanced vegetation index	EVI2	$\frac{2.5(R_{NIR}-R_{Red})}{R_{NIR}+2.4R_{Red}+1}$	[65]
4	Optimized soil-adjusted vegetation index	OSAVI	$\frac{1.16(R_{NIR}-R_{Red})}{R_{NIR}+R_{Red}+0.16}$	[66]
5	Renormalized difference vegetation index	RDVI	$\frac{R_{NIR}-R_{Red}}{\sqrt{R_{NIR}+R_{Red}}}$	[67]
6	Pigment-specific normalized difference index	PSND	$\frac{R_{NIR}-R_{Blue}}{R_{NIR}+R_{Blue}}$	[68]
7	Transformed chlorophyll absorption reflectance index/OSAVI	TCARI/OSAVI	$\frac{3\left[(R_{NIR}-R_{Red})-0.2(R_{NIR}-R_{Green})\frac{R_{NIR}}{R_{Red}}\right]}{(1+0.16)*\frac{R_{NIR}-R_{Red}}{R_{NIR}+R_{Red}+0.16}}$	[66,69]
8	Red-edge Transformed chlorophyll absorption reflectance index/OSAVI	TCARI/OSAVI$_{\text{red edge}}$	$\frac{3\left[(R_{RE1}-R_{Red})-0.2(R_{RE1}-R_{Green})\frac{R_{RE1}}{R_{Red}}\right]}{(1+0.16)\frac{R_{NIR}-R_{Red}}{R_{NIR}+R_{Red}+0.16}}$	[70]
9	MERIS terrestrialchlorophyll index	MTCI	$\frac{R_{RE2}-R_{RE1}}{R_{RE1}-R_{Red}}$	[27]
10	Normalized difference red-edge version 1	NDRE1	$\frac{R_{RE2}-R_{RE1}}{R_{RE2}+R_{RE1}}$	[28]
11	Normalized difference red-edge version 2	NDRE2	$\frac{R_{RE3}-R_{RE1}}{R_{RE3}+R_{RE1}}$	[71]
12	Red-edge chlorophyll index	CI$_{\text{red edge}}$	$\frac{R_{RE3}}{R_{RE1}}-1$	[72]
13	Ratio index	RI	$\frac{R_{\lambda1}}{R_{\lambda2}}$	[57]
14	Normalized difference index	NDI	$\frac{R_{\lambda1}-R_{\lambda2}}{R_{\lambda1}+R_{\lambda2}}$	[73]
15	Difference index	DI	$R_{\lambda1}-R_{\lambda2}$	[74]
16	Soil adjusted index	SAI	$\frac{1.5\,(R_{\lambda1}-R_{\lambda2})}{(R_{\lambda1}+R_{\lambda2}+0.5)}$	[75]
17	Modified simple ratio index	MSR	$\left[\frac{R_{\lambda1}}{R_{\lambda2}}-1\right]\times\left[\sqrt{\frac{R_{\lambda1}}{R_{\lambda2}}}+1\right]^{-1}$	[57]
18	Modified soil adjusted index	MSAI	$\frac{2R_{\lambda1}+1-\sqrt{(2R_{\lambda1}+1)^2-8(R_{\lambda1}-R_{\lambda2})}}{2}$	[76]

Table 2. *Cont.*

No	Index	Abbreviation	Formulation	Reference
19	Triangular index	TI	$0.5\,[(\lambda_2-\lambda_1)(R_{\lambda3}-R_{\lambda1})-(\lambda_3-\lambda_1)(R_{\lambda2}-R_{\lambda1})]$	[77]
20	Gitelson's three-band	Git	$\left(\frac{1}{R_{\lambda1}}-\frac{1}{R_{\lambda2}}\right)*R_{\lambda3}$	[78]
21	Tian's three-band spectral index	BSI-T	$\frac{R_{\lambda1}-R_{\lambda2}-R_{\lambda3}}{R_{\lambda1}+R_{\lambda2}+R_{\lambda3}}$	[79]
22	Verrelts's three-band spectral index	BSI-V	$\frac{R_{\lambda1}-R_{\lambda3}}{R_{\lambda2}+R_{\lambda3}}$	[80]
23	Wang's three-band spectral index	BSI-W	$\frac{R_{\lambda1}-R_{\lambda2}+2R_{\lambda3}}{R_{\lambda1}+R_{\lambda2}-2R_{\lambda3}}$	[81]

The bands used for the test vegetation index calculations for Sentinel-2 are R_{Red} (B4), R_{Green} (B3), R_{Blue} (B2), R_{RE1} (B5), R_{RE2} (B6), R_{RE3} (B7) and R_{NIR} (B8); for GaoFen-6 R_{Red} (B3), R_{Green} (B2), R_{Blue} (B1), R_{RE1} (B5), R_{RE2} (B6) and R_{NIR} (B4).

2.5. Statistical Analysis

The relationships between the CCC, MTA and vegetation indices were evaluated using the coefficients of determination (R^2). The R^2 between vegetation indices and CCC is indicated as $R^2{}_{\text{CCC}}$ and that relationship with MTA is indicated as $R^2{}_{\text{MTA}}$. The difference between $R^2{}_{\text{CCC}}$ and $R^2{}_{\text{MTA}}$ is used for the quantitative assessment of the CCC-sensitive and MTA-insensitive vegetation indices. The correlations between the CCC, MTA and individual band reflectance were also calculated.

3. Results

3.1. Responses of Satellite Broadband Reflectance to MTA

For illustration, the responses of individual broadband reflectance bands to MTA from PROSAIL model simulations are presented at four combinations of high and low LAI and Cab in Figure 3. At two low LAI conditions (LAI = 1), reflectance in the NIR region had a strong negative correlation with MTA for all the satellites. At the same time, MTA presented a medium to strong negative correlation with reflectance in the red edge depending on the satellite sensors. In the visible region, MTA had little effect on reflectance when Cab was high (Cab = 90). At two high LAI conditions (LAI = 5), MTA presented strong negative correlations with reflectance in NIR, and this correlation was enhanced when MTA varied between 60 and 70°. The determination coefficients between CCC, MTA and individual band reflectance using field-measured and model-simulated datasets were presented in Table A1. Generally, the bands with the strongest correlation to CCC appeared in visible regions, and those with the strongest correlations to MTA appeared in red edge and NIR regions.

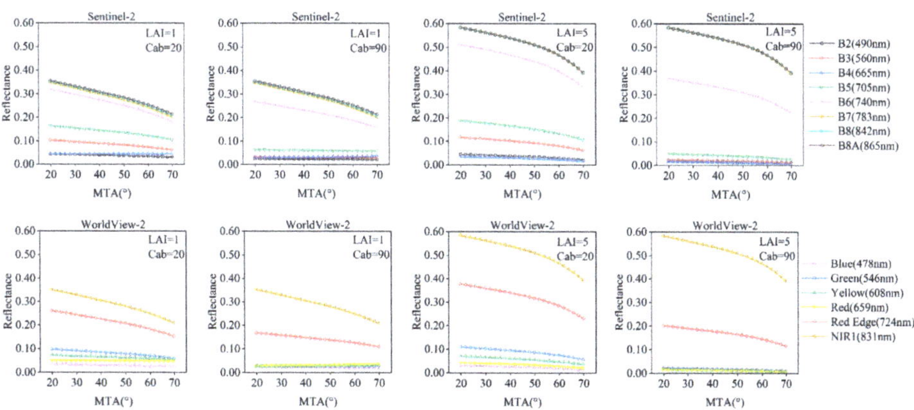

Figure 3. *Cont.*

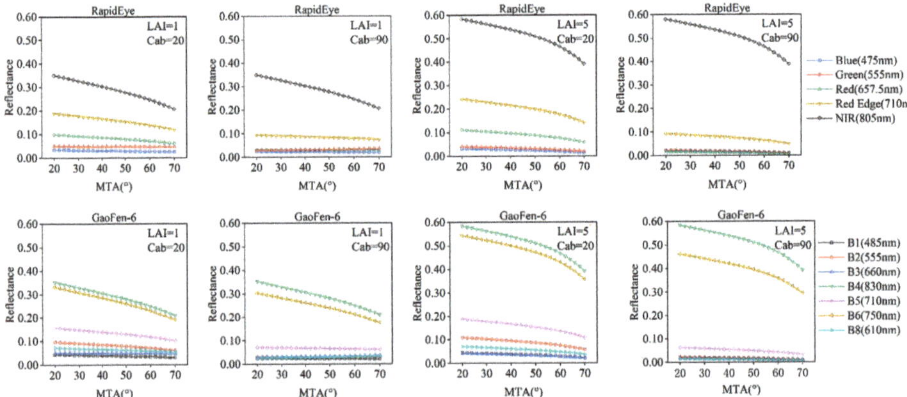

Figure 3. Responses of satellite broadband reflectance to leaf mean tilt angle (MTA) from PROSAIL model simulation for four combinations of high and low LAI and Cab: low LAI and low Cab (**left column**), low LAI and high Cab (**second column**), high LAI and low Cab (**third column**) and high LAI and high Cab (**right column**) for Sentinel-2 (**top row**), WorldView-2 (**second row**), RapidEye (**third row**), and GeoFen-6 (**bottom row**).

3.2. Performance of Existing Vegetation Indices

The relationships between CCC, MTA and the tested vegetation indices derived from four broadband satellites are presented in Table 3, including both the field-measured dataset and model simulations. In general, model-simulated dataset-derived VIs had stronger correlations with CCC than those of the field-measured dataset.

Table 3. Coefficient of determination (R^2) between canopy chlorophyll content (CCC), leaf mean tilt angle (MTA) and tested vegetation indices.

Dataset	Index	Sentinel-2		WorldView2		RapidEye		GaoFen-6	
		R^2_{CCC}	R^2_{MTA}	R^2_{CCC}	R^2_{MTA}	R^2_{CCC}	R^2_{MTA}	R^2_{CCC}	R^2_{MTA}
Measurement	NDVI	0.46	0.24	0.47	0.23	0.46	0.24	0.47	0.23
	EVI	0.16	0.65	0.18	061	0.17	0.63	0.17	0.62
	EVI2	0.19	0.63	0.19	0.60	0.18	0.62	0.19	0.60
	OSAVI	0.32	0.46	0.32	0.43	0.31	0.45	0.32	0.43
	RDVI	0.22	0.56	0.23	0.55	0.22	0.57	0.23	0.55
	PSND	0.52	0.17	0.50	0.18	0.49	0.19	0.52	0.17
	TCARI/OSAVI	0.31	0.40	0.32	0.38	0.29	0.41	0.33	0.37
	TCARI/OSAVI$_{red\ edge}$	0.31	0.18	0.20	0.48	0.27	0.31	0.36	0.08
	MTCI	0.12	0.14	—	—	—	—	0.48	0.21
	NDRE1	0.41	0.30	—	—	—	—	0.49	0.21
	NDRE2	0.64	0.07	—	—	—	—	—	—
	CI$_{red\ edge}$	0.68	0.05	—	—	—	—	—	—
Model	NDVI	0.50	0.01	0.57	0.01	0.56	0.01	0.56	0.01
	EVI	0.26	0.33	0.37	0.31	0.36	0.32	0.31	0.33
	EVI2	0.36	0.28	0.39	0.28	0.38	0.28	0.39	0.28
	OSAVI	0.41	0.18	0.46	0.17	0.45	0.17	0.46	0.17
	RDVI	0.37	0.26	0.40	0.26	0.39	0.26	0.40	0.26
	PSND	0.67	0.00	0.57	0.01	0.56	0.01	0.68	0.00
	TCARI/OSAVI	0.82	0.01	0.88	0.01	0.87	0.01	0.87	0.01
	TCARI/OSAVI$_{red\ edge}$	0.51	0.05	0.35	0.04	0.42	0.00	0.54	0.03
	MTCI	0.76	0.00	—	—	—	—	0.82	0.00
	NDRE1	0.76	0.00	—	—	—	—	0.79	0.00
	NDRE2	0.76	0.00	—	—	—	—	—	—
	CI$_{red\ edge}$	0.90	0.00	—	—	—	—	—	—

The transverse line ("—") denotes the sensor without band to calculate corresponding vegetation index.

In field measurements, for the tested VIs calculated using Sentinel-2 bands, the $CI_{red\ edge}$ had the strongest correlation with CCC (R^2_{CCC} = 0.68) and the smallest influence from MTA (R^2_{MTA} = 0.05). In model simulations, the $CI_{red\ edge}$ had the strongest correlation with CCC (R^2_{CCC} = 0.90) and a weak correlation with MTA (R^2_{MTA} = 0.00). For the other three satellite sensors, in the field-measured dataset analysis, PSND produced the strongest correlations with CCC (R^2_{CCC} = 0.49–0.52) and the weakest correlation with MTA (R^2_{MTA} = 0.17–0.19). Model-simulated PSND presented a medium-strong correlation with CCC (R^2_{CCC} = 0.57–0.67) and a weak correlation with MTA (R^2_{MTA} = 0.00–0.01). In model simulations, TCARI/OSAVI had the strongest correlation with CCC (R^2_{CCC} = 0.87–0.88) and the weakest correlation with MTA (R^2_{MTA} = 0.01). This index had medium-strong correlations with both CCC (R^2_{CCC} = 0.29–0.33) and MTA (R^2_{MTA} = 0.37–0.41). MTA had the largest effect on EVI in both the field-measured dataset (R^2_{MTA} = 0.61–0.64) and model simulations (R^2_{MTA} = 0.31–0.36).

3.3. Identification of New Indices

In addition to the twelve tested vegetation indices, the potential of six two-band and five three-band new vegetation indices of predefined type were investigated for CCC estimation using the four satellite bands. In Figures A2 and A3, for the six two-band types of indices, the matrices of determinations of coefficients between CCC (R^2_{CCC}), MTA (R^2_{MTA}) and vegetation indices using all possible combinations of field-measured datasets based on RI, NDVI, DI, SAI, MSR, MSAI formulations are presented. The corresponding difference matrices between R^2_{CCC} and R^2_{MTA} based on the six formulations are presented in Figure 4. The three best band sets for the three-band indices identified using simulated satellite bands in the field-measured dataset are presented in Table 4. These identified best bands for the two-band and three-band indices and the corresponding R^2_{CCC} and R^2_{MTA} using the field-measured data are presented in Tables 4 and 5, respectively. The identified best indices were validated with PROSAIL model simulations, and the results are presented in Table 6.

Table 4. Three best band configurations for the new three-band vegetation indices in the field measured dataset for each simulated satellite.

Index		Sentinel-2		WorldView-2		RapidEye		GaoFen-6	
		B1, B2, B3	R^2_{CCC}, R^2_{MTA}	**B1, B2, B3**	R^2_{CCC}, R^2_{MTA}	**B1, B2, B3**	R^2_{CCC}, R^2_{MTA}	**B1, B2, B3**	R^2_{CCC}, R^2_{MTA}
TI	1	B7, B4, B5	0.79, 0.05	NIR1, Green, Red Edge	0.77, 0.02	Blue, Green, Red Edge	0.22, 0.32	B1, B3, B8	0.14, 0.02
	2	B2, B6, B7	0.78, 0.06	NIR1, Blue, Red Edge	0.72, 0.03	Blue, Green, NIR	0.26, 0.45	B5, B1, B2	0.24, 0.20
	3	B3, B6, B7	0.66, 0.27	Red, Blue, Yellow	0.13, 0.06	Red Edge, Blue, NIR	0.25, 0.52	B4, B5, B8	0.31, 0.39
Git	1	B5, B8, B8A	0.76, 0.00	Green, Red Edge, NIR1	0.58, 0.10	Green, Red Edge, NIR	0.55, 0.11	B5, B6, B4	0.66, 0.07
	2	B5, B8A, B8	0.75, 0.00	Yellow, Red Edge, Red	0.46, 0.02	Green, Red Edge, Blue	0.38, 0.00	B2, B5, B8	0.55, 0.06
	3	B5, B7, B8A	0.74, 0.01	Green, Red Edge, Blue	0.33, 0.00	Green, NIR, Red Edge	0.48, 0.17	B2, B6, B4	0.58, 0.10
BSI-T	1	B7, B6, B2	0.78, 0.00	NIR1, Blue, Red Edge	0.76, 0.00	Red Edge, Green, NIR	0.76, 0.00	B5, B3, B4	0.78, 0.01
	2	B7, B5, B6	0.77, 0.00	NIR1, Green, Red Edge	0.73, 0.00	Red Edge, Blue, NIR	0.74, 0.00	B5, B4, B8	0.77, 0.00
	3	B8, B6, B4	0.76, 0.00	NIR1, Yellow, Red Edge	0.70, 0.02	Red Edge, Red, NIR	0.76, 0.09	B4, B3, B6	0.74, 0.00
BSI-V	1	B8, B6, B2	0.78, 0.02	NIR1, Red, Red Edge	0.78, 0.00	NIR, Blue, Red Edge	0.72, 0.03	B4, B6, B1	0.77, 0.01
	2	B8, B6, B5	0.78, 0.01	NIR1, Yellow, Red Edge	0.78, 0.01	NIR, Green, Red Edge	0.71, 0.03	B4, B6, B5	0.77, 0.00
	3	B2, B6, B8	0.76, 0.01	Red Edge, Red, NIR1	0.76, 0.00	Red Edge, Green, NIR	0.67, 0.04	B1, B6, B4	0.75, 0.00
BSI-W	1	B6, B8, B2	0.74, 0.01	Red Edge, Blue, NIR1	0.74, 0.03	Red Edge, Blue, NIR	0.64, 0.04	B6, B4, B1	0.72, 0.00
	2	B6, B5, B7	0.73, 0.01	Red Edge, Green, NIR1	0.72, 0.01	Red Edge, Green, NIR	0.62, 0.04	B5, B6, B4	0.68, 0.00
	3	B6, B3, B7	0.73, 0.01	Red Edge, NIR1, Blue	0.71, 0.00	Red Edge, NIR, Blue	0.62, 0.07	B6, B4, B2	0.65, 0.01

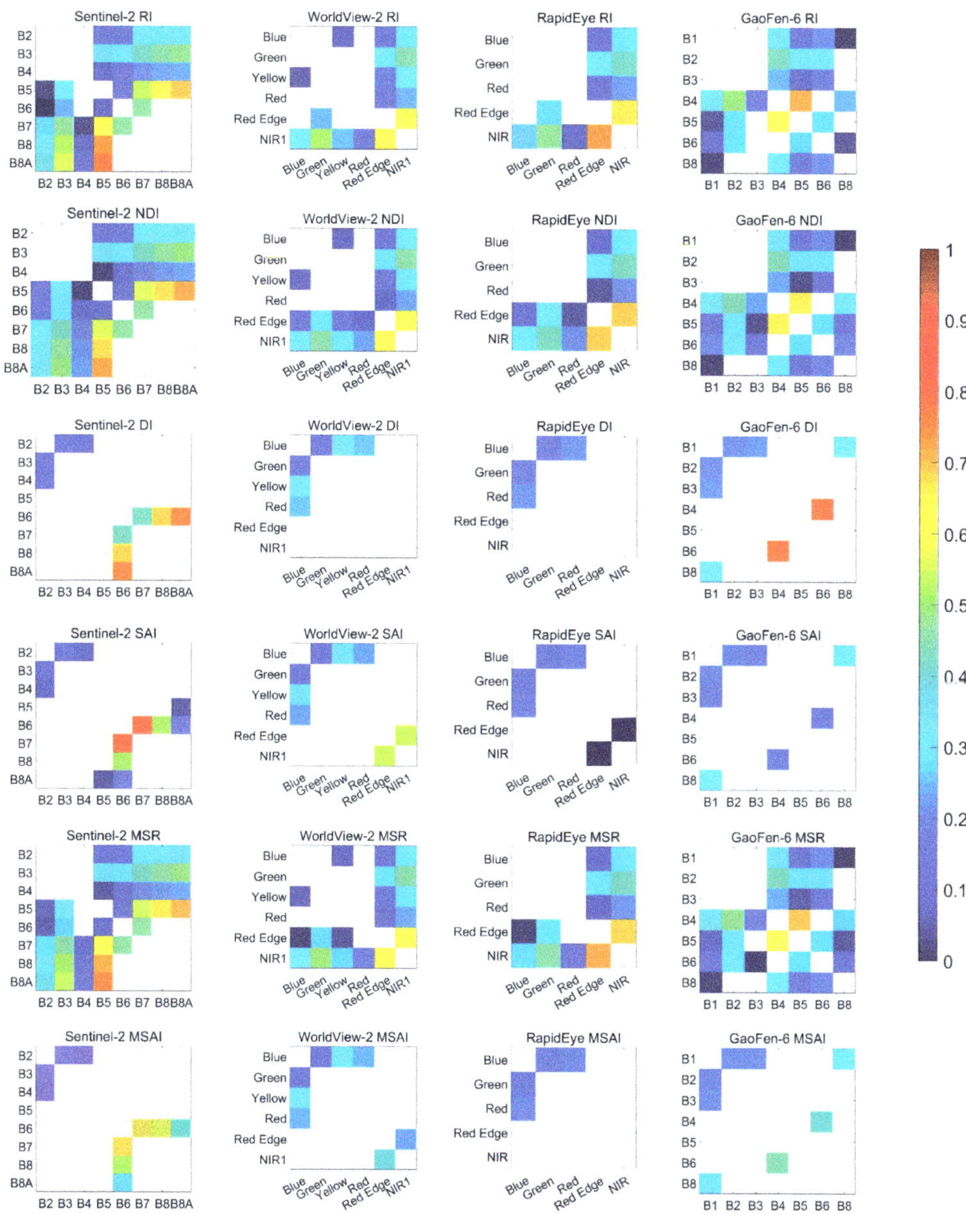

Figure 4. Matrices of difference between R^2_{CCC} and R^2_{MTA} in all possible two band combinations for RI, NDI, DI, SAI, MSR and MSAI formulations. The color indicates different R^2 values, blank negative values.

Table 5. Best band configurations for the two-band indices in the field measured dataset for each simulated satellite.

Index	Sentinel-2		WorldView-2		RapidEye		GaoFen-6	
	B1, B2	R^2_{CCC}, R^2_{MTA}	B1, B2	R^2_{CCC}, R^2_{MTA}	B1, B2	R^2_{CCC}, R^2_{MTA}	B1, B2	R^2_{CCC}, R^2_{MTA}
RI	B5, B8A	0.77, 0.00	NIR1, Red Edge	0.73, 0.10	Red Edge, NIR	0.74, 0.01	B5, B4	0.73, 0.02
NDVI	B5, B8A	0.73, 0.00	Red Edge, NIR1	0.74, 0.11	Red Edge, NIR	0.71, 0.02	B5, B4	0.69, 0.03
DI	B6, B8A	0.76, 0.00	Blue, Yellow	0.36, 0.03	Blue, Red	0.40, 0.18	B6, B4	0.78, 0.00
SAI	B6, B7	0.80, 0.00	Red Edge, NIR1	0.65, 0.09	Blue, Red	0.39, 0.19	B8, B1	0.36, 0.05
MSR	B5, B8A	0.75, 0.00	NIR1, Red Edge	0.74, 0.11	Red Edge, NIR	0.73, 0.01	B5, B4	0.72, 0.02
MSAI	B6, B7	0.78, 0.00	Red Edge, NIR1	0.56, 0.17	Blue, Red	0.40, 0.18	B4, B6	0.69, 0.23

Table 6. Performance of the best new indices of each type for the four simulated satellite sensors in model simulations.

Index	Sentinel-2		WorldView-2		RapidEye		GaoFen-6	
	Bands	R^2_{CCC}, R^2_{MTA}	Bands	R^2_{CCC}, R^2_{MTA}	Bands	R^2_{CCC}, R^2_{MTA}	Bands	R^2_{CCC}, R^2_{MTA}
RI	B5, B8A	0.89, 0.00	NIR1, Red Edge	0.80, 0.00	Red Edge, NIR	0.90, 0.00	B5, B4	0.90, 0.00
NDVI	B5, B8A	0.76, 0.00	Red Edge, NIR1	0.83, 0.01	Red Edge, NIR	0.80, 0.00	B5, B4	0.79, 0.00
DI	B6, B8A	0.93, 0.04	Blue, Yellow	0.51, 0.00	Blue, Red	0.61, 0.05	B6, B4	0.94, 0.04
SAI	B6, B7	0.95, 0.00	Red Edge, NIR1	0.90, 0.02	Blue, Red	0.62, 0.06	B8,B1	0.57, 0.00
MSR	B5, B8A	0.87, 0.00	NIR1, Red Edge	0.82, 0.00	Red Edge, NIR	0.87, 0.00	B5, B4	0.88, 0.00
MSAI	B6, B7	0.96, 0.01	Red Edge, NIR1	0.90, 0.04	Blue, Red	0.61, 0.06	B4, B6	0.95, 0.00
TI	B7, B4, B5	0.82, 0.05	NIR1, Green, Red Edge	0.92, 0.02	Blue, Green, Red Edge	0.43, 0.05	B1, B3, B8	0.36, 0.01
Git	B5, B8, B8A	0.89, 0.00	Green, Red Edge, NIR1	0.88, 0.00	Green, Red Edge, NIR	0.88, 0.00	B5, B6, B4	0.91, 0.00
BSI-T	B7, B6, B2	0.90, 0.01	NIR1, Blue, Red Edge	0.85, 0.01	Red Edge, Green, NIR	0.84, 0.00	B5, B3, B4	0.79, 0.00
BSI-V	B8, B6, B2	0.90, 0.01	NIR1, Red, Red Edge	0.90, 0.01	NIR, Blue, Red Edge	0.91, 0.01	B4, B6, B1	0.87, 0.02
BSI-W	B6, B8, B2	0.87, 0.01	Red Edge, Blue, NIR1	0.76, 0.00	Red Edge, Blue, NIR	0.72, 0.00	B6, B4, B1	0.83, 0.01

In the Sentinel-2 bands, all the best new indices presented strong correlations with CCC (R^2_{CCC} = 0.74–0.80) and no correlation with MTA (R^2_{MTA} = 0.00–0.02). SAI (B6, B7), was identified as the best (R^2_{CCC} = 0.80 and R^2_{MTA} = 0.00) among all the new indices in the field-measured dataset (Figure 5). This combination was found to have a strong correlation with CCC (R^2_{CCC} = 0.95) and a weak correlation with MTA (R^2_{MTA} = 0.00) in the model-simulated dataset (Figure 6), as shown in Table 6. In the simulated WorldView-2 data, the R^2_{CCC} varied between 0.44 and 0.78 and R^2_{MTA} varied between 0.00 and 0.11. The identified new three-band of indices performed better (R^2_{CCC} = 0.58–0.78 and R^2_{MTA} = 0.0–0.10) than the two-band indices (R^2_{CCC} = 0.44–0.74 and R^2_{MTA} = 0.02–0.11). BSI-V (NIR1, Red, Red Edge) was identified as the best new index (R^2_{CCC} = 0.78 and R^2_{MTA} = 0.00). In the model-simulated dataset, this combination was found to have a strong correlation with CCC (R^2_{CCC} = 0.90) and no correlation with MTA (R^2_{MTA} = 0.01). In the simulated RapidEye data, large variations on correlation were identified among the best new indices for CCC (R^2_{CCC} = 0.22–0.76) and MTA (R^2_{MTA} = 0.00–0.32). BSI-T (red edge, green, NIR) was the best-performing index (R^2_{CCC} = 0.76 and R^2_{MTA} = 0.00) and was found to have a strong correlation with CCC (R^2_{CCC} = 0.84) and no correlation with MTA (R^2_{MTA} = 0.00) in the model-simulated dataset. In the simulated GaoFen-6 data, the best new indices presented large variations in correlations with CCC (R^2_{CCC} = 0.14–0.78) and MTA (R^2_{MTA} = 0.00–0.23). DI (B6, B4) was identified as the best index (R^2_{CCC} = 0.78 and R^2_{MTA} = 0.00) and was found to have a strong correlation with CCC (R^2_{CCC} = 0.94) and almost no correlation with MTA (R^2_{MTA} = 0.04) in the model-simulated dataset.

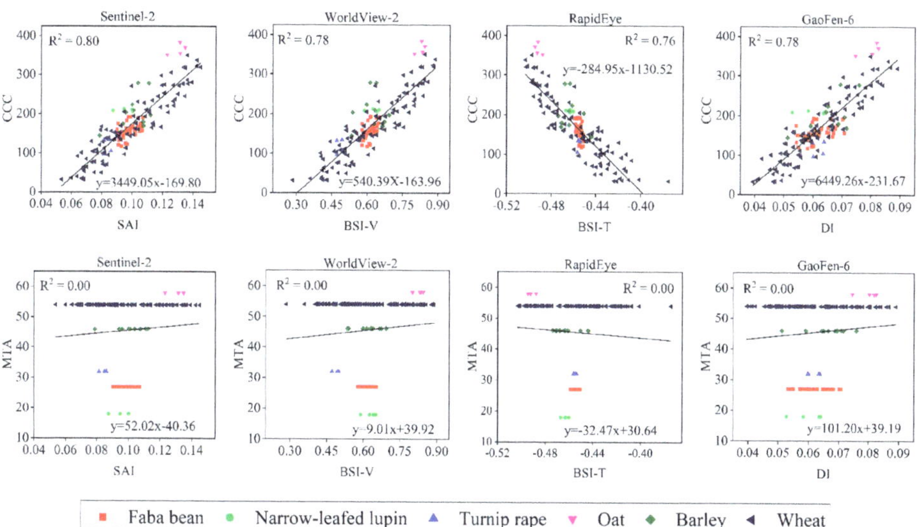

Figure 5. Correlation between the best vegetation indices, and CCC (**top row**) and MTA (**bottom row**) in Sentienl−2 (**left column**), WorldView−2 (**second column**), RapidEye (**third column**) and GaoFen-6 (**right column**) in the field measured dataset.

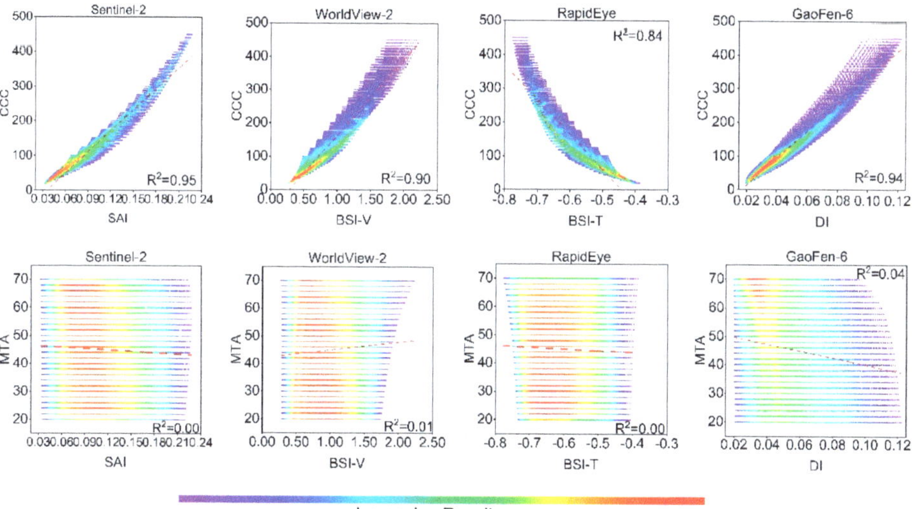

Figure 6. Correlation between the best vegetation indices, and CCC (**top row**) and MTA (**bottom row**) for Sentinel−2 (**left column**), WorldView−2 (**second column**), RapidEye (**third column**) and GaoFen-6 (**right column**) in model simulations.

4. Discussion

Potential CCC-sensitive but MTA-insensitive satellite broadband vegetation indices were developed. To our knowledge, this is among the few studies that have focused on specifically designing this type of vegetation index. The vegetation indices were calibrated with field measurements and validated with widely used PROSAIL model simulations. The canopy reflectance model can be used to accurately simulate the actual reflectance spectra without the inherent bias caused by the specific growth conditions at any study sites.

Actual field-measured datasets have limited ranges of variables of interest and specific data distributions (with possibly site-specific) internal correlations. This limits their generality for calibrating vegetation indices. While model-based fits are universal, they inevitably include simplifications, such as the absence of material other than leaves. Before application, all theoretical models need to be validated in the field. A compromise is to link an existing field-measured dataset with model simulations as suggested in a previous study [82]. An efficient vegetation index should be supported both by field measurements and model simulations. In this study, the identified best indices for each satellite presented a good match between measurements and simulations.

The newly developed indices performed better than the tested existing vegetation indices and are recommended to remotely estimate crop CCC from satellites across species and seasonality. Theoretically, three-band vegetation indices have a larger information content and flexibility than two-band combinations. However, in our study, the three-band vegetation indices did not show a great advantage over the simpler two-band formulations. For the simulated Sentinel-2 and GaoFen-6 bands, the best indices were two-band, while for the WorldView-2 and RapidEye, the identified best indices were three-band.

Regardless of the number of bands, all the best indices for each satellite were constructed from NIR and red edge bands. This agreed with previous studies performed by [33], who demonstrated that these two band combinations are minimally affected by crop phenology and can potentially be used as generic algorithms to crop CCC estimation. Red edge reflectance is strongly negatively correlated with MTA [44,46], and the addition of this channel can attenuate the sensitivity of vegetation indices to leaf angles [83]. Sentinel-2 MSI performed better than the other evaluated satellite sensors in both field-measured data and model simulations, indicating a more optimal spectral band combination. Similarly, in all tested vegetation indices, the $CI_{red\ edge}$ computed with Sentinel-2 data was the best vegetation index strongly correlated with CCC (R^2_{CCC} = 0.68 in field measured data and R^2_{CCC} = 0.90 in model simulated data) and no correlation with MTA (R^2_{MTA} = 0.05 in field measured data and R^2_{MTA} = 0.00 in model simulated data). In previous studies, the performance of $CI_{red\ edge}$ has been evaluated for single crop species either from real Sentinel-2 imagery or resampled from field canopy reflectance. The following relationships have been reported in the literature for $CI_{red\ edge}$ and CCC: R^2_{CCC} = 0.58 for potato [34], R^2_{CCC} = 0.86 and 0.94 for maize and soybean, respectively [33], and R^2_{CCC} = 0.74 for wheat [35]. These relationships agree with the results in this study, which can be explained by the fact that the $CI_{red\ edge}$ was suitable for crop CCC estimation under a mixed pixel scenario [3].

For the other vegetation indices derived from Sentinel-2 bands, such as NDVI, NDRE1, NDRE2, MTCI, TCARI/OSAVI and TCARI/$OSAVI_{red\ edge}$, R^2_{CCC} varied between 0.12 and 0.64 for field measured data and between 0.50 and 0.82 for model simulations. In a previous study, these correlations were between 0.66 and 0.78 for single wheat species [35], which are larger than that found in the field-measured data but within the range of our model simulations. Especially for the MTCI, which is specifically designed for the MERIS spectrometer, the correlation between CCC and real MERIS data-derived MTCI is R^2_{CCC} = 0.24 for soybean [26]. The value is better than that from Sentinel-2 data (R^2_{CCC} = 0.12) but lower than that from GaoFen-6 data (R^2_{CCC} = 0.48). The model-simulated MERIS-based MTCI presented a stronger correlation with CCC (R^2_{CCC} = 0.69) than real MERIS data [26], but this value is lower than the model simulation based on Sentinel-2 (R^2_{CCC} = 0.76) and GanFen-6 (R^2_{CCC} = 0.82) data in this study and even lower than that of proximal spectra-simulated Sentinel-2 data (R^2_{CCC} = 0.89) for maize and soybean [33].

Except for Sentinel-2, the three other satellites (WorldView2, RapidEye and GaoFen-6) have been widely used for remote sensing of vegetation. Surprisingly, there are few reports on their use for the estimation of CCC for field crops. In all tested vegetation indices, PSND had the strongest correlations with CCC in the field-measured data (R^2_{CCC} = 0.49–0.52), and similar results were found in PROSAIL model simulations (R^2_{CCC} = 0.56–0.68). TCARI/

OSAVI presented the best correlation with CCC in PROSAIL model simulations (R^2_{CCC} = 0.82–0.88) and no correlation with MTA (R^2_{MTA} = 0.01), but this good performance was not consistent in field measurements. The matrices of difference between R^2_{CCC} and R^2_{MTA} for the three two-band RI and NDI are similar (Figure 4), and identical bands were identified for the best vegetation indices of both types. This can be explained by their mathematical similarity [84]. However, comparing the four satellite sensors, large differences in performance were found among the best vegetation indices of each type in both field measurements (Tables 4 and 5) and model simulations (Table 6). Thus, finding the right type is also very important for optimizing vegetation indices.

For CCC estimation, it is essential to use band combinations. CCC effects on the responses of MTA to individual broadband reflectance varied with the combination of LAI and Cab. Even at similar CCC levels (CCC = 90–100 in Figure 3 in the second and third columns), this relationship can vary greatly. This is mainly because LAI and Cab determine the reflectance of different broadband separately. Generally, the MTA responses to NIR reflectance were determined by LAI and those to visible reflectance were determined by Cab.

Although the identified vegetation indices for the four satellite spectral configurations in this study produced good results in both field-measured and model-simulated data and are recommended for crop CCC estimation, there are some limitations in this study. First, the derived vegetation indices were not validated with real satellite imagery. Satellite sensor imaging needs to consider the atmospheric radiation and transmittance, geometric characteristics, spatial resolutions and signal-to-noise ratio, which limit the transferability of the vegetation indices developed in this study. Unfortunately, real satellite imagery could not be acquired simultaneously for the particular study area over a given time. In the future, more effort needs to be put into vegetation index evaluations using real satellite imagery.

The potential CCC-sensitive but MTA-insensitive satellite broadband vegetation indices developed in this study may provide a convenient method for accurately estimating crop CCC with diverse canopy architectures using satellite remote sensing data.

5. Conclusions

This research attempted to investigate the potential of satellite broadband vegetation indices for crop canopy chlorophyll content estimation with minimum effects from leaf inclination angle distribution. The broadband vegetation indices of four satellites (Sentinel-2, RapidEye, WorldView-2 and GaoFen-6) were resampled from canopy airborne imaging spectroscopy data of six crop species with various canopy structures. To obtain generic and robust crop CCC indices, both field-measured datasets and model simulations were used in this study. The best vegetation indices identified in this study are the soil-adjusted index type index SAI (B6, B7) for Sentinel-2, Verrelts's three-band spectral index type index BSI-V (NIR1, Red, Red Edge) for WorldView-2, Tian's three-band spectral index type index BSI-T (Red Edge, Green, NIR) for RapidEye and difference index type index DI (B6, B4) for GaoFen-6. The recommended indices produced strong correlations with CCC (R^2_{CCC} = 0.76–0.80 in field-measured data and R^2_{CCC} = 0.84–0.95 in model simulations) and no correlation with MTA (R^2_{MTA} = 0.00 for field-measured data and R^2_{MTA} = 0.00–0.04 for model simulations) and maintained consistent performance in both the field-measured dataset and model simulations. Thus, it is anticipated that more generic vegetation indices for crop CCC estimation can be derived from satellite broadband data. However, this is only a case study, and further studies are required to examine the suitability across more crop species and growth stages using real satellite imagery.

Author Contributions: X.Z. and J.J. conceived the research and implemented the data analysis. X.Z. prepared the original draft. M.M. revised the manuscript and supervised the research. All authors have read and agreed to the published version of the manuscript.

Funding: This research was supported by the National Science Foundation of China (grant No. 41801243) and the Academy of Finland (grant No. 317387).

Data Availability Statement: All data are presented within the article.

Acknowledgments: The authors would like to thank Priit Tammeorg, Clara Lizarazo Torres, Piia Kekkonen, F.L. Stoddard and Pirjo Mäkelä from the University of Helsinki, who kindly provided the SunScan and SPAD measurement data, and Petri Pellikka from the University of Helsinki for the hyperspectral acquisitions.

Conflicts of Interest: The authors declare no conflict of interest.

Appendix A

Table A1. The central wavelength, bandwidth and spatial resolution and R^2 values from field measured dataset between CCC, MTA and individual band reflectance of four satellite sensors.

Sensor	Central Wavelength (nm)	Band/Band Number	Bandwidth (nm)	Spatial Resolution (m)	Measurements		Model	
					R^2_{CCC}	R^2_{MTA}	R^2_{CCC}	R^2_{MTA}
Sentinel-2	490	2	65	10	0.58	0.00	0.39	0.25
	560	3	50	10	0.44	0.05	0.42	0.08
	665	4	30	10	0.53	0.08	0.54	0.07
	705	5	15	20	0.07	0.77	0.43	0.10
	740	6	15	20	0.00	0.87	0.00	0.45
	783	7	20	20	0.04	0.78	0.27	0.39
	842	8	115	10	0.04	0.77	0.26	0.39
	865	8A	20	20	0.04	0.76	0.26	0.40
Worldview-2	478	Blue	60	1.8	0.60	0.00	0.29	0.45
	546	Green	70	1.8	0.45	0.05	0.41	0.08
	608	Yellow	40	1.8	0.49	0.01	0.51	0.05
	659	Red	60	1.8	0.54	0.05	0.57	0.06
	724	Red Edge	40	1.8	0.00	0.87	0.10	0.33
	831	NIR1	125	1.8	0.04	0.77	0.26	0.39
RapidEye	475	Blue	70	5	0.60	0.00	0.29	0.47
	555	Green	70	5	0.45	0.04	0.42	0.08
	657.5	Red	55	5	0.53	0.07	0.57	0.07
	710	Red Edge	40	5	0.03	0.83	0.31	0.19
	805	NIR	90	5	0.04	0.78	0.26	0.39
GaoFen-6	485	1	70	16	0.58	0.01	0.39	0.26
	555	2	70	16	0.46	0.03	0.42	0.08
	660	3	60	16	0.55	0.05	0.57	0.06
	830	4	120	16	0.04	0.77	0.26	0.39
	710	5	40	16	0.08	0.76	0.39	0.15
	750	6	40	16	0.01	0.85	0.08	0.44
	610	8	40	16	0.49	0.01	0.51	0.05

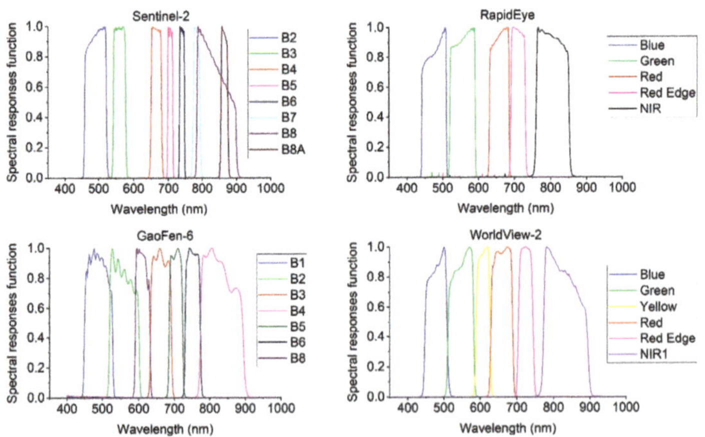

Figure A1. Spectral response functions of satellite sensors used for simulation of broadband reflectance.

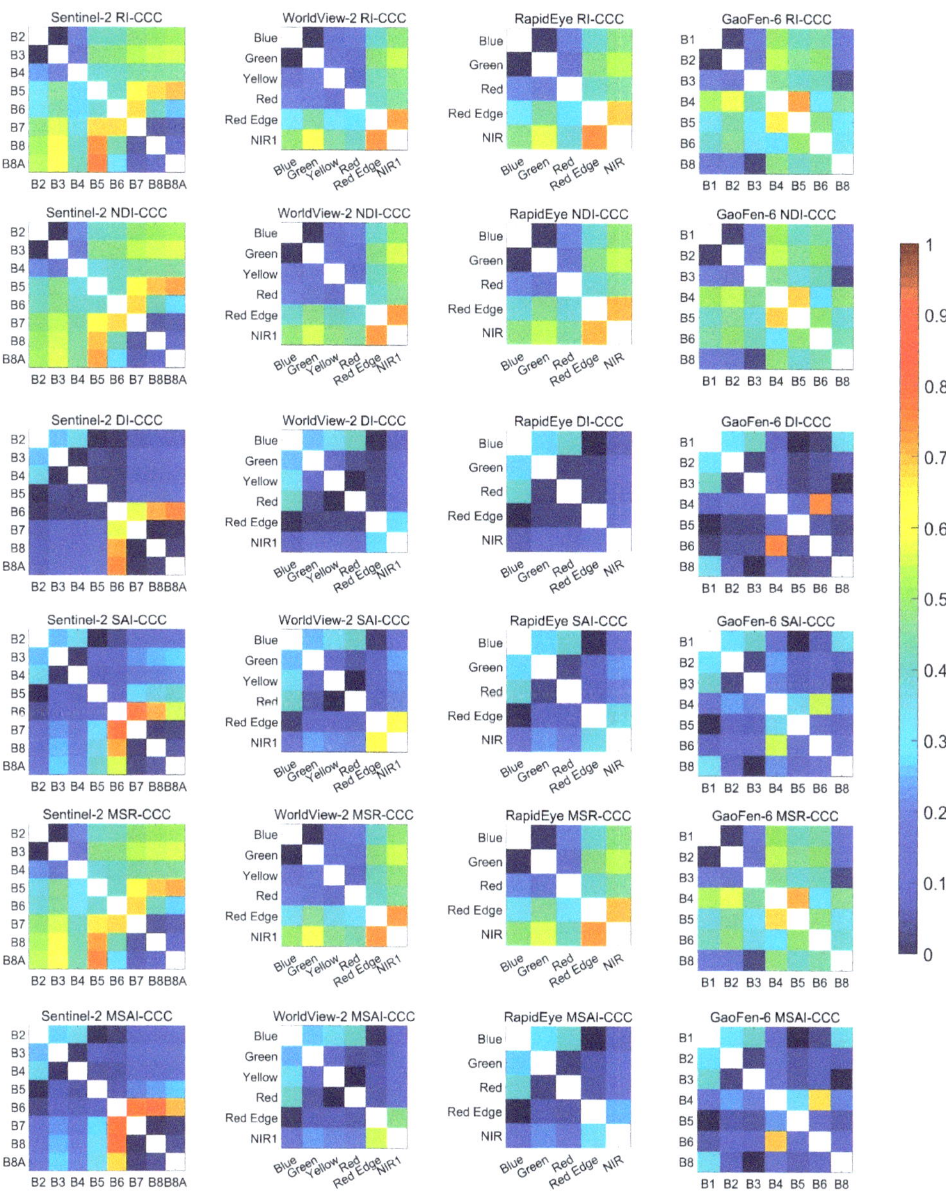

Figure A2. Map of the coefficient of determination between the CCC (R^2_{CCC}) and vegetation indices using all two band combinations based on the RI, NDVI, DI, SAI, MSR and MSAI formulations. The color indicates different R^2 values.

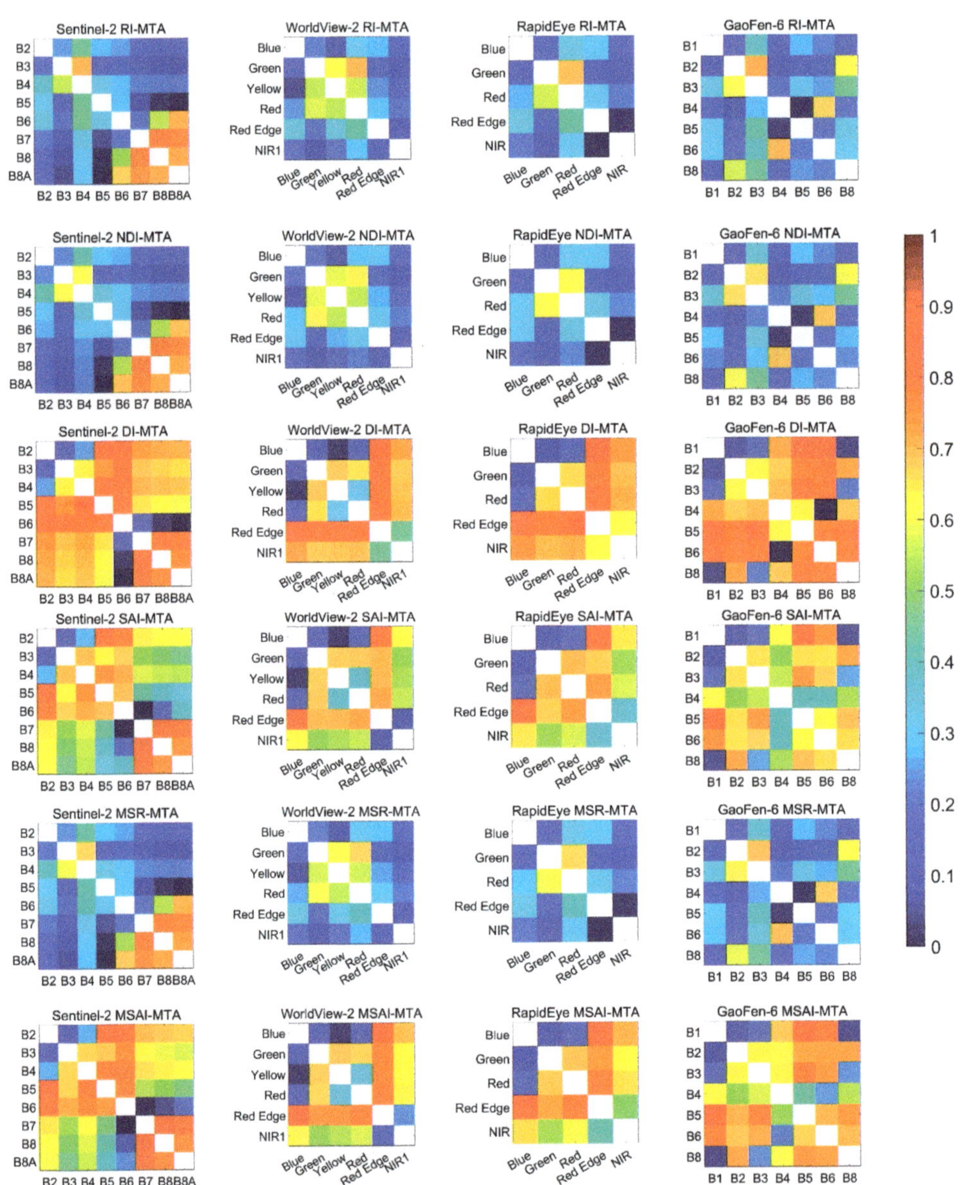

Figure A3. Map of the coefficient of determination between MTA (R^2_{MTA}) and vegetation indices using all two band combinations based on RI, NDVI, DI, SAI, MSR and MSAI formulations. The color indicates different R^2 values.

References

1. Richardson, A.D.; Duigan, S.P.; Berlyn, G.P. An Evaluation of Noninvasive Methods to Estimate Foliar Chlorophyll Content. *New Phytol.* **2002**, *153*, 185–194. [CrossRef]
2. Ustin, S.L.; Gitelson, A.A.; Jacquemoud, S.; Schaepman, M.; Asner, G.P.; Gamon, J.A.; Zarco-Tejada, P. Retrieval of Foliar Information about Plant Pigment Systems from High Resolution Spectroscopy. *Remote Sens. Environ.* **2009**, *113*, S67–S77. [CrossRef]
3. Gitelson, A.A. Remote Estimation of Canopy Chlorophyll Content in Crops. *Geophys. Res. Lett.* **2005**, *32*, L08403. [CrossRef]

4. Bacour, C.; Baret, F.; Béal, D.; Weiss, M.; Pavageau, K. Neural Network Estimation of LAI, FAPAR, FCover and LAI×Cab, from Top of Canopy MERIS Reflectance Data: Principles and Validation. *Remote Sens. Environ.* **2006**, *105*, 313–325. [CrossRef]
5. Wu, C.; Niu, Z.; Gao, S. The Potential of the Satellite Derived Green Chlorophyll Index for Estimating Midday Light Use Efficiency in Maize, Coniferous Forest and Grassland. *Ecol. Indic.* **2012**, *14*, 66–73. [CrossRef]
6. Murchie, E.H.; Lawson, T. Chlorophyll Fluorescence Analysis: A Guide to Good Practice and Understanding Some New Applications. *J. Exp. Bot.* **2013**, *64*, 3983–3998. [CrossRef] [PubMed]
7. Luo, X.; Croft, H.; Chen, J.M.; Bartlett, P.; Staebler, R.; Froelich, N. Incorporating Leaf Chlorophyll Content into a Two-Leaf Terrestrial Biosphere Model for Estimating Carbon and Water Fluxes at a Forest Site. *Agric. For. Meteorol.* **2018**, *248*, 156–168. [CrossRef]
8. Baret, F.; Houlès, V.; Guérif, M. Quantification of Plant Stress Using Remote Sensing Observations and Crop Models: The Case of Nitrogen Management. *J. Exp. Bot.* **2007**, *58*, 869–880. [CrossRef]
9. Wu, C.; Wang, L.; Niu, Z.; Gao, S.; Wu, M. Nondestructive Estimation of Canopy Chlorophyll Content Using Hyperion and Landsat/TM Images. *Int. J. Remote Sens.* **2010**, *31*, 2159–2167. [CrossRef]
10. Watson, D.J. Comparative Physiological Studies on the Growth of Field Crops: I. Variation in Net Assimilation Rate and Leaf Area between Species and Varieties, and within and between Years. *Ann. Bot.* **1947**, *11*, 41–76. [CrossRef]
11. Inoue, Y.; Guérif, M.; Baret, F.; Skidmore, A.; Gitelson, A.; Schlerf, M.; Darvishzadeh, R.; Olioso, A. Simple and Robust Methods for Remote Sensing of Canopy Chlorophyll Content: A Comparative Analysis of Hyperspectral Data for Different Types of Vegetation. *Plant Cell Environ.* **2016**, *39*, 2609–2623. [CrossRef] [PubMed]
12. Zhao, C.; Wang, Z.; Wang, J.; Huang, W.; Guo, T. Early Detection of Canopy Nitrogen Deficiency in Winter Wheat (*Triticum aestivum* L.) Based on Hyperspectral Measurement of Canopy Chlorophyll Status. *N. Z. J. Crop Hortic. Sci.* **2011**, *39*, 251–262. [CrossRef]
13. Korus, A. Effect of Preliminary and Technological Treatments on the Content of Chlorophylls and Carotenoids in Kale (*Brassica oleracea* L. Var. *Acephala). J. Food Process. Preserv.* **2013**, *37*, 335–344. [CrossRef]
14. Schlemmer, M.; Gitelson, A.; Schepers, J.; Ferguson, R.; Peng, Y.; Shanahan, J.; Rundquist, D. Remote Estimation of Nitrogen and Chlorophyll Contents in Maize at Leaf and Canopy Levels. *Int. J. Appl. Earth Obs. Geoinf.* **2013**, *25*, 47–54. [CrossRef]
15. Gitelson, A.A.; Viña, A.; Verma, S.B.; Rundquist, D.C.; Arkebauer, T.J.; Keydan, G.; Leavitt, B.; Ciganda, V.; Burba, G.G.; Suyker, A.E. Relationship between Gross Primary Production and Chlorophyll Content in Crops: Implications for the Synoptic Monitoring of Vegetation Productivity. *J. Geophys. Res. Atmos.* **2006**, *111*, D08S11. [CrossRef]
16. Peng, Y.; Gitelson, A.A.; Keydan, G.; Rundquist, D.C.; Moses, W. Remote Estimation of Gross Primary Production in Maize and Support for a New Paradigm Based on Total Crop Chlorophyll Content. *Remote Sens. Environ.* **2011**, *115*, 978–989. [CrossRef]
17. Prey, L.; Hu, Y.; Schmidhalter, U. High-Throughput Field Phenotyping Traits of Grain Yield Formation and Nitrogen Use Efficiency: Optimizing the Selection of Vegetation Indices and Growth Stages. *Front. Plant Sci.* **2020**, *10*, 1672. [CrossRef]
18. Dian, Y.; Le, Y.; Fang, S.; Xu, Y.; Yao, C.; Liu, G. Influence of Spectral Bandwidth and Position on Chlorophyll Content Retrieval at Leaf and Canopy Levels. *J. Indian Soc. Remote Sens.* **2016**, *44*, 583–593. [CrossRef]
19. Li, X.; Liu, X.; Liu, M.; Wang, C.; Xia, X. A Hyperspectral Index Sensitive to Subtle Changes in the Canopy Chlorophyll Content under Arsenic Stress. *Int. J. Appl. Earth Obs. Geoinf.* **2015**, *36*, 41–53. [CrossRef]
20. Darvishzadeh, R.; Matkan, A.A.; Dashti Ahangar, A. Inversion of a Radiative Transfer Model for Estimation of Rice Canopy Chlorophyll Content Using a Lookup-Table Approach. *IEEE J. Sel. Top. Appl. Earth Obs. Remote Sens.* **2012**, *5*, 1222–1230. [CrossRef]
21. Ali, A.M.; Darvishzadeh, R.; Skidmore, A.; Heurich, M.; Paganini, M.; Heiden, U.; Mücher, S. Evaluating Prediction Models for Mapping Canopy Chlorophyll Content Across Biomes. *Remote Sens.* **2020**, *12*, 1788. [CrossRef]
22. Dorigo, W.A.; Zurita-Milla, R.; de Wit, A.J.W.; Brazile, J.; Singh, R.; Schaepman, M.E. A Review on Reflective Remote Sensing and Data Assimilation Techniques for Enhanced Agroecosystem Modeling. *Int. J. Appl. Earth Obs. Geoinf.* **2007**, *9*, 165–193. [CrossRef]
23. Moulin, S. Impacts of Model Parameter Uncertainties on Crop Reflectance Estimates: A Regional Case Study on Wheat. *Int. J. Remote Sens.* **1999**, *20*, 213–218. [CrossRef]
24. Taddeo, S.; Dronova, I.; Depsky, N. Spectral Vegetation Indices of Wetland Greenness: Responses to Vegetation Structure, Composition, and Spatial Distribution. *Remote Sens. Environ.* **2019**, *234*, 111467. [CrossRef]
25. Mao, Z.H.; Deng, L.; Duan, F.Z.; Li, X.J.; Qiao, D.Y. Angle Effects of Vegetation Indices and the Influence on Prediction of SPAD Values in Soybean and Maize. *Int. J. Appl. Earth Obs. Geoinf.* **2020**, *93*, 102198. [CrossRef]
26. Sun, Q.; Jiao, Q.; Qian, X.; Liu, L.; Liu, X.; Dai, H. Improving the Retrieval of Crop Canopy Chlorophyll Content Using Vegetation Index Combinations. *Remote Sens.* **2021**, *13*, 470. [CrossRef]
27. Dash, J.; Curran, P.J. The MERIS Terrestrial Chlorophyll Index. *Int. J. Remote Sens.* **2004**, *25*, 5403–5413. [CrossRef]
28. Gitelson, A.; Merzlyak, M.N. Quantitative Estimation of Chlorophyll-a Using Reflectance Spectra: Experiments with Autumn Chestnut and Maple Leaves. *J. Photochem. Photobiol. B* **1994**, *22*, 247–252. [CrossRef]
29. Clevers, J.G.P.W.; Kooistra, L. Using Hyperspectral Remote Sensing Data for Retrieving Canopy Chlorophyll and Nitrogen Content. *IEEE J. Sel. Top. Appl. Earth Obs. Remote Sens.* **2012**, *5*, 574–583. [CrossRef]
30. He, R.; Li, H.; Qiao, X.; Jiang, J. Using Wavelet Analysis of Hyperspectral Remote-Sensing Data to Estimate Canopy Chlorophyll Content of Winter Wheat under Stripe Rust Stress. *Int. J. Remote Sens.* **2018**, *39*, 4059–4076. [CrossRef]

31. Li, L.; Ren, T.; Ma, Y.; Wei, Q.; Wang, S.; Li, X.; Cong, R.; Liu, S.; Lu, J. Evaluating Chlorophyll Density in Winter Oilseed Rape (*Brassica napus* L.) Using Canopy Hyperspectral Red-Edge Parameters. *Comput. Electron. Agric.* **2016**, *126*, 21–31. [CrossRef]
32. Okuda, K.; Taniguchi, K.; Miura, M.; Obata, K.; Yoshioka, H. Application of Vegetation Isoline Equations for Simultaneous Retrieval of Leaf Area Index and Leaf Chlorophyll Content Using Reflectance of Red Edge Band. In Proceedings of the Remote Sensing and Modeling of Ecosystems for Sustainability XIII, San Diego, CA, USA, 28 August–1 September 2016; Volume 9975, pp. 87–93.
33. Peng, Y.; Nguy-Robertson, A.; Arkebauer, T.; Gitelson, A.A. Assessment of Canopy Chlorophyll Content Retrieval in Maize and Soybean: Implications of Hysteresis on the Development of Generic Algorithms. *Remote Sens.* **2017**, *9*, 226. [CrossRef]
34. Clevers, J.; Kooistra, L.; van den Brande, M. Using Sentinel-2 Data for Retrieving LAI and Leaf and Canopy Chlorophyll Content of a Potato Crop. *Remote Sens.* **2017**, *9*, 405. [CrossRef]
35. Xie, Q.; Dash, J.; Huete, A.; Jiang, A.; Yin, G.; Ding, Y.; Peng, D.; Hall, C.C.; Brown, L.; Shi, Y.; et al. Retrieval of Crop Biophysical Parameters from Sentinel-2 Remote Sensing Imagery. *Int. J. Appl. Earth Obs. Geoinf.* **2019**, *80*, 187–195. [CrossRef]
36. Gausman, H.W.; Allen, W.A.; Cardenas, R.; Richardson, A.J. Effects of Leaf Nodal Position on Absorption and Scattering Coefficients and Infinite Reflectance of Cotton Leaves, *Gossypium hirsutum* L. *Agron. J.* **1971**, *63*, 87–91. [CrossRef]
37. Sellers, P.J. *Vegetation-Canopy Spectral Reflectance and Biophysical Processes*; Asrar, G., Ed.; John Wiley and Sons: New York, NY, USA, 1989.
38. Asner, G.P. Biophysical and Biochemical Sources of Variability in Canopy Reflectance. *Remote Sens. Environ.* **1998**, *64*, 234–253. [CrossRef]
39. Zou, X.; Hernández-Clemente, R.; Tammeorg, P.; Lizarazo Torres, C.; Stoddard, F.L.; Mäkelä, P.; Pellikka, P.; Mõttus, M. Retrieval of Leaf Chlorophyll Content in Field Crops Using Narrow-Band Indices: Effects of Leaf Area Index and Leaf Mean Tilt Angle. *Int. J. Remote Sens.* **2015**, *36*, 6031–6055. [CrossRef]
40. Zou, X.; Mõttus, M. Sensitivity of Common Vegetation Indices to the Canopy Structure of Field Crops. *Remote Sens.* **2017**, *9*, 994. [CrossRef]
41. Ross, J. Tasks for Vegetation Science. In *The Radiation Regime and Architecture of Plant Stands*; Springer: Dordrecht, The Netherlands, 1981; ISBN 978-90-6193-607-7.
42. McNeil, B.E.; Pisek, J.; Lepisk, H.; Flamenco, E.A. Measuring Leaf Angle Distribution in Broadleaf Canopies Using UAVs. *Agric. For. Meteorol.* **2016**, *218*, 204–208. [CrossRef]
43. Zou, X.; Mõttus, M. Retrieving Crop Leaf Tilt Angle from Imaging Spectroscopy Data. *Agric. For. Meteorol.* **2015**, *205*, 73–82. [CrossRef]
44. Zou, X.; Zhu, S.; Mõttus, M. Estimation of Canopy Structure of Field Crops Using Sentinel-2 Bands with Vegetation Indices and Machine Learning Algorithms. *Remote Sens.* **2022**, *14*, 2849. [CrossRef]
45. Jiao, Q.; Sun, Q.; Zhang, B.; Huang, W.; Ye, H.; Zhang, Z.; Zhang, X.; Qian, B. A Random Forest Algorithm for Retrieving Canopy Chlorophyll Content of Wheat and Soybean Trained with PROSAIL Simulations Using Adjusted Average Leaf Angle. *Remote Sens.* **2022**, *14*, 98. [CrossRef]
46. Zou, X.; Mõttus, M.; Tammeorg, P.; Torres, C.L.; Takala, T.; Pisek, J.; Mäkelä, P.; Stoddard, F.L.; Pellikka, P. Photographic Measurement of Leaf Angles in Field Crops. *Agric. For. Meteorol.* **2014**, *184*, 137–146. [CrossRef]
47. Ryu, Y.; Sonnentag, O.; Nilson, T.; Vargas, R.; Kobayashi, H.; Wenk, R.; Baldocchi, D.D. How to Quantify Tree Leaf Area Index in an Open Savanna Ecosystem: A Multi-Instrument and Multi-Model Approach. *Agric. For. Meteorol.* **2010**, *150*, 63–76. [CrossRef]
48. Pisek, J.; Ryu, Y.; Alikas, K. Estimating Leaf Inclination and G-Function from Leveled Digital Camera Photography in Broadleaf Canopies. *Trees* **2011**, *25*, 919–924. [CrossRef]
49. Pisek, J.; Sonnentag, O.; Richardson, A.D.; Mõttus, M. Is the Spherical Leaf Inclination Angle Distribution a Valid Assumption for Temperate and Boreal Broadleaf Tree Species? *Agric. For. Meteorol.* **2013**, *169*, 186–194. [CrossRef]
50. Campbell, G.S. Derivation of an Angle Density Function for Canopies with Ellipsoidal Leaf Angle Distributions. *Agric. For. Meteorol.* **1990**, *49*, 173–176. [CrossRef]
51. Markwell, J.; Osterman, J.C.; Mitchell, J.L. Calibration of the Minolta SPAD-502 Leaf Chlorophyll Meter. *Photosynth. Res.* **1995**, *46*, 467–472. [CrossRef]
52. Vohland, M.; Mader, S.; Dorigo, W. Applying Different Inversion Techniques to Retrieve Stand Variables of Summer Barley with PROSPECT+SAIL. *Int. J. Appl. Earth Obs. Geoinf.* **2010**, *12*, 71–80. [CrossRef]
53. Feret, J.-B.; François, C.; Asner, G.P.; Gitelson, A.A.; Martin, R.E.; Bidel, L.P.R.; Ustin, S.L.; le Maire, G.; Jacquemoud, S. PROSPECT-4 and 5: Advances in the Leaf Optical Properties Model Separating Photosynthetic Pigments. *Remote Sens. Environ.* **2008**, *112*, 3030–3043. [CrossRef]
54. Verhoef, W. Light Scattering by Leaf Layers with Application to Canopy Reflectance Modeling: The SAIL Model. *Remote Sens. Environ.* **1984**, *16*, 125–141. [CrossRef]
55. Kuusk, A. The Hot Spot Effect in Plant Canopy Reflectance. In *Photon-Vegetation Interactions*; Springer: Berlin/Heidelberg, Germany, 1991; pp. 139–159. [CrossRef]
56. Hosgood, B.; Jacquemoud, S.; Andreoli, G.; Verdebout, J.; Pedrini, G.; Schmuck, G. *Leaf Optical Properties EXperiment 93 (LOPEX93)*; Office for Official Publications of the European Communities: Luxembourg, 1994.

57. Haboudane, D.; Miller, J.R.; Pattey, E.; Zarco-Tejada, P.J.; Strachan, I.B. Hyperspectral Vegetation Indices and Novel Algorithms for Predicting Green LAI of Crop Canopies: Modeling and Validation in the Context of Precision Agriculture. *Remote Sens. Environ.* **2004**, *90*, 337–352. [CrossRef]
58. Mäkelä, P.; Kleemola, J.; Jokinen, K.; Mantila, J.; Pehu, E.; Peltonen-Sainio, P. Growth Response of Pea and Summer Turnip Rape to Foliar Application of Glycinebetaine. *Acta Agric. Scand. Sect. B Soil Plant Sci.* **1997**, *47*, 168–175. [CrossRef]
59. Dennett, M.D.; Ishag, K.H.M. Use of the Expolinear Growth Model to Analyse the Growth of Faba Bean, Peas and Lentils at Three Densities: Predictive Use of the Model. *Ann. Bot.* **1998**, *82*, 507–512. [CrossRef]
60. Pinheiro, C.; Rodrigues, A.P.; de Carvalho, I.S.; Chaves, M.M.; Ricardo, C.P. Sugar Metabolism in Developing Lupin Seeds Is Affected by a Short-Term Water Deficit. *J. Exp. Bot.* **2005**, *56*, 2705–2712. [CrossRef]
61. Vile, D.; Garnier, É.; Shipley, B.; Laurent, G.; Navas, M.L.; Roumet, C.; Lavorel, S.; Díaz, S.; Hodgson, J.G.; Lloret, F.; et al. Specific Leaf Area and Dry Matter Content Estimate Thickness in Laminar Leaves. *Ann. Bot.* **2005**, *96*, 1129–1136. [CrossRef] [PubMed]
62. Vermote, E.F.; Tanre, D.; Deuze, J.L.; Herman, M.; Morcette, J.-J. Second Simulation of the Satellite Signal in the Solar Spectrum, 6S: An Overview. *IEEE Trans. Geosci. Remote Sens.* **1997**, *35*, 675–686. [CrossRef]
63. Rouse, J.W. *Monitoring the Vernal Advancement and Retrogradation (Green Wave Effect) of Natural Vegetation*; NASA/GSFC, Type II; NASA: Greenbelt, MD, USA, 1973.
64. Huete, A.; Didan, K.; Miura, T.; Rodriguez, E.P.; Gao, X.; Ferreira, L.G. Overview of the Radiometric and Biophysical Performance of the MODIS Vegetation Indices. *Remote Sens. Environ.* **2002**, *83*, 195–213. [CrossRef]
65. Jiang, Z.; Huete, A.R.; Didan, K.; Miura, T. Development of a Two-Band Enhanced Vegetation Index without a Blue Band. *Remote Sens. Environ.* **2008**, *112*, 3833–3845. [CrossRef]
66. Rondeaux, G.; Steven, M.; Baret, F. Optimization of Soil-Adjusted Vegetation Indices. *Remote Sens. Environ.* **1996**, *55*, 95–107. [CrossRef]
67. Roujean, J.-L.; Breon, F.-M. Estimating PAR Absorbed by Vegetation from Bidirectional Reflectance Measurements. *Remote Sens. Environ.* **1995**, *51*, 375–384. [CrossRef]
68. Blackburn, G.A. Quantifying Chlorophylls and Caroteniods at Leaf and Canopy Scales: An Evaluation of Some Hyperspectral Approaches. *Remote Sens. Environ.* **1998**, *66*, 273–285. [CrossRef]
69. Daughtry, C.S.T.; Walthall, C.L.; Kim, M.S.; de Colstoun, E.B.; McMurtrey, J.E. Estimating Corn Leaf Chlorophyll Concentration from Leaf and Canopy Reflectance. *Remote Sens. Environ.* **2000**, *74*, 229–239. [CrossRef]
70. Wu, C.; Niu, Z.; Tang, Q.; Huang, W. Estimating Chlorophyll Content from Hyperspectral Vegetation Indices: Modeling and Validation. *Agric. For. Meteorol.* **2008**, *148*, 1230–1241. [CrossRef]
71. Barnes, E.; Clarke, T.R.; Richards, S.E.; Colaizzi, P.; Haberland, J.; Kostrzewski, M.; Waller, P.; Choi, C.; Riley, E.; Thompson, T.L. Coincident Detection of Crop Water Stress, Nitrogen Status, and Canopy Density Using Ground Based Multispectral Data. In Proceedings of the Fifth International Conference on Precision Agriculture, Bloomington, MN, USA, 16–19 July 2000.
72. Gitelson, A.A.; Gritz, Y.; Merzlyak, M.N. Relationships between Leaf Chlorophyll Content and Spectral Reflectance and Algorithms for Non-Destructive Chlorophyll Assessment in Higher Plant Leaves. *J. Plant Physiol.* **2003**, *160*, 271–282. [CrossRef] [PubMed]
73. Rouse, J.W.; Haas, R.H.; Deering, D.W.; Schell, J.A.; Harlan, J.C. *Monitoring the Vernal Advancement of Retrogradation (Green Wave Effect) of Natural Vegetation*; NASA/GSFC, Type III, Final Report; NASA: Greenbelt, MD, USA, 1974.
74. Tucker, C.J. Red and Photographic Infrared Linear Combinations for Monitoring Vegetation. *Remote Sens. Environ.* **1979**, *8*, 127–150. [CrossRef]
75. Huete, A.R. A Soil-Adjusted Vegetation Index (SAVI). *Remote Sens. Environ.* **1988**, *25*, 295–309. [CrossRef]
76. Qi, J.; Chehbouni, A.; Huete, A.R.; Kerr, Y.H.; Sorooshian, S. A Modified Soil Adjusted Vegetation Index. *Remote Sens. Environ.* **1994**, *48*, 119–126. [CrossRef]
77. Broge, N.H.; Leblanc, E. Comparing Prediction Power and Stability of Broadband and Hyperspectral Vegetation Indices for Estimation of Green Leaf Area Index and Canopy Chlorophyll Density. *Remote Sens. Environ.* **2001**, *76*, 156–172. [CrossRef]
78. Gitelson, A.A.; Keydan, G.P.; Merzlyak, M.N. Three-Band Model for Noninvasive Estimation of Chlorophyll, Carotenoids, and Anthocyanin Contents in Higher Plant Leaves. *Geophys. Res. Lett.* **2006**, *33*, 026457. [CrossRef]
79. Tian, Y.; Gu, K.; Chu, X.; Yao, X.; Cao, W.; Zhu, Y. Comparison of Different Hyperspectral Vegetation Indices for Canopy Leaf Nitrogen Concentration Estimation in Rice. *Plant Soil* **2014**, *376*, 193–209. [CrossRef]
80. Verrelst, J.; Rivera, J.P.; Veroustraete, F.; Muñoz-Marí, J.; Clevers, J.G.P.W.; Camps-Valls, G.; Moreno, J. Experimental Sentinel-2 LAI Estimation Using Parametric, Non-Parametric and Physical Retrieval Methods—A Comparison. *ISPRS J. Photogramm. Remote Sens.* **2015**, *108*, 260–272. [CrossRef]
81. Wang, W.; Yao, X.; Yao, X.; Tian, Y.; Liu, X.; Ni, J.; Cao, W.; Zhu, Y. Estimating Leaf Nitrogen Concentration with Three-Band Vegetation Indices in Rice and Wheat. *Field Crops Res.* **2012**, *129*, 90–98. [CrossRef]
82. le Maire, G.; François, C.; Soudani, K.; Berveiller, D.; Pontailler, J.Y.; Bréda, N.; Genet, H.; Davi, H.; Dufrêne, E. Calibration and Validation of Hyperspectral Indices for the Estimation of Broadleaved Forest Leaf Chlorophyll Content, Leaf Mass per Area, Leaf Area Index and Leaf Canopy Biomass. *Remote Sens. Environ.* **2008**, *112*, 3846–3864. [CrossRef]

83. Dong, T.; Liu, J.; Shang, J.; Qian, B.; Ma, B.; Kovacs, J.M.; Walters, D.; Jiao, X.; Geng, X.; Shi, Y. Assessment of Red-Edge Vegetation Indices for Crop Leaf Area Index Estimation. *Remote Sens. Environ.* **2019**, *222*, 133–143. [CrossRef]
84. Zeng, Y.; Hao, D.; Huete, A.; Dechant, B.; Berry, J.; Chen, J.M.; Joiner, J.; Frankenberg, C.; Bond-Lamberty, B.; Ryu, Y.; et al. Optical Vegetation Indices for Monitoring Terrestrial Ecosystems Globally. *Nat. Rev. Earth Environ.* **2022**, 3, 477–493. [CrossRef]

Article

Effect of Snow Cover on Detecting Spring Phenology from Satellite-Derived Vegetation Indices in Alpine Grasslands

Yiting Wang [1,2,*], Yuanyuan Chen [1], Pengfei Li [1], Yinggang Zhan [1], Rui Zou [1], Bo Yuan [2,3] and Xiaode Zhou [2]

1 College of Geomatics, Xi'an University of Science and Technology, Xi'an 710054, China
2 State Key Laboratory of Eco-Hydraulics in Northwest Arid Region, Xi'an University of Technology, Xi'an 710048, China
3 College of Geology and Environment, Xi'an University of Science and Technology, Xi'an 710054, China
* Correspondence: wyt_rs@163.com

Abstract: The accurate estimation of phenological metrics from satellite data, especially the start of season (SOS), is of great significance to enhance our understanding of trends in vegetation phenology under climate change at regional or global scales. However, for regions with winter snow cover, such as the alpine grasslands on the Tibetan Plateau, the presence of snow inevitably contaminates satellite signals and introduces bias into the detection of the SOS. Despite recent progress in eliminating the effect of snow cover on SOS detection, the mechanism of how snow cover affects the satellite-derived vegetation index (VI) and the detected SOS remains unclear. This study investigated the effect of snow cover on both VI and SOS detection by combining simulation experiments and real satellite data. Five different VIs were used and compared in this study, including four structure-based (i.e., NDVI, EVI2, NDPI, NDGI) VIs and one physiological-based (i.e., NIRv) VI. Both simulation experiments and satellite data analysis revealed that the presence of snow can significantly reduce the VI values and increase the local gradient of the growth curve, allowing the SOS to be detected. The bias in the detected SOS caused by snow cover depends on the end of the snow season (ESS), snow duration parameters, and the snow-free SOS. An earlier ESS results in an earlier estimate of the SOS, a later ESS results in a later estimate of the SOS, and an ESS close to the snow-free SOS results in small bias in the detected SOS. The sensitivity of the five VIs to snow cover in SOS detection is NDPI/NDGI < NIRv < EVI2 < NDVI, which has been verified in both simulation experiments and satellite data analysis. These findings will significantly advance our research on the feedback mechanisms between vegetation, snow, and climate change for alpine ecosystems.

Keywords: vegetation phenology; snow cover; vegetation index; SOS; Tibetan Plateau; remote sensing

Citation: Wang, Y.; Chen, Y.; Li, P.; Zhan, Y.; Zou, R.; Yuan, B.; Zhou, X. Effect of Snow Cover on Detecting Spring Phenology from Satellite-Derived Vegetation Indices in Alpine Grasslands. *Remote Sens.* **2022**, *14*, 5725. https://doi.org/10.3390/rs14225725

Academic Editors: Kenji Omasa, Shan Lu and Jie Wang

Received: 3 October 2022
Accepted: 8 November 2022
Published: 12 November 2022

1. Introduction

Land surface phenology is the assessment of seasonal vegetation growth at a large scale using satellite remote sensing and has been widely used to quantify the response of terrestrial ecosystems to climate change [1–3]. Alpine ecosystems, characterized by high elevations, low temperatures, snows, and short growing seasons, are very sensitive to climate change and are regarded as "climate change hot spots". The accurate estimation of phenological metrics from satellite data, especially the start of season (SOS), is critical for understanding the dynamics of alpine vegetation and climate change. As the third pole of the earth and the largest alpine pasture in Asia, the Tibetan Plateau is a research focus in land surface phenology [4]. However, the prevalent and seasonal snow cover, one of the major features of alpine ecosystems, increases the complexity of monitoring vegetation phenology from satellites [5]. Especially in the Tibetan Plateau, existing studies have yielded inconsistent results on the SOS changes, and snow cover has been attributed as a major cause [6,7].

Phenological transitions are generally detected from the seasonal dynamics of the satellite-derived vegetation index (VI). The VI measures the greenness of vegetation through algebraic combinations of the multiband reflectance of satellite data and is closely related to the biophysical and structural properties of the canopy [8]. From the VI's trajectory, the SOS is detected as the point in time when the VI reaches a threshold, the growth gradient reaches its maximum, or the VI exceeds the moving average VI curve, corresponding to the threshold method, the derivative method, and the moving average methods, respectively [9]. While each method has its own advantages and shortcomings, there is no consensus on which method performs best [10,11]. The dynamic threshold method, achieving a balance between simplicity, universality, and robustness to noise, is one of the most commonly used methods, especially in the latest MODIS phenology product MCD12Q2 C006 [12]. Among the various VIs, the normalized difference vegetation index (NDVI) [13] is the earliest and most commonly used VI in SOS detection due to its simplicity and long records of historical data [14–16], but it suffers from saturation in densely vegetated areas and interference from soil backgrounds. To reduce the sensitivity of VI to the soil background and atmosphere, the two-band enhanced vegetation index (EVI2) [17] was proposed and has been widely used in SOS detection, such as in the VIIRS phenology product VNP12Q2 [18].

Although various satellite-derived VIs have been successfully applied in phenology detection [19–21], they face major limitations in alpine grasslands due to snow's interference with satellite signals [22,23]. The presence of snow can significantly affect the VI's value and change the VI's trajectory, while snowmelt can cause a rapid increase in the VI's trajectory [24,25]. If the effect of snow cover is not considered, the detected SOS may be a snowmelt date instead of the SOS [26,27], which will further cause bias in our understanding of vegetation phenology trends and climate change [28]. For example, preseason snow was found to cause the SOS detected by NDVI to advance compared to snow-free cases [20,21].

Previous studies have attempted to eliminate the effect of snow cover on SOS detection from satellite data. Some studies introduced auxiliary information on snow, precipitation, and temperature to replace the SOS of snow-covered pixels with those of snow-free background pixels [18,29]. However, auxiliary data are not always available in large alpine areas, and additional data can also add bias and uncertainty [30]. Alternative approaches have attempted to propose new snow-free VIs, such as the normalized difference phenology index (NDPI) [31] and the normalized difference greenness index (NDGI) [21], which were recently developed to eliminate the effects of snow and soil. Both VIs were found to have better correlation with the in situ measurements and outperform the traditional VIs under snow conditions, such as NDVI and EVI [20,21]. In some studies, the SOS dates detected by NDPI or NDGI were used as the SOS detected under snow-free conditions to evaluate the advancement or delay of the SOS under snow conditions [20,32]. In addition, the near-infrared reflectance of vegetation (NIRv) [33] and solar-induced chlorophyll fluorescence (SIF), as direct indicators of vegetation photosynthesis, are promising indicators for phenological monitoring [33,34]. Both SIF and NIRv are physiological-based VIs and overcome the saturation problem of NDVI. Existing studies verified the good consistency of the SOS detected by NIRv and SIF with the SOS measured by flux towers [20,35].

Despite recent progress in eliminating the effect of snow cover on SOS detection, the mechanism of how snow cover affects VI values and subsequent SOS detection remains unclear. Existing studies have attempted to find evidence from satellite data or in situ measurements [36–39], yet it is challenging to compare snow-free and snow-covered areas directly. Vegetation growth on snow-free pixels cannot simply represent the growth on the snow-covered pixels due to the confounding effects of snow cover on SOS detection and on SOS itself. Furthermore, although several new snow-free VIs and SIF-related VIs have been proposed [31,35,40], there are no definitive answers as to how they are affected by snow cover and which VI performs best for alpine ecosystems. Direct evidence on how snow cover affects SOS detected from satellite-derived VIs is urgently needed to enhance our understanding of vegetation phenology changes on the Tibetan Plateau.

To address the above issues, this study combined simulation experiments and satellite data to investigate the effect of snow cover on VI values and subsequent SOS detection, aiming to clarify the mechanism of how snow cover affects SOS detection from satellite-derived VIs. Four snow parameters were adopted to describe the coverage and phonological characteristics of snow, including snow cover fractions (SCFs), snow cover duration (i.e., consecutive days with snow and the ratio of days with snow to total days, hereafter referred to as SCDc and SCDr), and the end of the snow season (ESS). Five different VIs were used and compared in this study, including four structure-based (i.e., NDVI, EVI2, NDPI, NDGI) VIs and one physiological-based (i.e., NIRv) VI. Simulation experiments were carefully designed to model the time series of different VIs under different snow scenarios to investigate the difference in SOS between snow and snow-free conditions. Then, the variations in the SOS under different snow conditions were analyzed using satellite data. The main objectives of this study are: (1) to elucidate the effect of snow cover on the five VIs and subsequent SOS detection through simulation experiments; (2) to analyze the spatial and temporal patterns of snow cover and investigate the effect of snow cover on the detected SOS using real satellite data from 2020; and (3) to compare the performance of the five VIs in detecting the SOS under snow-covered conditions.

2. Study Area and Data

2.1. Study Area

The study area is on the east of the Tibetan Plateau (29°35′30″N–35°48′06″N, 94°08′08″E–101°03′36″E, Figure 1), covering a large area of alpine grasslands at an average elevation of 4000 m. Our study focuses on alpine grasslands, which are dominated by alpine meadows and alpine steppe [41]. The study area is typical of alpine meadows, which normally grow from May to September. The mean temperature ranges from near 0 °C to above 20 °C, and precipitation ranges from 100 mm to over 1000 mm along a south–north gradient [4]. Seasonal snow covers the study area from October to May, which may contaminate the satellite signals and introduce bias in SOS detection.

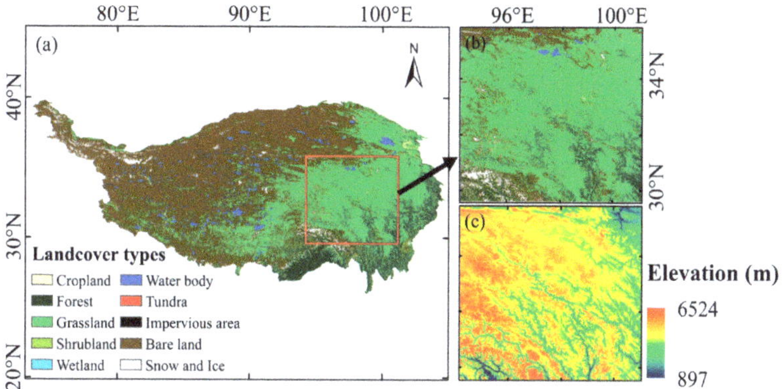

Figure 1. Map of the study area. (**a**) MODIS land cover map of the Qinghai–Tibet Plateau. (**b**) MODIS land cover map and (**c**) DEM of the study area.

2.2. Datasets

2.2.1. Satellite Reflectance Data

The MODIS MCD43A4 Nadir Bidirectional Reflectance Distribution Function (BRDF) Adjusted Reflectance (NBAR) product with a 500 m resolution was used in this study to remove the viewing angle effects from directional reflectivity [42]. A total of 366 images from 1 January 2020 to 31 December 2020 were freely downloaded from the NASA Land Processes Distributed Active Archive Center (LP DAAC, http://lpdaac.usgs.gov (accessed on 10 July 2021)), which were processed to UTM/WGS-84 projection and 500 m resolution. Based on repeated manual tests, missing values of 50 consecutive days would introduce significant errors in reconstructing the VI trajectory. To eliminate errors caused by missing values in the daily NBAR products, pixels with more than 50 consecutive days of missing data were flagged as poor-quality pixels and excluded from subsequent analysis, which accounted for only 3.34% of the study area. Reflectance in the red, green, near-infrared (NIR), and shortwave infrared (SWIR) bands was used to calculate the different VIs and to further determine the SOS dates.

2.2.2. Snow Cover Data

The MODIS snow cover data MOD10A2, with a 500 m spatial resolution and 8-day intervals, was used to derive snow phenology parameters in our study. This snow cover product has higher spatial and temporal resolutions [43]. Previous studies have verified that MOD10A2 is effective in reducing the effect of cloud contamination in most cases and has higher classification accuracy than another MODIS snow product (MOD10A1) [37,44]. A total of 46 images from 1 January 2020 to 31 December 2020 were freely downloaded from the National Snow and Ice Data Center (NSIDC, http://nsidc.org (accessed on 17 December 2021)). All images were processed to UTM/WGS-84 projection and 500 m resolution. Pixels coded as '200' were extracted as snow pixels according to the product's user guide [43].

2.2.3. Land Cover Type and DEM

A subset of the global land cover product from Tsinghua University at 10 m resolution in 2017 (available at http://data.ess.tsinghua.edu.cn (accessed on 15 July 2021)) was used to define our study area for its high spatial resolution and high overall accuracy [45], as shown in Figure 1. The SRTM digital elevation model (DEM) data at 90 m resolution were used to characterize the variations in snow phenology and vegetation phenology. The DEM data were freely downloaded from the Geospatial Data Cloud (http://www.gscloud.cn (accessed on 15 July 2021)) and processed to UTM/WGS-84 projection and 500 m resolution, as shown in Figure 1c.

3. Methods

3.1. Derivation of Vegetation Indices

Five VIs were selected and derived from MODIS daily NBAR data for phenology detection, including NDVI, EVI2, NDPI, NDGI, and NIRv. Their definitions and corresponding references are given in Table 1.

Table 1. Derivation of vegetation indices used in the study.

Index Acronym	Formula	Reference
NDVI	$NDVI = \frac{R_{NIR}-R_{Red}}{R_{NIR}+R_{Red}}$	[13]
EVI2	$EVI2 = 2.5\frac{R_{NIR}-R_{Red}}{R_{NIR}+2.4R_{Red}+1}$	[17]
NDPI	$NDPI_{MODIS} = \frac{R_{NIR}-[0.74\times R_{Red}+0.26\times R_{SWIR}]}{R_{NIR}+[0.74\times R_{Red}+0.26\times R_{SWIR}]}$	[31]
NDGI	$NDGI_{MODIS} = \frac{0.65\times R_{Green}+0.35\times R_{NIR}-R_{Red}}{0.65\times R_{Green}+0.35\times R_{NIR}+R_{Red}}$	[21]
NIRv	$NIRv = NDVI \times R_{NIR}$	[33]

Note: R_{Green}, R_{Red}, R_{NIR}, and R_{SWIR} are the surface reflectance values in the green, red, NIR, and SWIR bands, respectively.

Among the five VIs, NDVI is the most commonly used VI for monitoring land surface phenology. EVI2 was developed from the enhanced vegetation index (EVI) [46] to adapt to satellite sensors without a blue band [17], while both EVI and EVI2 were designed to overcome the saturation problem and sensitivity to the soil background with NDVI. The NDPI and NDGI and NIRv were recently developed and are rather new. Both NDPI and NDGI were proposed to maximize the contrast between vegetation and soil/snow [21]. The NDPI assumes that the reflectance of soil and snow increases or decreases monotonically from red to SWIR wavelengths. In contrast, the reflectance of vegetation is high in the NIR band and low in both the red and SWIR bands [40]. The NDPI is thus designed by replacing the red band in NDVI with a weighted sum of the red and SWIR bands. As a result, the NDPI is close to zero for soil and snow but high for vegetation [31,40]. NDGI is a semi-analytical snow-free VI based on a linear mixture model. It connects a straight line between the reflectance of the green and NIR bands, and the difference between this line and the reflectance of the red band is defined as the NDGI [21]. NDGI is positive for vegetation but is close to zero for snow, soil, and dry grass. NIRv has been proposed as a proxy for SIF, a very effective indicator of vegetation photosynthesis [47], and has been successfully used in phenology detection [33]. The comparison of these five typical VIs would provide a useful reference for the performance of both structural-based and physiological-based VIs, as well as the traditional VIs and snow-free VIs.

3.2. Spring Phenology Detection and Evaluation

3.2.1. Detection of SOS Dates

Vegetation growth in the alpine grasslands on the Tibetan Plateau has a distinct seasonal cycle, and the SOS can be detected from the time series of five different VIs. We first removed the poor-quality/filled values from the original MODIS datasets and then used a double-logistic fitting method to smooth the noisy time series of VIs in the TIMESAT program [48], since it is more robust in extracting phenological parameters [49–52]. We applied two upper-envelope iterations with an adaptation strength of 2 to reduce the bias of atmospheric effects in the smoothing process [20]. Then, the simple and intuitive dynamic threshold method [53,54] was used to detect the SOS (Figure 2) for its simplicity and robustness [55], and a 20% amplitude was used to determine the SOS, which was consistent with previous studies [20,55,56]. Figure 2 illustrates how the dynamic threshold method detects the SOS under snow and snow-free conditions. The SOS date was detected as the point in time when the VI increased to 20% of the amplitude plus the base value.

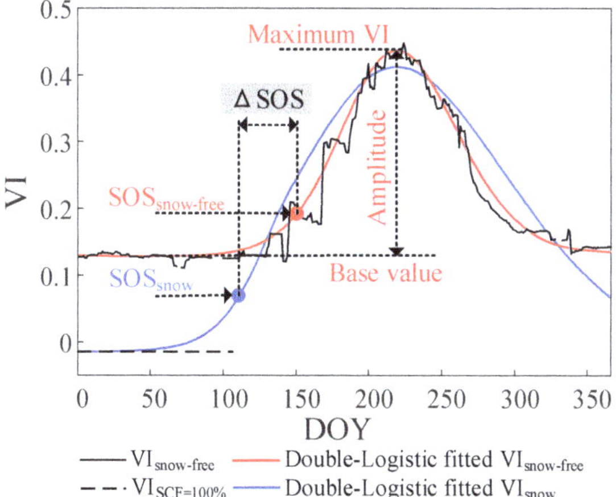

Figure 2. A schematic diagram of how the dynamic threshold method detects the start of the season (SOS). $\Delta SOS = SOS_{snow} - SOS_{snow\text{-}free}$.

3.2.2. Evaluation of SOS Dates

The SOS detected from the five VIs were compared pixel-by-pixel via scatterplots on each pair of image combinations. Three indicators, the correlation coefficient (R), the mean absolute error (MAE), and the root mean square error (RMSE), were calculated to evaluate the consistency between the SOS detected from different VIs. Then, considering the influence of elevation on the SOS, the study area was divided into four elevation zones, including < 3500 m, 3500–4000 m, 4000–4500 m, and ≥ 4500 m. The changes in the detected

SOS dates with varying elevations were analyzed through zonal statistics. The effect of snow on the detected SOS was further analyzed in each elevation zone.

3.3. *Effect of Snow Cover on Detecting SOS from VIs*

3.3.1. Snow Cover Analysis

Four parameters were used to describe the coverage and phenology of snow, including SCF, SCDc, SCDr, and ESS. The SCF is defined as the snow cover fraction at each location. The SCDc and SCDr are complementary parameters used to describe snow cover duration. The SCDc denotes the maximum number of consecutive days with snow cover and is more specific, while SCDr is defined as the ratio of days with snow to total days and is more general. The SCDc is more ready-to-use in simulating different snow scenarios if we only consider the longest snow duration, while the SCDr is difficult to simulate, as the spring snow cover might include several periods of snow duration. Thus, SCDc was used in both simulations and satellite data analysis, and SCDr was only used in satellite data analysis. The ESS is defined as the ending date of snow, i.e., the last day with snow cover. Since our focus is spring phenology, the SCDc, SCDr, and ESS were calculated for the period from the day of year (DOY) 001 to 208. The four parameters were calculated from the MOD10A2 snow cover data to reveal the spatial and temporal characteristics of snow in the study area. Furthermore, they were also used in simulation experiments to generate different snow scenarios.

3.3.2. Tests with Simulated Data

Since it is very difficult to find appropriate snow-free vegetation pixels as a reference, we designed simulation experiments to evaluate the effect of snow on SOS detection. A linear spectral mixture model was used to simulate the pixel reflectance with different SCF values. For simplicity, it is assumed that each pixel is composed of soil and vegetation; the snow layer is covered above and only absorbs and reflects the incident light, i.e., the transmittance of the snow layer equals 0. Assuming that the vegetation and soil components are both homogeneous, which is reasonable for alpine grasslands, the presence of snow will not affect the areal compositions of soil and vegetation. As a result, the spectral reflectance of a snow-covered pixel can be computed as:

$$R_{mixed} = (1 - f_{snow})\left[f_{veg} \cdot R_{veg} + (1 - f_{veg})R_{soil})\right] + f_{snow} \cdot R_{snow} \quad (1)$$

where R_{mixed} denotes the simulated mixed pixel reflectance; f_{veg}, f_{soil}, and f_{snow} are the coverage fractions of vegetation, soil, and snow, respectively; and R_{veg}, R_{soil}, and R_{snow} are the corresponding endmember reflectances.

To simulate the satellite-derived band reflectance, the spectral reflectance $\rho(\lambda)$ was convolved with the spectral response function (SRF) of the MODIS sensor $S(\lambda)$ as follows:

$$R = \int_{\lambda_1}^{\lambda_2} \rho(\lambda)S(\lambda)d\lambda \quad (2)$$

where λ1 and λ2 are the minimum and maximum wavelengths of each band, respectively. The $\rho(\lambda)$ for snow, soil, and vegetation were selected from the Johns Hopkins University Spectral Library [57], corresponding to medium granular snow, dark brown fine sandy loam, and green grass, respectively, as in Figure 3.

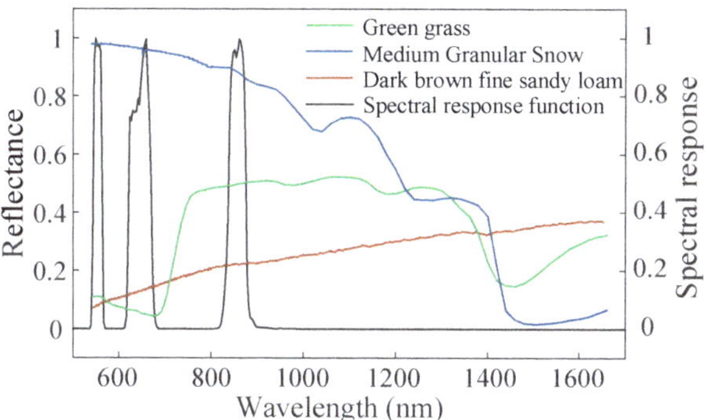

Figure 3. Schematic diagram of endmember reflectance spectra and the spectral response of the MODIS sensor.

Through Equations (1) and (2), the band reflectance and further VIs of a pixel can be computed using the endmember reflectance with varying FVC and SCF values. To investigate the effect of snow cover on SOS detection, simulation experiments were designed in the following two aspects.

First, we investigated the effect of snow cover on five VIs. We computed the VI values when FVC was varied from 0 to 1 for five cases of SCF = 0%, 25%, 50%, 75%, and 100%. The effect of snow cover on VI values can be described by the difference in VI values under specific snow scenarios and snow-free conditions (i.e., SCF = 0%), expressed as $\Delta VI = VI_{SCF > 0} - VI_{SCF = 0}$. As the values of ΔVI vary with SCF and FVC, we further define a quantitative indicator, the maximum impact of snow (MIS), to represent the maximum effect of snow on VI for a specific FVC, expressed as:

$$MIS = \frac{|VI_{SCF = 100\%} - VI_{SCF = 0\%}|}{VI_{max} - VI_{min}} \tag{3}$$

where $VI_{SCF = 100\%}$ and $VI_{SCF = 0\%}$ correspond to VI values when SCF = 100% and SCF = 0% for a given FVC, respectively; VI_{max} and VI_{min} are the maximum and minimum VI values, respectively. The denominator is the range of VI values and is used to eliminate the effect of different value ranges of VIs. Both VI and MIS vary with FVC. The numerator represents the absolute difference in VI between SCF = 100% and SCF = 0%, while the MIS represents the maximum percentage change in VI caused by snow relative to the range of VI values. The MIS provides a direct indication of the extent to which a VI is affected by snow and provides a basis to further investigate the effect of snow cover on SOS detection.

Second, to investigate the effect of snow cover on SOS detection, we designed a series of experiments to generate different VI time curves under different snow scenarios. We extracted the NDVI time curve of a typical snow-free vegetation pixel and converted the time series of NDVI to FVC using the dimidiate pixel model [58]. Using the derived time series of FVC, the time series of band reflectance and VI were calculated under different snow scenarios defined by snow parameters, including SCF, SCDc, and ESS. The time curves of five different Vis were then filtered, and the SOS was detected. As shown in Figure 2, given the same growth curve of FVC, the red and blue lines represent the VI trajectories for snow-free and preseason snow conditions, respectively. The preseason snow caused a bias in the detected SOS, expressed as $\Delta SOS = SOS_{snow} - SOS_{snow\text{-}free}$, which is defined as the effect of snow on SOS detection.

Three sets of experiments were designed, and the corresponding settings of snow scenarios are shown in Table 2. In all experiments, as we were only concerned with the SOS, only the snow season from DOY 001 to DOY 208 was considered. Experiment I corresponded to a completely snow-free case with SCF = 0% during the period, which served as the baseline to assess the effect of snow cover on SOS detection (Figure 4a).

Experiments II and III were snow cover conditions, where four cases of SCF = 25%, 50%, 75%, and 100% were considered, and in all cases, the SCF remained constant during the snow season. Experiment II referred to the cases of snow persisting from DOY 001 to ESS. Three cases of ESS at DOY 104, 136, and 168 were considered, which were the mean ESS plus or minus its standard deviation analyzed from snow cover data, as shown in Figure 4b. In experiment III, three cases of SCDc = 32, 64, and 96 days were considered. As the SCDc could be in any interval during the snow season, we simulated all cases by iterating the start of the snow season from DOY 001 at 16-day intervals while keeping the ESS no later than DOY 208. For example, a case with SCDc = 64 can generate 10 different time curves of a VI with a snow season ranging from DOY 1–64 to 145–208, as shown in Figure 4c.

Table 2. Experimental settings for the investigation of the effect of snow on spring vegetation phenology detection.

Experiments No.	Snow Scenarios *
I	Snow free with SCF = 0% constantly during the period from DOY 001 to 208.
II	Snow persists from DOY 001 to ESS (DOY 104, 136, and 168) with constant SCF.
III	Snow persists from DOY *t* to ESS with constant SCF, where *t* is iterated from DOY 001 at 16-day interval; ESS = *t* + SCDc − 1 ≤ 208; and SCDc = 32, 64, and 96 days.

* In experiments II and III, four cases with SCF = 25%, 50%, 75%, and 100% were considered.

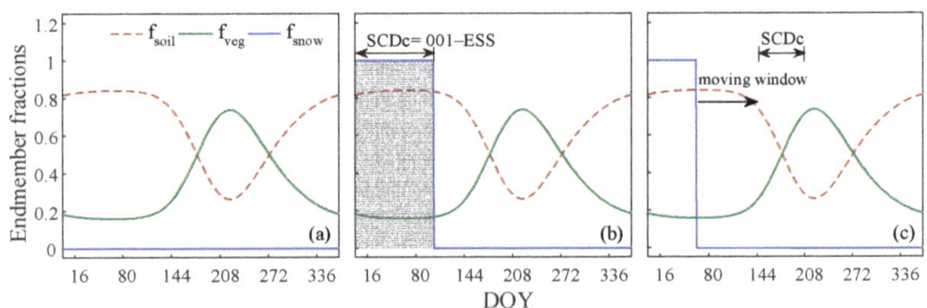

Figure 4. Temporal trajectories of coverage fractions of vegetation, snow, and soil for experiments I to III. (**a**) Experiment I with SCF = 0% constantly from DOY 001 to 208; (**b**) experiment II with snow persisting from DOY 001 to ESS with constant SCF; (**c**) experiment III with snow persisting from DOY *t* to ESS with constant SCF.

3.3.3. Tests with Satellite Data

Since snow is prevalent, a direct comparison of VI in snow-covered and snow-free pixels is difficult. Based on the simulation results, the effect of snow cover on SOS detection was analyzed from satellite data in two aspects. First, the statistical distribution of the minimum and maximum values of VI over time for each SCDr interval in each elevation zone was analyzed. This analysis could help reveal the difference in VI values between different SCDr intervals in the same elevation zone. Second, the statistical distribution of the SOS detected from real satellite data for different SCDc and ESS scenarios was analyzed, which could help reveal the variations in the detected SOS under different SCDc and ESS cases.

The flowchart of this study is shown in Figure 5.

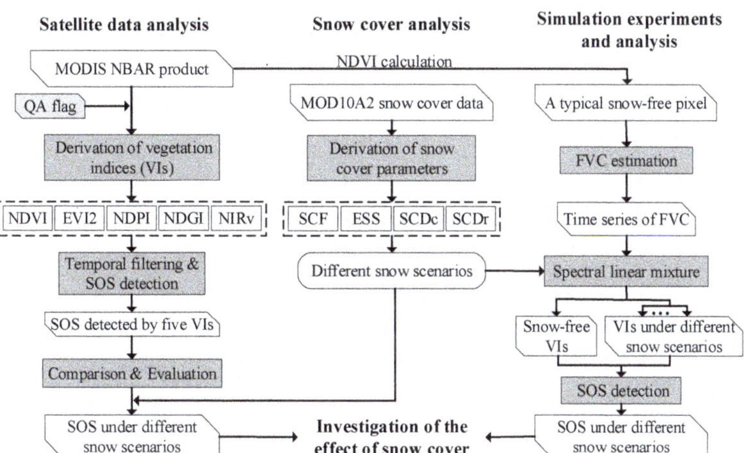

Figure 5. Flowchart of the investigation of the effect of snow cover on spring phenology detection.

4. Results

4.1. Snow Cover Analysis

The snow cover analysis from MOD10A2 data in Figure 6 shows that a large portion of the study area was covered by snow, while the SCDr varied across the area. Statistically, the SCDr in 52.27% of the study area was higher than 40%, and the SCDr was lower than 20% only in 21.25% of the study area.

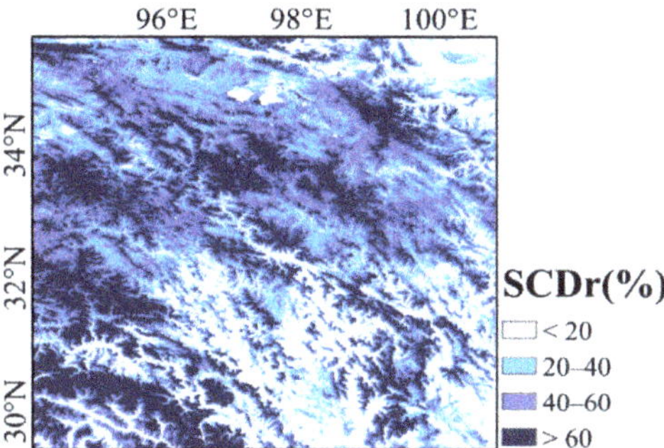

Figure 6. The derived SCDr map in the study area. SCDr is the ratio of days with snow to total days from DOY 001 to DOY 208.

Statistically, approximately 77% of the study area has an elevation $\geq$ 4000 m, while only approximately 6% of the area has an elevation < 3500 m. Figure 7 shows the statistical distributions of ESS, SCDr, and SCDc in each elevation zone. In all four elevation zones, the ESS values were concentrated on DOY 136. The increase in elevation led to the increase in ESS. When the elevation increased to $\geq$4000 m, a significant subpeak in the ESS appeared on DOY 160. Both SCDr and SCDc increased with increasing elevation. The peak values of SCDr were 0%, 8%, 36%, and 64%, while the peak values of SCDc were 8, 8, 16, and 88 days, respectively, for all four elevation zones from elevations <3500 m to $\geq$4500 m.

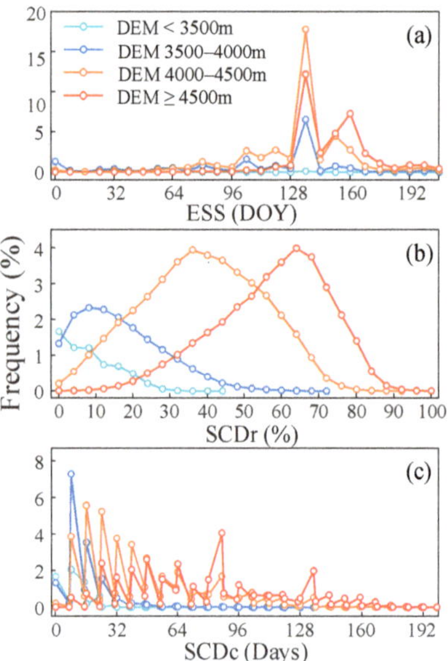

Figure 7. Statistical distributions of (**a**) ESS, (**b**) SCDr, and (**c**) SCDc in each elevation zone.

4.2. Spring Phenology Derived from Different VIs

Figure 8 shows the SOS detected from five different VIs. Nongrassland pixels (e.g., water bodies, forests, etc.) and the pixels whose SOSs were poorly detected were flagged as no data. SOS trends derived from different VIs were generally spatially consistent with the elevation variations. The SOS dates in the eastern part of the study area were earlier than those in the western parts. As shown in Figure 8, the SOS dates detected by NDVI were very similar to the SOS dates detected by EVI2, while those from NDPI and NDGI were highly consistent and those detected by NIRv were in between. The histograms revealed that the SOS dates detected by the five VIs were generally similar in terms of the value ranges and peak values. The detected SOS dates ranged from 105 to 175, while the average SOS dates derived from the five VIs followed the order of NDPI (DOY 154) > NDGI (DOY 152) > NIRV (DOY 150) > EVI2 (DOY 145) > NDVI (DOY 141), which was very consistent with previous studies in this study area [38,55,59]. This indicates that the SOS dates detected by the five VIs were generally consistent and reliable.

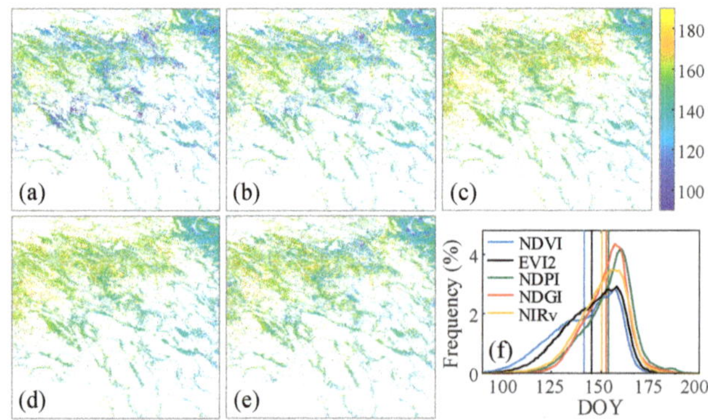

Figure 8. The SOS detected by (**a**) NDVI, (**b**) EVI2, (**c**) NDPI, (**d**) NDGI, (**e**) NIRV, and (**f**) the corresponding histograms for 2020.

Scatterplots in Figure 9 compare the SOS dates detected from different VIs on a pixel-by-pixel basis. According to the consistency between the detected SOS, the five different VIs were aggregated into two groups. One group included NDGI and NDPI (R = 0.849 and RMSE = 5.957), and the other group included EVI2, NDVI, and NIRv. The SOS dates detected by EVI2 were highly correlated with those detected by NDVI (R = 0.934 and RMSE = 6.416) and NIRv (R = 0.938 and RMSE = 6.870), while the SOS detected by NDVI and NIRv had lower correlations (R = 0.776 and RMSE = 12.722). The SOS detected by the VIs from different groups showed large discrepancies. For example, the SOS dates from NDVI and NDPI were poorly correlated (R = 0.445) and had a large bias (RMSE = 17.823). In addition to the differences in VI calculations, this different performance may also be attributed to the different effects of snow cover on different VIs.

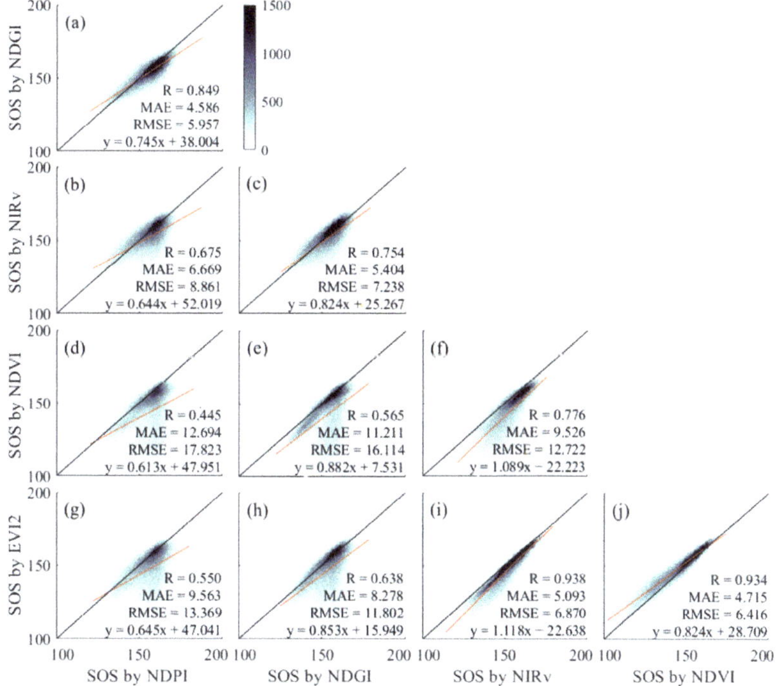

Figure 9. Comparison of the SOS detected by different VIs. The color from light gray to dark gray indicates increasing sample densities. (**a**–**j**) are scatterplots of the SOS dates detected from different VIs on a pixel-by-pixel basis.

4.3. Simulation Results

4.3.1. Effect of Snow Cover on VI

Figure 10 compares the changes in the five VIs under different SCF conditions as the FVC increases from 0 to 1. For all indices, the values of the five VIs generally decreased with increasing SCF. In the presence of snow cover, the larger the FVC was, the greater the decrease in the VI value. The deviation of the data points from the 1:1 line showed the effect of snow cover on VI values. The dashed line in each subplot of Figure 10 indicates the difference in VI values between SCF = 0% and SCF = 100%, denoted as $|\Delta VI|_{max}$. MIS values were further calculated for FVC = 0 and FVC = 1 to show the maximum possible influence of snow on VI values using Equation (3). Thus, the sensitivity of the five VIs to snow cover is NDPI (MIS range 0.0466–0.9534) < NDGI (MIS range 0.0122–1.0122) < NIRv (MIS range 0.2335–1.2335) < EVI2 (MIS range 0.3320–1.3320) < NDVI (MIS range 0.5280–1.5280).

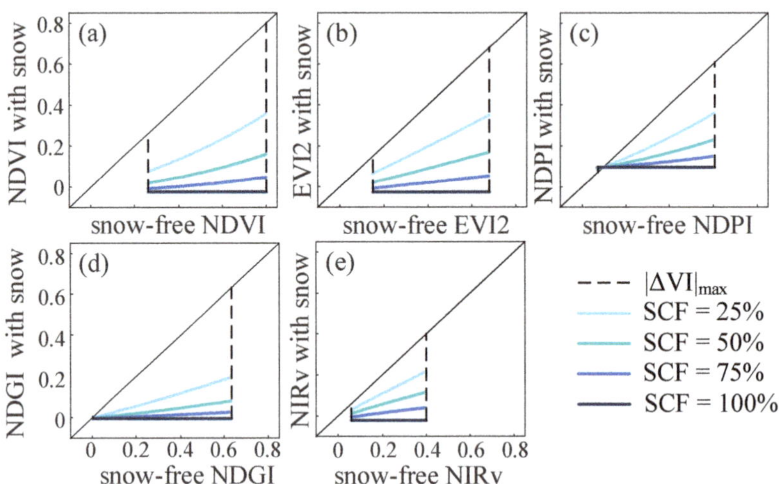

Figure 10. Comparison of simulated VIs under different SCFs with snow-free cases. (**a**–**e**) are the changes of the five VIs as the FVC increases from 0 to 1 under different SCF conditions.

4.3.2. Effect of Snow Cover on SOS Detection

Figure 11 shows the temporally filtered time curves of the five VIs and the detected SOS dates in simulation experiments I and II. Under snow-free conditions, the SOS detected by the different VIs ranged from DOY 144 to DOY 152, indicating small differences in snow-free SOS detected by the different VIs. Under snow conditions, the SOS dates detected by NDVI and EVI2 were earlier, while those detected by NDGI and NDPI were later, and those detected by NIRv were in between, as shown in all subplots of Figure 11.

Considering the differences in the SOS between SCF = 0 and SCF > 0 in Figure 11, the effect of snow on the detected SOS generally follows the order of NDPI/NDGI < NIRv < EVI2/NDVI. Generally, the presence of snow significantly reduced the VI values during the pregrowth period and advanced the SOS for all five VIs. For all five VIs, the greatest advances in the SOS were found for the earliest snow season (i.e., for ESS at DOY 104), which ranged from 16 to 56 days. As the ESS increased from DOY 104 to 168, the advances in the detected SOS decreased rapidly. For the ESS at DOY 168, the SOS estimated by NDVI, EVI2, and NIRv was only 4–6 days earlier than the snow-free SOS, while those estimated by NDPI and NDGI were 6 and 2 days later than the snow-free SOS, respectively. This indicates that the ending date of persisting snow is very important. When persisting snow ends earlier than the snow-free SOS, the presence of snow reduces the minimum VI value during the pregrowth period but does not affect the maximum VI value during the peak growth period, which would increase the gradient of the time curve of VI significantly and cause the SOS to be detected earlier. As analyzed in Section 4.3.1, the snow-induced decrease in the VI value is very small at small VI values and is relatively larger at large VI values. When the snow season ends later than the snow-free SOS, the decrease in VI values around the SOS is larger than that of the pregrowth period. This may locally smooth the time curve of VI and delay the detected SOS compared to the case of an early snow season.

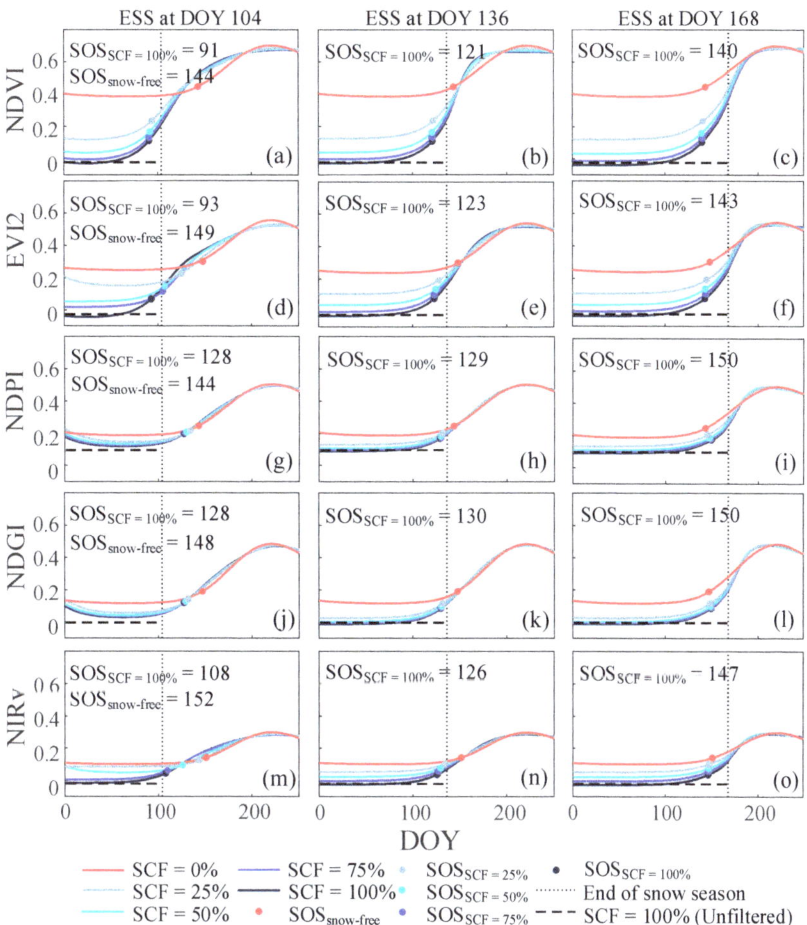

Figure 11. Time curves and the detected SOS of the five VIs under different SCF and ESS cases in experiments I and II. (**a–c**), (**d–f**), (**g–i**), (**j–l**), and (**m–o**) are the time curves of five VIs, for each of which three ESS cases at DOY 104, 136, and 168 and five cases of SCF = 0%, 25%, 50%, 75%, and 100% were plotted.

To further investigate the mechanism of how snow affects SOS detection, simulation experiment III was implemented to analyze the ΔSOS under different snow scenarios defined by SCDc, ESS, and SCF, and the results are shown in Figure 12. It clearly shows that ΔSOS changes with varying SCDc, ESS, and SCF values. Using the absolute values of ΔSOS as a standard, the effect of snow on SOS detection followed the order of NDPI/NDGI < NIRv < EVI2 < NDVI, which is consistent with the effect of snow on VI values analyzed in Section 4.3.1. Both SCDc and ESS are very important in determining the ΔSOS. In general, the larger the SCDc value was, the larger the absolute value of ΔSOS. This is reasonable because the reduction in VI values during a short snow period (i.e., small SCDc) can be better recovered by time series filtering performed prior to SOS detection. Specifically, for short snow with SCDc = 32, ΔSOS was very close to 0 for all VIs, except NDVI, for which ESS was earlier than the snow-free SOS. For longer snow with SCDc = 64 and 96, ΔSOS increases from negative to positive values as ESS increases. This also indicates that an earlier ESS generally advances the SOS, while an ESS much later than the snow-free

SOS delays the SOS, and ΔSOS approaches 0 when ESS approaches the snow-free SOS. These findings were consistent with the results of experiments I and II as analyzed above.

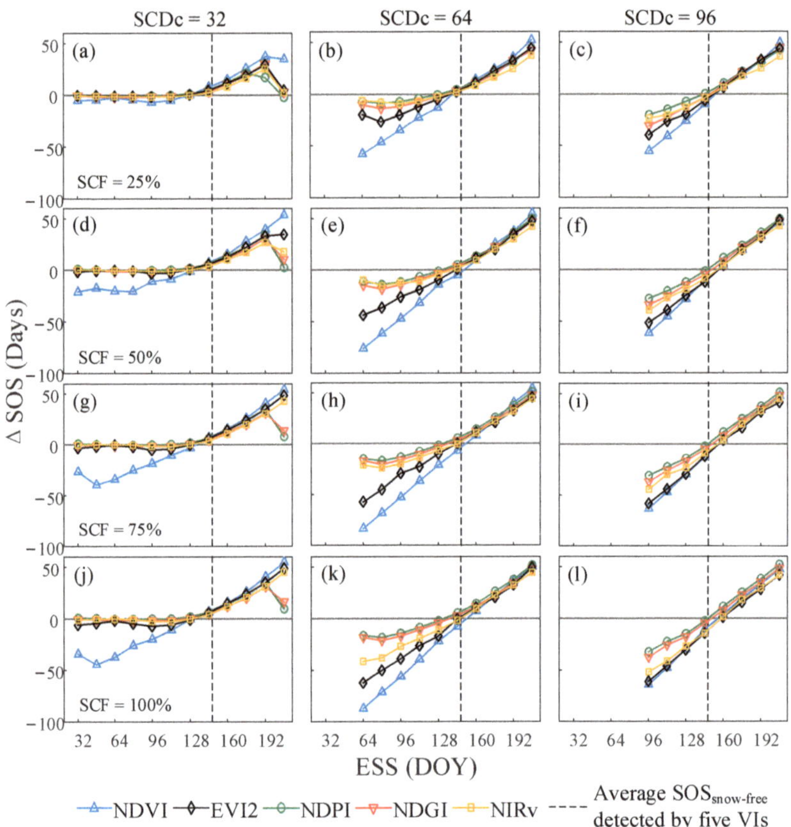

Figure 12. Changes in ΔSOS with increasing ESS under different SCF and SCDc cases derived from simulation experiment III. (**a–c**), (**d–f**), (**g–i**), and (**j–l**) are the changes in the ΔSOS with ESS for four cases of SCF = 25%, 50%, 75%, and 100%, in each group of which three cases of SCDc = 32, 64, and 96 were plotted.

To further investigate the different effects of ESS and SCDc on ΔSOS, we also analyzed the time curves of Vis under different ESS scenarios based on simulation experiment III. The medium SCDc = 64 was used. As the ΔSOS was close to 0 around the ESS at DOY 144, the ESS was varied as DOY 144 minus or plus three 16-day intervals, corresponding to three cases of ESS at DOY 96, 144, and 192. Figure 13 shows the temporally filtered time curves of VIs and the detected SOS. It clearly shows that the snow season ending much earlier than the snow-free SOS (ESS at DOY 96) advanced the SOS by up to 56 days, while the snow season ending much later than the snow-free SOS (ESS at DOY 192) delayed the SOS by up to 38 days. When the ESS approached the snow-free SOS (ESS at DOY 144), the changes in SOS caused by snow were very small. Therefore, the effect of snow cover on SOS detection depends on snow parameters, specifically SCDc, ESS, and the snow-free SOS. Because the presence of snow increases the local gradient of the VI growth curve and causes SOS to be detected, ESS and snow-free SOS determine where and to what extent the gradient of the VI growth curve increases.

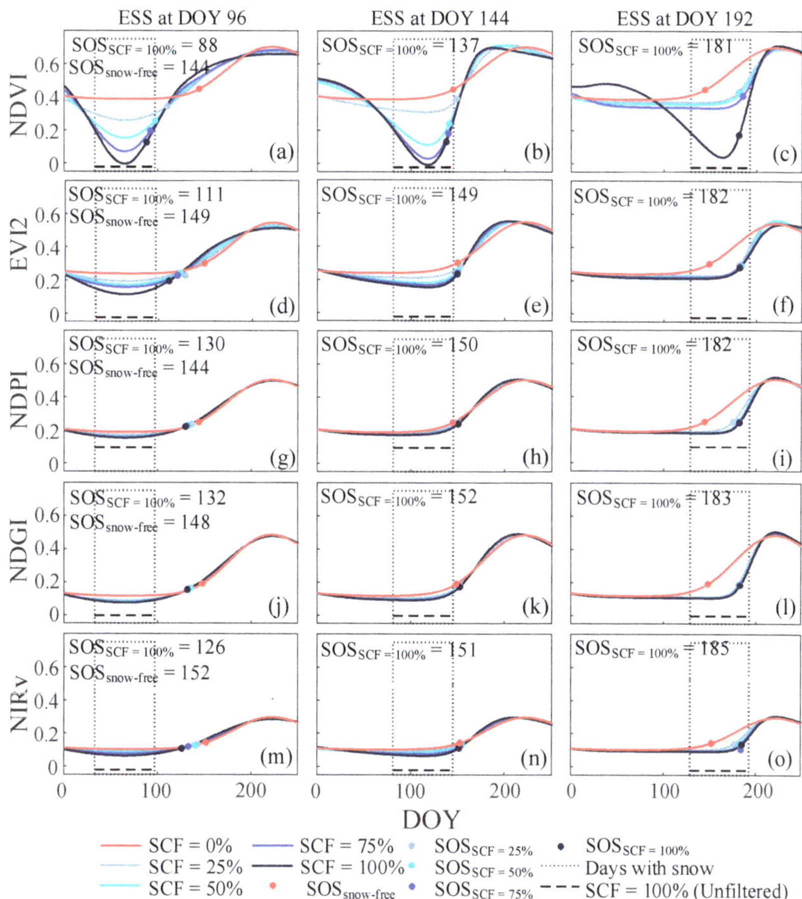

Figure 13. Time curves and the detected SOS of the five VIs under different SCF and ESS cases at SCDc = 64. (**a–c**), (**d–f**), (**g–i**), (**j–l**), and (**m–o**) are the time curves of five VIs, for each of which three ESS cases at DOY 96, 144, and 192 and five cases of SCF = 0%, 25%, 50%, 75%, and 100% were plotted.

4.4. Effect of Snow Cover on Spring Phenology Detection from Satellite Data

4.4.1. Effect of Snow Cover on VI

Figure 14 shows the statistical distributions of the maximum, minimum, and range of the time series of VIs for different SCDr intervals within the same elevation zone. To ensure a sufficient number of pixels for statistics, the SCDr intervals with too few pixels were excluded from the subsequent analysis, including the intervals of SCDr = 40–60% and SCDr > 60% in the zone of DEM < 3500 m. Generally, the decrease in minimum VI values with increasing SCDr was more significant than the decrease in maximum VI values. This is because snow always melts before the peak growth season and does not affect the maximum VI over time. Consequently, the range of VI values increased with the increasing SCDr for all five Vis. These findings were consistent with the simulation results. However, the reduction in VI values with increasing SCDr was not as significant as in the simulation results. This is probably because there were few snow-free pixels, and the minimum VI in the interval SCDr < 20% was affected by snow. Based on the decrease in the minimum VI value with increasing SCDr, the effect of snow on the VI value follows the order of NDPI/NDGI < NIRv < EVI2 < NDVI, which was consistent with the findings from the simulation experiments described in Section 4.3.1.

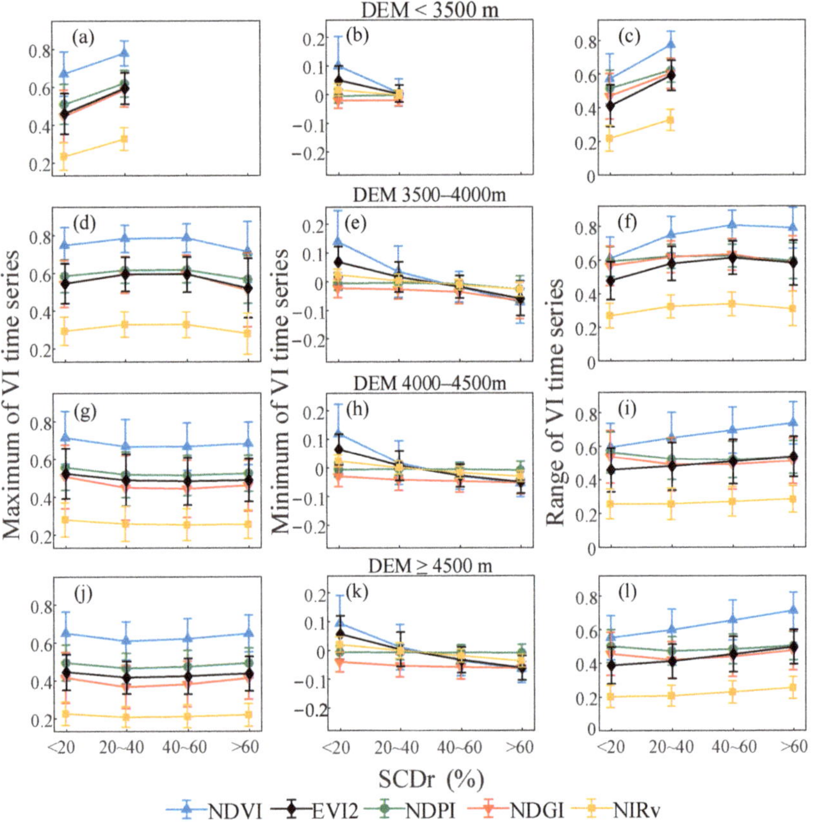

Figure 14. Statistical distribution of the maximum, minimum, and range of the time series of five VIs for different SCDr intervals and elevation zones. **(a–c)**, **(d–f)**, **(g–i)**, and **(j–l)** are statistics for elevation zones of DEM < 3500 m, 3500–4000 m, 4000–4500 m, and ≥ 4500 m, in each of which the maximum, minimum, and range of the VI values over time were plotted.

4.4.2. Effect of Snow Cover on SOS Detection

Figure 15 compares the statistical distribution of SOS detected from the five VIs at different SCDr and elevation zones. At the same SCDr, an increase in elevation caused a delay in the SOS for all five VIs. For less snowy areas with SCDr < 20%, an increase in elevation from < 3500 m to ≥ 4500 m caused a delay of approximately 19–25 days. Within the same elevation zone, the increase in SCDr caused different effects on the SOS detected by different VIs. At elevations ≥ 4500 m, an increase in SCDr from < 20% to > 60% delayed the SOS by 3 and 5 days for NDGI and NDPI, respectively, but advanced the SOS by 11, 9, and 4 days for NDVI, EVI2, and NIRv, respectively. Generally, the SOS detected by NDGI, NDPI, and NIRv was less affected by snow, while the SOS detected by NDVI and EVI2 was more affected by snow.

Based on the simulation results, the effect of snow on SOS detection depends on SCDc, ESS, and snow-free SOS. Figure 16 shows the statistical distribution of the SOS for different SCDc and ESS cases in each elevation zone using real satellite data. To ensure sufficient pixels for the statistical analysis, we considered three SCDc cases, including SCDc < 48, 48 ≤ SCDc < 80, and SCDc ≥ 80, and divided the ESS values into 12 cases with 16-day intervals from DOY 32 to 208. In the zone with DEM < 3500 m, the intervals of

48 ≤ SCDc < 80 and SCDc > 80 had a maximum of two pixels and were excluded from the statistical analysis.

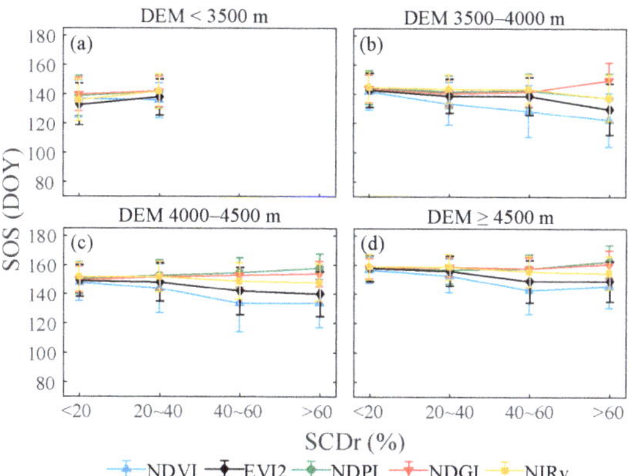

Figure 15. Error bars of the SOS detected by five different VIs for different SCDr intervals at each elevation zone. Error bars show the mean and standard deviation of the SOS in each case. (**a**–**d**) are the statistical distribution of SOS detected from the five VIs at different SCDr and elevation zones.

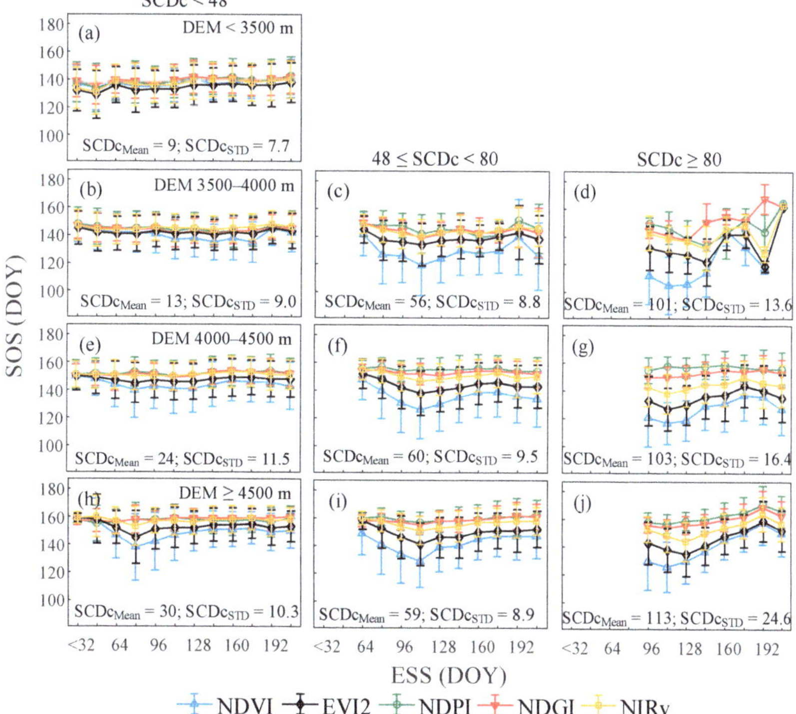

Figure 16. Changes in SOS with increasing ESS under different SCDc cases derived from satellite data. (**a**), (**b**–**d**), (**e**–**g**), and (**h**–**j**) are statistics for elevation zones of DEM < 3500 m, 3500–4000 m, 4000–4500 m, and ≥ 4500 m, in each of which three cases of SCDc < 48, 48 ≤ SCDc < 80, and SCDc ≥ 80 were plotted. $SCDc_{Mean}$ and $SCDc_{STD}$ are the mean and standard deviation of SCDc in each elevation zone.

As shown in Figure 16, the effect of snow on the detected SOS followed the order of NDPI/NDGI < NIRv < EVI2 < NDVI. Generally, the larger the SCDc was, the larger the effect of snow cover on the detected SOS. For short snow (i.e., SCDc < 48), the fluctuation in the SOS with varying ESS was the smallest for all five Vis, indicating a negligible effect of snow. Using the SOS detected for short snow (i.e., SCDc < 48) as a benchmark, the SOS detected under medium snow (i.e., 48 ≤ SCDc < 80) was generally advanced for NDVI and EVI2 and was slightly delayed for NDPI, NDGI, and NIRv for ESS later than DOY 144. For long snow (i.e., SCDc ≥ 80), the detected SOS dates were advanced for ESS earlier than DOY 160 and delayed for ESS later than DOY 160. For short and medium snows, the effect of snow on SOS detection was generally small for NDPI, NDGI, and NIRv. Generally, an earlier ESS results in earlier estimates of the SOS, while a later ESS results in later estimates of the SOS. These findings were highly consistent with the simulation results and verified the validity of the simulation experiments.

5. Discussion

5.1. Validity of the SOS Dates Detected by Different VIs

The SOS dates detected by the five different VIs all captured the spatial pattern of the SOS, which occurred earlier in the eastern areas and later in the western areas (Figure 8). The SOS dates detected by five different VIs also have very similar data distributions, ranging from 105 to 175 with peak values of approximately 157. Both the spatial details and data values were highly consistent with each other and with previous studies [38,55,59], verifying the validity of the detected SOS dates.

Statistical analysis in Figure 15 revealed that both the mean and standard deviation values of the SOS detected in less snowy areas (e.g., SCDr < 20%) by different VIs were very close to each other. However, for more snowy areas, such as for SCDr > 40%, discrepancies occurred in the SOS detected by different VIs. The SOS derived from NDVI and EVI2 were earlier, while those from NDPI and NDGI were later and those from NIRv were in between. These discrepancies could be attributed to the different sensitivities of various VIs to preseason snow cover.

For the entire study area, the SOS dates detected by different VIs have different correlations with each other, such as the results of NDGI/NDPI, which have relatively low correlations with those of NDVI/EVI2/NIRv. However, for less snowy areas, very high correlations were found for the results of all five VIs. Figure 17 compares the different SOS results for the areas with SCDr < 20%. The R values between the different SOS results ranged from 0.818 to 0.982, indicating a very high consistency of the SOS detected by different VIs.

Although previous studies confirmed the accuracy of satellite-derived SOS dates under snow-free conditions [20,21,31,40], a direct comparison with field data was lacking in this study due to the unavailability of in situ SOS measurements. However, we referred to previous studies [20,38,55,59,60] and found a high consistency of our results with the previous in situ measurements and satellite-derived SOS dates in terms of value ranges and spatial distribution. Moreover, a high consistency of the SOS results detected by different VIs was confirmed for less snowy conditions, which was also consistent with previous findings [19,20]. Since the satellite-derived SOS dates represent the macroscale spring phenology and cannot be simply equated with the in situ measurement [61,62], the absence of a direct comparison with in situ measurements does not affect the validity of our results, yet we hope more evidence from in situ measurements to support our findings in the future.

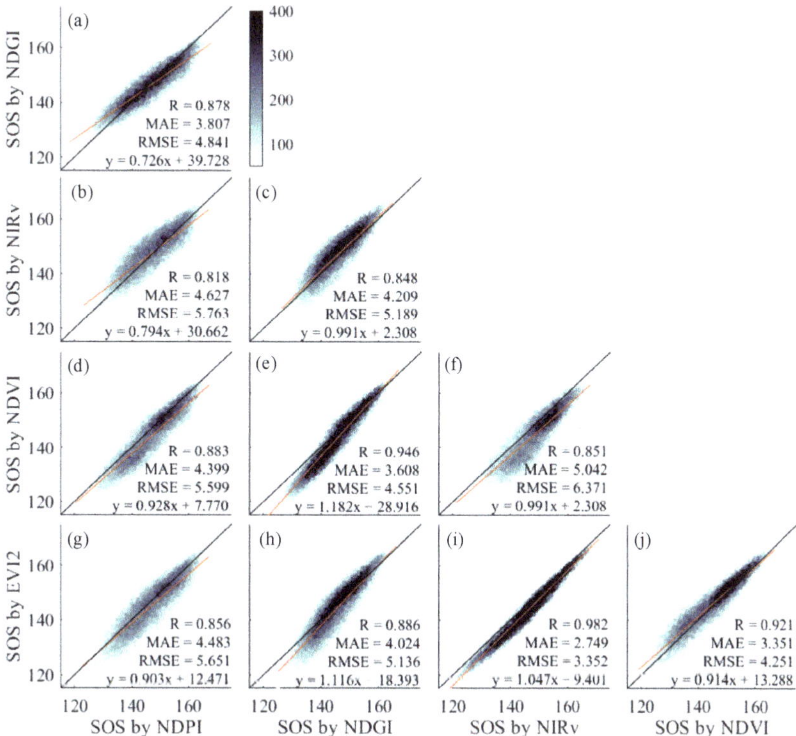

Figure 17. Comparison of the SOS detected by different VIs for SCDr < 20%. The color from light gray to dark gray indicates increasing sample densities. (**a**–**j**) are scatterplots of the SOS dates detected from different VIs on a pixel-by-pixel basis.

5.2. Influence of Snow Phenology Parameters

Both the simulation experiments and satellite data analysis showed that the presence of snow could significantly reduce the VI values, increase the local gradient of the time curve of VI during pregrowth periods, and cause the SOS to be detected. The bias in the SOS (ΔSOS) caused by snow cover depends on snow phenology parameters, especially the SCDc, ESS, and snow-free SOS.

In general, the SCDc represents the maximum duration of snow cover, corresponding to the length of the reduced values in the VI trajectory, which determines whether the reduced VI values could be recovered from temporal filtering. Thus, the larger the SCDc value was, the larger the absolute value of ΔSOS. Compared to the SCDc, the SCDr only represents overall snow-covered days. A high SCDr does not necessarily imply a high SCDc, while a high SCDc does lead to a high SCDr. Thus, the relationship between the SCDr and ΔSOS was rather indirect and was not discussed in detail.

Whether the ESS is earlier or later than the snow-free SOS determines whether ΔSOS is negative or positive, corresponding to the advancement or delay of the SOS detected under snow conditions. The further the temporal distance of ESS to the snow-free SOS was, the larger the absolute value of ΔSOS. These findings were consistent with previous studies [37,38,63,64]. An earlier ESS results in an earlier estimate of SOS, while a later ESS results in a later estimate of SOS. For example, Wang et al. [38] analyzed the correlation between the duration of snow cover and the SOS on the Qinghai–Tibet Plateau and found that snow cover can advance the SOS in the northeastern, central, and southwestern edges of the Qinghai–Tibet Plateau; however, in some areas, longer snow cover duration delayed the SOS. Xie et al. [64] found that a shortened snow duration advances the SOS, whereas a prolonged snow duration delayed the SOS in their study in the European Alps.

5.3. Performance of Five VIs under Snow Conditions

Our study showed that the effects of snow cover on the five VIs were NDPI < NDGI < NIRv < EVI2 < NDVI. For SOS detection, NDPI and NDGI were rather stable even with winter snow cover, which verified their abilities to minimize the effect of snow cover for alpine grasslands, such as those also found for the American prairie [65]. Our study further revealed the variations in the ΔSOS with snow phenology parameters for NDPI and NDGI. For short and medium snow (i.e., SCDc $\leq$ 64), preseason snow ending prior to the snow-free SOS caused insignificant biases in the SOS detected by NDPI and NDGI. For long snow (i.e., SCDc $\geq$ 96) or late snow that ends far later than the snow-free SOS, the biases in the SOS detected by NDPI and NDGI are significant. These findings increase our knowledge about the specific conditions under which the NDPI and NDGI are reliable for SOS detection with snow cover. The traditional NDVI and EVI2 are easily and heavily affected by snow cover. Either an early or a late snow season can cause significant bias in the detected SOS. The physiological-based NIRv could derive SOS dates highly consistent with those detected by NDVI and EVI2 and was less sensitive to snow cover than NDVI and EVI2, indicating its great potential for phenological detection in alpine grasslands. These findings with respect to the performance of different VIs under snow conditions were consistent with the study of Yang et al. [21].

5.4. Limitations and Future Improvements

There are several state-of-the-art methods to smooth the temporal profiles of VIs and extract phenological metrics [11,52,66–68]. Only one of them was used in this study. More methods can be used and evaluated in further studies. However, we assumed that the conclusion can hold for other SOS detection method because the affecting mechanism of snow in increasing the local gradient of the growth curve is still valid. A previous study using the derivative-based method for SOS detection achieved similar conclusions that snow cover would advance the SOS, but prolonged snow duration would delay the SOS date [38].

The design of the simulation experiments made a series of simplifications of the actual situation. One major simplification is that light cannot penetrate the snow layer. This assumption is representative of most cases with a thick snow layer but may not apply to thin snow layers. However, thin snow can melt or form into a thick snow layer quickly, which will cause negligible effects on the time curves of the VIs. This assumption is thus reasonable, yet further studies can consider the case of temporary thin snow. The other major simplification is that we only considered the presence or absence of snow cover without considering the snowmelt process. Snowmelt can also last for several days and affect vegetation phenology [39,69]. Although such simplification would cause a sharp increase in the time curve of the VIs at the ESS, temporal smoothing performed prior to the SOS detection can locally smooth the VI temporal curve and remedy the problem. In addition, the simulation experiments showed that a small SCF, such as 25%, can cause a large reduction in the VI value. Although the snowmelt process leads to a gradual decrease in SCF, the largest increase in the VI value is expected to be at the stage when SCF decreases from 25% to 0%, which is a relatively short time interval. Therefore, the presence and immediate melting of snow in our simulation experiments is reasonable, and more complex situations can be considered in future work.

For the debate on whether NDVI-based spring phenology trends are overestimated on the Tibetan Plateau [6,7,30,70], we suggest that snow phenology, particularly ESS, should be given much attention in related studies. Normally, NDPI, NDGI, and NIRv would be less affected by snow cover, and their performance in detecting long-term phenology trends can be further investigated. Based on the findings in this study, we also expect to decouple the effect of snow cover on satellite signals and on vegetation physiological phenology, which will enhance our understanding of vegetation–climate feedbacks.

6. Conclusions

Considering the difficulty in assessing the effect of snow cover on SOS detection, this study investigated the effect of snow cover on both VI and SOS detection by combining simulation experiments and real satellite data, aiming to determine how snow affects the different VIs and the subsequent SOS detection and how different VIs perform in capturing the SOS for alpine grasslands on the Tibetan Plateau. Five VIs, including NDVI, EVI2, NDPI, NDGI, and NIRv, were used for SOS detection, and their performance was compared.

Based on the simulation experiments, we found that the presence of snow, even at a low SCF, can significantly reduce the values of the five VIs and increase the local gradient of the growth curve, allowing the SOS to be detected. Thus, the bias in the detected SOS due to snow cover depends on both snow phenological parameters (i.e., ESS and SCDc) and the snow-free SOS. An earlier ESS results in an earlier estimate of SOS, while a later ESS results in a later estimate of SOS, and an ESS close to the snow-free SOS results in small bias in the detected SOS.

The analysis from satellite data showed consistent results with those from the simulations. The presence of snow especially reduced the minimum VI values over time, and the detected SOS within the same elevation zone varied with snow parameters such as SCDc and ESS. Generally, an earlier ESS led to an earlier estimate of SOS, while a later ESS led to a later estimate of SOS.

The sensitivity of the five VIs to snow cover in SOS detection is NDPI/NDGI < NIRv < EVI2 < NDVI, which has been tested in both simulation experiments and satellite data analysis. For SOS detection with winter snow cover, NDPI, NDGI, and the physiological-based NIRv were rather stable, while NDVI and EVI2 were easily and heavily affected by snow cover. However, the performance of a specific VI in SOS detection also depends on snow phenology parameters such as SCDc and ESS.

These findings will significantly advance our research on the feedback mechanisms between vegetation, snow, and climate change for alpine ecosystems.

Author Contributions: Conceptualization, Y.W.; Formal analysis, Y.W. and Y.C.; Funding acquisition, X.Z.; Investigation, Y.W. and P.L.; Methodology, Y.W. and P.L.; Resources, Y.W.; Validation, Y.C.; Visualization, Y.C. and R.Z.; Writing—original draft, Y.W. and Y.C.; Writing—review and editing, Y.W., Y.C., P.L., Y.Z., and B.Y. All authors have read and agreed to the published version of the manuscript.

Funding: This research was funded by the National Natural Science Foundation of China (NSFC), grant number 41901301; The Open Research Fund Program of State Key Laboratory of Eco-hydraulics in Northwest Arid Region, Xi'an University of Technology, grant number 2020KFKT-7; and The Natural Science Foundation of Shaanxi Province, grant number 2020JQ-739.

Data Availability Statement: Publicly available datasets were analyzed in this study. Satellite reflectance data can be found here: [LP DAAC, http://lpdaac.usgs.gov]; snow cover data can be found here: [NSIDC, http://nsidc.org]; land cover type data can be found here: [http://data.ess.tsinghua.edu.cn]; and DEM data can be found here: [http://www.gscloud.cn].

Conflicts of Interest: The authors declare no conflict of interest.

References

1. Richardson, A.D.; Keenan, T.F.; Migliavacca, M.; Ryu, Y.; Sonnentag, O.; Toomey, M. Climate change, phenology, and phenological control of vegetation feedbacks to the climate system. *Agric. For. Meteorol.* **2013**, *169*, 156–173. [CrossRef]
2. Zeng, Z.; Piao, S.; Li, L.Z.; Zhou, L.; Ciais, P.; Wang, T.; Li, Y.; Lian, X.; Wood, E.F.; Friedlingstein, P. Climate mitigation from vegetation biophysical feedbacks during the past three decades. *Nat. Clim. Change* **2017**, *7*, 432–436. [CrossRef]
3. Wang, Y.; Xie, D.; Hu, R.; Yan, G. Spatial scale effect on vegetation phenological analysis using remote sensing data. In Proceedings of the 2016 IEEE International Geoscience and Remote Sensing Symposium, Beijing, China, 10–15 July 2016; pp. 1329–1332. [CrossRef]
4. Shang, Z.H.; Gibb, M.; Leiber, F.; Ismail, M.; Ding, L.M.; Guo, X.S.; Long, R.J. The sustainable development of grassland-livestock systems on the Tibetan plateau: Problems, strategies and prospects. *Rangel. J.* **2014**, *36*, 267–296. [CrossRef]

5. Thompson, J.A.; Paull, D.J.; Lees, B.G. Using phase-spaces to characterize land surface phenology in a seasonally snow-covered landscape. *Remote Sens. Environ.* **2015**, *166*, 178–190. [CrossRef]
6. Wang, T.; Peng, S.; Lin, X.; Chang, J. Declining snow cover may affect spring phenological trend on the Tibetan Plateau. *Proc. Natl. Acad. Sci.* **2013**, *110*, E2854–E2855. [CrossRef]
7. Wang, X.; Xiao, J.; Li, X.; Cheng, G.; Ma, M.; Che, T.; Dai, L.; Wang, S.; Wu, J. No consistent evidence for advancing or delaying trends in spring phenology on the Tibetan Plateau. *J. Geophys. Res. Biogeosciences* **2017**, *122*, 3288–3305. [CrossRef]
8. Nyasha, M.T.; Onisimo, M.; Mbulisi, S.; John, O. Estimating and monitoring land surface phenology in rangelands: A review of progress and challenges. *Remote Sens.* **2021**, *13*, 2060. [CrossRef]
9. Wang, M.Y.; Luo, Y.; Zhang, Z.Y.; Xie, Q.Y.; Wu, X.D.; Ma, X.L. Recent advances in remote sensing of vegetation phenology: Retrieval algorithm and validation strategy. *Natl. Remote Sens. Bull.* **2022**, *26*, 431–455. [CrossRef]
10. Helman, D. Land surface phenology: What do we really 'see' from space? *Sci. Total Environ.* **2018**, *618*, 665–673. [CrossRef]
11. White, M.A.; De Beurs, K.M.; Didan, K.; Inouye, D.W.; Richardson, A.D.; Jensen, O.P.; O'Keefe, J.; Zhang, G.; Nemani, R.R.; Van Leeuwen, W.J.D.; et al. Intercomparison, interpretation, and assessment of spring phenology in North America estimated from remote sensing for 1982–2006. *Glob. Change Biol.* **2009**, *15*, 2335–2359. [CrossRef]
12. Friedl, M.; Gray, J.; Sulla-Menashe, D. User Guide to Collection 6 MODIS Land Cover Dynamics (MCD12Q2) Product [EB/OL]. 2021. Available online: https://modis.ornl.gov/documentation/guides/mcd12q2_v6_user_guide.pdf (accessed on 8 September 2022).
13. Tucker, C.J. Red and photographic infrared linear combinations for monitoring vegetation. *Remote Sens. Environ.* **1979**, *8*, 127–150. [CrossRef]
14. Pettorelli, N.; Vik, J.O.; Mysterud, A.; Gaillard, J.-M.; Tucker, C.J.; Stenseth, N.C. Using the satellite-derived NDVI to assess ecological responses to environmental change. *Trends Ecol. Evol.* **2005**, *20*, 503–510. [CrossRef] [PubMed]
15. Piao, S.L.; Wang, X.H.; Ciais, P.; Zhu, B.; Wang, T.A.O.; Liu, J.I.E. Changes in satellite-derived vegetation growth trend in temperate and boreal Eurasia from 1982 to 2006. *Glob. Change Biol.* **2011**, *17*, 3228–3239. [CrossRef]
16. Zhang, X.Y.; Jayavelu, S.; Liu, L.L.; Friedl, M.A.; Henebry, G.M.; Liu, Y.; Schaaf, C.B.; Richardson, A.D.; Gray, J. Evaluation of land surface phenology from VIIRS data using time series of PhenoCam imagery. *Agric. For. Meteorol.* **2018**, *256*, 137–149. [CrossRef]
17. Jiang, Z.; Huete, A.R.; Didan, K.; Miura, T. Development of a two-band enhanced vegetation index without a blue band. *Remote Sens. Environ.* **2008**, *112*, 3833–3845. [CrossRef]
18. Zhang, X.; Liu, L.; Liu, Y.; Jayavelu, S.; Wang, J.; Moon, M.; Henebry, G.M.; Friedl, M.A.; Schaaf, C.B. Generation and evaluation of the VIIRS land surface phenology product. *Remote Sens. Environ.* **2018**, *216*, 212–229. [CrossRef]
19. Chang, Q.; Xiao, X.; Jiao, W.; Wu, X.; Doughty, R.; Wang, J.; Du, L.; Zou, Z.; Qin, Y. Assessing consistency of spring phenology of snow-covered forests as estimated by vegetation indices, gross primary production, and solar-induced chlorophyll fluorescence. *Agric. For. Meteorol.* **2019**, *275*, 305–316. [CrossRef]
20. Huang, K.; Zhang, Y.; Tagesson, T.; Brandt, M.; Wang, L.; Chen, N.; Zu, J.; Jin, H.; Cai, Z.; Tong, X.; et al. The confounding effect of snow cover on assessing spring phenology from space: A new look at trends on the Tibetan Plateau. *Sci. Total Environ.* **2021**, *756*, 144011. [CrossRef]
21. Yang, W.; Kobayashi, H.; Wang, C.; Shen, M.; Chen, J.; Matsushita, B.; Tang, Y.; Kim, Y.; Bret-Harte, M.S.; Zona, D. A semi-analytical snow-free vegetation index for improving estimation of plant phenology in tundra and grassland ecosystems. *Remote Sens. Environ.* **2019**, *228*, 31–44. [CrossRef]
22. Shabanov, N.V.; Zhou, L.; Knyazikhin, Y.; Myneni, R.B.; Tucker, C.J. Analysis of interannual changes in northern vegetation activity observed in AVHRR data from 1981 to 1994. *IEEE Trans. Geosci. Remote Sens.* **2002**, *40*, 115–130. [CrossRef]
23. Delbart, N.; Kergoat, L.; Le Toan, T.; Lhermitte, J.; Picard, G. Determination of phenological dates in boreal regions using normalized difference water index. *Remote Sens. Environ.* **2005**, *97*, 26–38. [CrossRef]
24. De Beurs, K.M.; Henebry, G.M. Spatio-Temporal Statistical Methods for Modelling Land Surface Phenology. In *Phenological Research*; Hudson, I.L., Keatley, M.R., Eds.; Springer: Dordrecht, The Netherlands, 2010; Volume 22, pp. 177–208.
25. Migliavacca, M.; Galvagno, M.; Cremonese, E.; Rossini, M.; Meroni, M.; Sonnentag, O.; Cogliati, S.; Manca, G.; Diotri, F.; Busetto, L.; et al. Using digital repeat photography and eddy covariance data to model grassland phenology and photosynthetic CO_2 uptake. *Agric. For. Meteorol.* **2011**, *151*, 1325–1337. [CrossRef]
26. Gonsamo, A.; Chen, J.M.; Price, D.T.; Kurz, W.A.; Wu, C.Y. Land surface phenology from optical satellite measurement and CO_2 eddy covariance technique. *J. Geophys. Res. Biogeosciences* **2012**, *117*, G03032. [CrossRef]
27. Jönsson, A.M.; Eklundh, L.; Hellström, M.; Bärring, L.; Jönsson, P. Annual changes in MODIS vegetation indices of Swedish coniferous forests in relation to snow dynamics and tree phenology. *Remote Sens. Environ.* **2010**, *114*, 2719–2730. [CrossRef]
28. Shen, M.G.; Sun, Z.Z.; Wang, S.P.; Zhang, G.X.; Kong, W.D.; Chen, A.P.; Piao, S.L. No evidence of continuously advanced green-up dates in the Tibetan Plateau over the last decade. *Proc. Natl. Acad. Sci. USA* **2013**, *110*, E2329. [CrossRef] [PubMed]
29. Shen, M.; Tang, Y.; Chen, J.; Zhu, X.; Zheng, Y. Influences of temperature and precipitation before the growing season on spring phenology in grasslands of the central and eastern Qinghai-Tibetan Plateau. *Agric. For. Meteorol.* **2011**, *151*, 1711–1722. [CrossRef]
30. Shen, M.; Zhang, G.; Cong, N.; Wang, S.; Kong, W.; Piao, S. Increasing altitudinal gradient of spring vegetation phenology during the last decade on the Qinghai–Tibetan Plateau. *Agric. For. Meteorol.* **2014**, *189*, 71–80. [CrossRef]
31. Wang, C.; Chen, J.; Wu, J.; Tang, Y.; Shi, P.; Black, T.A.; Zhu, K. A snow-free vegetation index for improved monitoring of vegetation spring green-up date in deciduous ecosystems. *Remote Sens. Environ.* **2017**, *196*, 1–12. [CrossRef]

32. Wang, S.; Wang, X.; Chen, G.; Yang, Q.; Wang, B.; Ma, Y.; Shen, M. Complex responses of spring alpine vegetation phenology to snow cover dynamics over the Tibetan Plateau, China. *Sci. Total Environ.* **2017**, *593*, 449–461. [CrossRef]
33. Badgley, G.; Field, C.B.; Berry, J.A. Canopy near-infrared reflectance and terrestrial photosynthesis. *Sci. Adv.* **2017**, *3*, e1602244. [CrossRef]
34. Mohammed, G.H.; Colombo, R.; Middleton, E.M.; Rascher, U.; Tol, C.V.D.; Nedbal, L.; Goulas, Y.; Pérez-Priego, O.; Damm, A.; Meroni, M.; et al. Remote sensing of solar-induced chlorophyll fluorescence (SIF) in vegetation: 50 years of progress. *Remote Sens. Environ.* **2019**, *231*, 111177. [CrossRef] [PubMed]
35. Zhang, J.R.; Xiao, J.F.; Tong, X.J.; Zhang, J.S.; Meng, P.; Li, J.; Liu, P.R.; Yu, P.Y. NIRv and SIF better estimate phenology than NDVI and EVI: Effects of spring and autumn phenology on ecosystem production of planted forests. *Agric. For. Meteorol.* **2022**, *315*, 108819. [CrossRef]
36. Liu, Y.; Wei, Z.; Si, G.; Xuanlong, M.; Kai, Y. Phenological responses to snow seasonality in the qilian mountains is a function of both elevation and vegetation types. *Remote Sens.* **2022**, *14*, 3629. [CrossRef]
37. Wang, K.; Zhang, L.; Qiu, Y.; Ji, L.; Tian, F.; Wang, C.; Wang, Z. Snow effects on alpine vegetation in the Qinghai-Tibetan Plateau. *Int. J. Digit. Earth* **2015**, *8*, 58–75. [CrossRef]
38. Wang, X.; Wu, C.; Peng, D.; Gonsamo, A.; Liu, Z. Snow cover phenology affects alpine vegetation growth dynamics on the Tibetan Plateau: Satellite observed evidence, impacts of different biomes, and climate drivers. *Agric. For. Meteorol.* **2018**, *256*, 61–74. [CrossRef]
39. Chen, X.; An, S.; Inouye, D.W.; Schwartz, M.D. Temperature and snowfall trigger alpine vegetation green-up on the world's roof. *Glob. Change Biol.* **2015**, *21*, 3635–3646. [CrossRef]
40. Xu, D.; Wang, C.; Chen, J.; Shen, M.; Shen, B.; Yan, R.; Li, Z.; Karnieli, A.; Chen, J.; Yan, Y. The superiority of the normalized difference phenology index (NDPI) for estimating grassland aboveground fresh biomass. *Remote Sens. Environ.* **2021**, *264*, 112578. [CrossRef]
41. Zhang, X. A vegetation-climate classification system for global change studies in China. *Quat. Sci.* **1993**, *2*, 157–169.
42. Diem, P.K.; Diem, N.K.; Hung, H.V. Assessment of the efficiency of using modis MCD43A4 in Mapping of rice planting calendar in the Mekong Delta. *IOP Conf. Ser. Earth Environ. Sci.* **2021**, *652*, 012015. [CrossRef]
43. Hall, D.; Salomonson, V.; Riggs, G. *MODIS/Terra Snow Cover Daily L3 Global 500m Grid, Version 5*; NASA National Snow and Ice Data Center: Boulder, CO, USA, 2006. [CrossRef]
44. Huang, X.D.; Zhang, X.T.; Li, X.; Liang, T.G. Accuracy analysis for MODIS snow products of MOD10A1 and MOD10A2 in northern Xinjiang area. *J. Glaciol. Geocryol.* **2007**, *29*, 722–729. [CrossRef]
45. Gong, P.; Liu, H.; Zhang, M.; Li, C.; Wang, J.; Huang, H.; Clinton, N.; Ji, L.; Li, W.; Bai, Y.; et al. Stable classification with limited sample: Transferring a 30-m resolution sample set collected in 2015 to mapping 10-m resolution global land cover in 2017. *Sci. Bull.* **2019**, *64*, 370–373. [CrossRef]
46. Huete, A.; Didan, K.; Miura, T.; Rodriguez, E.P.; Gao, X.; Ferreira, L.G. Overview of the radiometric and biophysical performance of the MODIS vegetation indices. *Remote Sens. Environ.* **2002**, *83*, 195–213. [CrossRef]
47. Wang, S.; Zhang, Y.; Ju, W.; Qiu, B.; Zhang, Z. Tracking the seasonal and inter-annual variations of global gross primary production during last four decades using satellite near-infrared reflectance data. *Sci. Total Environ.* **2021**, *755*, 142569. [CrossRef] [PubMed]
48. Jönsson, P.; Eklundh, L. TIMESAT—A program for analyzing time-series of satellite sensor data. *Comput. Geosci.* **2004**, *30*, 833–845. [CrossRef]
49. Busetto, L.; Colombo, R.; Migliavacca, M.; Cremonese, E.; Meroni, M.; Galvagno, M.; Rossini, M.; Siniscalco, C. Remote sensing of larch phenological cycle and analysis of relationships with climate in the Alpine region. *Glob. Change Biol.* **2010**, *16*, 2504–2517. [CrossRef]
50. Beck, P.S.A.; Atzberger, C.; Høgda, K.A.; Johansen, B.; Skidmore, A.K. Improved monitoring of vegetation dynamics at very high latitudes: A new method using MODIS NDVI. *Remote Sens. Environ.* **2006**, *100*, 321–334. [CrossRef]
51. Fisher, J.I.; Mustard, J.F.; Vadeboncoeur, M.A. Green leaf phenology at Landsat resolution: Scaling from the field to the satellite. *Remote Sens. Environ.* **2006**, *100*, 265–279. [CrossRef]
52. Cai, Z.; Per, J.; Jin, H.; Lars, E. Performance of smoothing methods for reconstructing NDVI time-series and estimating vegetation phenology from modis data. *Remote Sens.* **2017**, *9*, 1271. [CrossRef]
53. Hufkens, K.; Friedl, M.; Sonnentag, O.; Braswell, B.H.; Milliman, T.; Richardson, A.D. Linking near-surface and satellite remote sensing measurements of deciduous broadleaf forest phenology. *Remote Sens. Environ.* **2012**, *117*, 307–321. [CrossRef]
54. Richardson, A.D.; Black, T.A.; Ciais, P.; Delbart, N.; Friedl, M.A.; Gobron, N.; Hollinger, D.Y.; Kutsch, W.L.; Longdoz, B.; Luyssaert, S.; et al. Influence of spring and autumn phenological transitions on forest ecosystem productivity. *Philos. Trans. R. Soc. B Biol. Sci.* **2010**, *365*, 3227–3246. [CrossRef]
55. Zu, J.; Zhang, Y.; Huang, K.; Liu, Y.; Chen, N.; Cong, N. Biological and climate factors co-regulated spatial-temporal dynamics of vegetation autumn phenology on the Tibetan Plateau. *Int. J. Appl. Earth Obs. Geoinf.* **2018**, *69*, 198–205. [CrossRef]
56. Jin, H.; Jönsson, A.M.; Bolmgren, K.; Langvall, O.; Eklundh, L. Disentangling remotely-sensed plant phenology and snow seasonality at northern Europe using MODIS and the plant phenology index. *Remote Sens. Environ.* **2017**, *198*, 203–212. [CrossRef]
57. Xie, B.S.; Zhou, S.Y.; Wu, L.X. An integrated mineral spectral library using shared data for hyperspectral remote sensing and geological mapping. *Int. Arch. Photogramm. Remote Sens. Spat. Inf. Sci.* **2020**, *43*, 69–75. [CrossRef]

58. Gutman, G.; Ignatov, A. The derivation of the green vegetation fraction from NOAA/AVHRR data for use in numerical weather prediction models. *Int. J. Remote Sens.* **1998**, *19*, 1533–1543. [CrossRef]
59. Zhang, Q.; Kong, D.; Shi, P.; Singh, V.P.; Sun, P. Vegetation phenology on the Qinghai-Tibetan Plateau and its response to climate change (1982–2013). *Agric. For. Meteorol.* **2018**, *248*, 408–417. [CrossRef]
60. Zeng, H.; Jia, G. Impacts of snow cover on vegetation phenology in the arctic from satellite data. *Adv. Atmos. Sci.* **2013**, *30*, 1421–1432. [CrossRef]
61. Liang, L.; Schwartz, M.D. Landscape phenology: An integrative approach to seasonal vegetation dynamics. *Landsc. Ecol.* **2009**, *24*, 465–472. [CrossRef]
62. Wang, Y.; Yan, G.; Xie, D.; Hu, R.; Zhang, H. Generating long time series of high spatiotemporal resolution FPAR images in the remote sensing trend surface framework. *IEEE Trans. Geosci. Remote Sens.* **2021**, *60*, 1–15. [CrossRef]
63. Huang, K.; Zu, J.; Zhang, Y.; Cong, N.; Liu, Y.; Chen, N. Impacts of snow cover duration on vegetation spring phenology over the Tibetan Plateau. *J. Plant Ecol.* **2018**, *12*, 583–592. [CrossRef]
64. Xie, J.; Kneubühler, M.; Garonna, I.; Notarnicola, C.; De Gregorio, L.; De Jong, R.; Chimani, B.; Schaepman, M.E. Altitude-dependent influence of snow cover on alpine land surface phenology. *J. Geophys. Res. Biogeosciences* **2017**, *122*, 1107–1122. [CrossRef]
65. Wang, H.; Liu, H.; Cao, G.; Ma, Z.; Li, Y.; Zhang, F.; Zhao, X.; Zhao, X.; Jiang, L.; Sanders, N.J.; et al. Alpine grassland plants grow earlier and faster but biomass remains unchanged over 35 years of climate change. *Ecol. Lett.* **2020**, *23*, 701–710. [CrossRef]
66. Zeng, L.; Wardlow, B.D.; Xiang, D.; Hu, S.; Li, D. A review of vegetation phenological metrics extraction using time-series, multispectral satellite data. *Remote Sens. Environ.* **2020**, *237*, 111511. [CrossRef]
67. Atkinson, P.M.; Jeganathan, C.; Dash, J.; Atzberger, C. Inter-comparison of four models for smoothing satellite sensor time-series data to estimate vegetation phenology. *Remote Sens. Environ.* **2012**, *123*, 400–417. [CrossRef]
68. Cong, N.; Piao, S.; Chen, A.; Wang, X.; Lin, X.; Chen, S.; Han, S.; Zhou, G.; Zhang, X. Spring vegetation green-up date in China inferred from SPOT NDVI data: A multiple model analysis. *Agric. For. Meteorol.* **2012**, *165*, 104–113. [CrossRef]
69. Mo, L.; Luo, P.; Mou, C.; Yang, H.; Wang, J.; Wang, Z.; Li, Y.; Luo, C.; Li, T.; Zuo, D. Winter plant phenology in the alpine meadow on the eastern Qinghai–Tibetan Plateau. *Ann. Bot.* **2018**, *122*, 1033–1045. [CrossRef]
70. Zhang, G.; Zhang, Y.; Dong, J.; Xiao, X. Green-up dates in the Tibetan Plateau have continuously advanced from 1982 to 2011. *Proc. Natl. Acad. Sci. USA* **2013**, *110*, 4309–4314. [CrossRef] [PubMed]

Article

Quantifying Temperate Forest Diversity by Integrating GEDI LiDAR and Multi-Temporal Sentinel-2 Imagery

Chunying Ren [1], Hailing Jiang [2], Yanbiao Xi [1,*], Pan Liu [1] and Huiying Li [3]

1 Key Laboratory of Wetland Ecology and Environment, Northeast Institute of Geography and Agroecology, Chinese Academy of Sciences, Changchun 130102, China
2 College of Tourism and Geographic Sciences, Jilin Normal University, Siping 136000, China
3 School of Environmental and Municipal Engineering, Qingdao University of Technology, Qingdao 266520, China
* Correspondence: xiyb@smail.nju.edu.cn; Tel.: +86-(209)-50270604

Abstract: Remotely sensed estimates of forest diversity have become increasingly important in assessing anthropogenic and natural disturbances and their effects on biodiversity under limited resources. Whereas field inventories and optical images are generally used to estimate forest diversity, studies that combine vertical structure information and multi-temporal phenological characteristics to accurately quantify diversity in large, heterogeneous forest areas are still lacking. In this study, combined with regression models, three different diversity indices, namely Simpson (λ), Shannon (H'), and Pielou (J'), were applied to characterize forest tree species diversity by using GEDI LiDAR data and Sentinel-2 imagery in temperate natural forest, northeast China. We used Mean Decrease Gini (MDG) and Boosted Regression Tree (BRT) to assess the importance of certain variables including monthly spectral bands, vegetation indices, foliage height diversity (FHD), and plant area index (PAI) of growing season and non-growing seasons (68 variables in total). We produced 12 forest diversity maps on three different diversity indices using four regression algorithms: Support Vector Machines (SVM), Random Forest (RF), K-Nearest Neighbors (KNN), and Lasso Regression (LR). Our study concluded that the most important variables are FHD, NDVI, NDWI, EVI, short-wave infrared (SWIR) and red-edge (RE) bands, especially in the growing season (May and June). In terms of algorithms, the estimation accuracies of the RF (averaged R^2 = 0.79) and SVM (averaged R^2 = 0.76) models outperformed the other models (R^2 of KNN and LR are 0.68 and 0.57, respectively). The study demonstrates the accuracy of GEDI LiDAR data and multi-temporal Sentinel-2 images in estimating forest diversity over large areas, advancing the capacity to monitor and manage forest ecosystems.

Keywords: forest diversity; GEDI LiDAR; Sentinel-2; machine Learning

Citation: Ren, C.; Jiang, H.; Xi, Y.; Liu, P.; Li, H. Quantifying Temperate Forest Diversity by Integrating GEDI LiDAR and Multi-Temporal Sentinel-2 Imagery. *Remote Sens.* **2023**, *15*, 375. https://doi.org/10.3390/rs15020375

Academic Editors: Kenji Omasa, Shan Lu and Jie Wang

Received: 20 November 2022
Revised: 18 December 2022
Accepted: 4 January 2023
Published: 7 January 2023

1. Introduction

Forests host unique tree species diversities, which support key ecosystem services such as nutrient cycles, head-water conservation, and biomass estimation [1]. Forest diversity is changing in response to climate change, soil erosion, species introductions and more [2]. In addition, forest productivity increases with tree species richness, and higher tree species diversity provides more food options for wildlife. Thus, developing effective technology is urgently needed for mapping forest diversity distribution over large areas to assess their current states and carrying capacity for animal populations [3].

Forest diversity is typically assessed by botanical surveys of the woods and metrics related to their species diversity (i.e., richness, Simpson, and Pielou diversity) [2,4]. Traditionally, forest diversity is calculated by counting the number and types of trees, which is an expensive, time-consuming process. Additionally, due to accuracy problems and difficulty in recognizing intertwined tree species, such a strategy is difficult to implement in large (e.g., hundreds of hectares) forest communities [5]. The challenges are more significant in natural forests with dense canopies. Remote sensing techniques have shown great potential for large-scale estimations of forest diversity and have been successfully used to estimate species diversity of subtropical and tropical forest ecosystems [6,7]. However, contemporary remote sensing-based approaches to estimate forest diversity vary with regard to the satellite data and machine learning models deployed. Plant richness of herbaceous ecosystems has been assessed using hyperspectral imagery by Oldeland et al. [3]. Nagendra et al. [7] used IKONOS and Landsat images to estimate forest species richness and diversity in central India. Stenzel et al. [8] used multi-seasonal, multi-spectral remote sensing data (RapidEye) to map ecological regions with high species richness. Almeida et al. [9] used hyperspectral images and airborne LiDAR data to assess the structure and diversity of restoration plantings. Clearly, rich spectral information plays an important role in species richness. However, these remote-sensing data are limited by area coverage, weather conditions, high costs, and acquisition time [10], making it challenging to develop detailed maps of forest diversity across large areas. Currently, commonly used methods for estimating forests diversity based on remote sensing data are extrapolated by using field data collected. Leutner et al. [11] examined the relationship between remotely sensed and field data, and mapped α- and β-diversity in the Yucatan Peninsula by using a regression kriging procedure. Hakkenberg et al. [12] predicted floristic diversity at different spatial scales using nonparametric models trained with spatially nested field plots and aerial LiDAR-hyperspectral data. Chrysafis et al. [13] developed a workflow to obtain tree diversity maps with machine learning algorithms using multispectral and multi-seasonal Sentinel-2 images and geodiversity data at the regional scale. The most important process in these methods is to extract features from remote sensing data, which are spectral indices or LiDAR-based metrics highly relevant to forest diversity, and then using these features as a set of mixed variables for regression analysis. Although these methods have achieved good prediction accuracy, it is unclear which types of algorithms are more effective in estimating forest diversity.

Sentinel-2 satellite data with 10 m spatial resolution has large spatial coverage, short acquisition time, and rich spectral bands that offer unprecedented opportunities to estimate tree species diversity [14]. The phenological differences of plant communities can be captured by their high temporal resolution and used as metrics to calculate plant diversity [15]. Detailed spectral information is related to plant biochemical composition, canopy structure, and leaf morphology characteristics, specifically for red-edge wavelengths [16]. Then, being available for free, they can be used to process large areas and complement field surveys at a reduced cost [17]. Sentinel-2 imagery has achieved good performance in mapping tree species classification [15], vegetation phenology monitoring [18], and forest aboveground biomass [19]. However, estimating tree species diversity is still lacking, especially in temperate mixed forests. Additionally, since April 2019, the NASA Global Ecosystem Dynamics Investigation (GEDI), a spaceborne LiDAR sensor in the International Space Station, has acquired footprint data with an average diameter of 25 m [20]. GEDI is a full waveform LiDAR that was created with the purpose of detecting vegetation structure [21] and provides an unprecedented sampling density, which could be an ideal structure parameter for estimating forest diversity [22]. Potapov et al. [23] combined GEDI LiDAR and Landsat to produce a global tree height map at a 30 m resolution. Liang et al. [24] quantified aboveground biomass dynamics of charcoal degradation in Mozambique using GEDI LiDAR and Landsat. These studies provide promising examples for the potential of GEDI-Sentinel data fusion to estimate forest diversity continuously across large extents.

In this study, GEDI LiDAR data and multi-temporal Sentinel-2 images were integrated to estimate forest diversity at the pixel level within natural forests in northeast China. Specifically, this study aims to: (1) quantify the relationships between forest diversity and variables from Sentinel-2 and GEDI LiDAR, (2) explore the effective algorithm for high precision mapping of forest diversity, and (3) map forest diversity by using GEDI LiDAR data and Sentinel-2 images for forest ecological assessment.

2. Materials

2.1. Study Area

The study area is located in the southeast region of Jilin Province, northeast China (Figure 1). It covers approximately 311,000 ha with an average elevation of 500 m. The average annual temperature ranges from nearly −3 to 7 °C, and the precipitation ranges from 500 mm to 1400 mm [25]. The forest types are mainly temporal mixed broadleaf-conifer woodlands, which are dominated by *Juglans mandshurica, Pinus koraiensis, Betula costata, Larix gmelinii, Quercus mongolica,* and *Populus tremula*.

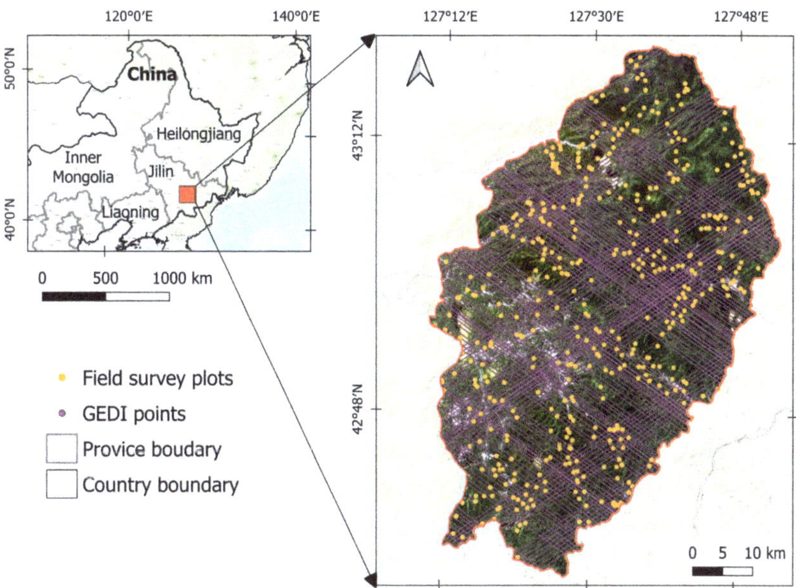

Figure 1. Location of the study area. The yellow dots on Sentinel-2 imagery are the sampling plots, while purple dots indicate the GEDI footprints used in this study.

2.2. Field Botanical Surveys

Compared with other forest parameters, forest diversity is related to spatial variability. Prior to field excursion, one would need to determine what size plots can achieve a stable range of spatial variability. In this study, we used the semi-variogram to determine the investigated plot size, which quantifies the spatial variability due to distance change [2]. Specifically, we calculated the square deviation between adjacent pixel values to test spatial variability with the Sentinel-based NDVI band. Semi-variance gradually increases with the distance between pixels until it starts to level off. Our findings indicated that lag distances of 50 m correspond to the scale for tree species variability in the study area (Appendix A, Figure A1). Thus, the plot size of 50 m × 50 m was identified as optimal in terms of capturing spatial variation in tree species diversity. From June to July 2019, field surveys were conducted. Based on spatial distribution randomness and road accessibility principles, a total of 452 plots were designed; The Global Positioning System was used to record each plot of position. In this study, based on the Chinese Forest Biodiversity Monitoring Network (CForBio) [26], all trees with a diameter at breast height (DBH) greater than 10 cm were identified, while trees with DBH less than 10 cm, shrubs and grasslands were not investigated considering the effect of dense canopy. In addition, the spatial distribution of different forest types was also obtained through the 9th National Forest Inventory of China (2018).

2.3. Data Source and Processing

2.3.1. Diversity Index Data

Based on sample data obtained from the field survey, tree species diversity for each plot was calculated using the three commonly used plant diversity indices, namely Shannon (H'), Simpson (λ) and Pielou (J') (Table 1). Specifically, we first counted the species (i) and proportion (P_i) of trees in each plot, and then input the statistical parameters into the equations of diversity index to calculate diversity values of each plot (see Figure A2). Finally, the diversity values of each plot were used as dependent variables, and multi-variables from remote sensing data corresponding to the plot location were used as the prediction variables for the next step.

Table 1. Three diversity indexes and corresponding equations were used in the study. Note: S is the total number of tree species in a plot; P_i is the proportional abundance of species i relative to the total abundance of all species S in a plot; InP_i is the natural logarithm of this proportion.

Diversity Index	Equation	Reference	Description
Shannon index (H', based e)	$H' = -\sum_{i=1}^{s} P_i InP_i$	[27]	Species richness and equitability in distribution in a plot
Simpson index (λ form)	$\lambda = \sum_{i=1}^{s} P_i^2$	[28]	The dominance of a species in a plot
Pielou evenness index (J')	$J' = \frac{-\sum_{i=1}^{s} P_i InP_i}{InS}$	[29]	How close in numbers each species in a plot

2.3.2. Sentinel-2 Images

Multi-temporal Sentinel-2 imagery was obtained from the Copernicus open access (COA) Hub [30]. We extracted 4 tiles of Sentinel-2 images which corresponded to different phenological phases and covered the study areas from May, June, September, and October in 2020. Using the Sen2Cor plug-in provided by the ESA [31], the Sentinel images were atmospherically corrected. In the sentinel application platform [32], bilinear interpolation method is used to resample all bands to 50 m, and then multiple vegetation indices were also calculated using Sentinel-2 bands (Table 2).

Table 2. Vegetation indices extracted from Sentinel-2 satellite imagery.

Vegetation Indices	Expression	References
Normalized Difference Vegetation Index (NDVI)	$\frac{R_{nir}-R_{red}}{R_{nir}+R_{red}}$	[33]
Normalized Difference Water Index (NDWI)	$\frac{R_{nir}-R_{swir}}{R_{nir}+R_{swir}}$	[34]
Difference Vegetation Index (DVI)	$R_{nir} - R_{red}$	[35]
Enhanced Vegetation Index (EVI)	$2.5\left[\frac{R_{nir}-R_r}{L+R_{nir}+C_1R_r}\right]$	[36]
Soil Adjusted Vegetation Index (SAVI)	$\frac{R_{nir}-R_{red}}{L+R_{nir}+R_{red}} * (1+L)$	[37]

2.3.3. GEDI LiDAR Data

GEDI LiDAR L2B data were obtained from NASA Land Processes Distributed Active Archive Center (https://search.earthdata.nasa.gov/search, accessed on 21 October 2022) in 2019–2021, matching the region of study. The GEDI instrument acquired structural information, such as canopy height metrics, vertical profiles, and surface topography, by analyzing the amount of energy returned by various tree components at different heights above the ground [38]. In this study, the foliage height diversity (FHD) and plant area index (PAI) were extracted from 154,371 observations from GEDI L2B. The FHD index is a plant structural measure that describes the vertical heterogeneity of the foliage profile (Table 3) [39]. The PAI, which comprises various plant components (stem, branches, and leaves), is the one-sided area of plant material surface per unit ground surface area [39]. Considering the changes in forest structure caused by phenological differences, we differentiated the two metrics as growing season and non-growing season. Considering the signal-to-noise ratio of the waveform, the sensitivity of a GEDI footprint shows the dense canopy cover that can be penetrated. Thus, we excluded footprints with sensitivity less than 0.9. After filtering out these invalid observations, 62,593 pairs of FHD and PAI were used for further processing.

Table 3. Characteristics of data used in this study.

Data Type	Variables	Time	Description
Sentinel-2	B1	May. Jun. Sep. and Oct.	Coastal aerosol, 443 nm
	B2	May. Jun. Sep. and Oct.	Blue, 490 nm
	B3	May. Jun. Sep. and Oct.	Green, 560 nm
	B4	May. Jun. Sep. and Oct.	Red, 665 nm
	B5	May. Jun. Sep. and Oct.	Red edge, 705 nm
	B6	May. Jun. Sep. and Oct.	Red edge, 740 nm
	B7	May. Jun. Sep. and Oct.	Red edge, 783 nm
	B8	May. Jun. Sep. and Oct.	Near infrared, 842 nm
	B8A	May. Jun. Sep. and Oct.	Near infrared, 865 nm
	B11	May. Jun. Sep. and Oct.	Short-wave infrared, 1610 nm
	B12	May. Jun. Sep. and Oct.	Short-wave infrared, 2190 nm
Vegetation indices	NDVI	May. Jun. Sep. and Oct.	Normalized Difference Vegetation Index
	NDWI	May. Jun. Sep. and Oct.	Normalized Difference Water Index
	EVI	May. Jun. Sep. and Oct.	Enhanced Vegetation Index
	DVI	May. Jun. Sep. and Oct.	Difference Vegetation Index
	SAVI	May. Jun. Sep. and Oct.	Soil Adjusted Vegetation Index
GEDI LiDAR	FHD_NGS	Non-growing season	Foliage height diversity in non-growing season
	FHD_GS	Growing season	Foliage height diversity in growing season
	PAI_NGS	Non-growing season	Plant area index in non-growing season
	PAI_GS	Growing season	Plant area index in growing season

To obtain spatially continuous FHD and PAI, we used inverse distance weighting (IDW) interpolation to achieve wall-to-wall diversity mapping. The IDW, as a global interpolation, is usually used for sample datasets that are uniformly distributed and dense enough to reflect local differences [40]. Measured values closest to the predicted location have a greater effect on the predicted value than those farther away, resulting in sensitivity of IDW interpolation to outliers and sampling configurations (i.e., clustering and isolation points) [41]. Thus, we randomly select dense GEDI points until these points are uniformly distributed throughout the study area. Then, we selected 80% of GEDI points for interpolation and parameter optimization and applied the remaining sample data (20%) for validation until the correlation coefficient was higher than 0.8.

3. Methods

3.1. Variable Importance Assessment

Selecting the most important variables from high-dimensional datasets is beneficial in improving efficiency and reducing model overfitting. In this study, Boosted Regression Tree (BRT) and Mean Decrease Gini (MDG) algorithms were used to evaluate the importance of independent variables. MDG indicates the contribution of each variable to the homogeneity of the nodes and leaves in the resulting random forest, while BRT evaluates variable performance by iteratively fitting and combining multiple regression tree models [42]. Both algorithms are capable of ingesting multiple classes of predicted variables to model complex interactions without making assumptions about variable interactions and have been widely used in ecological and remote sensing research [43].

3.2. Algorithms for Forest Diversity Mapping

In this study, four machine-learning algorithms with various setups were employed: Lasso Regression (LR), Random Forest (RF), K-Nearest Neighbors (KNN), and Support Vector Machine (SVM). Non-parametric, non-linear algorithms including KNN, RF, and SVM have been applied successfully in a variety of remote sensing applications [44]. KNN and SVM represent distance-based and kernel-based models, respectively, while RF represents tree-based models. Specifically, KNN finds similarities between the new data and available results and puts the new results into the category most similar to those available. SVM can hold regression problems with multidimensional data by separating positive and negative samples to identify the optimum decision hyperplane [45]. RF is a classifier containing a large number of decision tree classifiers [46], and each tree is trained with randomly selected training samples to solve a single problem [47]. All algorithms were implemented using the Scikit-learn python library, and the hyperparameters of LR, K-NN, SVM, and RF methods were fitted through cross-validation (Table 4) [48].

Table 4. Description of the regression models used in this study, including the parameters considered and the criteria used to rank the feature importance.

Model	Abbr.	Parameters	Feature Rank Criteria
Lasso regression	LR	—	Absolute value of coefficients
K-Nearest Neighbors	KNN	K values = 3, 5, 7, 9, 11	Minimum error rate
Support Vector Machine	SVM	cost = 0.1, 0.5, 1, 2, 4, 10 kernel = linear, radial, sigmoid, rbf.	Squared weights
Random Forest	RF	ntree = 200, 500, 800, 1000 mtry = 2, 5, 10, 20, or k/3	Increase in mean squared error by permuting a variable

3.3. Accuracy Assessment

The coefficient of determination (R^2), root-mean-square error (RMSE) and mean absolute error (MAE) were applied to assess the accuracy of tree species diversity estimation. The following equations were used to calculated R^2, RMSE, and MAE:

$$R^2 = 1 - \frac{\sum_{i=1}^{n}(y_i - x_i)^2}{\sum_{i=1}^{n}(y_i - \overline{y})^2} \quad (1)$$

$$RMSE = \sqrt{\frac{\sum_{i=1}^{n}(x_i - y_i)^2}{n}} \quad (2)$$

$$MAE = \frac{\sum_{i=1}^{n}|x_i - y_i|}{n} \quad (3)$$

where x_i and y_i are the estimated and measured values, respectively. $\overline{y}$ is the average measured values, and n is the sample number.

All samples were randomly assigned to one of the two sets of training and validation, following the ratio of 70%:30%. Then, k-fold cross validation was also employed. The generalization error of a given method is directly estimated by cross-validation: The data is divided into K folds of almost equal size, and K folds are used to fit the model. Additionally, the estimated generalization error is the average error over the K folds.

4. Results

4.1. Optimal Features from SENTINEL-2 Images and GEDI LiDAR Data

MDG and BRT algorithms were applied to analyze the 68 features obtained by Sentinel-2 images and GEDI LiDAR data to find the optimal features for diversity mapping. Cross-validation is further used to score several feature subsets and choose the best scoring feature collection. Figure 2 shows the ranking results of key features for three diversity indices, other detailed results are displayed in Appendix A, Table A1. Using the FHD and PAI of GEDI LiDAR in growing season, the vegetation indices of NDVI, NDWI, and EVI, and the spectral bands of B7, B8A, B11, and B12 were identified. Compared with individual spectral bands, GEDI feature and vegetation indices have a stronger explanation on the variations of forest diversity.

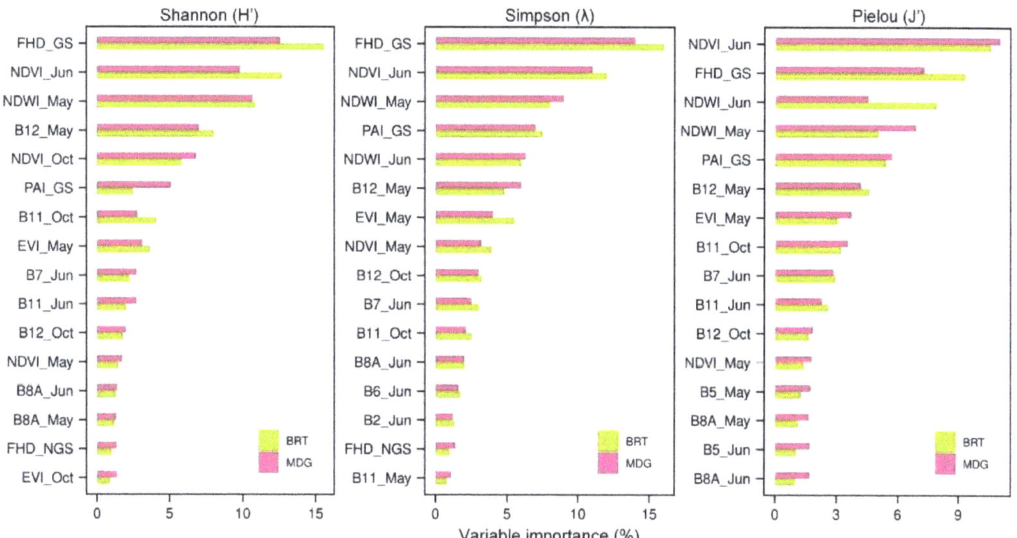

Figure 2. Relative importance of the features selected for estimations of forest diversity indices.

After feature selection, we applied mixed features from GEDI LiDAR data and Sentinel-2 images to estimate forest diversity. For comparison, we selected RF model and applied only GEDI LiDAR data or Sentinel-2 images for forest diversity estimation. Our results show that the Sentinel-2 data alone (averaged $R^2 = 0.62$) gives better prediction accuracies than the GEDI LiDAR data alone (averaged $R^2 = 0.51$), but both are lower than that of combined data sources (Table 5). Specifically, the Sentinel-2&VIs has a good performance on the prediction of H' and J' indices, with R^2 values of 0.66 and 0.63, RMSE of 0.56 and 0.18, although the result of λ index is slightly lower than other indices ($R^2 = 0.57$, RMSE = 0.15). The GEDI data alone is observed to have a relatively high prediction on H' and λ indices ($R^2 = 0.51$; $R^2 = 0.54$ respectively), but a lower prediction on J' index ($R^2 = 0.48$).

Table 5. Estimated accuracy for different data combinations in three diversity indices.

Combined Variables	H' Index		λ Index		J' Index	
	R^2	RMSE	R^2	RMSE	R^2	RMSE
GEDI	0.51	0.78	0.54	0.26	0.48	0.35
Sentinel-2 &VIs	0.66	0.56	0.57	0.15	0.63	0.18
GEDI & Sentinel-2 &VIs	0.72	0.46	0.78	0.14	0.86	0.11

4.2. Diversity Indices Modelling Using Machine Learning Algorithms

Based on selected optimal predictor variables from Sentinel-2 and GEDI data, three diversity indices were characterized using LR, K-NN, RF, and SVM models. Our results showed that the R^2 values of all models are above 0.45 in all the diversity indices (Figure 3). Specifically, the RF model exhibited the best performance with R^2 = 0.86 (RMSE = 0.11) for the J' index, 0.78 (RMSE = 0.15) for the λ index, and 0.73 (RMSE = 0.47) for H' index (Figure 3a,e,i). The SVM also had positive results on the H' and λ indices, with R^2 values of 0.80 and 0.72, RMSE of 0.37 and 0.16, although the result of the J' index was lower than the other models (R^2 = 0.57, RMSE = 0.21) (Figure 3b,f,j). The KNN and LR models showed relatively low results on the λ index (R^2 = 0.46 and 0.57, respectively) (Figure 3c,d) but higher results on the J' index (R^2 = 0.81 and 0.71, respectively) (Figure 3k,l). Overall, the main trend was that lower values of the three indices were a bit overestimated (above the 1:1 line) while high values were underestimated (below the 1:1 line).

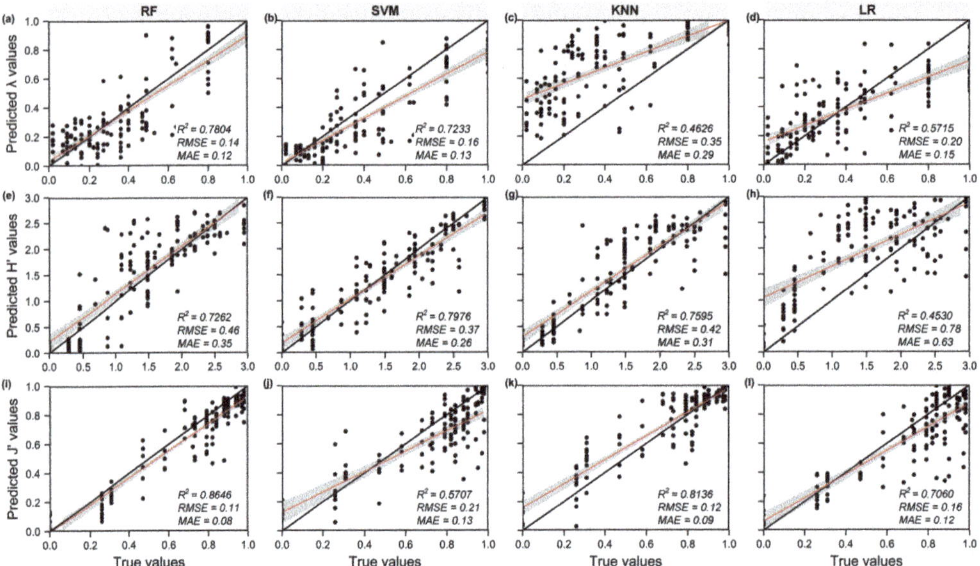

Figure 3. Scatterplot matrix of true values and predicted values by using RF, SVM, KNN, and LR models in three diversity indices. Shannon indices (**a–d**), Simpson (**e–h**), and Pielou (**i–j**). **: Significant correlation ($p < 0.01$).

4.3. Spatial Variability of the Predicted Diversity Indices

Based on the four regression models, we plotted the spatial variation of diversity indices and predicted variables within forests. The spatial distribution of the three diversity indices for the RF result is displayed in Figure 4, while the other results are displayed in Appendix A, Figure A3. Visually, the predicted maps show strong spatial agreements between the H' and J' indices, which are negatively related to the λ index in most parts of a forest. The H' and λ indices account for species richness (i.e., number of different species) and abundance (i.e., number of individual trees per species), while J' index accounts for species evenness (i.e., the numerical dominance of a few abundant tree species). Generally speaking, forest diversity was higher in the north than in the south, especially in the northeast. It is worth noting that the forest diversity of the sparse woods in the southwest area is significantly lower than that of other regions.

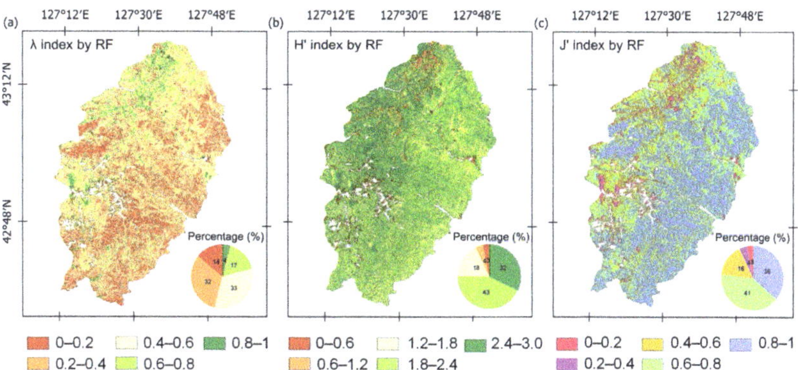

Figure 4. Predicted maps and pixel statistics of forest diversity by the three indices (**a**) λ index, (**b**) H' index, (**c**) J' index using the RF model.

There are notable differences in tree species diversity according to the various forest types obtained by the 9th National Forest Inventory of China (2018). The diversity of the secondary forest regions (the right part of Figure 5a) could be easily distinguished from the natural forests based on predicted variables (Figure 5b–d). However, the performance varies amongst the three indices. Compared with other regions, the diversity of areas along rivers and roads did not significantly differ, but the J' index along rivers expressed relatively low values (Figure 5f–h,j–l). Although the best prediction results were obtained by testing four regression models, we found that a single indicator does not adequately characterize diversity. For example, on the right side of the road in Figure 5f–h, there are significant differences in the three diversity indices, which forced us to obtain a more comprehensive assessment of diversity.

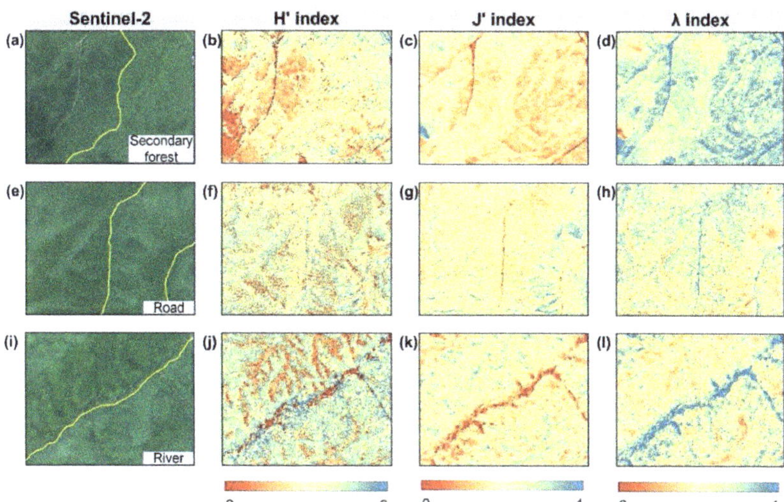

Figure 5. Zoom-in examples of true color Sentinel-2 images (RGB = bands 4, 3, 2) and forest diversity predictions under different forest environments. Sentinel-2 image (**a**) contains two forest types, secondary forest (right) and natural forest (lift); Images (**e**,**i**) show natural forests traversed by rivers and roads. The (**b**–**d**), (**f**–**h**), and (**j**–**l**) indicate three diversity index results corresponding to images (**a**,**e**,**i**).

5. Discussion

5.1. Prospects of GEDI LiDAR and Sentinel-2 Data on Forest Diversity

In this study, we succeeded in estimating forest diversity in a mixed broadleaf-conifer forest, using multi-temporal Sentinel-2 and GEDI LiDAR data. This suggests promising potential for LiDAR data and optical images, combined with machine-learning approach, to estimate forest species diversity over large areas. Such a method would greatly improve conservation and management of forest resources. GEDI LiDAR data uses the reflected laser energy within ~25 m footprints to determine the height, canopy cover, and vertical distribution of plant material. This study is the first to apply the GEDI-derived FHD metrics to forest diversity estimation, our results demonstrate the importance of FHD metrics in future diversity studies. In forest ecology, a high FHD value typically indicates a more complex forest structure (e.g., caused by multiple canopy layers). Structure differences across tree species provide a different directional gap probability, which underlies the LiDAR-based estimations of forest diversity and were confirmed by the direct correlations between tree species diversity by indices (H', λ and J') and GEDI-derived FHD and PAI indices (Figure 6). Therefore, GEDI LiDAR data will become one of the most important parameters in forest diversity estimation. Nonetheless, we argue that it is difficult to achieve good performance using only GEDI data. Our study demonstrated that combined remote sensing data sources were better than GEDI LiDAR data or Sentinel-2 images alone in explaining tree species diversity. The higher explanatory power of the combined data sources was attributed to the full utilization of vegetation properties (vegetation structure information, biochemical properties, and phenological variability).

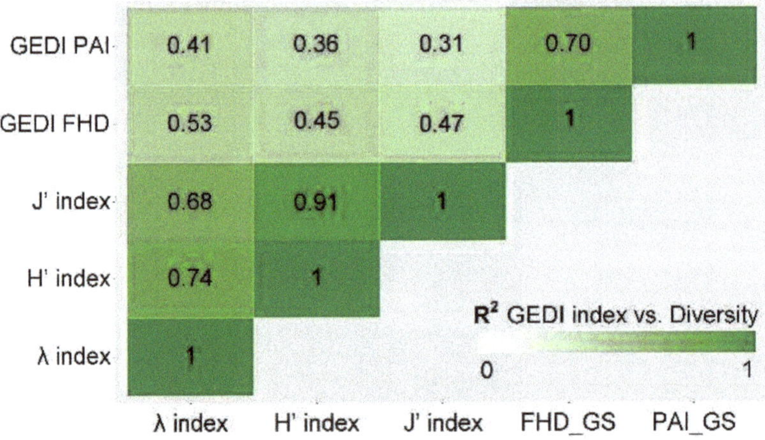

Figure 6. Coefficients of determination (R^2) between measured diversity indices and GEDI LiDAR indices.

Unique spectral responses are caused by differences in the physical and chemical characteristics of various tree species, which is the main driver of forest diversity estimation. Compared to band features, vegetation indices (NDVI, NDWI, EVI, and SAVI) were more significantly correlated with forest diversity (H', λ and J'). These results coincide with those reported by Madonsela et al. [2]. Vegetation indices enhance the spectral information from vegetation while limiting the spectral reflectance from non-vegetative characteristics [49]. This is also proven in Figure 7: The correlation coefficient between predicted H' index and vegetation indices in the fall season is significantly higher than that of the band features. Variability in vegetation indices is caused by a variety of vegetation properties, such as photosynthetic pigments, biomass, and structural carbohydrates [50]. Thus, it is unsurprising that vegetation indices have a significant relationship with forest diversity indices (H', λ and J'). Additionally, the value of Red-Edge, NIR, and SWIR bands for estimating plant diversity has been demonstrated in previous studies by Sothe et al. [51] and Grabska et al. [52]. This study also confirmed the importance of these bands using the BRT and MDG algorithm (see Figure 2). This success is attributed to the rich spectral band setting in Sentinel-2, for example, NIR and SWIR bands are sensitive to water content, lignin, starch, and nitrogen [53]. In addition, we noticed that the correlation coefficients of growing season and non-growing seasons showed a great gap, especially for spectral features. Seasonal variations in canopy structure and biochemical characteristics among several tree species were captured by the spectral values and vegetation indices. These differences provide important references for estimating forest diversity in various forest environments.

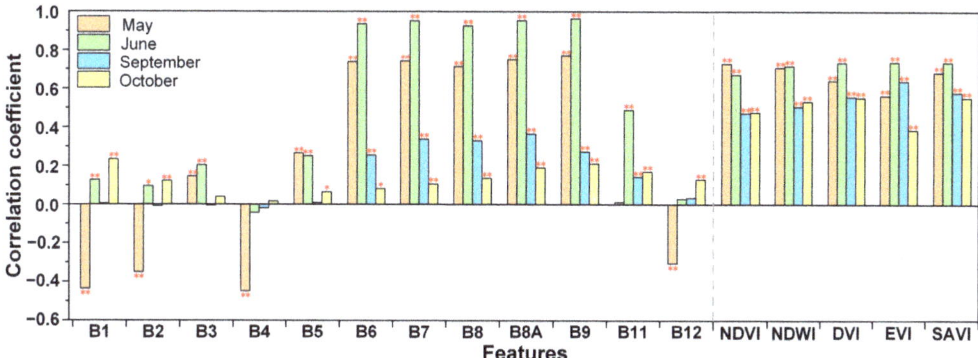

Figure 7. The correlation coefficient between predicted H' index and Sentinel-2 derived feature variables. **: Significant correlation ($p < 0.01$), *: Significant correlation ($p < 0.05$).

5.2. Machine Learning Algorithms for Forest Diversity Mapping

Four different types of machine-learning algorithms were used to estimate forest diversity indices, with three of the diversity indices used having their own variable selection. Our results showed that RF and SVM models provided the highest estimation accuracy in terms of the highest R^2, the lower RMSE, and MAE. This was confirmed by the KNN and LR models. The RF classifier, as an ensemble approach, consisted of a number of tree classifiers, which reduces overfitting impacts and has been the most often used in remote sensing tasks [54]. Similarly, SVMs are a high-performance method designed to solve nonlinear problems using various kernel functions, such as the radial basis function [55]. The solid performance of RF and SVM models were confirmed in other studies [56,57]. For λ and J' indices, RF has the best prediction result, while in the H' index, the SVM model is best. The kernel-based algorithm (e.g., SVM) is prone to overfitting when presented with an extreme value that cannot be identified in the sample [57]. In contrast, tree-based algorithms (e.g., RF) seem to be more resistant to overfitting, though they do not fit as well as kernel-based algorithms [58].

5.3. Prediction Performance and Uncertainty for Forest Diversity

Among the three diversity indices, J' index has the highest correlation coefficient ($R^2 = 0.86$), followed by H' index ($R^2 = 0.80$) and λ index ($R^2 = 0.78$). The three diversity indices, being different representations of plant diversity, varied in spatial distribution (Figure 5). λ index, which accounts for the proportion of species in a sample, is considered to be a dominance indicator [59]. H' index reflects both species richness and equitable distribution of those species within a sample [3]. Moreover, Oldeland et al. [3] emphasized that the H' index better mirrors what one could call "vegetation structure", which is a subset of habitat heterogeneity and thus better reflects spectral variability. The spatial difference between the two indices has been well demonstrated in natural forests and secondary forests (Figure 5b,d). The J' index is an indication of dominance and distribution of individuals across the community within a sample. Relatively few studies have reported this index in remote sensing studies, but it is still of great significance, especially considering the landscape scale [60].

While we derived the forest diversity map with high accuracy, several issues that may limit further estimations still exist. The first is the uncertainty of on-site measurements. In this study, we used semi-variance to determine a spatial scale for forest diversity investigation. Although fixed spatial scales are highly efficient in field surveys, they do not adequately represent the diversity values of the survey region [61]. Secondly, the presence of rare tree species in the understory and trees with DBH less than 10 cm may bring uncertainty on the estimation of forest diversity. Our study area is primarily composed of protected pristine natural forests [27], and the DBH of most trees exceeds 10 cm, which is also confirmed in field surveys. Thus, these trees have no impact on the experimental design and analysis, especially under dense canopy [49]. Finally, errors already exist in the process of forest diversity prediction. For example, the background, including the shading caused by tree canopy, topography, and/or soil color, could cause biased reflectance captured by Sentinel-2 [62].

6. Conclusions

In this study, we applied machine-learning-based regression models to map the spatial patterns of forest diversity in a temperate mixed forest in northeast China. We did this by coupling the newly available diversity product from GEDI LiDAR and multi-temporal Sentinel-2 imagery. Our results showed that a variety of diversity indices can be predicted accurately through combining forest vertical structure information, plant biochemistry, and phenological variability. More accurately, utilizing the FHD index from GEDI, vegetation indices (NDVI, NDWI and EVI), and shortwave infrared band from Sentinel-2 imagery enhanced our ability to estimate forest diversity better than other variables, especially during the growing season. Moreover, comparing four regression algorithms, the study confirmed that the RF model, combined with GEDI LiDAR and Sentinel-2 data, showed strong performance on forest diversity estimation (R^2 = 0.79) and outperformed SVM, KNN, and LR models (R^2 = 0.76, 0.68 and 0.57, respectively). Our results also stressed the great potential of GEDI LiDAR and Sentinel-2 images as explanatory variables for the prediction of forest biodiversity indices. From a forest management perspective, our study developed a reproducible workflow, based on free and openly available GEDI LiDAR and Sentinel-2, that can potentially be used in a routine manner to map forest diversity distribution with a high-resolution, advancing biodiversity conservation and forest ecological restoration.

Author Contributions: Conceptualization, Y.X. and C.R.; methodology, Y.X.; validation, P.L. and H.J.; formal analysis, P.L.; writing—original draft preparation, Y.X.; writing—review and editing, C.R.; visualization, H.L.; project administration, C.R.; funding acquisition, All authors have read and agreed to the published version of the manuscript.

Funding: The research was funded by the National Natural Science Foundation of China (No. 42171367), and Science & Technology Fundamental Resources Investigation Program (No. 2022FY101902).

Data Availability Statement: The remote sensing data were downloaded from Land Processes Distributed Active Archive Center (https://lpdaac.usgs.gov/, accessed on 21 October 2022), and the code used in this study are openly available at https://github.com/xiyanbiao (accessed on 21 October 2022).

Acknowledgments: The authors appreciate the colleagues for cooperation on field campaign and measurements.

Conflicts of Interest: The authors declare no conflict of interest.

Appendix A

Table A1. Detailed results of variables importance.

Rank	Simpson			Shannon			Pielou		
	Variables	BRT (%)	MDG (%)	Variables	BRT (%)	MDG (%)	Variables	BRT (%)	MDG (%)
1	FHD_GS	16.49	14.07	FHD_GS	15.59	12.14	NDVI_Jun	11.43	11.34
2	NDVI_Jun	12.16	12.37	NDVI_Jun	12.78	8.52	FHD_GS	9.33	7.48
3	NDWI_May	7.74	8.91	NDWI_May	10.48	9.38	NDWI_Jun	8.51	3.97
4	PAI_GS	7.55	6.33	B12_May	7.59	7.09	NDWI_May	5.38	5.07
5	NDWI_Jun	5.25	5.66	NDVI_Oct	6.40	6.84	PAI_GS	4.72	3.77
6	B12_May	4.58	5.55	PAI_GS	2.10	4.73	B12_May	3.97	2.94
7	EVI_May	5.18	3.84	B11_Oct	4.11	3.40	EVI_May	2.95	2.38
8	NDVI_May	3.75	3.18	EVI_May	3.52	2.70	B11_Oct	2.83	2.42
9	B12_Oct	3.21	2.76	B7_Jun	2.77	2.37	B7_Jun	2.77	2.40
10	B7_Jun	2.91	2.47	B11_Jun	2.36	2.25	B11_Jun	2.70	2.06
11	B11_Oct	2.58	2.25	B12_Oct	2.07	2.23	B12_Oct	1.66	2.02
12	B8A_Jun	1.97	2.18	NDVI_May	1.83	1.94	NDVI_May	1.59	1.68
13	B6_Jun	1.71	1.80	B8A_Jun	1.43	1.85	B5_May	1.59	1.45
14	B2_Jun	1.46	1.56	B8A_May	1.32	1.46	B8A_May	1.50	1.41
15	FHD_NGS	1.27	1.10	FHD_NGS	1.20	1.46	B5_Jun	1.49	1.40
16	B11_May	1.15	1.02	EVI_Oct	1.26	1.34	B8A_Jun	1.44	1.29
17	B1_Jun	0.92	0.99	B1_May	1.14	1.29	B5_Sep	1.29	1.18
18	NDWI_Sep	0.85	0.87	B8_Jun	1.08	1.19	NDVI_Oct	1.16	1.17
19	B4_May	0.83	0.85	DVI_Sep	1.02	1.05	B3_Sep	1.15	1.17
20	B3_May	0.78	0.84	B5_Oct	1.02	1.03	B3_Oct	1.11	1.16
21	B1_May	0.76	0.79	EVI_Jun	0.98	1.00	B4_Oct	1.09	1.13
22	B7_Sep	0.75	0.78	DVI_Oct	0.96	0.82	B7_Sep	1.07	1.12
23	B1_Oct	0.74	0.77	B3_May	0.89	0.80	B6_Sep	1.05	1.11
24	EVI_Oct	0.74	0.71	B4_May	0.89	0.78	B1 May	1.05	1.11
25	B6_Sep	0.71	0.69	B5_Jun	0.84	0.75	B1_Oct	1.04	1.09
26	B5_Jun	0.70	0.67	B1_Oct	0.78	0.75	B4_Sep	0.99	1.06
27	DVI_Sep	0.69	0.60	B12_Jun	0.75	0.72	B2_Oct	0.97	1.05
28	B12_Jun	0.68	0.60	B4_Sep	0.66	0.72	B1_Jun	0.97	1.05
29	B5_Sep	0.65	0.59	B2_Jun	0.65	0.69	DVI_Sep	0.97	1.04
30	B11_Jun	0.65	0.58	B3_Sep	0.62	0.68	B2_Sep	0.96	1.03
31	DVI_Oct	0.60	0.58	B5_May	0.61	0.68	PAI_NGS	0.95	1.03
32	B8_Jun	0.56	0.55	B1_Jun	0.61	0.65	DVI_Jun	0.93	1.00
33	NDVI_Oct	0.56	0.55	B2_Sep	0.57	0.64	DVI_Oct	0.93	0.99
34	PAI_NGS	0.55	0.49	B2_Oct	0.55	0.63	B6_May	0.85	0.98
35	DVI_Jun	0.54	0.49	B7_Sep	0.50	0.61	B6_Jun	0.84	0.98
36	B2_Oct	0.51	0.49	NDWI_Sep	0.48	0.61	B4_May	0.82	0.96
37	B8_Oct	0.48	0.48	B8_May	0.45	0.61	EVI_Oct	0.82	0.94
38	B9_Oct	0.48	0.47	NDWI_Oct	0.44	0.56	B9_Oct	0.80	0.92
39	B3_Jun	0.47	0.44	B6_May	0.44	0.54	B3_Jun	0.77	0.91
40	B3_Sep	0.47	0.43	B8A_Sep	0.43	0.52	B1_Sep	0.73	0.88
41	B3_Oct	0.46	0.42	NDWI_Jun	0.41	0.51	B5_Oct	0.73	0.88
42	B5_Oct	0.46	0.41	B11_May	0.40	0.51	B8_Jun	0.72	0.88
43	B2_Sep	0.31	0.40	B1_Sep	0.39	0.50	B2_May	0.71	0.86
44	B1_Sep	0.31	0.40	B6_Oct	0.37	0.49	EVI_Jun	0.71	0.85
45	B8A_Sep	0.30	0.39	B12_Sep	0.37	0.48	B12_Jun	0.70	0.85
46	B4_Oct	0.30	0.39	B6_Jun	0.36	0.48	NDVI_Sep	0.70	0.85
47	B2_May	0.28	0.39	DVI_Jun	0.36	0.47	B8A_Sep	0.70	0.84
48	B8_May	0.28	0.39	B8_Oct	0.31	0.47	B11_May	0.68	0.83
49	EVI_Jun	0.27	0.38	B5_Sep	0.30	0.46	B3_May	0.65	0.83
50	B4_Sep	0.26	0.38	NDVI_Sep	0.30	0.45	SAVI_May	0.56	0.83
51	NDWI_Oct	0.24	0.37	B6_Sep	0.30	0.45	SAVI_Sep	0.55	0.81
52	B5_May	0.24	0.37	B2_May	0.29	0.45	B8_Oct	0.53	0.81
53	B7_Oct	0.21	0.37	B11_Sep	0.28	0.43	FHD_NGS	0.52	0.81
54	B8A_May	0.20	0.37	PAI_NGS	0.26	0.42	B6_Oct	0.52	0.81
55	B6_Oct	0.19	0.37	B7_May	0.23	0.42	B11_Sep	0.50	0.79
56	B11_Sep	0.19	0.36	B9_Oct	0.22	0.42	B8_Sep	0.48	0.79
57	B8_Sep	0.18	0.36	B3_Oct	0.21	0.41	B4_Jun	0.48	0.77
58	B12_Sep	0.18	0.35	B4_Jun	0.19	0.41	B7_May	0.48	0.77

Table A1. *Cont.*

Rank	Simpson			Shannon			Pielou		
	Variables	BRT (%)	MDG (%)	Variables	BRT (%)	MDG (%)	Variables	BRT (%)	MDG (%)
59	NDVI_Sep	0.16	0.33	SAVI_Sep	0.15	0.41	B2_Jun	0.47	0.77
60	B4_Jun	0.15	0.33	B4_Oct	0.07	0.40	B8_May	0.45	0.75
61	B6_May	0.13	0.32	DVI_May	0.02	0.40	DVI_May	0.33	0.73
62	DVI_May	0.03	0.31	B7_Oct	0.02	0.39	B12_Sep	0.30	0.72
63	SAVI_May	0.01	0.29	B8_Sep	0.02	0.39	B7_Oct	0.22	0.72
64	B7_May	0.00	0.27	SAVI_May	0.01	0.38	NDWI_Oct	0.17	0.67
65	SAVI_Sep	0.00	0.27	B3_Jun	0.00	0.37	SAVI_Jun	0.00	0.66
66	SAVI_Jun	0.00	0.25	SAVI_Jun	0.00	0.34	EVI_Sep	0.00	0.65
67	EVI_Sep	0.00	0.24	EVI_Sep	0.00	0.33	SAVI_Oct	0.00	0.65
68	SAVI_Oct	0.00	0.20	SAVI_Oct	0.00	0.32	NDWI_Sep	0.00	0.65

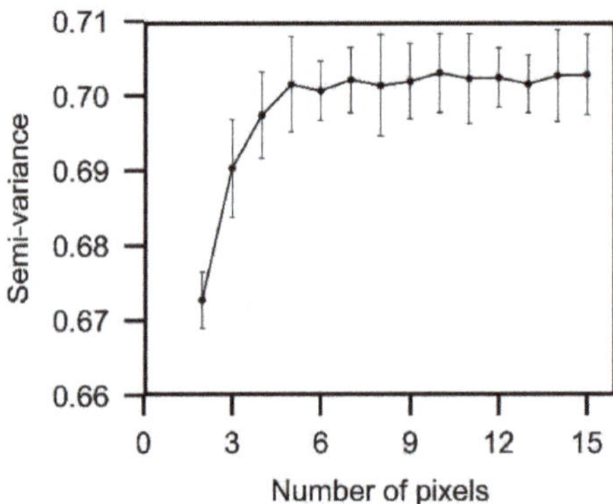

Figure A1. The scale of tree species variability by semi-variogram analysis.

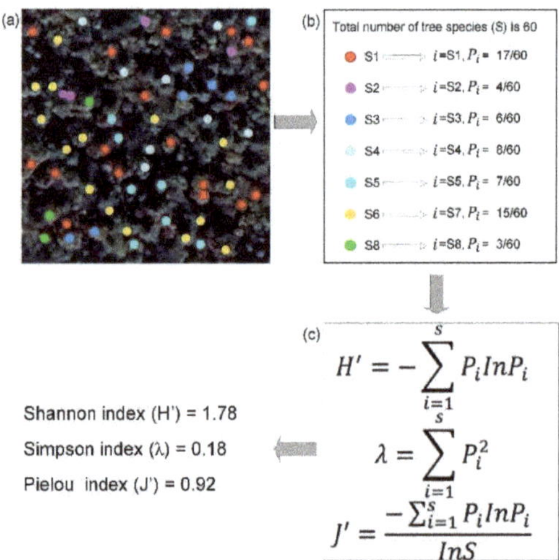

Figure A2. Illustrations of forest tree species diversity calculated by the three indices (H', J' and λ). (**a**) Plot with tree species information; (**b**) Statistics on the number and types of tree species in the plot; (**c**) Calculation equation of tree species diversity. In Figure A2c, S is the total number of tree species in a plot; P_i is the proportional abundance of species i relative to the total abundance of all species S in a plot; InP_i is the natural logarithm of this proportion.

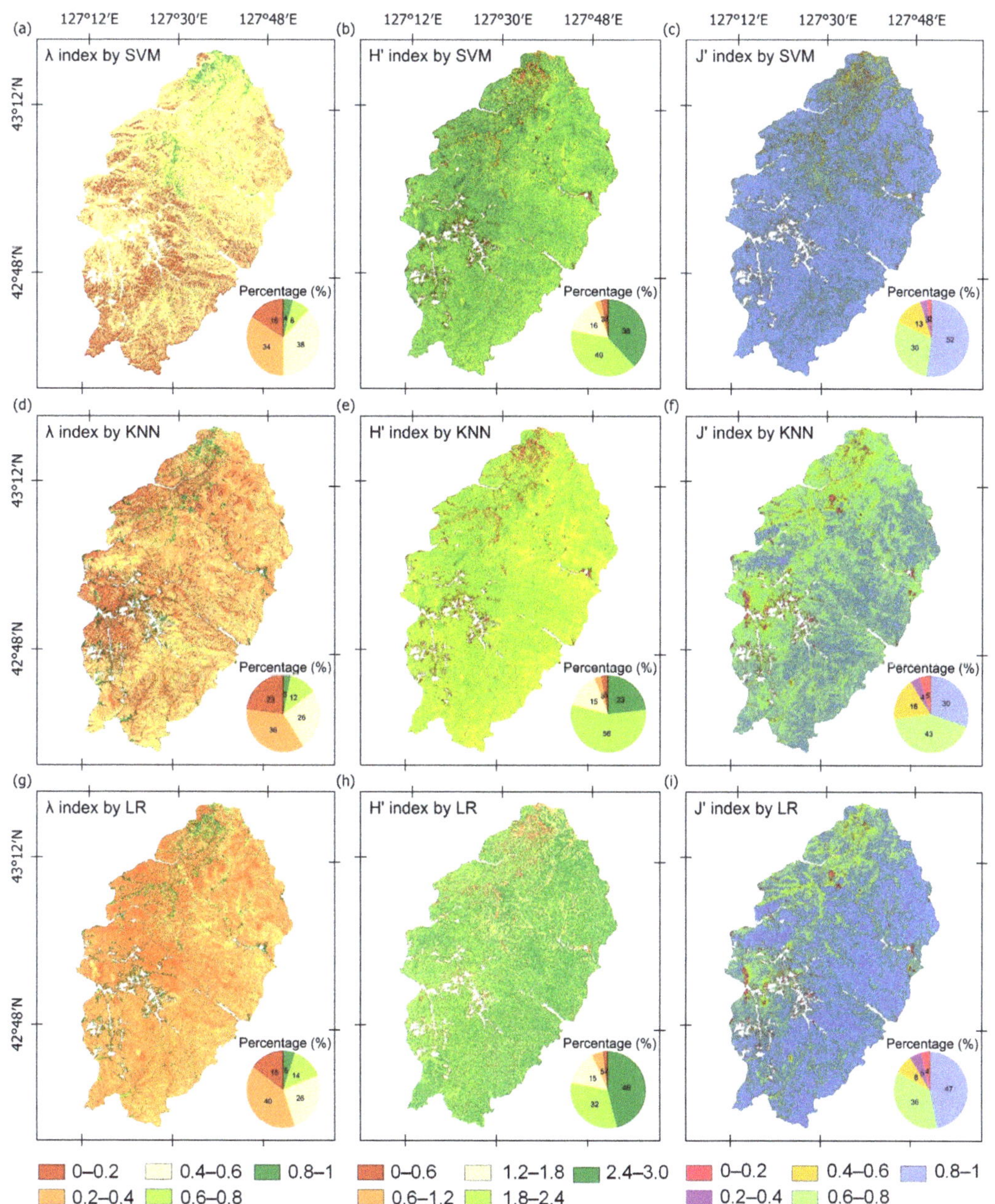

Figure A3. Predicted maps of three diversity indices (λ, H' and J') by SVM (**a**–**c**), KNN (**d**–**f**) and LR (**g**–**i**) models.

References

1. Qi, W.; Saarela, S.; Armston, J.; Ståhl, G.; Dubayah, R. Forest biomass estimation over three distinct forest types using TanDEM-X InSAR data and simulated GEDI lidar data. *Remote Sens. Environ.* **2019**, *232*, 111283. [CrossRef]
2. Madonsela, S.; Cho, M.A.; Ramoelo, A.; Mutanga, O. Remote sensing of species diversity using Landsat 8 spectral variables. *ISPRS J. Photogramm. Remote Sens.* **2017**, *133*, 116–127. [CrossRef]
3. Oldeland, J.; Wesuls, D.; Rocchini, D.; Schmidt, M.; Jürgens, N. Does using species abundance data improve estimates of species diversity from remotely sensed spectral heterogeneity? *Ecol. Indic.* **2010**, *10*, 390–396. [CrossRef]
4. Brown, K.A.; Gurevitch, J. Long-term impacts of logging on forest diversity in Madagascar. *Proc. Natl. Acad. Sci. USA* **2004**, *101*, 6045–6049. [CrossRef] [PubMed]
5. Asner, G.P.; Martin, R.E.; Knapp, D.E.; Tupayachi, R.; Anderson, C.B.; Sinca, F.; Vaughn, N.R.; Llactayo, W. Airborne laser-guided imaging spectroscopy to map forest trait diversity and guide conservation. *Science* **2017**, *355*, 385–389. [CrossRef] [PubMed]
6. Zhang, J.; Zhang, Z.; Lutz, J.A.; Chu, C.; Hu, J.; Shen, G.; Li, B.; Yang, Q.; Lian, J.; Zhang, M.; et al. Drone-acquired data reveal the importance of forest canopy structure in predicting tree diversity. *For. Ecol. Manag.* **2022**, *505*, 119945. [CrossRef]
7. Nagendra, H.; Rocchini, D.; Ghate, R.; Sharma, B.; Pareeth, S. Assessing Plant Diversity in a Dry Tropical Forest: Comparing the Utility of Landsat and Ikonos Satellite Images. *Remote Sens.* **2010**, *2*, 478–496. [CrossRef]
8. Stenzel, S.; Fassnacht, F.E.; Mack, B.; Schmidtlein, S. Identification of high nature value grassland with remote sensing and minimal field data. *Ecol. Indic.* **2017**, *74*, 28–38. [CrossRef]
9. Almeida, D.R.A.d.; Broadbent, E.N.; Ferreira, M.P.; Meli, P.; Zambrano, A.M.A.; Gorgens, E.B.; Resende, A.F.; de Almeida, C.T.; do Amaral, C.H.; Corte, A.P.D.; et al. Monitoring restored tropical forest diversity and structure through UAV-borne hyperspectral and lidar fusion. *Remote Sens. Environ.* **2021**, *264*, 112582. [CrossRef]
10. Fauvel, C.; Weizman, O.; Trimaille, A.; Mika, D.; Pommier, T.; Pace, N.; Douair, A.; Barbin, E.; Fraix, A.; Bouchot, O.; et al. Pulmonary embolism in COVID-19 patients: A French multicentre cohort study. *Eur. Heart J.* **2020**, *41*, 3058–3068. [CrossRef]
11. Leutner, B.F.; Reineking, B.; Müller, J.; Bachmann, M.; Beierkuhnlein, C.; Dech, S.; Wegmann, M. Modelling forest α-diversity and floristic composition—On the added value of LiDAR plus hyperspectral remote sensing. *Remote Sens.* **2012**, *4*, 2818–2845. [CrossRef]
12. Hakkenberg, C.R.; Zhu, K.; Peet, R.K.; Song, C. Mapping multi-scale vascular plant richness in a forest landscape with integrated Li DAR and hyperspectral remote-sensing. *Ecology* **2018**, *99*, 474–487. [CrossRef]
13. Chrysafis, I.; Korakis, G.; Kyriazopoulos, A.P.; Mallinis, G. Predicting Tree Species Diversity Using Geodiversity and Sentinel-2 Multi-Seasonal Spectral Information. *Sustainability* **2020**, *12*, 9250. [CrossRef]
14. Hauser, L.T.; Féret, J.-B.; An Binh, N.; van der Windt, N.; Sil, Â.F.; Timmermans, J.; Soudzilovskaia, N.A.; van Bodegom, P.M. Towards scalable estimation of plant functional diversity from Sentinel-2: In-situ validation in a heterogeneous (semi-)natural landscape. *Remote Sens. Environ.* **2021**, *262*, 112505. [CrossRef]
15. Xi, Y.; Ren, C.; Tian, Q.; Ren, Y.; Dong, X.; Zhang, Z. Exploitation of Time Series Sentinel-2 Data and Different Machine Learning Algorithms for Detailed Tree Species Classification. *IEEE J. Sel. Top. Appl. Earth Obs. Remote Sens.* **2021**, *14*, 7589–7603. [CrossRef]
16. Chen, C.; Ma, Y.; Ren, G.; Wang, J. Aboveground biomass of salt-marsh vegetation in coastal wetlands: Sample expansion of in situ hyperspectral and Sentinel-2 data using a generative adversarial network. *Remote Sens. Environ.* **2022**, *270*, 112885. [CrossRef]
17. Shamshiri, R.; Eide, E.; Høyland, K.V. Spatio-temporal distribution of sea-ice thickness using a machine learning approach with Google Earth Engine and Sentinel-1 GRD data. *Remote Sens. Environ.* **2022**, *270*, 112851. [CrossRef]
18. Kowalski, K.; Senf, C.; Hostert, P.; Pflugmacher, D. Characterizing spring phenology of temperate broadleaf forests using Landsat and Sentinel-2 time series. *Int. J. Appl. Earth Obs. Geoinf.* **2020**, *92*, 102172. [CrossRef]
19. Li, H.; Kato, T.; Hayashi, M.; Wu, L. Estimation of Forest Aboveground Biomass of Two Major Conifers in Ibaraki Prefecture, Japan, from PALSAR-2 and Sentinel-2 Data. *Remote Sens.* **2022**, *14*, 468. [CrossRef]
20. Coyle, D.B.; Stysley, P.R.; Poulios, D.; Clarke, G.B.; Kay, R.B. Laser transmitter development for NASA's Global Ecosystem Dynamics Investigation (GEDI) lidar. In *Lidar Remote Sensing for Environmental Monitoring XV*; SPIE: Bellingham, WA, USA, 2015; pp. 19–25. [CrossRef]
21. Qi, W.; Dubayah, R.O. Combining Tandem-X InSAR and simulated GEDI lidar observations for forest structure mapping. *Remote Sens. Environ.* **2016**, *187*, 253–266. [CrossRef]
22. Lang, N.; Kalischek, N.; Armston, J.; Schindler, K.; Dubayah, R.; Wegner, J. Global canopy height regression and uncertainty estimation from GEDI LIDAR waveforms with deep ensembles. *Remote Sens. Environ.* **2022**, *268*, 112760. [CrossRef]
23. Potapov, P.; Li, X.; Hernandez-Serna, A.; Tyukavina, A.; Hansen, M.C.; Kommareddy, A.; Pickens, A.; Turubanova, S.; Tang, H.; Silva, C.E.; et al. Mapping Global Forest Canopy Height Through Integration of GEDI and Landsat Data. *Remote Sens. Environ.* **2020**, *253*, 112165. [CrossRef]
24. Liang, M.; Duncanson, L.; Silva, J.A.; Sedano, F. Quantifying aboveground biomass dynamics from charcoal degradation in Mozambique using GEDI Lidar and Landsat. *Remote Sens. Environ.* **2023**, *284*, 113367. [CrossRef]
25. Zeng, W.; Tomppo, E.; Healey, S.P.; Gadow, K.V. The national forest inventory in China: History—results—international context. *For. Ecosyst.* **2015**, *2*, 23. [CrossRef]
26. Mi, X.; Guo, J.; Hao, Z.; Xie, Z.; Guo, K.; Ma, K. Chinese forest biodiversity monitoring: Scientific foundations and strategic planning. *Biodivers. Sci.* **2016**, *24*, 1203. [CrossRef]
27. Shannon, C.E. A mathematical theory of communication. *Bell Syst. Tech. J.* **1948**, *27*, 379–423. [CrossRef]

28. Simpson, E.H. Measurement of diversity. *Nature* **1949**, *163*, 688. [CrossRef]
29. Pielou, E.C. The measurement of diversity in different types of biological collections. *J. Theor. Biol.* **1966**, *13*, 131–144. [CrossRef]
30. Bereta, K.; Caumont, H.; Daniels, U.; Goor, E.; Koubarakis, M.; Pantazi, D.-A.; Stamoulis, G.; Ubels, S.; Venus, V.; Wahyudi, F. *The Copernicus App Lab Project: Easy Access to Copernicus Data*; EDBT: Lisbon, Portugal, 2019; pp. 501–511.
31. Main-Knorn, M.; Pflug, B.; Louis, J.; Debaecker, V.; Müller-Wilm, U.; Gascon, F. Sen2Cor for sentinel-2. In *Image and Signal Processing for Remote Sensing XXIII*; SPIE Remote Sensing: Warsaw, Poland, 2017; pp. 37–48.
32. Zuhlke, M.; Fomferra, N.; Brockmann, C.; Peters, M.; Veci, L.; Malik, J.; Regner, P. SNAP (sentinel application platform) and the ESA sentinel 3 toolbox. In *Sentinel-3 for Science Workshop*; ESA: Venice, Italy, 2015; p. 21.
33. Rouse, J.W., Jr.; Haas, R.H.; Deering, D.W.; Schell, J.A.; Harlan, J.C. *Monitoring the Vernal Advancement and Retrogradation (Green Wave Effect) of Natural Vegetation*; Texas A&M university: College Station, TX, USA, 1974; No. E75-10354.
34. Gao, G.; Ting-Toomey, S.; Gudykunst, W.B. *Chinese Communication Processes*; Oxford University Press: Oxford, UK, 1996.
35. Tucker, C.J.; Elgin Jr, J.H.; McMurtrey Iii, J.E.; Fan, C.J. Monitoring corn and soybean crop development with hand-held radiometer spectral data. *Remote Sens. Environ.* **1979**, *8*, 237–248. [CrossRef]
36. Huete, A.; Didan, K.; Miura, T.; Rodriguez, E.P.; Gao, X.; Ferreira, L.G. Overview of the radiometric and biophysical performance of the MODIS vegetation indices. *Remote Sens. Environ.* **2002**, *83*, 195–213. [CrossRef]
37. Huete, A.R. A soil-adjusted vegetation index (SAVI). *Remote Sens. Environ.* **1988**, *25*, 295–309. [CrossRef]
38. Hancock, S.; Armston, J.; Hofton, M.; Sun, X.; Tang, H.; Duncanson, L.I.; Kellner, J.R.; Dubayah, R. The GEDI Simulator: A Large-Footprint Waveform Lidar Simulator for Calibration and Validation of Spaceborne Missions. *Earth Space Sci.* **2019**, *6*, 294–310. [CrossRef] [PubMed]
39. MacArthur, R.H.; Horn, H.S. Foliage profile by vertical measurements. *Ecology* **1969**, *50*, 802–804. [CrossRef]
40. Lu, G.Y.; Wong, D.W. An adaptive inverse-distance weighting spatial interpolation technique. *Comput. Geosci.* **2008**, *34*, 1044–1055. [CrossRef]
41. Bartier, P.M.; Keller, C.P. Multivariate interpolation to incorporate thematic surface data using inverse distance weighting (IDW). *Comput. Geosci.* **1996**, *22*, 795–799. [CrossRef]
42. Hallman, T.A.; Robinson, W.D. Comparing multi-and single-scale species distribution and abundance models built with the boosted regression tree algorithm. *Landsc. Ecol.* **2020**, *35*, 1161–1174. [CrossRef]
43. Valbuena, R.; Eerikäinen, K.; Packalen, P.; Maltamo, M. Gini coefficient predictions from airborne lidar remote sensing display the effect of management intensity on forest structure. *Ecol. Indic.* **2016**, *60*, 574–585. [CrossRef]
44. Maxwell, A.E.; Warner, T.A.; Fang, F. Implementation of machine-learning classification in remote sensing: An applied review. *Int. J. Remote Sens.* **2018**, *39*, 2784–2817. [CrossRef]
45. Vafaei, S.; Soosani, J.; Adeli, K.; Fadaei, H.; Naghavi, H.; Pham, T.; Tien Bui, D. Improving Accuracy Estimation of Forest Aboveground Biomass Based on Incorporation of ALOS-2 PALSAR-2 and Sentinel-2A Imagery and Machine Learning: A Case Study of the Hyrcanian Forest Area (Iran). *Remote Sens.* **2018**, *10*, 172. [CrossRef]
46. Mitchell, M.W. Bias of the Random Forest Out-of-Bag (OOB) Error for Certain Input Parameters. *Open J. Stat.* **2011**, *1*, 205–211. [CrossRef]
47. Chen, L.; Ren, C.; Zhang, B.; Wang, Z.; Liu, M.; Man, W.; Liu, J. Improved estimation of forest stand volume by the integration of GEDI LiDAR data and multi-sensor imagery in the Changbai Mountains Mixed forests Ecoregion (CMMFE), northeast China. *Int. J. Appl. Earth Obs. Geoinf.* **2021**, *100*, 102326. [CrossRef]
48. Hastie, T.; Tibshirani, R.; Friedman, J.H.; Friedman, J.H. *The Elements of Statistical Learning: Data Mining, Inference, and Prediction*; Springer: New York, NY, USA, 2009; Volume 2, pp. 1–758.
49. Fauvel, M.; Lopes, M.; Dubo, T.; Rivers-Moore, J.; Frison, P.-L.; Gross, N.; Ouin, A. Prediction of plant diversity in grasslands using Sentinel-1 and -2 satellite image time series. *Remote Sens. Environ.* **2020**, *237*, 111536. [CrossRef]
50. Yue, J.; Tian, Q.; Dong, X.; Xu, N. Using broadband crop residue angle index to estimate the fractional cover of vegetation, crop residue, and bare soil in cropland systems. *Remote Sens. Environ.* **2020**, *237*, 111538. [CrossRef]
51. Sothe, C.; Almeida, C.; Liesenberg, V.; Schimalski, M. Evaluating Sentinel-2 and Landsat-8 Data to Map Sucessional Forest Stages in a Subtropical Forest in Southern Brazil. *Remote Sens.* **2017**, *9*, 838. [CrossRef]
52. Grabska, E.; Frantz, D.; Ostapowicz, K. Evaluation of machine learning algorithms for forest stand species mapping using Sentinel-2 imagery and environmental data in the Polish Carpathians. *Remote Sens. Environ.* **2020**, *251*, 112103. [CrossRef]
53. Fassnacht, F.E.; Latifi, H.; Stereńczak, K.; Modzelewska, A.; Lefsky, M.; Waser, L.T.; Straub, C.; Ghosh, A. Review of studies on tree species classification from remotely sensed data. *Remote Sens. Environ.* **2016**, *186*, 64–87. [CrossRef]
54. Belgiu, M.; Drăguţ, L. Random forest in remote sensing: A review of applications and future directions. *ISPRS J. Photogramm. Remote Sens.* **2016**, *114*, 24–31. [CrossRef]
55. Wessel, M.; Brandmeier, M.; Tiede, D. Evaluation of Different Machine Learning Algorithms for Scalable Classification of Tree Types and Tree Species Based on Sentinel-2 Data. *Remote Sens.* **2018**, *10*, 1419. [CrossRef]
56. Kuter, S. Completing the machine learning saga in fractional snow cover estimation from MODIS Terra reflectance data: Random forests versus support vector regression. *Remote Sens. Environ.* **2021**, *255*, 112294. [CrossRef]
57. Poorazimy, M.; Shataee, S.; McRoberts, R.E.; Mohammadi, J. Integrating airborne laser scanning data, space-borne radar data and digital aerial imagery to estimate aboveground carbon stock in Hyrcanian forests, Iran. *Remote Sens. Environ.* **2020**, *240*, 111669. [CrossRef]

58. Latifi, H.; Nothdurft, A.; Koch, B. Non-parametric prediction and mapping of standing timber volume and biomass in a temperate forest: Application of multiple optical/LiDAR-derived predictors. *Forestry* **2010**, *83*, 395–407. [CrossRef]
59. Cesarz, S.; Ruess, L.; Jacob, M.; Jacob, A.; Schaefer, M.; Scheu, S. Tree species diversity versus tree species identity: Driving forces in structuring forest food webs as indicated by soil nematodes. *Soil Biol. Biochem.* **2013**, *62*, 36–45. [CrossRef]
60. Peng, Y.; Fan, M.; Song, J.; Cui, T.; Li, R. Assessment of plant species diversity based on hyperspectral indices at a fine scale. *Sci. Rep.* **2018**, *8*, 4776. [CrossRef] [PubMed]
61. Cabrero-González, C.; Garrido-Almonacid, A.; Esquivel, F.J.; Cámara-Serrano, J.A. A model of spatial location: New data for the Gor River megalithic landscape (Spain) from LiDAR technology and field survey. *Archaeol. Prospect.* **2022**, 1–15. [CrossRef]
62. Lechner, A.M.; Foody, G.M.; Boyd, D.S. Applications in remote sensing to forest ecology and management. *One Earth* **2020**, *2*, 405–412. [CrossRef]

Article

Cropland Productivity Evaluation: A 100 m Resolution Country Assessment Combining Earth Observation and Direct Measurements

Nándor Csikós [1,*], Brigitta Szabó [1], Tamás Hermann [2], Annamária Laborczi [1], Judit Matus [1], László Pásztor [1], Gábor Szatmári [1], Katalin Takács [1] and Gergely Tóth [1,2]

1 Centre for Agricultural Research, Institute for Soil Sciences, Hermann Otto 15., H-1022 Budapest, Hungary
2 Institute of Advanced Studies, Chernel u. 14., H-9730 Kőszeg, Hungary
* Correspondence: csikos.nandor@atk.hu

Abstract: A methodology is presented for the quantitative assessment of soil biomass productivity at 100 m spatial resolution on a national scale. The traditional land evaluation approach—where crop yield is the dependent variable—was followed using measured yield and net primary productivity data derived from satellite images, together with digital soil and climate maps. In addition to characterizing of soil biomass productivity based on measured data, the weight of soil properties on productivity was also quantified to provide measured soil health and soil quality indicators as an information base for designing sustainable land management practices. To produce these results, we used only the Random Forest method for our calculations. The study considers high-input agriculture, which is predominant in the country. Biomass productivity indices for the main crops (wheat, maize and sunflowers) and general productivity indices were calculated for the whole agricultural area of Hungary. Results can be implemented in cadastral systems, in applied in agricultural and rural development programs. The assessment can be repeated for monitoring purposes to support general monitoring objectives as well as for reporting in relation to the United Nations Sustainable Development Goals. However, on the basis of the results, we also propose a method for periodically updating the assessment, which can also be used for monitoring biomass productivity in the context of climate change, land degradation and the development of cultivation technology.

Keywords: random forest; land evaluation; soil; biomass; Hungary; gross primary productivity; soil health; soil quality

Citation: Csikós, N.; Szabó, B.; Hermann, T.; Laborczi, A.; Matus, J.; Pásztor, L.; Szatmári, G.; Takács, K.; Tóth, G. Cropland Productivity Evaluation: A 100 m Resolution Country Assessment Combining Earth Observation and Direct Measurements. *Remote Sens.* **2023**, *15*, 1236. https://doi.org/10.3390/rs15051236

Academic Editors: Kenji Omasa, Shan Lu and Jie Wang

Received: 6 December 2022
Revised: 21 February 2023
Accepted: 22 February 2023
Published: 23 February 2023

1. Introduction

A key natural resource that ensures food security, ecological security and sustainable development is cultivable land. Recently, the importance of soil has been increasingly put into focus as the general public also become more aware of it as a non-renewable resource that can be lost quickly if improperly used or managed with very little chance of regeneration. Despite the critical importance of soil productivity, not only as indicator, but also in sustaining life on Earth, knowledge of the spatial and temporal variability of soil from regional to global scales is limited or fragmented. For the creation of effective agricultural and food policies at the regional levels, accurate soil productivity predictions are essential. The limited information on soil productivity hinders national (Farmers' Soil Conservation Programme, National Rural Development Programme) and international (EU Soil Mission) programs to monitoring its changes and build future scenarios on it.

The Sustainable Development Goals (SDGs) of the United Nations' Agenda 2030 framework include targets that recommend direct consideration of land and soil resources [1–3], which were adopted by all United Nations member states in 2015. Soil resources are linked to the SDGs through several soil functions [2], of which the biomass productivity function is at the core of SDGs 2.3 and 2.4., which explicitly target the sustainable increases in agricultural productivity. Furthermore, biomass productivity is proposed as an indicator of land degradation [4], which is linked to SGD 15.3 [5].

Biomass productivity is conditioned by inherent soil properties, climatic and management factors, thus variable in both space and time [6]. Spatial variability of soil productivity is traditionally assessed within the broad framework of land evaluation [7]. However, land evaluation should also include socio-economic components [8], which are not necessary for soil productivity evaluation. Nevertheless, soil is an integral part of the land with a distinct spatial location and therefore biophysical characteristics of the studied sites, such as climate and relief conditions, need to be taken into account when assessing its productivity [9].

The aim of classical quantitative land evaluation is to establish productivity indices based on actual yields in order to reflect production potentials for taxation and planning purposes [10–18]. A similar quantitative approach can be applied to reveal soil biomass productivity, its drivers and changes for monitoring purposes.

Dynamic and simulation models [7,16,19–22] can provide an alternative to classical productivity evaluation, but their validation still requires measured biomass or yield data. Advantages of the classical data-driven assessment, i.e., where yield is the dependent variable and biophysical factors are independent inputs, are high reliability, explicit spatial validity and easy interpretation. Process-based modeling and statistical modeling are also two frequently employed techniques for forecasting crop yield responses to climate variability. Process-based crop models are effective for predicting crop yields because they simulate physiological processes of crop growth and development in response to environmental factors and management techniques, especially at the field scale [23]. Traditional regression techniques have some drawbacks that can be addressed by statistical modeling techniques based on machine-learning algorithms. Machine-learning techniques have been used increasingly in recent years as niche-based classification modeling tools [24–26]. For our analysis we selected the Random Forest (RF) technique [27,28], which uses the Classification and Regression Trees method as the basis for growing multiple classification trees. The study considers high-input agriculture, which is predominant in the country and uses time series information (measured crop yield statistics and satellite-derived biomass productivity indicators).

A scientific-based biomass productivity assessment should be based on a numerical assessment of production potential based on statistical studies. Previous national land evaluation techniques were estimation procedures, which inevitably introduced classification errors. Since the only objective measure of land quality is yield over time, our method is designed with yield as the dependent variable and environmental factors (soil, climate, topography) that affect yield as the independent variables. The method must be designed in such a way that the parameterization process can be repeated as the amount of available data increases, so that the land classification system can be easily revised and refined at any time on the basis of changes in production conditions.

Based on the above considerations, we performed a detailed study with country coverage with the following aims: (i) to identify main soil and climatic determinants of biomass productivity, (ii) to quantify the weights of soil and climatic factors of productivity for the main crop types (wheat, maize, sunflowers), (iii) to produce crop-specific and general productivity maps for all agricultural land of the country, and (iv) to propose a methodology for integrated monitoring of biomass productivity.

2. Materials and Methods

Soil biomass productivity evaluation must be based on biomass data and the assessment of the environmental and management factors influencing it. This requires biomass data, and geographical and management data, including soil, topography, climate and fertilizer data. Country-wide implementation of the agricultural biomass productivity model can only be based on information that is available for the full agricultural area of the country. To ensure the best possible spatial detail to develop and implement a new productivity model, data of dependent variables (measured yield and remotely sensed biomass indicators) as well as independent variables on soil properties were collected at parcel scale and implemented at soil property maps of 100 m resolution.

2.1. Study Area

Hungary is located in Central Europe and the Carpathian Basin, which is a part of the Pannonian biogeographic region (45°43′ to 48°35′N and 16°06′ to 22°53′E). The country is 93,033 km^2 and has an elevation range between 77 and 1014 m above sea level, and agricultural lands are typically located between 77 and 350 m altitude. Agriculture is the dominant land use, with non-irrigated arable land (Figure 1) accounting for 61% of the country's total area [29]. Winter wheat (Triticum aestivum), maize (Zea mays) and sunflowers (Helianthus annuus), which have been sown on up to 80% of Hungary's arable land in recent decades, were selected for the productivity assessment.

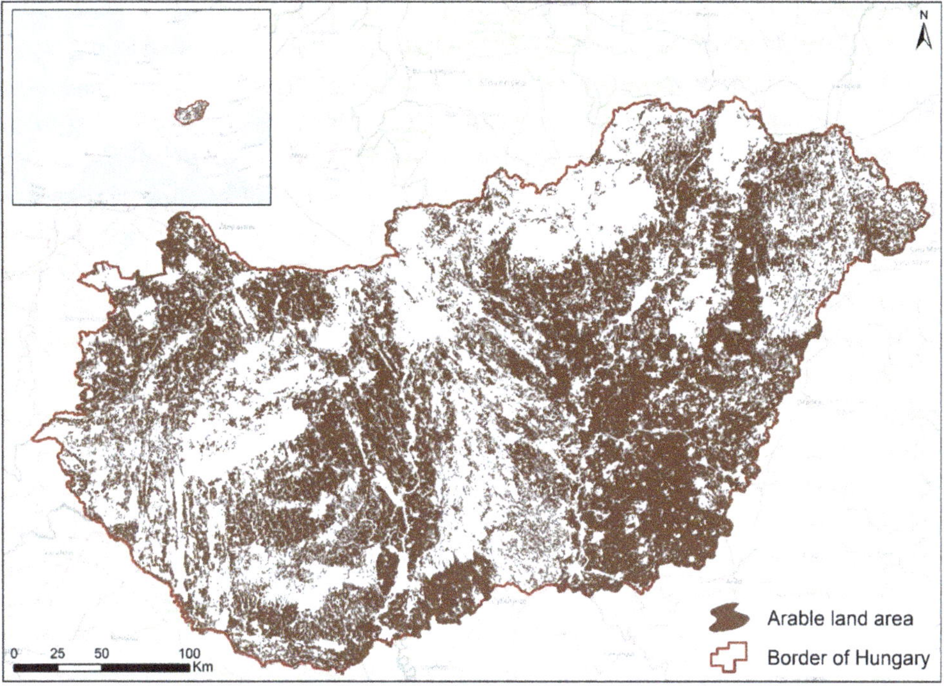

Figure 1. Arable land areas of Hungary (study area), based on Corine Land Cover 2018 dataset [30].

2.2. Databases

2.2.1. National Plot-Level Field Soil, Fertilization and Yield Databases (AIIR Field Database)

The AIIR database [31] contains crop type, yield, fertilization and soil information for each cultivated parcel, summing up to 80,000 cultivated parcels of Hungary for 5 years (1985–1989). The data were provided by the Central Plant and Soil Conservation Service (Budapest) for the purpose of land evaluation research. The sampling for the soil tests was carried out in such a way that the parcels were divided into 12 ha sections and then, along the diagonals of the selected sections, soil samples were taken from at least 20 locations using the so-called parallel sampling method. The subsamples were taken homogenized, so that an average sample was taken from the subplots of each agricultural field. For areas with a slope greater than 12%, average samples were taken separately for each (upper, middle, lower) section of the slope, taking into account erosion and different soil nutrient supply. The database was digitized in 2000 and in 2014 was upgraded to a modern geospatial database (point data with coordinates). We have selected the points that still fall on arable land at the time of our study. The database includes the following three major types of data:

- Basic data of the parcels (location, size, land user);
- Soil taxonomical and laboratory analysis data (soil type and subtype, pH, texture, organic matter, nitrogen, phosphorus and potassium content);
- Agricultural management data (crop type, yield, date of sowing, fertilization and harvest, fertilizer doses);
- Crop type and yield data.

Distribution of data by soil types is presented in Table 1.

Table 1. Main features of the AIIR dataset, based on Hungarian [32] and World Reference Base for Soil Resources [33] classification.

Soil Taxonomical Unit of Major Agricultural Soils		No. of Parcels Covered	Area (ha)	Area (%)
Hungarian classification	WRB 2014			
Lessivated brown forest soil (non-podzolic)	*Haplic Luvisol*	11,062	385,048	10.06
Raman-type brown forest soil	*Haplic Cambisol*	6567	270,239	7.06
Rust-brown sandy forest soil	*Arenic Cambisol*	2988	114,872	3
Typical calcareous chernozem	*Haplic Chernozems*	3792	228,240	5.96
Great Plains calcareous chernozem	*Haplic Chernozems*	2042	120,123	3.14
Carbonated meadow chernozem	*Gleyic Chernozems*	5540	330,200	8.63
Non-carbonated meadow chernozem	*Luvic Chernozems*	2021	108,149	2.83
Carbonated meadow soil	*Calcic Vertisols*	3952	184,853	4.83
Non-carbonated meadow soil	*Haplic Vertisols*	3460	151,394	3.96
Carbonated alluvial meadow soil	*Gleyic Fluvisols*	3129	142,535	3.73
Non-carbonated alluvial meadow soil	*Dystric Fluvisols*	4658	179,101	4.68
Carbonated humic alluvial soil	*Calcaric Fluvisols*	1210	51,720	1.35
Non-carbonated humic alluvial soil	*Dystric Fluvisols*	1584	50,789	1.33
Carbonated humic sandy soil	*Calcaric Cambisols*	3714	138,044	3.61
Non-carbonated humic sandy soil	*Dystric Cambisols*	2458	75,656	1.98
major soils in total		58,177	2,530,963	66.2
other soils		28,517	1,295,467	33.8
Σ		86,695	3,826,430	100

2.2.2. Remote Sensing Derived Biomass Productivity Indicators

Long term (2003–2018) time series remote sensing data were used to derive mean gross primary productivity (GPP) values as proposed by Jin and Eklundh (2014) [34]. The MODIS dataset (MOD17) [35] was used at a nominal 500 m spatial resolution to produce GPP datasets for the whole country. It is important to note that crop yields and GPP represent different aspects of productivity. However, in managed cropland there is a strong correlation between the two [36]. We used the normalized productivity (value range 1–100) as the target variable, and all of our results were normalized between 1 and 100, making it easier to integrate into our model.

2.2.3. Time Series Meteorological Data

The Central-European FORESEE meteorological database [37], which covers the whole area of the country with a 0.1 × 0.1 degree grid, was used to derive mean temperature and total precipitation at monthly scales (between 1951 and 2013). Mean temperature and precipitation values were linked to the spatial units (100 m pixels) of the assessment. The downscaling was performed by the bilinear resampling method.

2.2.4. Topographic Data

The Shuttle Radar Topography Mission [38,39] provides a dataset of 30 m resolution grid cells as the basis for the digital elevation model (DEM). SRTM mapped Earth's topography between 56 degrees south and 60 degrees north of the equator. SRTM has a vertical accuracy of 5.3 m (RMSE) in Hungary [40]. The SRTM-derived DEM was used to include a topographic component to the land evaluation model.

2.2.5. Land Use Data

The Hungarian coverage of the CORINE [30] land cover database for the year 2018 was used to delineate croplands in the country. The 1:100,000 scale datasets have a minimum mapping unit of 25 ha for patches and a minimum width of 100 m for linear elements. A total of 44 land cover and land use categories are included in the dataset, 28 of which are appropriate for Hungary [30]. All assessments and the map visualization of the results were based on the cropland areas (see Figure 1).

2.2.6. Map Series of Soil Types and Soil Properties

The unified national soil type and soil property maps of Pásztor et al. (2020, 2018, 2017, 2015) [41–44] provided the soil information base for the assessment. A total of 41 soil types, belonging to 9 main soil type groups of the Hungarian Soil Classification System [45], are covered by the dataset. Soil chemical and physical data include pH, calcium carbonate content, organic matter content and texture. The map series are all produced at a 100 m resolution and can be viewed on the dosoremi.hu website. The 100 m resolution of the soil maps was considered to be sufficiently detailed for parcel-scale productivity evaluation, and therefore this spatial resolution defined the resolution of the assessment. There is a slight difference in the semantic component of the soil type maps and the soil type information in the AIIR dataset (Table 1). There are soil types in the national soil map with areas covering <1% of the country that are not available in the AIIR dataset, or which are available only with a very limited sample size. These were not sufficient for statistical tests. This minor inconsistency required an expert-based modification of the final evaluation system.

2.3. Data Preparation

A quality and consistency check of the AIIR dataset was carried out in the first phase of the data preparation to filter out typos and false records. Inconsistent records (outliers), such as soil samples with high carbonate content and low pH, were excluded from the dataset. We then selected those records from the AIIR dataset that corresponded to agricultural parcels of intensive (i.e., high fertilizer use) cultivation. The selection was made based on the amount of fertilizers applied, and records containing at least 125 kg $\times$ ha^{-1} of nitrogen and 30 kg $\times$ ha^{-1} of active phosphorus input were kept. In this way, the analysis of the current assessment focused on data from intensively cultivated fields.

Winter wheat (*Triticum aestivum*), maize (*Zea mays*) and sunflowers (*Helianthus annuus*), which have been sown on up to 80% of the croplands in Hungary [46] in recent decades, were selected for the productivity evaluation. In order to establish a common basis for the analysis, the yield data of these three main crops from each parcel of the dataset were normalized to a scale of 1 to 100. For the same reason, the GPP values were also normalized to a scale of 1 to 100. Normalization was applied to all wheat, maize and sunflower yield data in the five years covered by the AIIR database and to all cropland pixels in the GPP dataset.

The AIIR database with normalized yield data and the normalized GPP dataset were integrated with the climate geodatabase into a single geodatabase using geographical coordinates as unique identifiers. The result was a georeferenced dataset created to include all soil, climate, management and yield data. Productivity analysis was carried out using information of the georeferenced pixels, including their geographical coordinates.

The GPP data, originally produced at 500 m resolution, were downscaled to 100 m resolution and normalized to values between 1 and 100. The downscaling was performed by the nearest neighbor resampling method The SRTM data, which were originally produced at 30 m resolution were generalized to 100 m resolution using the bilinear interpolation technique.

All datasets were converted to the Uniform National Projection System (EOV) to create a coherent geodatabase.

2.4. Assessment and Implementation Methods

2.4.1. Model Development

Soil biomass productivity assessment is the process of establishing relationships between soil properties and yields. Data-mining methods are tools for revealing hidden relationships in datasets structured by input variables. In soil assessment, data mining can help to identify the most important factors in yield formation and establish the weights of these factors. For our analysis we chose the Random Forest technique [27,28], which uses the Classification and Regression Trees method as a basis for growing multiple classification trees. For this operation, the database is divided into a series of training and test datasets to establish and validate relationships, respectively. Each training dataset (80% of the dataset) is a randomly selected subset that is used to develop a tree model using randomly selected predictors. The remaining data (10%) after the random selection of the subset (test data, 10% of dataset) are used to validate the developed model [47]. We used the createDataPartition function from the caret package to select data randomly. The generalized error of the forest depends on two parameters: how accurate each individual classifier is and how independent the different classifiers are from each other (i.e., the strength of each tree in the forest and the correlation between them). The Random Forest analysis was performed with the ranger R package [48]. The long-term means normalized productivity index (MNPI), taking into account both the measured AIIR data and the GPP data, was computed by taking the average of the two normalized datasets. The Random Forest operation was performed with the MNPI as the dependent variable and the environmental (soil, climate) variables as explanatory variables (Figure 2). First, the assessment was carried out separately for winter wheat, maize and sunflowers in order to evaluate crop-specific productivity of Hungarian croplands using the MNPI data of these crops. As a result, crop-specific productivity indices were produced for the three main crops. As our overall interest was to establish the MNPI for each Hungarian parcel at 100 m resolution, three parallel models were developed for the three major crops (wheat, maize, sunflowers) based on the crop-specific entries of the normalized yield data, and a fourth, a general productivity model, was developed based on the MNPI. As a result, both crop-specific (weighted means, wheat 40%, maize 40% and sunflowers 20%) and general productivity indices were assigned to climate and soil property combinations. Due to the limited information for some minor soil types (i.e., occupying area < 0.5% of agricultural lands), statistical testing could not be successfully performed for these soils. To assess the productivity evaluation of these soils, two evaluation approaches were applied and their results were combined. Firstly, an expert-based judgement was carried out. Productivity indices were established considering those of closely related soils in the Hungarian soil taxonomy using information from previous land evaluation systems [49], related literature [50–54] and expert knowledge. Secondly, a statistical test based solely on the GPP data was carried out to evaluate the effect of soil properties and climate, although without statistically significant results, but for orientation purposes. The relative importance of the explanatory variables was calculated. We analyzed the importance of all variables using the imp function of bclust package in R [55]. Relative importance was calculated by dividing the importance score of each variable by the largest importance score of the variable, and then multiplying by 100. Harmonizing the results of the two approaches ensured the consistency across the system, even for parcels with soils that make up a small proportion of the country's croplands. The theoretical range of the final productivity indices was set between 1 and 100, corresponding to the normalized yield values of the test dataset and following the indexing approach of traditional soil productivity evaluation of Hungary [51]. Model validation was performed using normalized yield data as independent variables of the test dataset. The test dataset included a randomly selected 10% of the data and a predict function of the ranger package was used. We calculated the correlation coefficient to show the relationship between the observed and the predicted values, the mean absolute error (MAE) to show the distance of the predicted values from the observed values [56], and the mean absolute percentage error (MAPE) to show the percentage of error between observed and predicted values [57].

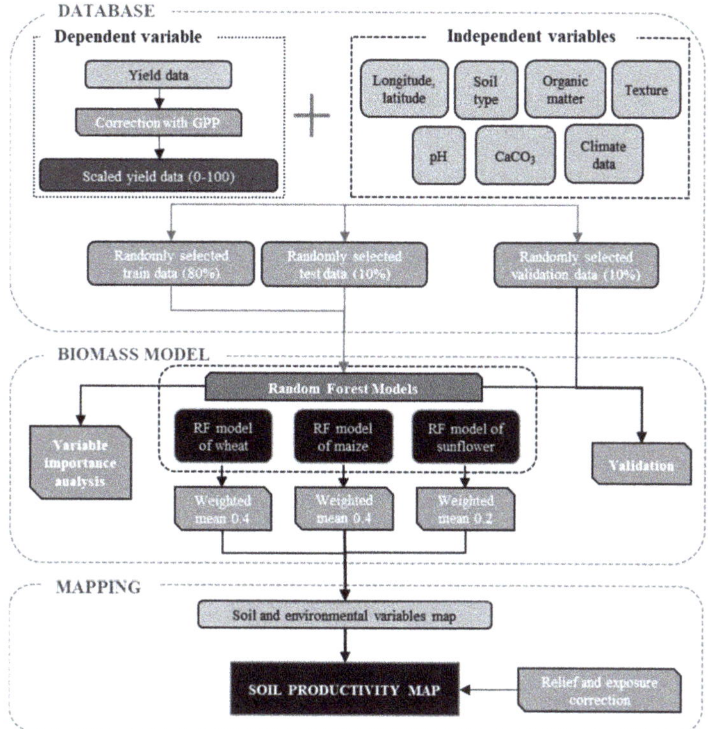

Figure 2. Flowchart of land evaluation modeling process.

2.4.2. Spatial Implementation

Soil, meteorological and digital terrain maps were used for the spatial implementation of the soil biomass productivity model, i.e., to produce soil productivity maps. The developed model provides productivity indices on a scale of 1 to 100 for several combinations of climate and soil properties in the country. Basic input layers for the spatial implementation include detailed (100 m) soil type and soil property maps and climate data. Slope correction coefficients (see Appendix A Table A1) from the previous official Hungarian land evaluation model [49] were applied to produce the final productivity indices. The coefficients reflect the effect of slope angle and slope direction on productivity. The SRTM digital topographic data were used to implement the correction coefficients and to produce the final maps. Presentation of the results covers all cropland areas of the country at a 100 m resolution.

3. Results

3.1. Model Development and Estimation Efficiency

The general, country-wide productivity model using biophysical explanatory variables explains up to 40% of the biomass productivity in the country (R^2 = 0.402). This model fit can be considered adequate for a country scale assessment, especially for a country with a wide variety of soil types from salt-affected soils to Arenosols, Luvisols and chernozems. The efficiency of the crop-specific models is best for wheat, followed by maize and sunflowers, in the order of the available sample size, respectively (Figure 3 and Table 2). Results were statistically significant at the 0.01 level.

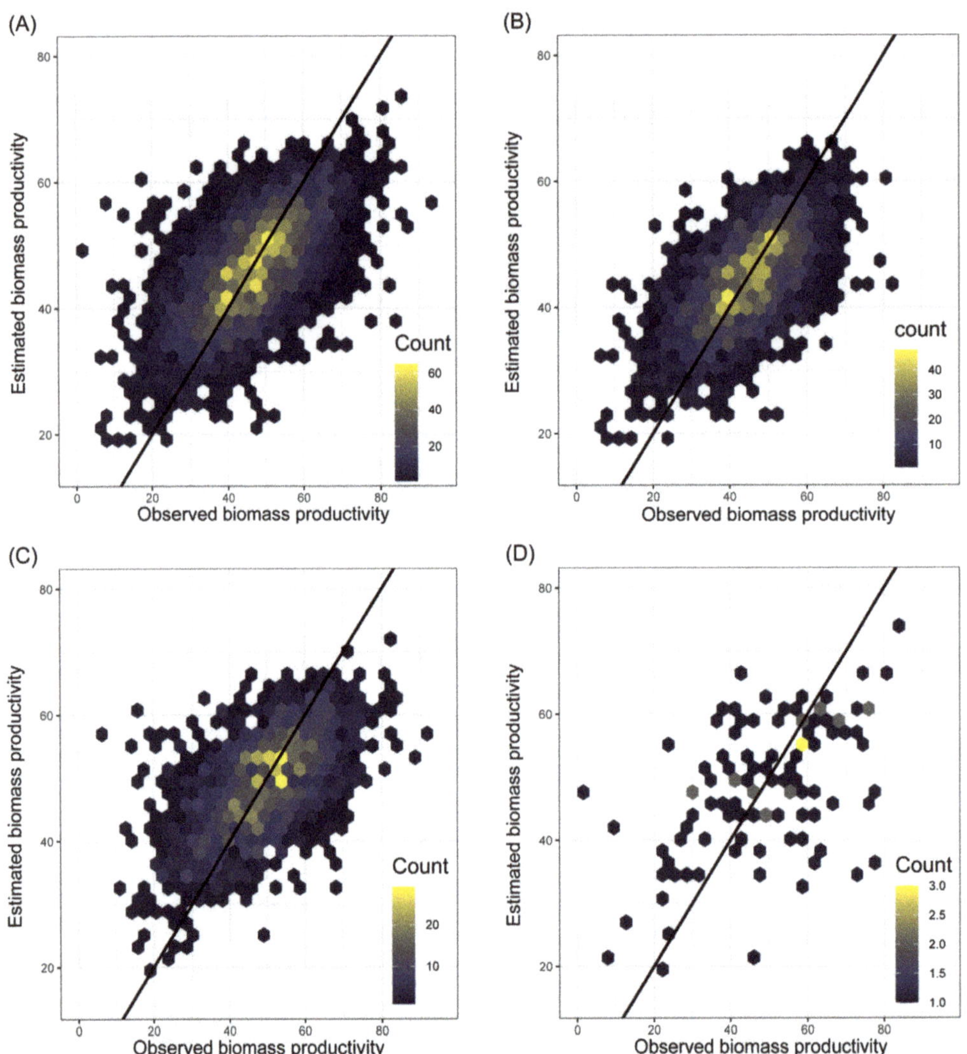

Figure 3. Scatter plot of observed vs estimated biomass productivity of total cropland area (**A**), wheat (**B**), maize (**C**) and sunflowers (**D**). Results were significant at the 0.01 level.

Table 2. Test validation results of all cropland, wheat, maize and sunflowers. R^2: correlation coefficient, R: Pearson correlation, MAPE: mean absolute percentage error, MAE: mean absolute error, N: number of pairs.

	R^2	R	MAPE (%)	MAE	N
All cropland	0.4	0.63	19.28	7.33	4381
Wheat	0.41	0.64	18.06	6.78	2631
Maize	0.35	0.59	19.17	7.93	1646
Sunflower	0.27	0.52	29.81	11.7	104

The combination of measured and satellite-driven data for the general productivity model development gave almost the same model fit as the crop-specific one for wheat, which was based on a large sample size of measured yields. The descriptive power of sunflower productivity estimation was not as strong (Figure 3D). The MAPE results are as follows: all cropland 19.28%, wheat 18.07%, maize 19.17% and sunflowers 29.81%. The most accurate prediction based on the MAPE and MAE results was for wheat followed by the maize and sunflower predictions.

3.2. Baseline Biomass Productivity Indices and Map for Croplands of Hungary

By implementing the biomass productivity model on the national soil, climate and topographic geodatabase, a new soil biomass productivity map was produced (Figure 4). The map shows the general productivity of croplands. In the same process, crop-specific productivity maps were also produced. While the crop-specific productivity indices and maps can be used for planning land use and cropping, the general productivity map provides an overview of the spatial pattern of biomass potential of agricultural parcels in the country.

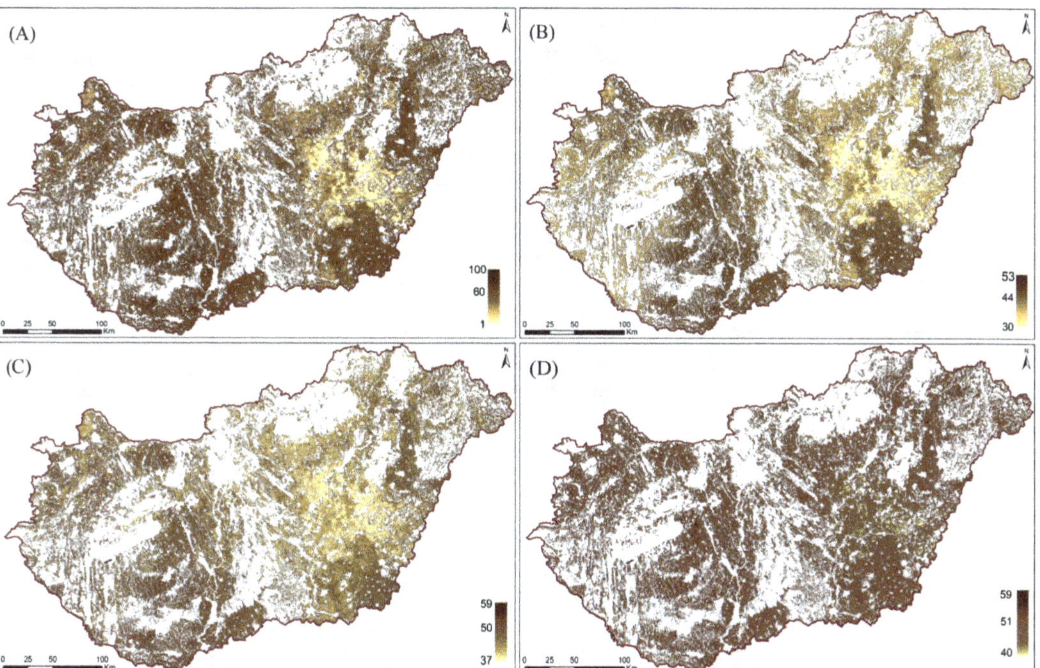

Figure 4. Croplands' land evaluation values range between 0 and 100 in the case of general biomass productivity of arable lands without slope correction coefficients (**A**) and separately wheat (**B**), maize (**C**) and sunflowers (**D**).

The map confirms the empirical knowledge that the most fertile areas are on chernozem soils in the east and on various loamy soils in the west of the country. Sandy soils, whether in the western, the central or the eastern part of the country, perform rather poorly. This phenomenon is typical of a country where water supply is the main climatic factor limiting crop production.

The mean productivity index for all the croplands is 64.7, with a standard deviation of 13.4, reflecting the dominancy of medium-to-good land within the agricultural areas of the country in terms of the spatial extent (Figure 4A).

The mean productivity index after slope coefficient correction for all the croplands of the country is 58.9, with a standard deviation of 18.5, reflecting the dominancy of medium-to-good land within the agricultural areas of the country (Figure 5).

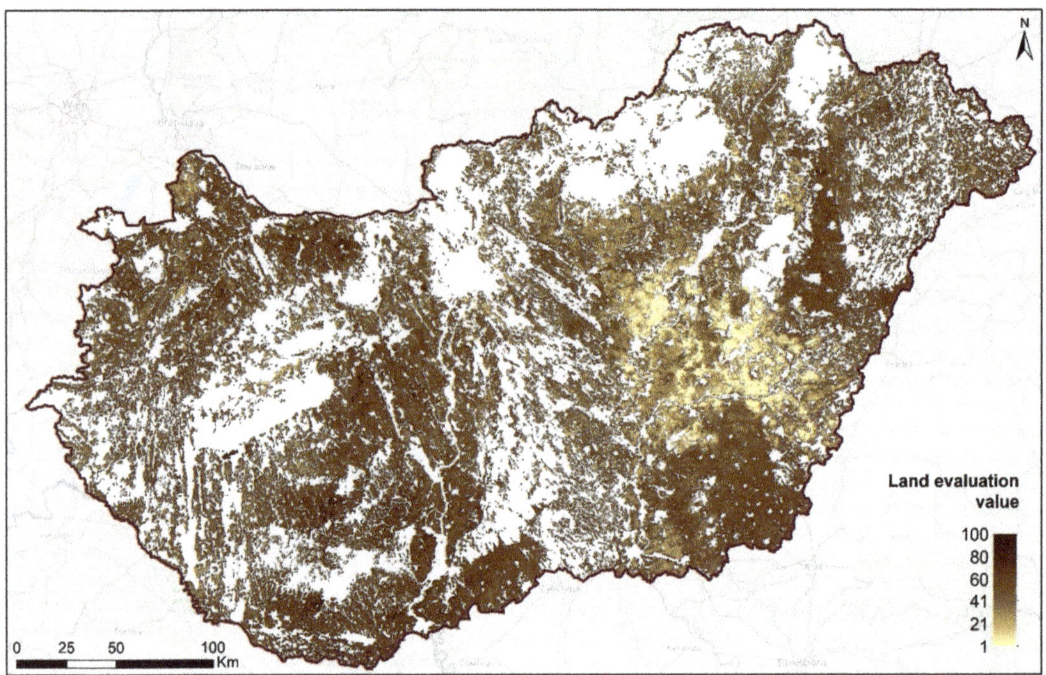

Figure 5. Croplands' land evaluation values range between 0 and 100 in the case of general biomass productivity of arable lands after slope correction.

3.3. Soil and Climatic Determinants of Biomass Productivity in Hungary

In general, humus content (which also reflects organic matter content), pH, $CaCO_3$ content, soil type and soil texture were the most important soil-based input parameters for predicting wheat productivity (Figure 6A). However, the geographical location was found to be an even more important explanatory variable. While this information suggests the importance of climate, the measured climatic variables ranked lower in the importance list. The mean temperatures of January, February, December and June (in the order of importance) are the most important thermal parameters for productivity. Regarding precipitation, the amounts in November, August, June and December are the most important determinants.

In the case of maize, on the other hand, the measured climatic variables were found to be of high importance, together with humus, pH, $CaCO_3$ and soil type, while the location was not considered to be relevant. These differences suggest the appropriateness of the crop-specific evaluation approach. There are also differences in the mean temperature and precipitation. Figure 6B shows that the precipitation (July, November, May, August and October) is a more important factor than temperature (most important in January and February) in the case of maize.

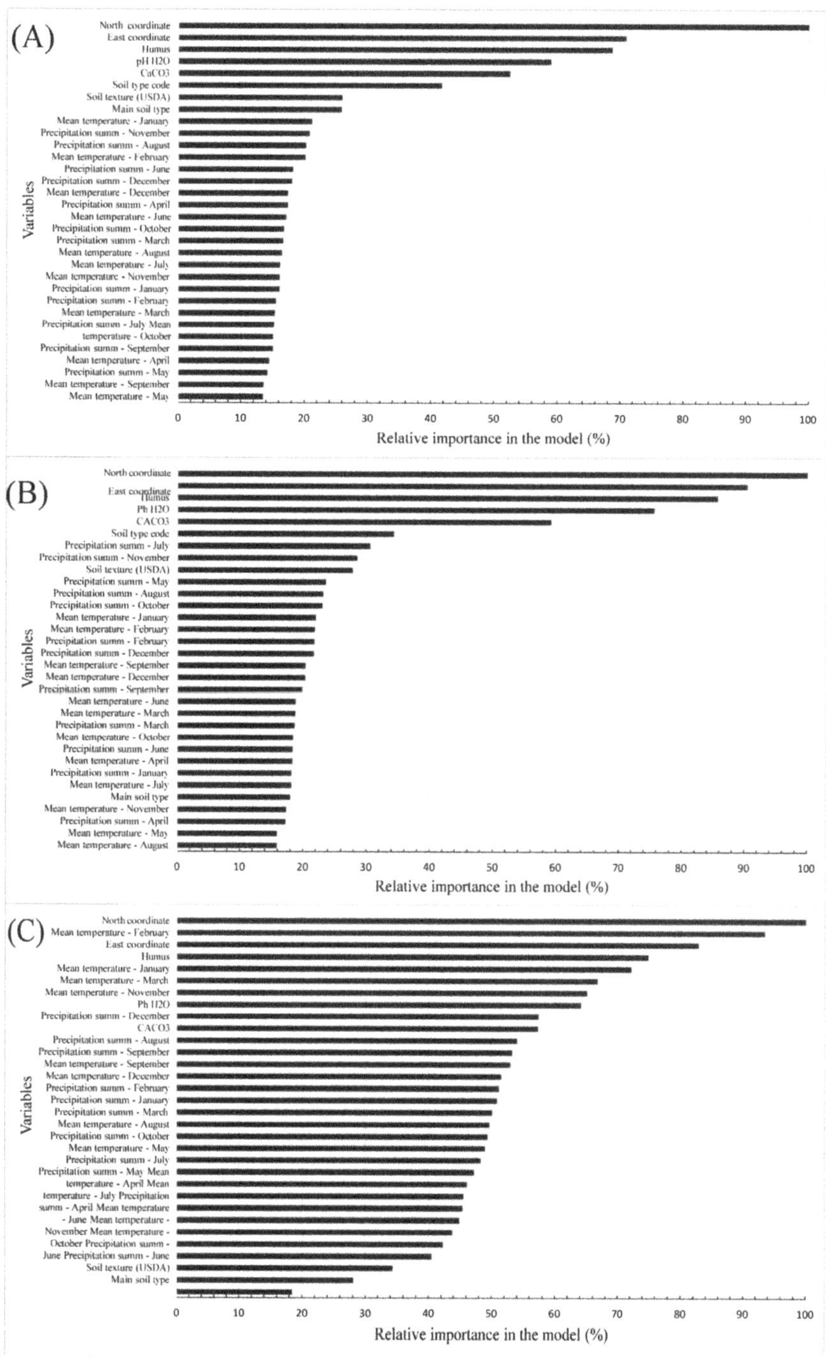

Figure 6. Overall importance of explanatory variables in predicting wheat (**A**), maize (**B**) and sunflower (**C**) biomass productivity.

Based on Figure 6C, the variables of the sunflower prediction show a completely different pattern. The most important variables are the location and mean temperature in February, followed by humus content, mean temperature in January, March and November, pH, precipitation in November and $CaCO_3$. Soil type and texture are the least important variables. Climatic variables (mainly amount of precipitation) have a more significant effect on the sunflower yield amount than the soil type and texture.

4. Discussion

Biomass productivity is a dynamic property that changes over time, partly due to changing climatic conditions (within and between years) and changing soil properties (pH, organic matter, soil nutrients content, etc.), but also due to new crop varieties and advances in crop management having an important influence. The effect of the changes in the biophysical factors may be synergistic or in the opposite direction. Nevertheless, it is possible to estimate the weight of the factors in productivity on a reasonable time scale. Twenty to thirty years seem to be an adequate time scale for estimating soil biomass productivity and for identifying the weights of different factors in it. A moving timeframe with intervals of 3 to 6 years can be proposed for the updating of the biomass productivity indices. If the system is to be used for monitoring purposes, soil biomass productivity will need to be compared on the basis of different time periods, e.g., on moving time windows or trends and supplemented by the monitoring of soil properties, subject to degradation. The moving time window for biomass productivity monitoring can be harmonized with the periodic assessments in soil monitoring, i.e., 3–6 years.

Our validation results (predicted data vs. measured data in the test set) showed that there is a significant difference in the prediction accuracy between the different crops. The sunflower model has a lower performance in calculating biomass productivity, which may be due to the fewer number of data available for model training. Furthermore, sunflowers are a cash crop grown on very diverse soils and not so much linked to bioclimatic factors and soil parameters [58,59], while the R^2 value (0.41) for the biomass productivity map can be regarded as adequate for a national estimate. The MAPE value indicates that only 18.06% of the model is inaccurate, that is, above the accuracy of results published in other studies [60,61]. The MAE indicator values indicate an average deviation of 6.78, which is not outstanding on a scale of 1 to 100. Cheng et al. (2022) [25] found similar R^2 values in the case of maize and wheat based on MODIS GPP values, with stronger correlation in the case of maize. However, other studies in the case of wheat presented lower values [24,61,62]. Although the performance of the sunflower productivity model is rather low, its inclusion in the assessment provides a more comprehensive overview of plant-specific productivities and their differences, including major factors and the varying weights of the factors in plant-specific productivities. Furthermore, the inclusion of an additional plant-specific model extends the potential of the applied method to provide a general soil productivity assessment considering multiple crops, which is often needed in land use planning.

A new soil biomass productivity map was created by applying the biomass productivity model using national soil, climate and topography geodatabase. Crop-specific productivity maps, which shall be the ultimate source of multicriteria land use planning [63], were also produced using the same technique. The spatial distribution of biomass potential is shown on the general productivity map, which can be used to plan land use and crop production. Further to that, weights of individual soil and climate parameters of crop-specific productivity indices were also derived. For our final model results (soil biomass productivity) we applied a correction that takes into account the topography. Slope angle and orientation both matter for solar radiation to be taken into account. Our solution for incorporating terrain indices from an earlier national land evaluation model [51] was tailored for Hungarian conditions (average slope under 2.3%), instead of applying more complex methods [64,65] used in other pedoclimatic conditions. Random Forest is considered to be an appropriate method to predict crop-specific biomass productivity, as proven by Jeon et al. [23] and as also highlighted in our country assessment. Results of

RF-based models can be applied to plan agricultural land use in order to increase the yield and make it sustainable, without environmental side effects. One of the most important and interesting results in our perspective is the quantification of the relative importance of explanatory variables, which best reflects the different edaphic and climatic needs of the observed crop species. For wheat, soil characteristics are the most important factors, while temperature and precipitation are less important [66]. In case of maize, soil parameters are still important but temperature and precipitation have more importance than in the case of wheat, highlighting that, even in a relatively small country like Hungary, climate tolerance of plants is a differentiating factor. This observation becomes more evident when studying sunflowers, where the importance of mean temperature and precipitation outweighs those of soil type and soil textures as earlier presented by Kern et al. (2018) in case studies from Hungary. Nevertheless climatic variables, such as precipitation in October and November and temperature in January and February are also important for winter wheat [66]. Our results also show the importance of summer rainfall totals (May, June, July) for maize, while for sunflowers the most important parameters are spring and autumn temperatures. We have to emphasize that it is often difficult to compare our results with those of other researchers, because the bioclimatic variables of the study area differ. For example, the work of Vannoppen and Gobin (2018) from northern Belgium, investigating the importance of climatic variables in winter wheat yield estimation, found similar parameters to be important, but in a different order. While in Hungary, the mean temperature in January and the amount of precipitation in November are the most important, in Belgium, winter precipitation is the most important [67]. The model fit can be further improved by adding information on management factors such as nutrient levels and fertilizer inputs [52,68].

Soil plays an important role in increasing crop production. The soil science community is trying to define the appropriate indicators. The presented analysis on the importance of variables in calculating productivity also provides a good basis for SDG indicators, as the related target of SDG is to improve land and soil quality progressively. Addressing soil health and soil quality are the main criteria for achieving sustainable agriculture. Climate change largely affects the minimum and maximum temperatures and the amount of precipitation per month [69–72]. Our results suggest that these variables are also important for winter wheat, maize and sunflowers, and that changes in these variables could change soil productivity in the future.

We established a baseline prediction model for biomass productivity applicable for Hungarian croplands using Earth observation data and yield statistics, identified the importance of soil and climatic determinants of biomass productivity, and proposed a methodology for integrated monitoring of biomass productivity.

5. Conclusions

Our present assessment shows the long-term productivity of soils in Hungary. Long-term productivity in this context means the mean productivity of the last three decades. A period of 20 to 30 years was found to be an adequate time scale for estimating the productivity of soil biomass and for identifying the weights of different factors in it, and also as prospective baseline and threshold values of soil health and soil quality indicators, which can be used in land degradation and soil improvement assessment. A new generalized biomass productivity map was created on a 100 m resolution, which can be implemented in the cadastral system and in multipurpose land use planning programs. The general map of productivity was produced from crop-specific productivity maps by applying biomass productivity models on the country-scale soil, climate and topography geodatabase. Soil properties and characteristics play the most important roles in wheat biomass productivity, while maize has a more significant relationship with precipitation. In the case of sunflowers, soil type and texture are less important factors. The spatial pattern of biomass potential is shown on the general productivity map at 100 m resolution. This map can be used to plan land use in general and agricultural production in croplands. Climate change largely affects the minimum and maximum temperatures, their variability and the amount of precipitation and its temporal distribution, which all have considerable impact on soil biomass productivity. The most important climatic variables for crops deserve particular attention in the next decade, particularly in developing adaptation strategies. We believe that our soil–climate-based land productivity models will help in developing new methods for such adaptation. However, in order to measure changes in biomass production potential, further assessment is required, including trend analysis and the analysis of the effects of changing combinations of soil properties. Nevertheless, the proposed methodology, in addition to possible applications in cadastral systems and in land use planning and agricultural development programs, is also applicable to the integrated monitoring of biomass productivity, which is in line with the goals related to the UN SDGs.

Author Contributions: Conceptualization, G.T., N.C. and B.S.; methodology, G.T., B.S., N.C. and G.S.; resources, L.P. and A.L.; writing—original draft preparation, B.S., G.T. and N.C.; writing—review and editing, N.C., B.S., T.H., A.L., J.M., L.P., G.S., K.T. and G.T.; visualization, N.C.; supervision, B.S. and G.T. All authors have read and agreed to the published version of the manuscript.

Funding: This research received no external funding.

Data Availability Statement: The data presented in this study are available on request from the corresponding author. The data are not publicly available due to private property.

Acknowledgments: We thank the anonymous referees for their valuable recommendations and suggestions.

Conflicts of Interest: The authors declare no conflict of interest.

Appendix A

Table A1. Coefficients to modify the computed productivity based on slope relief and orientation [49].

Slope (%)	South, South-West	West, South-East	East, North-West	North-East	North
1	1	1	1	1	0.98
2	1	1	1	0.98	0.96
3	1	1	0.98	0.96	0.94
4	1	0.98	0.96	0.94	0.92
5	0.98	0.96	0.94	0.92	0.9
6	0.96	0.94	0.92	0.9	0.88
7	0.94	0.92	0.9	0.88	0.86
8	0.92	0.9	0.88	0.86	0.84
9	0.9	0.88	0.86	0.84	0.82
10	0.88	0.86	0.84	0.82	0.8
11	0.86	0.84	0.82	0.8	0.78
12	0.84	0.82	0.8	0.78	0.76
13	0.82	0.8	0.78	0.76	0.74
14	0.8	0.78	0.76	0.74	0.72
15	0.78	0.76	0.74	0.72	0.7
16	0.76	0.74	0.72	0.7	0.68
17	0.74	0.72	0.7	0.68	0.66
18	0.72	0.7	0.68	0.66	0.64
19	0.7	0.68	0.66	0.64	0.62
20	0.68	0.66	0.64	0.62	0.6
21	0.66	0.64	0.62	0.6	0.58
22	0.64	0.62	0.6	0.58	0.56
23	0.62	0.6	0.58	0.56	0.54
24	0.6	0.58	0.56	0.54	0.52
25	0.58	0.56	0.54	0.52	0.5
25	0.56	0.54	0.52	0.5	0.48

References

1. United Nations' Agenda Sustainable Development Goals (SDGs). Available online: https://www.undp.org/sustainable-development-goals?utm_source=EN&utm_medium=GSR&utm_content=US_UNDP_PaidSearch_Brand_English&utm_campaign=CENTRAL&c_src=CENTRAL&c_src2=GSR&gclid=Cj0KCQjw2MWVBhCQARIsAIjbwoPU19Uvs4z3V0arAu_3QfBuDppDDaLFi5wfsb9husx4Hdj7FZ62lk (accessed on 4 December 2022).
2. Keesstra, S.D.; Bouma, J.; Wallinga, J.; Tittonell, P.; Smith, P.; Cerdà, A.; Montanarella, L.; Quinton, J.N.; Pachepsky, Y.; Van Der Putten, W.H.; et al. The significance of soils and soil science towards realization of the United Nations sustainable development goals. *SOIL* **2016**, *2*, 111–128. [CrossRef]
3. Bouma, J.; Montanarella, L.; Evanylo, G. The challenge for the soil science community to contribute to the implementation of the UN Sustainable Development Goals. *Soil Use Manag.* **2019**, *35*, 538–546. [CrossRef]
4. Sims, N.C.; Newnham, G.J.; England, J.R.; Guerscham, J.; Cox, S.J.D.; Roxburgh, S.H.; Viscara Rossel, R.A.; Fritz, S.; Wheeler, I. *Good Practice Guidance*; SDG Indicator 15.3.1, Proportion of Land That Is Degraded Over Total Land Area Version 2.0; United Nations Conventions to Combat Desertification: Bonn, Germany, 2021.
5. UNCCD SDG Indicator 15.3. Available online: https://knowledge.unccd.int/ldn/ldn-monitoring/sdg-indicator-1531 (accessed on 4 December 2022).
6. McBride, R.A.; Bober, M.L. Quantified evaluation of agricultural soil capability at the local scale: A GIS-assisted case study from Ontario, Canada. *Soil Use Manag.* **1993**, *9*, 58–65. [CrossRef]
7. Rossiter, D.G. A theoretical framework for land evaluation. *Geoderma* **1996**, *72*, 165–190. [CrossRef]
8. FAO. *A Framework for Land Evaluation*; Food and Agriculture Organization: Rome, Italy, 1976; p. 32.
9. Kumar, R.; Kalbende, A.R.; Landey, R.J. Soil Evaluation for Agricultural Land Use—II. Productivity Potential Appraisal. *J. Indian Soc. Soil Sci.* **1984**, *32*, 467–472.
10. Schroers, J.O. *Zur Entwicklung der Landnutzung auf Grenzstandorten in Abhängikeit Agrarmarktpolitischer, Agrarstrukturpolitischer und Produktions-Technologischer Rahmenbedingungen*; Justus-Liebig-Universität: Gießen, Germany, 2006.
11. Esch, E.; Mccann, K.; Kamm, C.; Arce, B.; Carroll, O.; Dolezal, A.; Mazzorato, A.; Noble, D.; Fraser, E.; Fryxell, J.; et al. Rising farm costs, marginal land cropping, and ecosystem service markets. *Preprint* **2021**. [CrossRef]
12. Esch, E.; MacDougall, A.S.; Esch, E.; MacDougall, A.S. *More at the Margin: Leveraging ECOSYSTEM Services on Marginal Lands to Improve Agricultural Sustainability and Slow Trends of Farming Costs Outpacing Yield Gains*; American Geophysical Union: Washington, DC, USA, 2018.
13. Gopalakrishnan, G.; Negri, M.C.; Snyder, S.W. A Novel Framework to Classify Marginal Land for Sustainable Biomass Feedstock Production. *J. Environ. Q.* **2011**, *40*, 1593–1600. [CrossRef]
14. Riquier, J.; Bramao, D.L.; Cornet, J.P. A new system of soil appraisal in terms of actual and potential productivity. *FAO Soil Resour. Dev. Conserv.* **1970**, *38*, 31–33.
15. Sys, C. *Land Evaluation. I-II-III*; State University of Ghent: Ghent, Belgium, 1985.
16. Van Lanen, H.A.J.; Broeke, M.J.D.H.; Bouma, J.; de Groot, W.J.M. A mixed qualitative/quantitative physical land evaluation methodology. *Geoderma* **1992**, *55*, 37–54. [CrossRef]
17. Tóth, G. Evaluation of cropland productivity in Hungary with the D-e-Meter land evaluation system. *Agrokémia Talajt* **2011**, *60*, 161–174.
18. Tóth, G.; Gardi, C.; Bódis, K.; Ivits, É.; Aksoy, E.; Jones, A.; Jeffrey, S.; Petursdottir, T.; Montanarella, L. Continental-scale assessment of provisioning soil functions in Europe. *Ecol. Process.* **2013**, *2*, 32. [CrossRef]
19. Burrough, P.A. Fuzzy mathematical methods for soil survey and land evaluation. *J. Soil Sci.* **1989**, *40*, 477–492. [CrossRef]
20. Godev, G.; Klestov, V. Statistical evaluation of soil fertility at given plant environment system. In Proceedings of the UNDP/FAO 472 Meeting of Panel of Experts on Land Productivity Evaluation, Sofia, Bulgaria, 27 September 1971.
21. Trashliev, H.; Godev, G.; Krastanov, S.; Klevstov, A.; Kabakchiev, I.; Hershkovich, E.; Dilkov, D. Assessment of ecological conditions for wheat and maize in Bulgaria by means of multivariate regression analysis. In Proceedings of the UNDP/FAO 472 Meeting of Panel of Experts on Land Productivity Evaluation, Sofia, Bulgaria, 27 September 1971.
22. Bonfante, A.; Terribile, F.; Bouma, J. Refining physical aspects of soil quality and soil health when exploring the effects of soil degradation and climate change on biomass production: An Italian case study. *SOIL* **2019**, *5*, 1–14. [CrossRef]
23. Jeong, J.H.; Resop, J.P.; Mueller, N.D.; Fleisher, D.H.; Yun, K.; Butler, E.E.; Timlin, D.J.; Shim, K.M.; Gerber, J.S.; Reddy, V.R.; et al. Random Forests for Global and Regional Crop Yield Predictions. *PLoS ONE* **2016**, *11*, e0156571. [CrossRef]
24. Roell, Y.E.; Beucher, A.; Møller, P.G.; Greve, M.B.; Greve, M.H. Comparing a Random Forest Based Prediction of Winter Wheat Yield to Historical Yield Potential. *Agronomy* **2020**, *10*, 395. [CrossRef]
25. Cheng, M.; Jiao, X.; Shi, L.; Penuelas, J.; Kumar, L.; Nie, C.; Wu, T.; Liu, K.; Wu, W.; Jin, X. High-resolution crop yield and water productivity dataset generated using random forest and remote sensing. *Sci. Data* **2022**, *9*, 641. [CrossRef] [PubMed]
26. Guo, Y.; Xia, H.; Zhao, X.; Qiao, L.; Du, Q.; Qin, Y. Early-season mapping of winter wheat and garlic in Huaihe basin using Sentinel-1/2 and Landsat-7/8 imagery. *IEEE J. Sel. Top. Appl. Earth Obs. Remote Sens.* **2022**, 1–10. [CrossRef]
27. Breiman, L. Random Forests. *Mach. Learn.* **2001**, *45*, 5–32. [CrossRef]
28. Cutler, A.; Cutler, D.R.; Stevens, J.R. Random Forests. In *Ensemble Machine Learning*; Springer: Boston, MA, USA, 2012; pp. 157–175. [CrossRef]

29. Farkas, J.Z.; Lennert, J. Modelling and predicting of the land use change in Hungary. In *Climate Change—Society—Economy: Long-Term Processes and Trends in Hungary*; Czirfusz, M., Hoyk, E., Suvák, A., Eds.; Publikon: Pécs, Hungary, 2015; pp. 193–221. ISBN 978-615-5457-62-3.
30. EEA. *Corine Land Cover (CLC) 2018*; EEA: Copenhagen, Denmark, 2018.
31. Kocsis, M.; Tóth, G.; Berényi-Üveges, J.; Makó, A. Presentation of soil data from the National Pedological and Crop Production Database (NPCPD) and investigations on spatial representativeness. *Agrokémia Talajt.* **2014**, *63*, 223–248. [CrossRef]
32. Stefanovits, P.; Michéli, E. *Talajgenetika, Talajosztályozás II*; Gödöllői Agrártudományi Egyetem: Gödöllő, Hungary, 1989.
33. FAO. World Reference Base for Soil Resources. In *International Soil Classification System for Naming Soils and Creating Legends for Soil Maps*; FAO: Rome, Italy, 2015.
34. Jin, H.; Eklundh, L. A physically based vegetation index for improved monitoring of plant phenology. *Remote Sens. Environ.* **2014**, *152*, 512–525. [CrossRef]
35. NASA Modis Dataset. Available online: https://modis.gsfc.nasa.gov/ (accessed on 4 December 2022).
36. He, M.; Kimball, J.S.; Maneta, M.P.; Maxwell, B.D.; Moreno, A.; Beguería, S.; Wu, X. Regional Crop Gross Primary Productivity and Yield Estimation Using Fused Landsat-MODIS Data. *Remote Sens.* **2018**, *10*, 372. [CrossRef]
37. Dobor, L.; Barcza, Z.; Hlásny, T.; Havasi; Horváth, F.; Ittzés, P.; Bartholy, J. Bridging the gap between climate models and impact studies: The FORESEE Database. *Geosci. Data J.* **2015**, *2*, 1–11. [CrossRef]
38. SRTM Shuttle Radar Topography Mission (SRTM) Global. Available online: https://portal.opentopography.org/datasetMetadata?otCollectionID=OT.042013.4326.1 (accessed on 4 December 2022).
39. Rabus, B.; Eineder, M.; Roth, A.; Bamler, R. The shuttle radar topography mission—A new class of digital elevation models acquired by spaceborne radar. *ISPRS J. Photogramm. Remote Sens.* **2003**, *57*, 241–262. [CrossRef]
40. Józsa, E.; Fábián, S.Á.; Kovács, M. An evaluation of EU-DEM in comparison with ASTER GDEM, SRTM and contour-based DEMs over the Eastern Mecsek Mountains. *Hungarian Geogr. Bull.* **2014**, *63*, 401–423. [CrossRef]
41. Pásztor, L.; Laborczi, A.; Bakacsi, Z.; Szabó, J.; Illés, G. Compilation of a national soil-type map for Hungary by sequential classification methods. *Geoderma* **2018**, *311*, 93–108. [CrossRef]
42. Pásztor, L.; Laborczi, A.; Takács, K.; Szatmári, G.; Dobos, E.; Illés, G.; Bakacsi, Z.; Szabó, J. Compilation of novel and renewed, goal oriented digital soil maps using geostatistical and data mining tools. *Hungarian Geogr. Bull.* **2015**, *64*, 49–64. [CrossRef]
43. Pásztor, L.; Laborczi, A.; Takács, K.; Szatmári, G.; Fodor, N.; Illés, G.; Farkas-Iványi, K.; Bakacsi, Z.; Szabó, J. Compilation of Functional Soil Maps for the Support of Spatial Planning and Land Management in Hungary. In *Soil Mapping and Process Modeling for Sustainable Land Use Management*; Pereira, P., Brevik, E.C., Munoz-Rojas, M., Miller, B.A., Eds.; Elsevier: Amsterdam, The Netherlands, 2017; pp. 293–317.
44. Pásztor, L.; Laborczi, A.; Takács, K.; Illés, G.; Szabó, J.; Szatmári, G. Progress in the elaboration of GSM conform DSM products and their functional utilization in Hungary. *Geoderma Reg.* **2020**, *21*, e00269. [CrossRef]
45. Baranyai, F.; Fekete, A.; Kovács, I. *A Magyarországi Tápanyag-Vizsgálatok Eredményei*; Mezőgazdasági Kiadó: Budapest, Hungary, 1987; pp. 8–79.
46. KSH Központi Statisztikai Hivatal (Central Statistical Office). Available online: https://www.ksh.hu (accessed on 4 December 2022).
47. Farmanov, N.; Amankulova, K.; Szatmari, J.; Sharifi, A.; Abbasi-Moghadam, D.; Mirhossein-Nejad, M.; Mucsi, L. Crop Type Classification by DESIS Hyperspectral Imagery and Machine Learning Algorithms. *IEEE J. Sel. Top. Appl. Earth Obs. Remote Sens.* **2023**, *16*, 1576–1588. [CrossRef]
48. Wright, M.N.; Ziegler, A. Ranger: A Fast Implementation of Random Forests for High Dimensional Data in C++ and R. *J. Stat. Softw.* **2017**, *77*, 1–17. [CrossRef]
49. *Mezőgazdasági és Élelmezésügyi Minisztérium (MÉM) Táblázatok a Földértékelés Végrehajtásához*; MÉM: Budapest, Hungary, 1982.
50. Fekete, Z. *Direktívák a Gyakorlati Földértékeléshez*; Mezőgazdasági Kiadó: Budapest, Hungary, 1965.
51. Fórizsné, J.; Máté, F.; Stefanovits, P. Talajbonitáció—Földértékelés. *MTA Agrártudományok Osztályának Közleményei* **1972**, *30*, 359–378.
52. Hermann, T.; Kismányoky, T.; Tóth, G. A humuszellátottság hatása a kukorica (*Zea mays* L.) termésére csernozjom és barna erdőtalajú termőhelyeken, különböző évjáratokban. *Növénytermelés* **2014**, *63*, 1–18.
53. Máté, F. Megjegyzések a talajok termékenységük szerinti osztályozásához. *Agrokémia Talajt.* **1960**, *9*, 419–426.
54. Tóth, G.; Máté, F. Jellegzetes dunántúli talajok főbb növényenkénti relatív termékenysége. *Agrokémia Talajt.* **1999**, *48*, 172.
55. Nia, V.P.; Davison, A.C. High-Dimensional Bayesian Clustering with Variable Selection: The R Package bclust. *J. Stat. Softw.* **2012**, *47*, 1–22. [CrossRef]
56. Willmott, C.; Matsuura, K. Advantages of the mean absolute error (MAE) over the root mean square error (RMSE) in assessing average model performance. *Clim. Res.* **2005**, *30*, 79–82. [CrossRef]
57. De Myttenaere, A.; Golden, B.; Grand, B.L.; Rossi, F. Mean Absolute Percentage Error for regression models. *Neurocomputing* **2016**, *192*, 38–48. [CrossRef]
58. Seddaiu, G.; Iocola, I.; Farina, R.; Orsini, R.; Iezzi, G.; Roggero, P.P. Long term effects of tillage practices and N fertilization in rainfed Mediterranean cropping systems: Durum wheat, sunflower and maize grain yield. *Eur. J. Agron.* **2016**, *77*, 166–178. [CrossRef]

59. Sadras, V.O.; Calviño, P.A. Quantification of Grain Yield Response to Soil Depth in Soybean, Maize, Sunflower, and Wheat. *Agron. J.* **2001**, *93*, 577–583. [CrossRef]
60. Feng, P.; Wang, B.; Liu, D.L.; Waters, C.; Xiao, D.; Shi, L.; Yu, Q. Dynamic wheat yield forecasts are improved by a hybrid approach using a biophysical model and machine learning technique. *Agric. For. Meteorol.* **2020**, *285–286*, 107922. [CrossRef]
61. Kheir, A.M.S.; Ammar, K.A.; Amer, A.; Ali, M.G.M.; Ding, Z.; Elnashar, A. Machine learning-based cloud computing improved wheat yield simulation in arid regions. *Comput. Electron. Agric.* **2022**, *203*, 107457. [CrossRef]
62. Reeves, M.C.; Zhao, M.; Running, S.W. Usefulness and limits of MODIS GPP for estimating wheat yield. *Int. J. Remote Sens.* **2007**, *26*, 1403–1421. [CrossRef]
63. Mishra, U.; Torn, M.S.; Fingerman, K. Miscanthus biomass productivity within US croplands and its potential impact on soil organic carbon. *GCB Bioenergy* **2013**, *5*, 391–399. [CrossRef]
64. Xie, X.; Li, A.; Tian, J.; Wu, C.; Jin, H. A fine spatial resolution estimation scheme for large-scale gross primary productivity (GPP) in mountain ecosystems by integrating an eco-hydrological model with the combination of linear and non-linear downscaling processes. *J. Hydrol.* **2023**, *616*, 128833. [CrossRef]
65. Sabetraftar, K.; Mackey, B.; Croke, B. Sensitivity of modelled gross primary productivity to topographic effects on surface radiation: A case study in the Cotter River Catchment, Australia. *Ecol. Modell.* **2011**, *222*, 795–803. [CrossRef]
66. Kern, A.; Barcza, Z.; Marjanović, H.; Árendás, T.; Fodor, N.; Bónis, P.; Bognár, P.; Lichtenberger, J. Statistical modelling of crop yield in Central Europe using climate data and remote sensing vegetation indices. *Agric. For. Meteorol.* **2018**, *260–261*, 300–320. [CrossRef]
67. Vannoppen, A.; Gobin, A. Estimating Farm Wheat Yields from NDVI and Meteorological Data. *Agron.* **2021**, *11*, 946. [CrossRef]
68. Debreczeni, B.; Németh, T.; Tóth, G. Nutrient factor of land quality. In *Land Evaluation and Land Use Information*; Gaál, Z., Máté, F., Tóth, G., Eds.; Veszprémi Egyetem: Keszthely, Hungary, 2003; pp. 39–48. ISBN 963949525.
69. Farooq, M.; Bramley, H.; Palta, J.A.; Siddique, K.H.M. Heat Stress in Wheat during Reproductive and Grain-Filling Phases. *Crit. Rev. Plant Sci.* **2011**, *30*, 491–507. [CrossRef]
70. Hatfield, J.L.; Boote, K.J.; Kimball, B.A.; Ziska, L.H.; Izaurralde, R.C.; Ort, D.; Thomson, A.M.; Wolfe, D. Climate Impacts on Agriculture: Implications for Crop Production. *Agron. J.* **2011**, *103*, 351–370. [CrossRef]
71. Lobell, D.B.; Schlenker, W.; Costa-Roberts, J. Climate trends and global crop production since 1980. *Science* **2011**, *333*, 616–620. [CrossRef]
72. Lobell, D.B.; Cahill, K.N.; Field, C.B. Historical effects of temperature and precipitation on California crop yields. *Clim. Change* **2007**, *81*, 187–203. [CrossRef]

Article

Convolutional Neural Network Maps Plant Communities in Semi-Natural Grasslands Using Multispectral Unmanned Aerial Vehicle Imagery

Maren Pöttker [1,*], Kathrin Kiehl [2], Thomas Jarmer [1] and Dieter Trautz [2]

[1] Remote Sensing Group, Institute of Computer Science, University of Osnabrück, 49074 Osnabrück, Germany
[2] Faculty of Agricultural Sciences and Landscape Architecture, Osnabrück University of Applied Sciences, 49090 Osnabrück, Germany
* Correspondence: maren.poettker@uos.de

Abstract: Semi-natural grasslands (SNGs) are an essential part of European cultural landscapes. They are an important habitat for many animal and plant species and offer a variety of ecological functions. Diverse plant communities have evolved over time depending on environmental and management factors in grasslands. These different plant communities offer multiple ecosystem services and also have an effect on the forage value of fodder for domestic livestock. However, with increasing intensification in agriculture and the loss of SNGs, the biodiversity of grasslands continues to decline. In this paper, we present a method to spatially classify plant communities in grasslands in order to identify and map plant communities and weed species that occur in a semi-natural meadow. For this, high-resolution multispectral remote sensing data were captured by an unmanned aerial vehicle (UAV) in regular intervals and classified by a convolutional neural network (CNN). As the study area, a heterogeneous semi-natural hay meadow with first- and second-growth vegetation was chosen. Botanical relevés of fixed plots were used as ground truth and independent test data. Accuracies up to 88% on these independent test data were achieved, showing the great potential of the usage of CNNs for plant community mapping in high-resolution UAV data for ecological and agricultural applications.

Keywords: convolutional neural networks (CNNs); remote sensing; unmanned aerial vehicles (UAVs); semi-natural grasslands; plant communities

Citation: Pöttker, M.; Kiehl, K.; Jarmer, T.; Trautz, D. Convolutional Neural Network Maps Plant Communities in Semi-Natural Grasslands Using Multispectral Unmanned Aerial Vehicle Imagery. *Remote Sens.* **2023**, *15*, 1945. https://doi.org/10.3390/rs15071945

Academic Editors: Kenji Omasa, Shan Lu and Jie Wang

Received: 20 February 2023
Revised: 30 March 2023
Accepted: 3 April 2023
Published: 6 April 2023

1. Introduction

In Central Europe, semi-natural grasslands (SNGs) are an essential part of ancient cultural landscapes. They have developed over centuries of anthropogenic land use by grazing and mowing [1,2]. Until the 19th century, most European SNGs were used as pastures, whereas hay meadows developed mainly over the last 100 to 150 years [1]. The highest diversity of species and plant communities in grasslands was reached in the middle of the 19th century [2]. Increasing intensification of land use, however, has led to decreasing species richness, especially since the 1950s [3,4]. Furthermore, the area used as grasslands in Germany decreased continuously from the 1970s until 2013. Since then, a reform of the common agricultural policy of the European Union (EU) regulates the transformation of grasslands into arable land [5]. Furthermore, subsidies for biodiversity-friendly use of grasslands were included as *greening* in the subsidy scheme of the EU [1]. For example, in Lower Saxony subsidies were granted for low-intensity use of high-nature-value grasslands [6]. This included a ban on mineral nitrogen fertilizers or pesticides and a prescribed earliest date for mowing.

Contrastingly, agriculturally improved grasslands are used, e.g., for dairy farming. Here, a high energy and protein concentration in the forage is required for increasing the milk production of the individual animal [7]. This is achieved by special grass cultivars and fertilizer application, which increase the number of mowings possible per year. Yield from SNG is not always processed into silage for milk production but can also be cut once or twice a year to produce hay in the traditional way, which maintains species richness [8]. If this hay is not fed to cattle or sheep but to horses, special importance must be paid to its plant species composition. Horses do not tolerate some *Lolium* or *Festuca* species due to their high fructose content [9,10]. Furthermore, these grass species may contain endophytic fungi that make them highly resistant to environmental conditions [11] and are harmful to horses but not ruminants [12]. Apart from their usage as fodder for meat, dairy, and wool production, SNGs' multiple ecosystem services include good groundwater quality and quantity, water flow regulation, carbon storage, mitigation of greenhouse gas fluxes, and erosion prevention, as well as cultural and health values. [13]. Furthermore, they are a habitat for many plant and animal species [13]. Both ecosystem services and habitat conditions of grasslands cannot be determined by mapping land use or land cover type only, because of the spatial variability in the biophysical variables [14]. Ecosystem services can vary over land use or land cover types [15], as species abundance and diversity in grassland plant communities influence their provision [16]. The composition of plant communities can change due to spatiotemporal dynamics, like water balance in the soil, light availability, or management [17].

To monitor vegetation structure and species composition, field-based methods in the form of phytosociological relevés are commonly used but are rather time-consuming [18,19]. In contrast, remote sensing is a cost-effective and non-destructive alternative, which is increasingly applied to get vegetation data of large-scale areas or areas showing spatiotemporal dynamics [20–23]. On a large scale, various remote sensing systems can be used to classify plant communities in grasslands. The authors of [24,25] used spaceborne data as a combination of multispectral and/or radar time series, whereas [20] analyzed airborne LiDAR. Over the last years, UAVs are increasingly used for ecological tasks on a smaller scale [26]. As an example, they were used in grasslands for the estimation of biodiversity [27], species and vegetation functional groups classification [23,28,29], forage quality, and biomass prediction [30,31] as well as for the detection of weed plants [32,33]. Various methodological approaches are suitable for the classification of plant communities in remote sensing data. To use the influence of phenology, some studies use multitemporal data for species and plant community classification [23,29]. The authors of [24,29] used machine learning techniques such as support vector machine and random forest for the classification of species and plant communities in grasslands. The authors of [34,35] tested the suitability of convolutional neural networks (CNNs) for their classification of plant communities in shrublands and forests. Recently, CNNs have been increasingly applied for the analysis of remote sensing data [36], but also specifically in vegetation remote sensing [22]. CNNs are particularly suitable for the detection of spatial patterns. As plant communities in grasslands are formed by plants of different heights and shapes, the spatial pattern is, in addition to spectral information, a strong feature for separation.

In our study, plant communities in a semi-natural hay meadow in northwestern Germany were classified with UAV imagery using CNNs. The aim is to use CNNs (1) to analyze the spatial distribution of the plant community composition before the first and second cut of the grassland vegetation and (2) to map the distribution of weed species with low forage value. Thereby, (3) the usage of mono- and multitemporal data for the mapping of plant communities with respect to the phenological phases is compared.

2. Material and Methods

2.1. Study Site

This study focuses on a 2.3 ha semi-natural meadow in the Osnabrück district, Lower Saxony, Germany (52.18°N, 8.10°E), as visible in Figure 1. According to the official soil survey map [37], the soil is predominantly gley, with part of the northern area being plaggic anthrosol. The climate is temperate oceanic, with an annual precipitation of 835 mm and a mean air temperature of 8.8 °C [38].

This survey covers the first growth (G_1) of plants from the beginning of May 2021 until the first mowing in the middle of June 2021, and the second growth (G_2) until the second mowing at the end of August 2021. The SNG can be assigned to the class *Molino-Arrhenatheretea* and the order *Arrhenatheretalia* [2]. Over the past 5 years, the study site was used twice a year for hay production according to the agri-environmental measure for low-intensity use of grasslands in Lower Saxony GL11 [6]. Before that, it had been used as a cattle pasture for about 30 years and a heterogeneous structure of plant communities had developed.

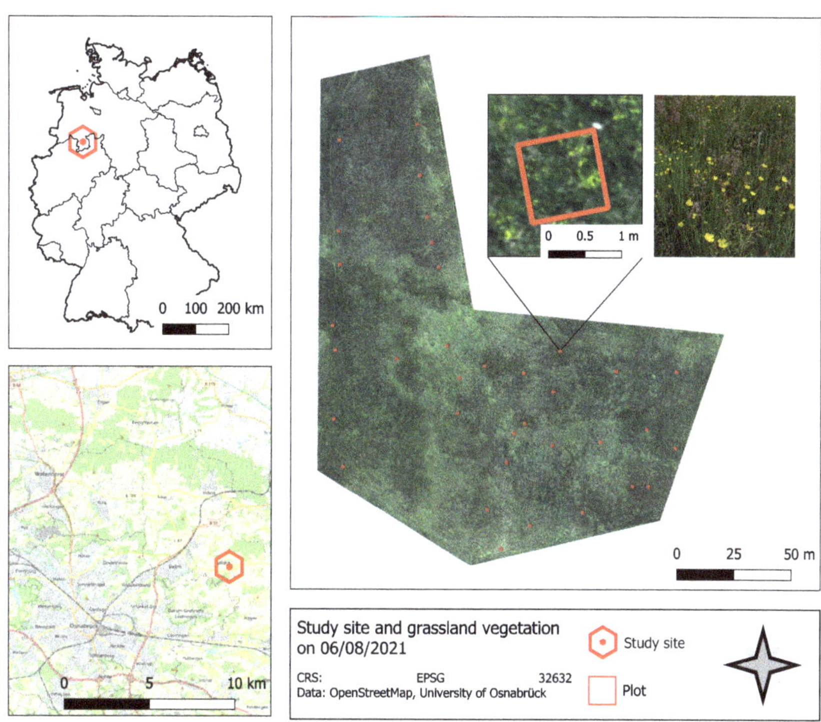

Figure 1. Location of the study site in Germany (**top left**) and the district of Osnabrück (**bottom left**). Orthomosaic and grassland vegetation of one plot of 06/08/2021 (**right**).

2.2. Data and Preprocessing

2.2.1. UAV Image Data

UAV data for this study were captured using a DJI Phantom 4 multispectral. The camera system offers five single-spectral cameras (blue (450 ± 16 nm), green (560 ± 16 nm), red (650 ± 16 nm), red edge (730 ± 16 nm), and infrared (840 ± 26 nm)) as well as an RGB camera. Each sensor has a resolution of 2.08 Megapixels and a focal length of 5.74 mm. Due to the flight altitude of 35 m a resolution of less then two centimeters was achieved. Images were taken on four dates during the first growth G_1 (Table 1, G_1T_0-G_1T_3), and four dates during the second growth G_2 (Table 1, G_2T_0-G_2T_3). Flights took place during noon (between 11 am and 3 pm) to minimize the influence of shadows. Each flight took about 30 to 35 min. The weather conditions on the observation days were inconsistent (see Table 1). Eight to ten field targets were placed before the flights and used as ground control points (GCPs). The center of each target was located using a differential GPS (bi-frequent GNSS receiver). On each observation day, around 350 images per channel were made with a front and side overlap of 70%. The images were stitched to a multispectral orthomosaic using

Agisoft Metashape software (version V1.7.2), georeferenced using the GCPs, and clipped to the extent of the study site.

Table 1. Growth, observation dates and times, number of botanically observed plots, weather conditions, and wind speed during the flights.

Growth	ID	Date	Time of Flight	No. of Plots	Weather Conditions	Wind Speed
Growth 1	G_1T_0	05/03/2021	10:58 a.m.–11:24 a.m.	0	closed cloud cover	2 m/s
Growth 1	G_1T_1	05/12/2021	2:47 p.m.–3:11 p.m.	30	closed cloud cover	5.7 m/s
Growth 1	G_1T_2	05/28/2021	1:59 p.m.–2:28 p.m.	30	sunny with a few clouds	6.4 m/s
Growth 1	G_1T_3	06/08/2021	2:06 p.m.–2:30 p.m.	29	closed cloud cover	2.9 m/s
Growth 2	G_2T_0	06/25/2021	12:45 a.m.–1:22 p.m.	0	sunny and cloudless	3.9 m/s
Growth 2	G_2T_1	07/12/2021	11:22 a.m.–11:47 a.m.	35	sunny and cloudless	1.6 m/s
Growth 2	G_2T_2	07/27/2021	11:22 a.m.–11:49 a.m.	35	first sunny, then cloudy	2.2 m/s
Growth 2	G_2T_3	08/06/2021	11:13 a.m.–11:44 a.m.	35	closed cloud cover	2.2 m/s

2.2.2. Vegetation Surveys in the Field

A total of 30 plots were stratified randomly distributed and marked during the first growth. For this, homogeneous areas were visually identified based on dominant species and 1 m × 1 m plots were placed. The four corner points of a plot were captured using a differential GPS. The area of one square meter is less than the minimum area of 10–25 m^2 recommended for botanical examinations in pastures [18], but to generate a variety of training data, a smaller plot size was chosen. At date G_1T_3, plot 26 was damaged and some of the vegetation was removed, leaving only 29 plots to be recorded (Table 1). After the first mowing, the existing plots were marked again and extended by five more plots. On six observation days, T_1–T_3 in each growth, vegetation relevés were recorded by visual cover estimation after the UAV flight using the scale of [39]. As many characteristic and indicative species were not fully grown at both G_1T_0 (early in the vegetation period) and G_2T_0 (immediately after mowing), no botanical data were recorded at these times.

2.3. *Methodology*

2.3.1. Analysis of Vegetation Data

We used the nomenclature for plant species according to [40]. Vegetation units (VUs) were formed by sorting the relevés in each growth by similar composition. The species in these VUs were sorted to form species groups. These groups show dominant species within the VUs. Four VUs were formed in the first growth, and three in the second. The plant species of a VU were listed in terms of their frequency to validate the separation into plant communities with the help of Ellenberg indicator values (EIV): soil moisture number (M, 1 = strong soil dryness, 5 = moist, 9 = wet, 12 = underwater), soil reaction number (R, 1 = extremely acidic, 5 = mildly acidic, 9 = alkaline) and nutrient number (N, 1 = least, 5 = average, 9 = excessive supply) [41]. The weighted means were calculated using the indicator values presented. The forage value, considering for example the protein and mineral content of the VUs, was determined using the values of [42].

2.3.2. Training and Test Data

The data used to train the CNN were obtained from the orthomosaics by visual interpretation and knowledge of the vegetation composition and regarding the time series. Since the plots were placed in homogeneous areas, it was assumed that the adjacent areas were dominated by the same plant community. Further training data for the VU dominated by *Rumex obtusifolius* could be obtained on the whole area, as this plant was easily identifiable. For each VU except for the one dominated by *Rumex obtusifolius*, 100 non-overlapping samples were taken in the homogeneous area around the observed plots. Only 30 samples of *Rumex obtusifolius* were taken because the plants in the study site were limited. Each training sample had an actual size of 1 × 1 m, according to the size of the plots, which corresponds to a size of 53 ± 1 × 53 ± 1 pixels. Following common standards to enhance the number of training samples [43], they were augmented as follows: Resampling to 64 × 64 pixels with nearest neighbor, rotating and flipping, and sporadic application of a median filter (kernel size 3) to add blur [44]. For use in the CNN, a random 75% (random state = 42) of the training data were used for training, the remaining 25% was used as a dependent test set for validation.

The spectral data of the observed plots were clipped and used for independent validation. Since the plot orientation does not correspond to the raster, the clipped plot samples were rotated and resampled. To avoid misclassification, a CNN with the same structure as shown in Table 2 was trained to binary classify objects that are not part of the vegetation. For this, training data were collected from fence posts, bare soil, fawns, molehills, and targets and augmented as described above.

Table 2. Architecture of the used CNN.

Layer	Parameter
Input	64 × 64 × 5
Conv2D_1	Filter: 32, Kernel: 3 × 3, Strides: 2 × 2, Activation: ReLU
BatchNormalization	-
Dropout	0.1
Conv2D_2	Filter: 128, Kernel: 3 × 3, Strides: 2 × 2, Activation: ReLU
BatchNormalization	-
Dropout	0.1
Reshape	-
FullyConnected_1	Dense: 64, Activation: ReLU
BatchNormalization	-
Dropout	0.2
FullyConnected_2	Dense: n, Activation: Softmax

2.3.3. CNN

The structure of CNNs is inspired by the biological structure of a brain. Both consist of repeating layers of simple and complex cells to solve segmentation, detection, and localization tasks [36]. The first CNNs were presented in the late 1980s, e.g., by [45] for the recognition of handwriting digits. Nowadays, they are the leading model for image classification, detection, and recognition tasks [36]. Each convolutional layer of a CNN extracts features and local conjunctions of the previous layer with weighted neurons. For this, kernels of a certain size are used to pass over the feature map or filter, and forwarded to a nonlinear activation function, e.g., rectified linear units (ReLU) [46]. There are two commonly applied techniques to simplify and aggregate the outputs of a convolutional layer. The first is to insert pooling layers. For this, features are merged (e.g., using the maximum or average value) with a pooling kernel to reduce the spacial resolution and decorrelate the features [47]. The second is the use of strides instead of pooling. Strides describe the step size of the kernel, and by increasing their size, the spatial resolution can be reduced. They are useful when input sizes are small [48] and are also utilized in more complex architectures such as ResNet to achieve higher accuracy and increase the training and classification speed [49]. Several convolutional layers in series can derive abstract features of the input. Fully connected layers of neurons and weights, as in standard neural networks, are attached to this to interpret these abstract features. For classification problems, in general a softmax function is used as the activation function in the last fully connected layer [46].

The CNN applied in this study was created with *TensorFlow's Keras* Python API (version 2.3.1). Its structure is shown in Table 2. Two convolutional layers, the first with 32 filters, the second with 128 filters, and two fully connected layers, the first of size 64, the second of size n, which is the number of output classes, were implemented. A softmax activation function in combination with a cross-entropy loss function (also known as categorical cross-entropy loss [50]) was used in this last layer to give a probability for the predicted output. The model utilizes *Adam* as an optimizer because it showed good results for CNNs [51]. Strides are applied within the convolutional layers to aggregate the features. A ReLU activation function is used for the two convolutional layers and the first dense layer. The performance of the CNN is improved via batch normalization [52]. To reduce overfitting and improve generalization, the L2 kernel regularizer and dropouts are applied as regularization methods [22,53].

2.3.4. Classification

Five different training sets were independently used to train CNNs with the structure described in Table 2: first, a binary training set for the identification of non-vegetation objects; second, a multispectral training set with the identified four vegetation units for G_1T_3; third, a multitemporal training set for G_1; fourth, a multispectral training set with the three vegetation units for G_2T_3 and last a multitemporal training set for G_2. For the monotemporal classification, both G_1T_3 and G_2T_3 were chosen, as they are closest to the harvest date in each growth and therefore most relevant for agricultural purposes. The models trained on vegetation units were used to classify the whole orthomosaic via a moving window approach to select and classify squared subimages. For both monotemporal models, each subimage was first classified with the object identification model to exclude misclassifications and then classified by the monotemporal model. The subimages of the multitemporal models were not pre-classified with the object identification model, since it was assumed that misclassifications of objects that only appear at a specific date can be avoided by the multitemporal features. The classification results of the subimages were aligned and rasterized with n channels. This workflow is depictured in Figure 2.

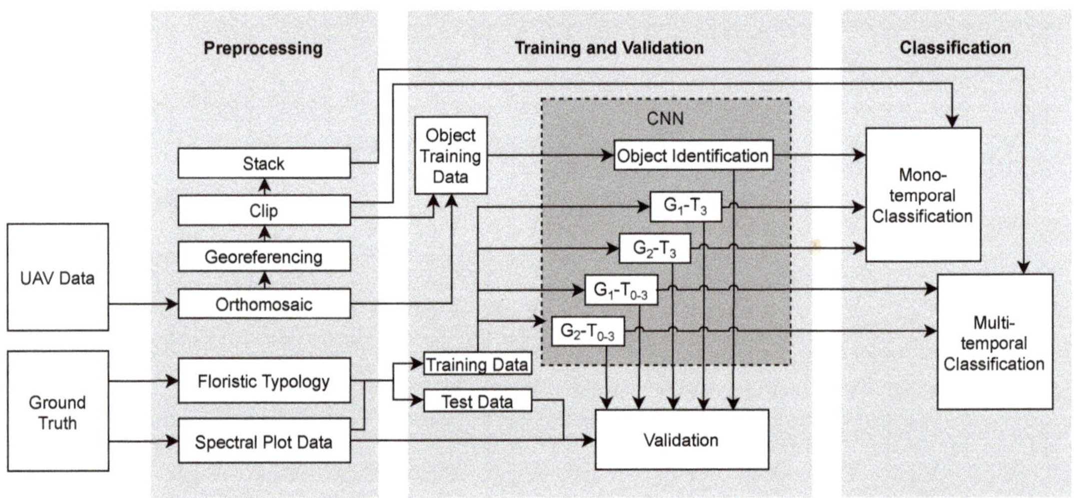

Figure 2. Schematic workflow of preprocessing, training, validation, and classification.

2.3.5. Validation Metrics

For evaluation of the classification model and the generated maps both dependent test data, which were 25% of the augmented samples set aside prior to training, and independent data, which were the resampled spectral information of the observed plots, were used. The number of true positives (t_p), true negatives (t_n), false negatives (f_n), and false positives (f_p) were calculated by using confusion matrices for each classification and for both the dependent and independent test data. The threshold for class probability was set to 50%; classification results below this threshold were listed as misclassification. The following metrics were used to estimate the performance of the models [54]:

$$Precision = \frac{t_p}{t_p + f_p} \tag{1}$$

$$Recall = \frac{t_p}{t_p + f_n} \tag{2}$$

$$Overall\ Accuracy = \frac{t_p + t_n}{t_p + t_n + f_p + f_n} \tag{3}$$

3. Results

3.1. Floristic Typology

We grouped the vegetation relevés of the first growth in three plant communities (see Appendix A) plus the VU dominated by *Rumex obtusifolius*. In both growths, VUs of a *Lolium perenne*-community and a *Alopecurus pratensis*-community could be found. In the first growth, we also identified a *Bromus hordeaceus* community. No dominant stands of this community could be found in the second growth. Common species of *Arrhenatheretalia* occur in all VUs (Appendix A, other species). Species groups highlighted in Appendix A were used to differentiate the individual VUs and to show phenological differences between the growths. Appendix B shows the VUs with their mean forage value and EIV. All values for both M, R, and N are in the moderate range (5–7).

3.2. Phenological Change in Species Spectrum

The influence of phenology is indicated by the shifting species spectrum of the species groups between the two growths and the percentage frequency of individual species (Appendix A). Although *Holcus lanatus* was found over the entire study site in the first growth, it was suppressed by other species such as *Alopecurus pratensis* or *Lolium perenne* in the second growth. During the first growth, the *Bromus hordeaceus*-community was present in some subareas, but in the second growth *Bromus hordeaceus* was only found sporadically in areas of the *Alopecurus pratensis*-community. Other grasses, such as *Phalaris arundinacea* or *Cynosurus cristatus*, were more abundant in the second growth. The flowering spectrum of the study site also changes with the seasons, following the phenological phases. In the first growth, all three plant communities showed a prominent flowering aspect with *Taraxacum officinale*, *Cerastium fontanum*, *Ranunculus repens*, and *Cardamine pratensis*. In the first growth, flowers of *Trifolium repens*, *Veronica chamaedrys*, and *Ajuga reptans* appeared in the *Lolium perenne*-community and in the *Alopecurus pratensis*-community some *Lychnis flos-cuculi*. In the second growth, the flowering aspect of the *Lolium perenne*-community was dominated by *Centaurea jacea*, *Trifolium repens* and *Crepis biennis* (species group D_3), whereas the *Alopecurus pratensis*-community showed barely any flowering plants. Not only the flowering aspect of the herbs but also the flowering of the grasses was a relevant feature differentiating the two growths. Mowing in the first growth took place during the flowering of *Holcus lanatus*, *Poa pratensis*, and *Poa trivialis*, and their flowering aspect is therefore prominent. In the second growth, barely any flowering grasses were present; flowering *Phleum pratense*, *Cynosurus cristatus* and *Agrostis capillaris* were found sporadically, but not, or only weakly, visible in the orthomosaics.

3.3. Separability of Training Data

The mean values for training set and plot samples for G_1T_3 and G_2T_3 in blue vs. green and red vs. infrared band combinations were shown in Figure 3.

The samples of the VUs formed clusters which partially overlap. In particular, the spectral samples of the *Rumex obtusifolius* plants could not be well separated. In blue vs. green band combinations, the clusters were better separated than in the red vs. infrared combination. It was noticeable that the spectral values of the *Lolium perenne*-community and the *Alopecurus pratensis*-community show higher variance and mean values at G_2T_3

than at G_1T_3. Furthermore, the samples at date G_2T_3 showed a higher reflectance in the green and infrared band than the samples at G_1T_3. This was caused by the prominent flower aspect of the grasses at G_1T_3.

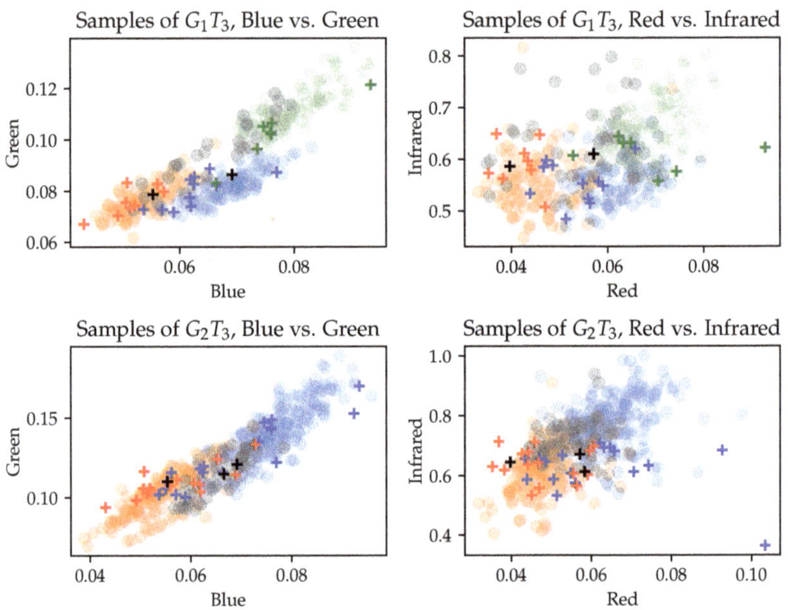

Figure 3. Scatter plots of the samples of the dependent (○) and independent (+) test data in blue vs. green and red vs. infrared. Colors are used as follows: **grey**: *Rumex obtusifolius* plants, **blue**: *Lolium perenne*-community, **red**: *Alopecurus pratensis*-community, **green**: *Bromus Hordeaceus*-community.

3.4. Classification Results

In Table 3, a summary of the validation of the monotemporal VU classification (G_1T_3 and G_2T_3), the multitemporal VU classification (G_1 and G_2) and the object identification (*OI*) can be found. All five classification models reached overall accuracies $> 91\%$ on the dependent test data. On the independent test data, the overall accuracy of the VU classifications reached from 68% to 88%. On both the dependent and independent test data, the multitemporal classification of G_1 got the lowest overall accuracy. In this, worse accuracies appeared for the classification of *Rumex obtusifolius* (precision and recall of 0%) and the *Alopecurus pratensis*-community (precision: 70%, recall: 58.33%).

The result maps of the classifications for G_1 and G_2 are shown in Figure 4. Both the monotemporal and the multitemporal classifications highlight similar spatial vegetation patterns. In both dates, the transition ranges between VUs were smaller in the multitemporal classification. In the multitemporal classification of G_1, more homogeneous areas could be found than in the monotemporal classification. In G_2, the results show strong similarities, but differ mainly at the western edge. Subsets of a *Rumex obtusifolius*-dominated area of the classification results are depicted in Figure 4. *Rumex obtusifolius* was mainly recognized in the multitemporal classification result of G_2. The areas eliminated by the object identification appear as white areas in the results of the monotemporal classifications. In G_1T_3, especially the area of open ground in the center of the subset was not classified. In G_2T_3, individual molehills were not included in the classification. In the multitemporal classification, these areas were assigned to the surrounding VUs.

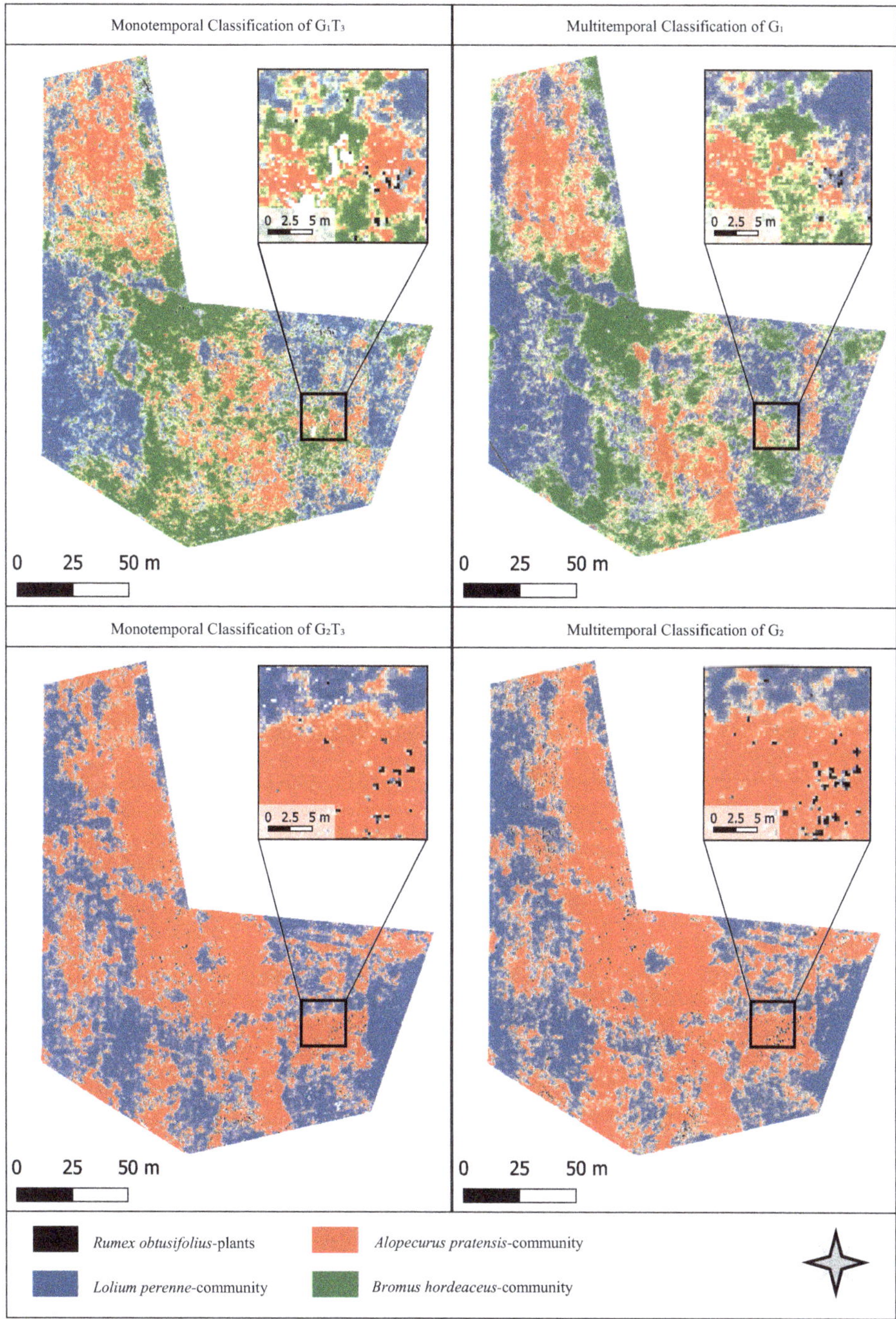

Figure 4. Subsets of the classification results of the mono- and multitemporal model and orthomosaics in RGB-color of G_1T_3 and G_2T_3.

Table 3. Precision, recall, and overall accuracy (OA) (in %) for dependent and **independent** test data of the four vegetation classifications for *Rumex obtusifolium* plants, the *Lolium perenne-*, *Alopecurus pratensis-*, and *Bromus hordeaceus-*community and overall accuracy for the object identification (OI).

	Precision in %									
	Rumex obtusifolius plants		*Lolium perenne-* community		*Alopecurus pratensis-* community		*Bromus hordeaceus-* community		**OA**	
G_1T_3	87.72	**100**	97.98	**81.81**	99.00	**80.00**	96.51	**83.33**	97.06	**82.75**
G_2T_3	96.25	**33.33**	95.51	**78.95**	97.11	**81.82**			96.01	**71.43**
G_1	83.33	**0.00**	93.94	**72.73**	95.81	**70.00**	87.34	**83.33**	91.14	**68.97**
G_2	96.62	**100**	95.12	**94.11**	97.68	**86.67**			95.72	**88.57**
	Recall in %									
G_1T_3	98.04	**71.42**	96.37	**100**	94.75	**75**	98.04	**100**	97.06	**82.75**
G_2T_3	95.06	**33.33**	97.50	**78.95**	94.39	**69.23**			96.01	**71.43**
G_1	79.71	**0.00**	96.44	**100**	83.40	**58.33**	99.07	**71.43**	91.14	**68.97**
G_2	98.85	**66.67**	97.82	**84.21**	92.32	**100**			95.72	**88.57**
OI									97.71	

4. Discussion

4.1. Usability of the Presented Methodology in an Agricultural Context

To estimate the forage value of the mown plant material, it is useful to know its species composition [42]. Since this varies spatially, a map is useful for yield estimation. However, it must be considered that the identified plant communities are not static in their composition and vary spatially and temporally [17]. The EIV of the VUs helps to understand the characteristics of an area and to identify potentially more humid, acidic, or nutrient-rich areas. Based on the EIV, few differences can be deduced, both for different observation dates and between the three communities of *Alopecurus pratensis*, *Lolium perenne*, and *Bromus hordeaceus* (see Appendix B). For assessment of forage quality, it is also helpful to estimate the forage value of a VU (see Appendix B), and spatially identify weeds [55]. The species *Bromus hordeaceus* and *Rumex obtusifolius* mentioned here as weeds are characterized by a low forage value. As can be seen in Appendix A, *Bromus hordeaceus* is represented over the entire area in G_1. *Bromus hordeaceus* is a perennial, self-seeding grass that is found primarily in patchy rich pastures [56]. If it exceeds 10% of the vegetation, it can be considered a weed [55]. The areas dominated by *Bromus hordeaceus* during G_1 were classified as *Alopecurus pratensis*-community in G_2.

Rumex obtusifolius occurs as a nitrogen and intensification indicator, as can be seen by N = 9, but due to its high content of oxalic acids and tannins, it is not fed fresh or in hay [42]. Due to its high seed potential, even a single plant should be controlled [55,57]. However, the occurrence of individual grass species that may be harmful to horses is only partially demonstrated by monitoring plant communities. The abundance of individual species within the plant community varies, possibly occurring only in sub-areas. To cover this issue, a classification of more detailed vegetation units is necessary.

4.2. Comparison of Mono- and Multitemporal Data for Plant Community Mapping

In comparison of the mono- and multitemporal VU classification, it was noticeable that larger homogeneous areas are found in both multitemporal classifications. Furthermore, class boundaries could be better delimited in the multitemporal results, and the transition areas were smaller. This could be explained by the expanded feature space of the multitemporal training data. As described in Section 3.2, both the flowering aspect and the occurrence of individual species changed with the phenological phases. It could therefore be assumed that the flowering aspect and the change in vegetation structure had a positive influence on the multitemporal classification, as they should vary the same or similar within a plant community over the vegetation period. However, the validation showed that the monotemporal model for G_1 had a higher accuracy on the independent plot data

(82.75% to 68.97%, Table 3). For G_2, the multitemporal model had a higher accuracy on the independent plot data (88.57% to 71.43%, Table 3).

The authors of [23] showed an improvement of 5–10% in the accuracy of the classification of vegetation functional groups by using multitemporal data. The influence of shadows and flowering was reduced when using data of different phenological stages. In our work, this improvement was only visible in the validation of independent plot data of G_2, but in general, the multitemporal models showed a weaker overall accuracy than the monotemporal models. It is possible that the multitemporal models could be improved with extended training data. These models have more input neurons than the monotemporal models and therefore need more data to properly learn the relevant features. The classifier of the multitemporal classification of G_1 showed problems, especially in the detection of *Rumex obtusifolius*. This plant is small and barely detectable at early observation dates of G_1 and later overgrown by tall grass, whereas it was present in G_2 from the beginning of the observation. The multitemporal classification of G_1 showed problems in the detection of the *Alopecurus pratensis*-community. At early dates, this class was dominated by *Alopecurus pratensis*, but at later dates the flowering of *Holcus lanatus* was also visible, especially in the transition areas to the other plant communities. Possibly, these plants caused a decreased accuracy in the multitemporal classification because the borders of the plant communities were less clear at G_1T_3. Some plots in the northwest of the study site lay in the transition area between the *Lolium perenne*- and the *Alopecurus pratensis*-community, which influences the separability.

Object identification showed good results in the monotemporal models (97.72% accuracy, Table 3). In the multitemporal models it was not necessary, because most objects (e.g., molehills) were not temporally stable. Areas that were not classified in the monotemporal models are replaced by the surrounding VU in the multitemporal models (see subfigures of Figure 4). So, areas removed by the object identification did not affect the applicability and interpretability of the result map.

4.3. CNNs for Plant Community Classification in Grasslands

The spectral classes of the VUs could not be separated linearly. Although there were correlations between class membership and spectral information (see Figure 3), these were not sufficient for a separation. The samples of *Rumex obtusifolius* extended across the other VUs and had no distinctive spectral signature. However, due to their size and structure in rosettes [55], they could be easily distinguished from the surrounding grasses and herbs. The detection of *Rumex obtusifolius* in grasslands with CNNs was already shown by [32]; the authors achieved an accuracy of over 91% on a monotemporal model. The accuracy of the identification of *Rumex obtusifolius* with the models presented here varies. The multitemporal model for G_1 achieved the worst accuracy with 79.71% on the test set (0% on the plot data). The best accuracy was achieved by the monotemporal model of G_1 with 98.04% on the test set (100% on the plot data). The other classes are characterized not only by different spectral values but also by a distinctive spatial structure. The *Alopecurus pratensis* community is dominated by tall grasses, which are no longer upright because of wind at later observation dates. Thus, a wavy structure becomes visible, which is less apparent in the *Lolium perenne*-community, where mainly herbs and low grasses are found (see Appendix A).

It was shown by other studies [34,35] that CNNs are suitable for the classification of different plant communities. In this work, individual plants of the species *Rumex obtusifolius* were identified in addition to the *Lolium perenne-*, *Alopecurus pratensis-*, and *Bromus hordeaceus*-community. Different requirements for classifications of VUs show the great potential of CNNs. A single network can infer and combine multiple spatial and spectral nonlinear features. In this complex problem, good accuracies in separating multiple plant communities and individual plants could be achieved. Even though only a single study site was observed in two growing periods within this study, it can be assumed that the presented methodology can be used in other grasslands with different or differently separated plant communities. For this, a database should be created from grasslands in various expressions at the same or similar phenological phases. With this database, plant communities in various grasslands could be classified with little effort and no deep ecological and botanical knowledge.

5. Conclusions

This work presents a method for the detection of plant communities in grasslands based on CNNs and UAV data. For this, UAV imagery and botanical data were collected at regular intervals in a hay meadow during two growths. Four VUs, a *Alopecurus pratensis*-community, a *Lolium perenne*-community, a *Bromus hordeaceus*-community, and *Rumex obtusifolius* plants were identified and classified with CNNs. It was investigated whether a multitemporal classification offers added value compared to a monotemporal classification. However, it was shown that not all models trained for this purpose achieved the same accuracy and the classification quality was influenced by phenology. For the preparation of phytosociological relevés, expert knowledge is essential. This complicates the generation of suitable training data for the presented models. Furthermore, only one study site with two different plant communities and two weed species was observed. To transfer the presented methodology to other grasslands to estimate the composition of the vegetation and thus the forage quality, a database of additional grassland plant communities in different variants at the same phenological phase would be necessary. The monotemporal model can give a good impression of the spatial distribution of the different plant communities from a single observation. It should further be investigated whether the accuracy of the multitemporal model can be improved with additional training data.

Author Contributions: M.P. conceptualized the study, captured and processed the botanical and UAV data, built and trained the CNN, and wrote the paper. K.K. added botanical and ecological aspects, reviewed the grouping of vegetation units, and guided the identification of plants and draft versions of the manuscript. T.J. and D.T. advised on remote sensing and agricultural issues, respectively, and commented on draft versions on the manuscript. All authors have read and agreed to the published version of the manuscript.

Funding: We acknowledge support by Deutsche Forschungsgemeinschaft (DFG) and Open Access Publishing Fund of Osnabrück University.

Data Availability Statement: The orthomosaics and botanical data generated and analyzed during the study are available from the corresponding author on reasonable request.

Acknowledgments: We would like to thank the Remote Sensing Group Osnabrück for providing a UAV and equipment for the field work. Special thanks goes to Nadine Molitor, who made her meadow available for the study and provided background information about its use.

Conflicts of Interest: The authors declare no competing interest.

Abbreviations

The following abbreviations are used in this manuscript:

SNG	Semi-Natural Grasslands
UAV	Unmanned Aerial Vehicle
CNN	Convolutional Neural Network
EIV	Ellenberg Indicator Values
VU	Vegetation Unit

Appendix A

Table A1. Frequency values (in %) of species in the plant communities of *Lolium perenne*, *Alopecurus pratensis*, and *Bromus hordeaceus*. Identified species groups indicating plant communities are marked. *Other species* include common *Arrhenatheretalia* species not differentiating between vegetation types. Note the changed order of the growths of the *Alopecurus pratensis* community for better visualization.

		***Lolium perenne*-Community**						***Alopecurus pratensis*-Community**						***Bromus hordeaceus*-Community**		
		G_1			G_2			G_2			G_1			G_1		
		T1	T2	T3	T1	T2	T3	T1	T2	T3	T1	T2	T3	T1	T2	T3
Species Group	No. of Plots		8			13			19			12			7	
D_1	*Anthoxanthum odoratum*	38	38	38												
	Ranunculus acris		38	50												
	Veronica chamedrys	13	13	13												
	Ajuga reptans		13	13												
D_2	*Lolium perenne*	100	100	100	100	100	100	25	25	25	25	58	58			
	Centaurea jacea	13	13		13	13	13									
	Galium mollugo	13	13	13	13	13	13							29		
D_3	*Crepis biennis*				31	31	25									
	Agrostis capillaris				25	25	25									
	Trifolium pratense				25	25	25									
D_4	*Cynosurus cristatus*				38	38	38	44	44	44						
D_5	*Phleum pratense*							19	19	0						
	Stellaria media							19	13							
	Rumex obtusifolius							13	13	13						
	Lamium album							6	6	13						
	Capsella bursa-pastoris							6	6	6						
D_6	*Alopecurus pratensis*	38	50	63	13	13	13	100	100	100	100	100	100	28	71	71
D_7	*Phalaris arundinaea*							38	38	38	8	8	8	14	14	14
	Cirsium arvense							13	13	13				14	14	14
D_8	*Bromus hordeaceus*	25	62	62				19	19	19	11	58	58	100	100	100
Other species	*Holcus lanatus*	100	100	100	56	56	56	81	81	81	100	100	100	100	100	100
	Poa pratensis	100	100	100	25	25	25	13	13	13	25	58	58	43	43	43
	Plantago laneolata	100	100	100	100	100	100	68	65	43	44	8	8			
	Taraxacum officinale agg.	100	100	100	87	68	44	68	62	38	67	41		29		
	Cerastium fontanum	88	100	75	43	56	31	31	31	19	67	58	33	43	71	14
	Ranunculus repens	38	50	75	68	62	56	38	31	31		16		29	29	29
	Trifolium repens	63	13	13	38	31	13	19	13	6						
	Rumex acetosa	63	13	13	43	31	25	19	19	13		8	16			
	Poa trivialis	100	100	100							25	67	67	100	100	100
	Festuca rubra agg.	67	100	100	13	13	13				42	67	67		57	57
	Molinia caerulea			13									16			14
	Cardamine pratensis	50	13								42	8				
	Lychnis flos-cuculi											16	16			

Appendix B

Table A2. EIV and forage values for *Rumex obtusifolius* plants, the *Lolium perenne*-, and the *Alopecurus pratensis*-communities in both growths and the *Bromus hordeaceus*-community in the first growth.

	***Rumex obtusifolius* Plants**	***Lolium perenne*-Community**		***Alopecurus pratensis*-Community**		***Bromus hordeaceus*-Community**
	G_1 & G_2	G_1	G_2	G_1	G_2	G_1
Ellenberg M	6	5.76	5.44	5.98	5.76	6.4
Ellenberg R	X	6.2	6.42	6.25	6.02	6.0
Ellenberg N	9	7.17	6.41	6.68	6.37	4.97
Forage Value	2	6.26	6.59	6.67	6.4	5.26

References

1. Dengler, J.; Tischew, S. Grasslands of Western and Northern Europe—Between intensification and abandonment. In *Grasslands of the World: Diversity, Management and Conservation*; Squires, V.S., Dengler, J., Hua, L., Feng, H., Eds.; CRC Press: Boca Raton, FL, USA, 2018.
2. Leuschner, C.; Ellenberg, H. *Ecology of Central European Non-Forest Vegetation: Coastal to Alpine, Natural to Man-Made Habitats: Vegetation Ecology of Central Europe*; Springer: Berlin/Heidelberg, Germany, 2018; Volume II.
3. Plantureux, S.; Peeters, A.; McCracken, D. Biodiversity in intensive grasslands: Effect of management, improvement and challenges. *Agron. Res.* **2005**, *3*, 153–164.
4. Wesche, K.; Krause, B.; Culmsee, H.; Leuschner, C. Fifty years of change in Central European grassland vegetation: Large losses in species richness and animal-pollinated plants. *Biol. Conserv.* **2012**, *150*, 76–85. [CrossRef]
5. European Parliament and the Council. *OJ L 347/608*; Regulation (EU) No 1307/2013 of the European Parliament and of the Council of 17 December 2013 Establishing Rules for Direct Payments to Farmers Under Support Schemes within the Framework of the Common Agricultural Policy and Repealing Council Regulation (EC) No 637/2008 and Council Regulation (EC) No 73/2009; European Parliament and the Council: Brussels, Belgium, 2013.
6. Niedersächsisches Ministerium für Ernährung, Landwirtschaft und Verbraucherschutz (ML). Merkblatt zu den Besonderen Förderbestimmungen GL 1—Extensive Bewirtschaftung von Dauergrünland GL 11—Grundförderung. Available online: https://www.ml.niedersachsen.de/download/85100/GL_12_-_Merkblatt_Zusatzfoerderung_nicht_vollstaendig_barrierefrei_.pdf (accessed on 7 October 2021).
7. Johansen, M.; Lund, P.; Weisbjerg, M. Feed intake and milk production in dairy cows fed different grass and legume species: A meta-analysis. *Animal* **2018**, *12*, 66–75. [CrossRef] [PubMed]
8. Sturm, P.; Zehm, A.; Baumbauch, H.; von Brackel, W.; Verbücheln, G.; Stock, M.; Zimmermann, F. *Grünlandtypen*; Quelle & Meyer Verlag: Wiebelsheim, Germany, 2018.
9. Gräßler, J.; von Borstel, U. Fructan content in pasture grasses. *Pferdeheilkunde* **2005**, *21*, 75. [CrossRef]
10. van Eps, A.; Pollitt, C. Equine laminitis induced with oligofructose. *Equine Vet. J.* **2006**, *38*, 203–208. [CrossRef] [PubMed]
11. Malinowski, D.; Belesky, D.; Lewis, G. Abiotic stresses in endophytic grasses. In *Neotyphodium in Cool-Season Grasses*; Blackwell Publishing: Hoboken, NJ, USA, 2005. [CrossRef]
12. Bourke, C.A.; Hunt, E.; Watson, R. Fescue-associated oedema of horses grazing on endophyte-inoculated tall fescue grass (Festuca arundinacea) pastures. *Aust. Vet. J.* **2009**, *87*, 492–498. [CrossRef] [PubMed]
13. Bengtsson, J.; Bullock, J.M.; Egoh, B.; Everson, C.; Everson, T.; O'Connor, T.; O'Farrell, P.; Smith, H.; Lindborg, R. Grasslands—More important for ecosystem services than you might think. *Ecosphere* **2019**, *10*, e02582. [CrossRef]
14. Le Clec'h, S.; Finger, R.; Buchmann, N.; Gosal, A.S.; Hörtnagl, L.; Huguenin-Elie, O.; Jeanneret, P.; Lüscher, A.; Schneider, M.K.; Huber, R. Assessment of spatial variability of multiple ecosystem services in grasslands of different intensities. *J. Environ. Manag.* **2019**, *251*, 109372. [CrossRef]
15. Lavorel, S.; Grigulis, K.; Lamarque, P.; Colace, M.P.; Garden, D.; Girel, J.; Pellet, G.; Douzet, R. Using plant functional traits to understand the landscape distribution of multiple ecosystem services. *J. Ecol.* **2011**, *99*, 135–147. [CrossRef]
16. Díaz, S.; Lavorel, S.; de Bello, F.; Quétier, F.; Grigulis, K.; Robson, T.M. Incorporating plant functional diversity effects in ecosystem service assessments. *Proc. Natl. Acad. Sci. USA* **2007**, *104*, 20684–20689. [CrossRef]
17. Smith, T.; Huston, M. A theory of the spatial and temporal dynamics of plant communities. In *Progress in Theoretical Vegetation Science*; Springer: Berlin/Heidelberg, Germany, 1990; pp. 49–69. [CrossRef]
18. Dierschke, H. *Pflanzensoziologie: Grundlagen und Methoden; 55 Tabellen*; Eugen Ulmer KG: Darmstadt, Germany, 1994.
19. Mueller-Dombois, D.; Ellenberg, H. *Aims and Methods of Vegetation Ecology*; The Blackburn Press: West Caldwell, NJ, USA, 2003.
20. Zlinszky, A.; Schroiff, A.; Kania, A.; Deák, B.; Mücke, W.; Vári, Á.; Székely, B.; Pfeifer, N. Categorizing grassland vegetation with full-waveform airborne laser scanning: A feasibility study for detecting Natura 2000 habitat types. *Remote Sens.* **2014**, *6*, 8056–8087. [CrossRef]
21. Reinermann, S.; Asam, S.; Kuenzer, C. Remote sensing of grassland production and management—A review. *Remote Sens.* **2020**, *12*, 1949. [CrossRef]
22. Kattenborn, T.; Leitloff, J.; Schiefer, F.; Hinz, S. Review on Convolutional Neural Networks (CNN) in vegetation remote sensing. *ISPRS J. Photogramm. Remote Sens.* **2021**, *173*, 24–49. [CrossRef]
23. Wood, D.J.; Preston, T.M.; Powell, S.; Stoy, P.C. Multiple UAV Flights across the Growing Season Can Characterize Fine Scale Phenological Heterogeneity within and among Vegetation Functional Groups. *Remote Sens.* **2022**, *14*, 1290. [CrossRef]
24. Rapinel, S.; Mony, C.; Lecoq, L.; Clement, B.; Thomas, A.; Hubert-Moy, L. Evaluation of Sentinel-2 time-series for mapping floodplain grassland plant communities. *Remote Sens. Environ.* **2019**, *223*, 115–129. [CrossRef]
25. Fauvel, M.; Lopes, M.; Dubo, T.; Rivers-Moore, J.; Frison, P.L.; Gross, N.; Ouin, A. Prediction of plant diversity in grasslands using Sentinel-1 and-2 satellite image time series. *Remote Sens. Environ.* **2020**, *237*, 111536. [CrossRef]
26. Cruzan, M.B.; Weinstein, B.G.; Grasty, M.R.; Kohrn, B.F.; Hendrickson, E.C.; Arredondo, T.M.; Thompson, P.G. Small unmanned aerial vehicles (micro-UAVs, drones) in plant ecology. *Appl. Plant Sci.* **2016**, *4*, 1600041. [CrossRef]
27. Gholizadeh, H.; Gamon, J.A.; Townsend, P.A.; Zygielbaum, A.I.; Helzer, C.J.; Hmimina, G.Y.; Yu, R.; Moore, R.M.; Schweiger, A.K.; Cavender-Bares, J. Detecting prairie biodiversity with airborne remote sensing. *Remote Sens. Environ.* **2019**, *221*, 38–49. [CrossRef]

28. Lu, B.; He, Y. Species classification using Unmanned Aerial Vehicle (UAV)-acquired high spatial resolution imagery in a heterogeneous grassland. *ISPRS J. Photogramm. Remote Sens.* **2017**, *128*, 73–85. [CrossRef]
29. Weisberg, P.J.; Dilts, T.E.; Greenberg, J.A.; Johnson, K.N.; Pai, H.; Sladek, C.; Kratt, C.; Tyler, S.W.; Ready, A. Phenology-based classification of invasive annual grasses to the species level. *Remote Sens. Environ.* **2021**, *263*, 112568. [CrossRef]
30. Wijesingha, J.; Astor, T.; Schulze-Brüninghoff, D.; Wengert, M.; Wachendorf, M. Predicting forage quality of grasslands using UAV-borne imaging spectroscopy. *Remote Sens.* **2020**, *12*, 126. [CrossRef]
31. Pecina, M.V.; Bergamo, T.F.; Ward, R.; Joyce, C.; Sepp, K. A novel UAV-based approach for biomass prediction and grassland structure assessment in coastal meadows. *Ecol. Indic.* **2021**, *122*, 107227. [CrossRef]
32. Valente, J.; Doldersum, M.; Roers, C.; Kooistra, L. Detecting Rumex Obtusifolius weed plants in grasslands from UAV RGB imagery using deep learning. *ISPRS Ann. Photogramm. Remote Sens. Spat. Inf. Sci.* **2019**, *4*, 179–185. [CrossRef]
33. Lam, O.H.Y.; Dogotari, M.; Prüm, M.; Vithlani, H.N.; Roers, C.; Melville, B.; Zimmer, F.; Becker, R. An open source workflow for weed mapping in native grassland using unmanned aerial vehicle: Using Rumex obtusifolius as a case study. *Eur. J. Remote Sens.* **2021**, *54*, 71–88. [CrossRef]
34. Kattenborn, T.; Eichel, J.; Fassnacht, F.E. Convolutional Neural Networks enable efficient, accurate and fine-grained segmentation of plant species and communities from high-resolution UAV imagery. *Sci. Rep.* **2019**, *9*, 17656. [CrossRef] [PubMed]
35. Kattenborn, T.; Eichel, J.; Wiser, S.; Burrows, L.; Fassnacht, F.E.; Schmidtlein, S. Convolutional Neural Networks accurately predict cover fractions of plant species and communities in Unmanned Aerial Vehicle imagery. *Remote Sens. Ecol. Conserv.* **2020**, *6*, 472–486. [CrossRef]
36. Zhu, X.X.; Tuia, D.; Mou, L.; Xia, G.S.; Zhang, L.; Xu, F.; Fraundorfer, F. Deep learning in remote sensing: A comprehensive review and list of resources. *IEEE Geosci. Remote Sens. Mag.* **2017**, *5*, 8–36. [CrossRef]
37. Niedersächsisches Landesamt für Bergbau, Energie und Geologie (LBEG). *Bodenübersichtskarte im Maßstab 1:50 000 (BÜK50)*; LBEG: Hannover, Germany, 1999.
38. Meteostat. Belm. Available online: https://meteostat.net/en/station/D0342 (accessed on 7 October 2021).
39. Reichelt, G.; Wilmanns, O. *Vegetationsgeographie*; Westermann: Braunschweig, Germany, 1973.
40. Jäger, E.J. *Rothmaler-Exkursionsflora von Deutschland. Gefäßpflanzen: Grundband*, 21st ed.; Springer-Verlag: Berlin/Heidelberg, Germany, 2017.
41. Ellenberg, H.; Weber, H.E.; Düll, R.; Wirth, V.; Werner, W.; Paulißen, D. Zeigerwerte von Pflanzen in Mitteleuropa, 3, durch gesehene Aufl. *Scr. Geobot.* **2001**, *18*, 1–261.
42. Dierschke, H.; Briemle, G. *Kulturgrasland*, 2nd ed.; Eugen Ulmer KG: Darmstadt, Germany, 2008.
43. Shorten, C.; Khoshgoftaar, T.M. A survey on image data augmentation for deep learning. *J. Big Data* **2019**, *6*, 1–48. [CrossRef]
44. Pawara, P.; Okafor, E.; Schomaker, L.; Wiering, M. Data augmentation for plant classification. In Proceedings of the International Conference on Advanced Concepts for Intelligent Vision Systems, Antwerp, Belgium, 18–21 September 2017; pp. 615–626. [CrossRef]
45. LeCun, Y.; Boser, B.E.; Denker, J.S.; Henderson, D.; Howard, R.E.; Hubbard, W.E.; Jackel, L.D. Handwritten digit recognition with a back-propagation network. In Proceedings of the Advances in Neural Information Processing Systems, Denver, CO, USA, 26–29 November 1990; pp. 396–404.
46. LeCun, Y.; Bengio, Y.; Hinton, G. Deep learning. *Nature* **2015**, *521*, 436–444. [CrossRef]
47. Rawat, W.; Wang, Z. Deep convolutional neural networks for image classification: A comprehensive review. *Neural Comput.* **2017**, *29*, 2352–2449. [CrossRef] [PubMed]
48. Springenberg, J.T.; Dosovitskiy, A.; Brox, T.; Riedmiller, M. Striving for simplicity: The all convolutional net. *arXiv* **2014**, arXiv:1412.6806.
49. He, K.; Zhang, X.; Ren, S.; Sun, J. Deep residual learning for image recognition. In Proceedings of the IEEE Conference on Computer Vision and Pattern Recognition, Las Vegas, NV, USA, 27–30 June 2016; pp. 770–778. [CrossRef]
50. Zhang, Z.; Sabuncu, M. Generalized cross entropy loss for training deep neural networks with noisy labels. *Adv. Neural Inf. Process. Syst.* **2018**, *31*, 8778–8788. [CrossRef]
51. Kingma, D.P.; Ba, J. Adam: A method for stochastic optimization. *arXiv* **2014**, arXiv:1412.6980. https://doi.org/10.48550/arXiv.1412.6980.
52. Thakkar, V.; Tewary, S.; Chakraborty, C. Batch Normalization in Convolutional Neural Networks—A comparative study with CIFAR-10 data. In Proceedings of the 2018 5th International Conference on Emerging Applications of Information Technology (EAIT), West Bengal, India, 12–13 January 2018; pp. 1–5. [CrossRef]
53. Phaisangittisagul, E. An analysis of the regularization between L2 and dropout in single hidden layer neural network. In Proceedings of the 2016 7th International Conference on Intelligent Systems, Modelling and Simulation (ISMS), Bangkok, Thailand, 25–27 January 2016; pp. 174–179. [CrossRef]
54. Sokolova, M.; Lapalme, G. A systematic analysis of performance measures for classification tasks. *Inf. Process. Manag.* **2009**, *45*, 427–437. [CrossRef]
55. Elsäßer, M.; Engel, S.; Roßberg, R.; Thumm, U. *Unkräuter im Grünland. Erkennen - Bewerten - Handeln*, 2nd ed.; DLG-Verlag: Frankfurt am Main, Germany, 2018.

56. Klapp, E.; Opitz von Boberfeld, W. *Taschenbuch der Gräser*, 12th ed.; Eugen Ulmer KG: Darmstadt, Germany, 2013.
57. Stilmant, D.; Bodson, B.; Vrancken, C.; Losseau, C. Impact of cutting frequency on the vigour of Rumex obtusifolius. *Grass Forage Sci.* **2010**, *65*, 147–153. [CrossRef]

Article

A Novel Vegetation Index Approach Using Sentinel-2 Data and Random Forest Algorithm for Estimating Forest Stock Volume in the Helan Mountains, Ningxia, China

Taiyong Ma [1], Yang Hu [2,3,4,5,*], Jie Wang [6], Mukete Beckline [7], Danbo Pang [2,3,4,5], Lin Chen [2,3,4,5], Xilu Ni [2,3,4,5] and Xuebin Li [2,3,4,5]

1 School of Agriculture, Ningxia University, Yinchuan 750021, China
2 Breeding Base for State Key Laboratory of Land Degradation and Ecological Restoration in Northwestern China, Yinchuan 750021, China
3 Key Laboratory of Restoration and Reconstruction of Degraded Ecosystems in Northwestern China of Ministry of Education, Yinchuan 750021, China
4 School of Ecology and Environment, Ningxia University, Yinchuan 750021, China
5 Ningxia Helan Mountain Forest Ecosystem Orientation Observation Research Station, Yinchuan 750021, China
6 College of Grassland Science and Technology, China Agricultural University, Beijing 100093, China
7 Research and Development Unit, Agrosystems Group, Tiko P.O. Box 76, Southwest Region, Cameroon
* Correspondence: huyang@nxu.edu.cn

Abstract: Forest stock volume (FSV) is a major indicator of forest ecosystem health and it also plays an important part in understanding the worldwide carbon cycle. A precise comprehension of the distribution patterns and variations of FSV is crucial in the assessment of the sequestration potential of forest carbon and optimization of the management programs of the forest carbon sink. In this study, a novel vegetation index based on Sentinel-2 data for modeling FSV with the random forest (RF) algorithm in Helan Mountains, China has been developed. Among all the other variables and with a correlation coefficient of r = 0.778, the novel vegetation index ($NDVI_{RE}$) developed based on the red-edge bands of the Sentinel-2 data was the most significant. Meanwhile, the model that combined bands and vegetation indices (bands + VIs-based model, BVBM) performed best in the training phase (R^2 = 0.93, RMSE = 10.82 m^3ha^{-1}) and testing phase (R^2 = 0.60, RMSE = 27.05 m^3ha^{-1}). Using the best training model, the FSV of the Helan Mountains was first mapped and an accuracy of 80.46% was obtained. The novel vegetation index developed based on the red-edge bands of the Sentinel-2 data and RF algorithm is thus the most effective method to assess the FSV. In addition, this method can provide a new method to estimate the FSV in other areas, especially in the management of forest carbon sequestration.

Keywords: forest stock volume; $NDVI_{RE}$; Sentinel-2; random forest; Helan mountains

Citation: Ma, T.; Hu, Y.; Wang, J.; Beckline, M.; Pang, D.; Chen, L.; Ni, X.; Li, X. A Novel Vegetation Index Approach Using Sentinel-2 Data and Random Forest Algorithm for Estimating Forest Stock Volume in the Helan Mountains, Ningxia, China. *Remote Sens.* **2023**, *15*, 1853. https://doi.org/10.3390/rs15071853

Academic Editor: Jochem Verrelst

Received: 28 February 2023
Revised: 29 March 2023
Accepted: 29 March 2023
Published: 30 March 2023

1. Introduction

Forest stock volume (FSV) refers to the total volume of tree trunks growing within a certain area of a forest, and it is thus an important indicator for measuring the total forest resources within that area [1]. It is also an important parameter to measure forest quality, forest carbon sequestration potentials, and an evaluation of the effectiveness of forest management [2]. Around the globe and ever since the Chinese government formally proposed a strategic plan for carbon peaking and carbon neutrality in 2020, global warming has drawn widespread attention [3–5]. This is because the carbon sink capacity of forests is an effective measure to mitigate global warming. Through the change in FSV [6], the dynamic change in carbon storage can be understood and the carbon sink capacity of the forest ecosystem can be obtained. Therefore, FSV studies are not only paramount in the global carbon cycle, but also practically significant in the realization of China's dual-carbon objectives.

The traditional FSV estimation method is mainly based on the manual measurement of the diameter at breast height (DBH) and tree height on the ground [7]. For fine-scale FSV estimation, it is indeed possible to obtain higher-precision estimation results [8]. However, if extended to a large-scale forest area, the small size and small number of sample plots will make it hard to obtain results close to the actual level [9]. Furthermore, forest ecosystems generally exhibit high spatial heterogeneity and inaccessibility [10,11]. Therefore, at this stage, it is not recommended to estimate FSV purely by manual field surveys. The advent of remote sensing has provided a solution to the challenge of large-scale FSV estimation [1,8,12]. By utilizing satellite images, it is now possible to obtain information about forest structures and compositions across vast areas, without the need for extensive ground measurements [13]. This technology has revolutionized the field of forest inventory, allowing for a more efficient and accurate estimation of FSV at a large-scale. Remote sensing images can be used in combination with a small number of ground samples to obtain highly accurate estimates of FSV or biomass [10]. By calibrating remote sensing data with ground-based measurements, it is possible to create statistical models that can accurately predict FSV at a much larger scale [14]. This combined approach has significant advantages over traditional manual field surveys, as it allows for a more efficient and cost-effective estimation of FSV across large areas. Furthermore, the use of remote sensing data can provide a more comprehensive understanding of forest ecosystems, allowing for more informed management decisions.

However, as more and more optical remote sensing images are applied to FSV studies, researchers have focused on the light saturation phenomenon that affects FSV estimation results [15–17]. Using the band reflectance of optical remote sensing images, all kinds of vegetation indices can be calculated. These traditional indices are usually used to estimate the corresponding FSV or biomass [18–22]. However, as the forest ages, the traditional vegetation indices will no longer respond accordingly to the decrease or increase in tree age [15,16]. This is the phenomenon of overestimation of low values and underestimation of high values that often occurs in FSV estimation studies. This is a result of the insensitivity of spectral variables to changes in FSV, especially in forest areas with high vegetation coverage. Previous studies have explored a variety of methods to decrease the influence of light saturation phenomena on remote sensing estimation. These studies include the utilization of spatial regression models and multi-source remote sensing image fusion [15,17]. Unfortunately, being an FSV study solely on a specific region, it has generalized limitations and it does not apply to other regions.

The present study proposes a novel vegetation index aimed at improving the ability to estimate FSV from remote sensing images. According to the literature, it is known that the Sentinel-2 imagery covers 13 spectral bands [23–26], from visible light to short-wave infrared, and each band has different spatial resolutions. Among all optical satellites, Sentinel-2 is the only satellite that includes three spectral bands in the red-edge range [24,26]. These bands are very effective in monitoring vegetation change information. Such as to estimate the FSV of the Helan Mountains, the vegetation reflectance of these three red-edge bands was used to calculate the novel vegetation index [27]. Similarly, by setting the step size, the optimal weighting coefficient of each red-edge band was determined. As this study was carried out the a typical semi-arid montane forest ecosystem of the Helan Mountains, this study may serve as a knowledge base for related research in similar areas across the globe.

Furthermore, the present study aims at developing a novel vegetation index based on Sentinel-2 multiple red-edge bands. It also combines the original band information and traditional vegetation indices to estimate the FSV of the Helan Mountains under the machine learning algorithm. The study will accomplish the following three goals: (1) to explore the potentials of the novel vegetation index developed based on Sentinel-2 data to estimate the FSV; (2) to compare the ability of the different variable combinations to estimate FSV and determine the best model among the three models developed in this study; (3) to map the FSV distribution of the study area by the best variable combination obtained in objective (2).

2. Materials and Methods

2.1. Study Area

This study focused on the forest resources in the Helan Mountains National Nature Reserve (38°19′–39°22′ N, 105°49′–106°41′ E) in Ningxia Province (Figure 1). The Helan Mountains belongs to the temperate arid climate zone with typical continental monsoon climate characteristics. The lack of rain and snow all year round leads to a dry climate. Although the average annual temperature is −0.7 °C, there is a wide seasonal variation in precipitation. For instance, the average precipitation from June to September, which accounts for over 62% of the annual precipitation, reaches 260.2 mm. Due to the steep mountain and complex terrain, the Helan Mountains are an important dividing line between climate and vegetation in western China. To the east is the grassland climate and grassland vegetation, and to the west is the desert climate and desert vegetation. It is located at the junction of the Qinghai-Tibetan Plateau, the Mongolian Plateau, and the North China Plain. The special geographical environment has shaped the unique biological groups of the Helan Mountains, making it the only biodiversity hotspot in northern China. Furthermore, the Helan Mountains National Nature Reserve in Ningxia Province has played a major role in studies on the virtuous cycle of vegetation development, succession, and restoration of ecosystems in semi-arid areas.

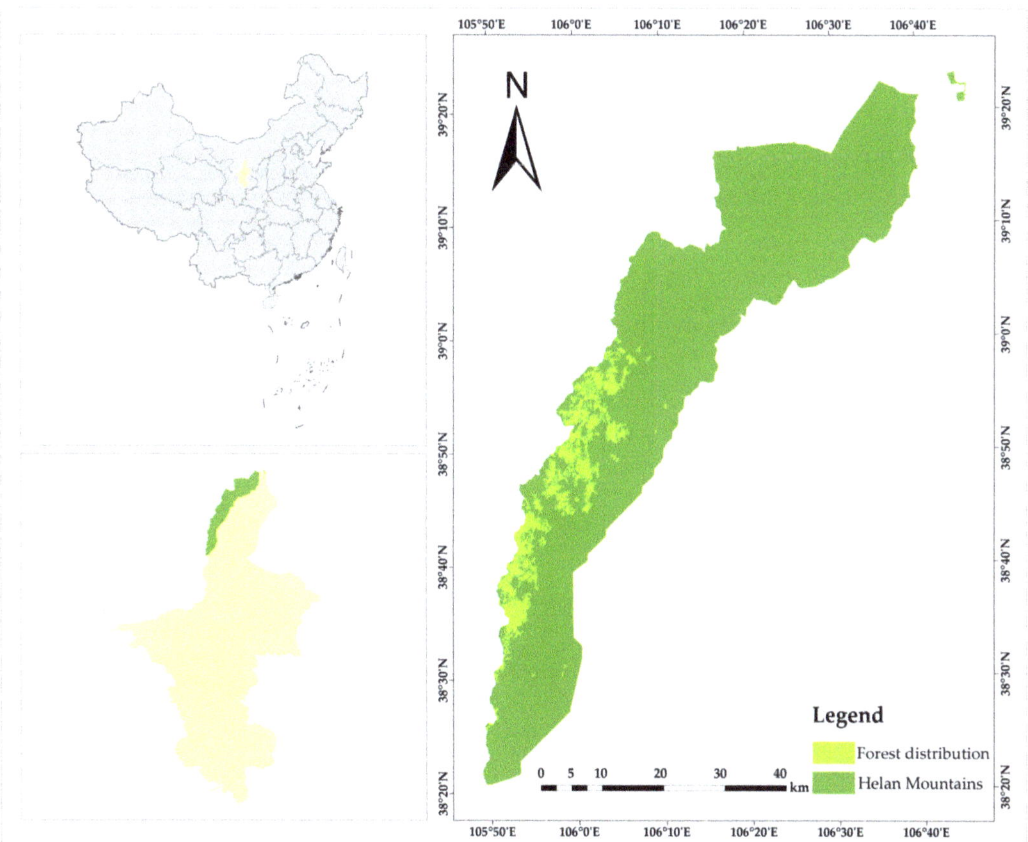

Figure 1. The geographical location of the Helan Mountains.

2.2. Field Data Collection

The field data were obtained from the 2020 forest resources management "one map" annual update data released by the Ningxia Forestry Survey and Planning Institute. Using these data reduces the workload of field surveys, and it provides access to a large amount of information on ground sample plots. Due to the wide distribution of national surface survey plots, not all sample plots can be surveyed on the spot, and there is a certain degree of uncertainty in these data. Therefore, based on previous studies, the NDVI obtained from Sentinel-2 data was used to screen plots and remove outlier data (NDVI < 0.2) [28,29]. In the end, 881 small class data were extracted for the modeling analysis, and took the hectare stock volume of living trees as the unit area FSV of each sample plot.

Random grouping was used to divide the training data and the testing data. Among the 881 sample plots, 530 (about 60%) were used as the training data and 351 (about 40%) were used as the testing data. Table 1 counts the characteristics of the field FSV training data and testing data, respectively.

Table 1. Descriptive statistics of the FSV training and testing data.

Statistical Category	Training Data (m^3ha^{-1})	Testing Data (m^3ha^{-1})
Minimum	3.30	6.40
Maximum	163.20	162.30
Median	45.15	48.80
Mean	56.66	63.84
Number of sample plots	530	351

2.3. The Acquiring and Processing of Sentinel-2 Data

Sentinel-2 covers spectral information in 13 bands, including visible light, near-infrared, red-edge, and short-wave infrared. The Sentinel-2 images are directly extracted from the processed surface reflectance product (COPERNICUS/S2_SR) through the Google Earth Engine (GEE) platform. To match the date of field data and consider the influence of cloud coverage of remote sensing images in the study area, the product date is selected from 1 July 2020 to 31 August 2020. The declouding process uses the method officially announced by the GEE to directly mask out the pixels whose pixel_QA band pixel attributes are 3 and 5. Following cloud removal, the overlaid images were medianized using the median function, followed by coordinate system matching and resampling to 30 m resolution.

2.3.1. Original Band Information

Then, the band information was extracted from the processed images using the vector file of the ground sample. Eight bands (Table 2) of the Sentinel-2 data were selected for this study [30–32], excluding bands 1, 9, 10, 11, and 12 because these bands are mainly associated with the atmosphere or water vapor.

Table 2. Selected band information of Sentinel-2.

Sentinel-2 Bands	Description	Central Wavelength (nm)	Bandwidth (nm)	Resolution (m)	Resampling Resolution (m)
B2	Blue	492.4	66	10	30
B3	Green	559.8	36	10	30
B4	Red	664.6	31	10	30
B5	Red Edge 1	704.1	15	20	30
B6	Red Edge 2	740.5	15	20	30
B7	Red Edge 3	782.8	20	20	30
B8	NIR	832.8	106	10	30
B8A	Narrow NIR	864.7	21	20	30

2.3.2. Traditional Vegetation Indices

The potential of six traditional vegetation indices for estimating FSV, calculated from the band reflectance extracted from the Sentinel-2 data (Table 3) were initially tested. Normalized Difference Vegetation Index (NDVI) reflects the background influence of plant canopy and is concerned with vegetation coverage. It is a vegetation index frequently utilized for detecting the growth status of plants. The difference vegetation index (DVI) can also reflect changes in vegetation coverage very well, and within a certain range of vegetation coverage, the DVI rises with the growth of biomass. The ratio vegetation Index (RVI) is a highly sensitive indicator parameter for monitoring green plants, which can be used to detect vegetation status and estimate the FSV. This index is the ratio of light scattered in the near-infrared to light absorbed in the red band, which lessens the effect of the atmosphere and terrain. The perpendicular vegetation index (PVI) represents the vertical distance from the vegetation pixel to the soil brightness line in the two-dimensional coordinate system of R—NIR and is less sensitive to the atmosphere than other vegetation indices. The transformed vegetation index (TVI) is based on the NDVI and introduces a constant of 0.5 to convert the negative value that the NDVI may take into a positive value. The EVI not only inherits the advantages of the NDVI, but also improves the saturation of high vegetation areas, incomplete correction of atmospheric effects, and soil background. The enhanced vegetation index (EVI) can improve the sensitivity of vegetation in high biomass areas and reduce the influence of soil background and atmosphere.

Table 3. Several traditional vegetation indices calculated based on Sentinel-2 data.

Original Vegetation Indices	Formulas	References
NDVI	$(\rho_{NIR} - \rho_{Red})/(\rho_{NIR} + \rho_{Red})$	[20]
DVI	$\rho_{NIR} - \rho_{Red}$	[33]
RVI	ρ_{NIR}/ρ_{Red}	[34]
PVI	$0.939\rho_{NIR} - 0.344\rho_{Red} + 0.9$	[34]
TVI	$\sqrt{(\rho_{NIR} - \rho_{Red})/(\rho_{NIR} + \rho_{Red}) + 0.5}$	[34]
EVI	$2.5(\rho_{NIR} - \rho_{Red})/(\rho_{NIR} + 6\rho_{Red} - 7.5\rho_{Blue} + 1)$	[20]

2.3.3. Novel Vegetation Index Based on Red-Edge Bands

The accuracy of traditional vegetation indices to estimate FSV is severely affected by the light saturation phenomenon. While the three red-edge bands in the Sentinel-2 data have been proven to be an effective way to improve the estimation of the forest parameters, unfortunately only one or two of the red-edge bands were used in existing indices. Therefore, to maximize the ability to estimate FSV using the three red-edge bands in the Sentinel-2 data, a novel vegetation index based on existing NDVI construction principles, the 4-band red-edge NDVI ($NDVI_{RE}$), such as Formula (1) was developed. According to the novel index construction rules, as elaborated in previous studies, in the $NDVI_{RE}$ formula, instead of using the NIR band, the reflectance values of *RE3* and *RE2* are averaged using weights and are substituted. Similarly, the Red band is replaced with a weighted average of the reflectance values of *RE1* and *RE2* [27]. The weighting coefficients "α" and "β" are designed to define the optimal proportion of each band in the construction of the novel index.

$$NDVI_{RE} = \frac{(\alpha \cdot R_{RE3} + (1-\alpha) \cdot R_{RE2}) - (\beta \cdot R_{Red} + (1-\beta) \cdot R_{RE1})}{(\alpha \cdot R_{RE3} + (1-\alpha) \cdot R_{RE2}) + (\beta \cdot R_{Red} + (1-\beta) \cdot R_{RE1})} \quad (1)$$

where R_{RE1}, R_{RE2}, R_{RE3}, and R_{Red} are the reflectance of B5, B6, B7, and B4, respectively. "α" and "β" represent weighting coefficients. The value range of "α" and "β" is (0,1), and the step size is 0.1.

2.4. Acquisition of the Forest Distribution Pattern in the Helan Mountains

The Global PALSAR-2/PALSAR Forest/Non-Forest Map product utilizes synthetic aperture radar (SAR) images obtained from the phased array type L-band synthetic aperture radar (PALSAR) on the ALOS-2 satellite to generate a global map of forest and non-forest areas. The classification accuracy of this map, in terms of forest and non-forest information, can reach 90%. This product is widely used for monitoring forest changes, assessing forest carbon storage, and providing information for forest management decisions. We downloaded this product using the GEE and extracted the pixels defined as forest areas within the Helan Mountains region, ultimately obtaining the forest distribution pattern of the Helan Mountains.

2.5. Machine Learning Algorithm of Modeling FSV

The random forest (RF) is a machine learning algorithm that uses multiple decision tree classifiers for classification and prediction. In recent years, studies on RF algorithms have rapidly developed accompanied by large numbers of applied research carried out in many fields. The RF algorithm is an efficient bagging-based integrated learning algorithm, and numerous prior studies have shown that the RF algorithm performs well in regression prediction [35–38]. Therefore, this study chooses the RF algorithm for modeling and analysis. The RF algorithm operates by utilizing the bootstrap method, that involves randomly sampling from the original population to create multiple samples. These samples are then used to generate a set of decision trees (ntree). The RF algorithm achieves higher accuracy and robustness by increasing the number of decision trees. At each splitting node, the RF algorithm randomly selects a subset of predictors (mtry) to build each tree. Additionally, there is no need to prune each tree. The RF algorithm employs the "out-of-bag" (OOB) error procedure to independently build each tree based on the training data. This procedure allows for the calculation of variable importance (VI) and OOB error for each tree grown by the RF algorithm. An estimation of the OOB error can be obtained using the following formula:

$$\mathrm{OOB}_{\mathrm{error}} = \frac{1}{n}\sum\nolimits_{i=1}^{n}(y_i - \hat{y}_i)^2 \quad (2)$$

where y_i is the measured FSV, $\hat{y}_i$ is the predicted FSV, and n is the total number of OOB samples.

In this study, three RF-based models composed of bands and vegetation indices (VIs) to estimate FSV, namely the bands-based model (BBM), VIs-based model (VBM), and bands + VIs-based model (BVBM) have been used.

2.6. Selecting Variables Using the VSURF Package

The VSURF package is a powerful tool for variable selection in regression problems using the RF algorithm. It is a three-step process that involves eliminating irrelevant variables, selecting relevant variables for interpretation, and improving prediction accuracy by removing redundant variables. To begin, the first step of the process involves identifying and eliminating irrelevant variables from the dataset. In the second step, all variables that are associated with the response variable are selected for interpretation. Finally, in the third step, redundant variables are removed to enhance the model's prediction performance. Once the relevant variables have been selected, the minimum mean square error (MSE) is used to determine the optimal number of decision trees (ntree) and the number of variables (mtry) to be used in the RF model. Initially, the ntree parameter is set to 500 and mtry parameter is set to the total number of variables. Once the optimal parameters are calculated, the RF regression model is established and tested.

2.7. Assessment of the Modeling Performance

This study utilized two metrics to assess the effectiveness of the RF model. The first metric was the coefficient of determination (R^2, Formula (3)), that indicates the extent to which the independent variable can account for the variability in the dependent variable. The second metric was the root mean square error (RMSE, Formula (4)), that represents the standard deviation of the difference between the observed data and the fitted model. A higher R^2 and a lower RMSE are indicative of a well-fitting model. The model is trained on 60% of the total samples, and the remaining 40% are used for testing. This approach allows for accurate predictions while reducing the risk of over-fitting.

$$R^2 = 1 - \frac{\sum_{i=1}^{n}(y_i - \hat{y}_i)^2}{\sum_{i=1}^{n}(y_i - \overline{y})^2} \tag{3}$$

$$\text{RMSE} = \sqrt{\frac{\sum_{i=1}^{n}(y_i - \hat{y}_i)^2}{n}} \tag{4}$$

where y_i is the measured FSV, $\hat{y}_i$ is the predicted FSV, $\overline{y}$ is the mean measured FSV, i is the same index, and n is the number of sample plots.

3. Results

3.1. Determination of the Optimal Novel Vegetation Index

According to the calculation formula of the novel vegetation index ($NDVI_{RE}$), the value range of the weighting coefficients "α" and "β" is (0,1), and the step size is 0.1, so 121 $NDVI_{RE}$ can be obtained. Python 3.10 software was used to calculate each $NDVI_{RE}$ value of all small class data, and the Pearson correlation coefficient of each $NDVI_{RE}$ with the FSV per unit area was also calculated. Results of the analysis are shown in Figure 2 (correlation is significant at the 0.01 level (two-tailed). In addition, the Pearson correlation coefficient was also put between the traditional NDVI and unit area FSV in the figure for comparison. Results showed the 47th $NDVI_{RE}$ to have the highest correlation coefficient (r = 0.778), which is better than the traditional NDVI (r = 0.767), and its corresponding values of "α" and "β" were 0.4 and 0.2, respectively. Therefore, the optimal $NDVI_{RE}$ was determined and used for the subsequent modeling analysis.

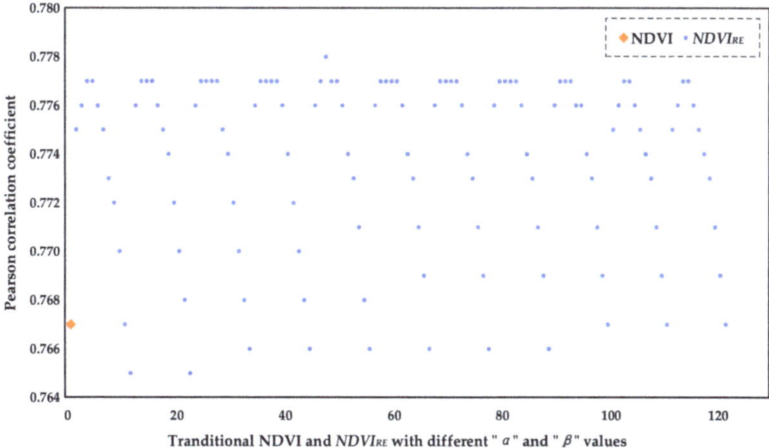

Figure 2. Pearson correlation coefficients of the NDVI and $NDVI_{RE}$ with FSV per unit area.

3.2. Major Variables Selection and the Importance Related to the FSV Data

Two types of variables, the band (B2, B3, B4, B5, B6, B7, B8, and B8A) and vegetation index (NDVI, DVI, RVI, PVI, TVI, EVI, and $NDVI_{RE}$) were selected to participate in the modeling. Figures 3–5, represent the variable selection process of the three models (BBM, VBM, and BVBM). Meanwhile, Table 4 shows the final variable selection results of each model.

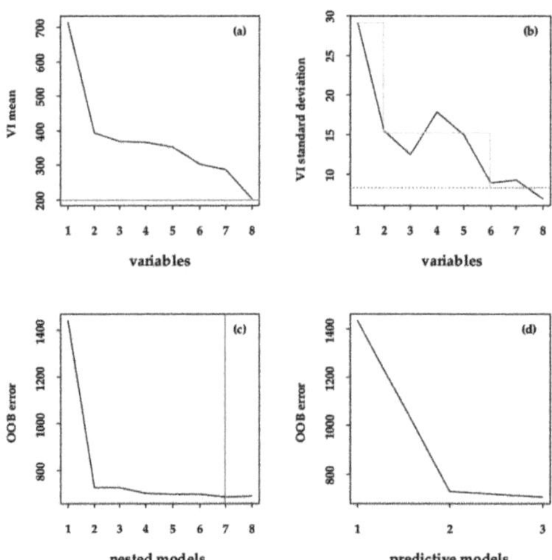

Figure 3. The variables selection of BBM. (**a**,**b**) Removes the negatively important variables based on the variable importance (VI) mean and standard deviation, respectively ((**a**), the threshold position is represented by a solid red line that runs horizontally, and (**b**), the green segmented line represents the predicted value given by the CART model, and the red line with dashes running horizontally represents the minimum predicted value). (**c**) Gradually builds a random forest from only the most important variables to all variables selected in the first step, and selects the corresponding variables according to the average OOB error (the vertical solid red line indicates the minimum error position). (**d**) Gives the number of variables meeting the requirements.

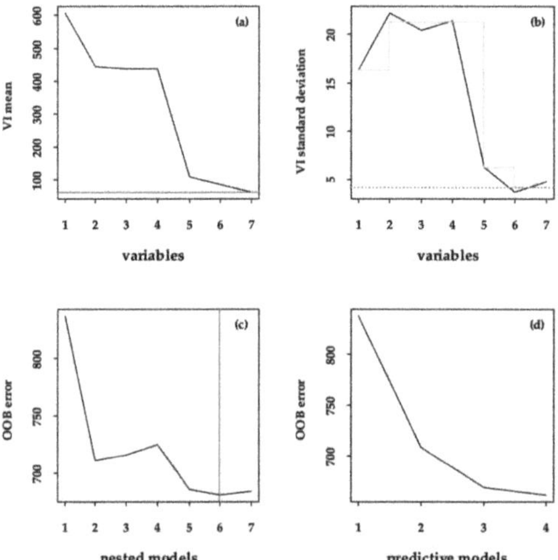

Figure 4. The variables selection of VBM. (**a**,**b**) Removes the negatively important variables based on the VI mean and standard deviation, respectively ((**a**), the threshold position is represented by a solid red line that runs horizontally, and (**b**), the green segmented line represents the predicted value given by the CART model, and the red line with dashes running horizontally represents the minimum predicted value). (**c**) Gradually builds a random forest from only the most important variables to all variables selected in the first step, and selects the corresponding variables according to the average OOB error (the vertical solid red line indicates the minimum error position). (**d**) Gives the number of variables meeting the requirements.

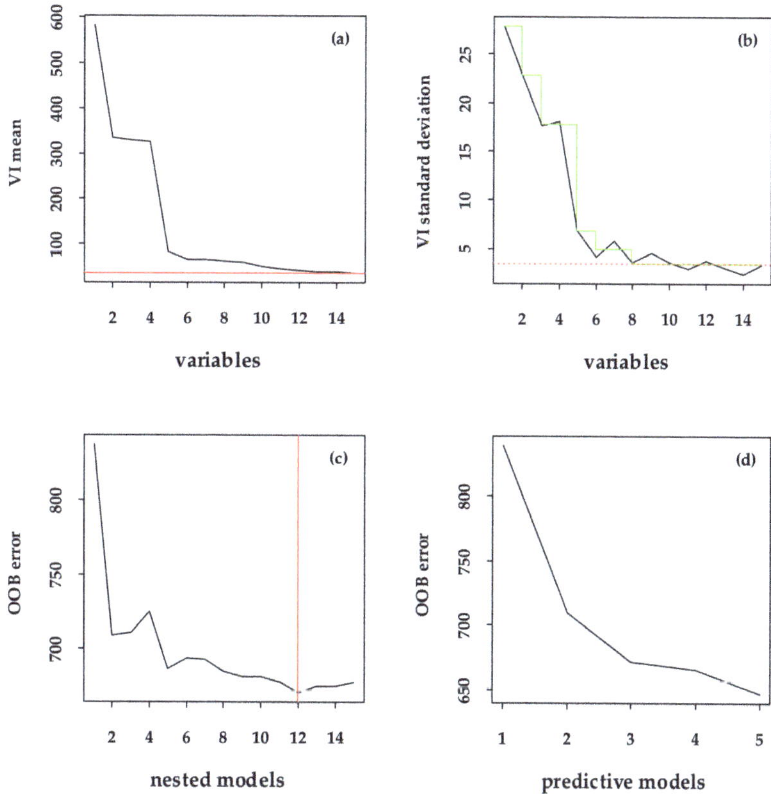

Figure 5. The variables selection of BVBM. (**a**,**b**) Removes the negatively important variables based on the VI mean and standard deviation, respectively ((**a**), the threshold position is represented by a solid red line that runs horizontally, and (**b**), the green segmented line represents the predicted value given by the CART model, and the red line with dashes running horizontally represents the minimum predicted value). (**c**) Gradually builds a random forest from only the most important variables to all variables selected in the first step, and selects the corresponding variables according to the average OOB error (the vertical solid red line indicates the minimum error position). (**d**) Gives the number of variables meeting the requirements.

Table 4. The variables selection results using the VSURF package.

RF Models	Variables Selected
BBM	B4, B8, B2
VBM	$NDVI_{RE}$, TVI, EVI, DVI
BVBM	$NDVI_{RE}$, NDVI, EVI, DVI, B2

Furthermore, all predictor variables were ranked based on their ability to estimate FSV using PercentIncMSE and IncNodePurity estimated from the OOB data. The greater the value, the greater the significance of the variable (Figure 6). It is worth noting that the novel vegetation index $NDVI_{RE}$ ranks first in importance under the two evaluation criteria.

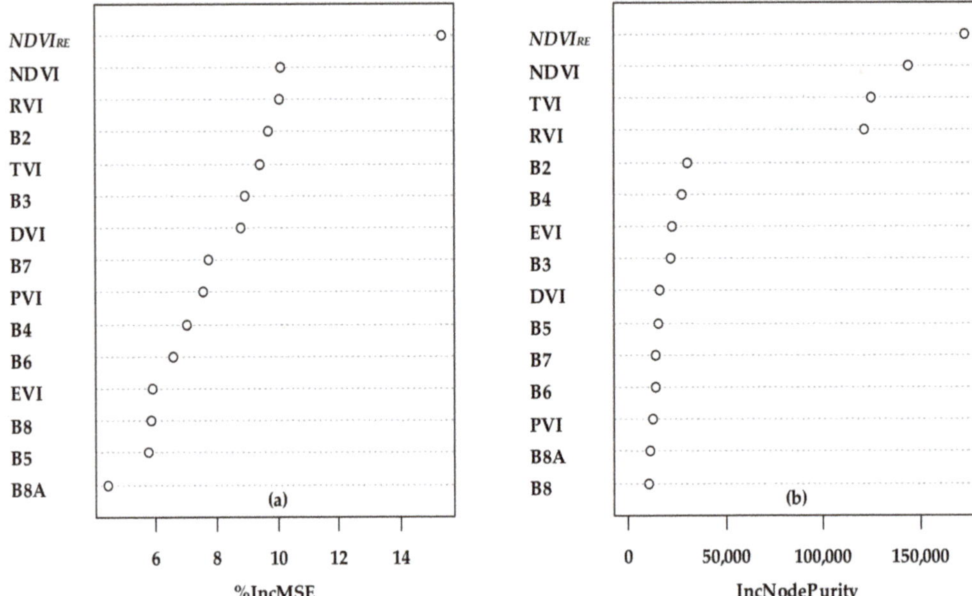

Figure 6. Importance ranking plot of all variables. Left, %IncMSE (percentage increase in the mean square error, (**a**)), and right, IncNodePurity (increase in NodePurity, (**b**)).

3.3. Optimal Regression Model for the Three Models

To optimize the RF regression model, we need to find the optimal values for two key parameters: "mtry", which determines the number of variables randomly selected as candidates for each split in the decision tree, and "ntree", which determines the total number of trees in the forest that have grown. To calculate the minimum error rate, an iterative algorithm was used, known as an "error rate loop", according to the number of variables participating in the modeling in the three models. Figure 7 shows the determination process of the optimal mtry and ntree of the three models. The values of mtry, ntree, and other performances of each model are summarized in Table 5.

3.4. Comparison of the Three Models Predicting FSV

In the training phase, BBM (Figure 8a) with an R^2 = 0.92 is slightly better than VBM (Figure 8c) with an R^2 = 0.91. However, the RMSE = 11.90 m^3ha^{-1} of the VBM is lower than the RMSE = 12.23 m^3ha^{-1} of the BBM. The BVBM (Figure 8e) has the highest R^2 = 0.93 and the smallest RMSE = 10.82 m^3ha^{-1}. In the testing phase, the BBM (Figure 8b) with an R^2 = 0.59 and RMSE = 27.72 m^3ha^{-1} performed almost the same as VBM (Figure 8d) with an R^2 = 0.59 and RMSE = 27.32 m^3ha^{-1}. Similarly, the BVBM (Figure 8f) had the best performance with an R^2 = 0.60 and RMSE = 27.05 m^3ha^{-1}. Obviously, the BVBM is the optimal model in this study, and its predicted FSV is used as the final estimation result to map the FSV. A summary of the data characteristics of FSV as predicted by the three models is presented in Table 6.

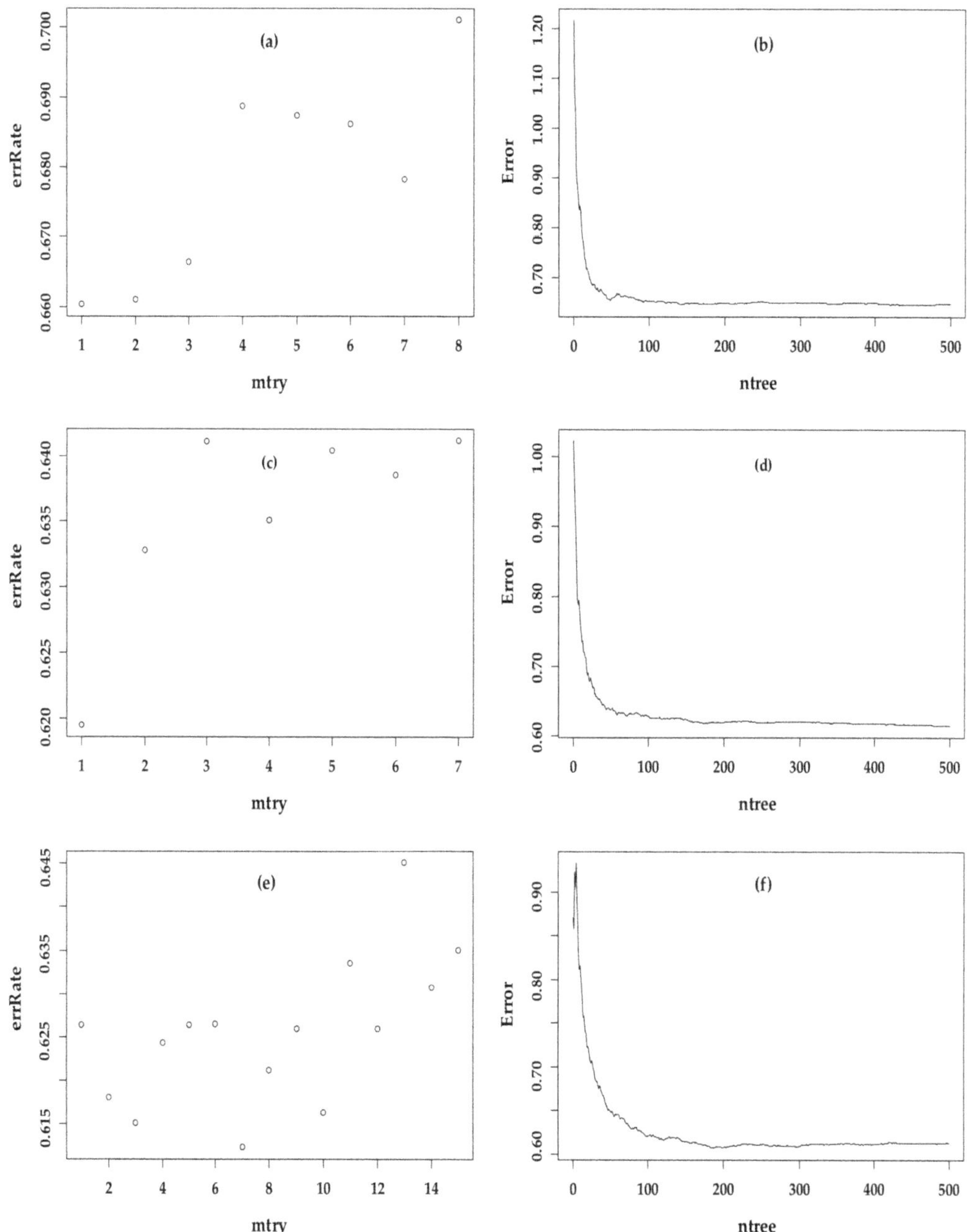

Figure 7. (**a**,**c**,**e**) are the distribution of error rate versus mtry; (**b**,**d**,**f**) are the distribution of the error versus ntree; (**a**,**b**) are related to the BBM; (**c**,**d**) are related to the VBM; and (**e**,**f**) are related to the BVBM.

Table 5. The best mtry, ntree, and performance of the three models.

RF Models	mtry	ntree	Mean of Squared Residuals	% Var Explained
BBM	1	468	636.68	56.77
VBM	1	494	612.33	58.42
BBVM	7	188	609.55	58.61

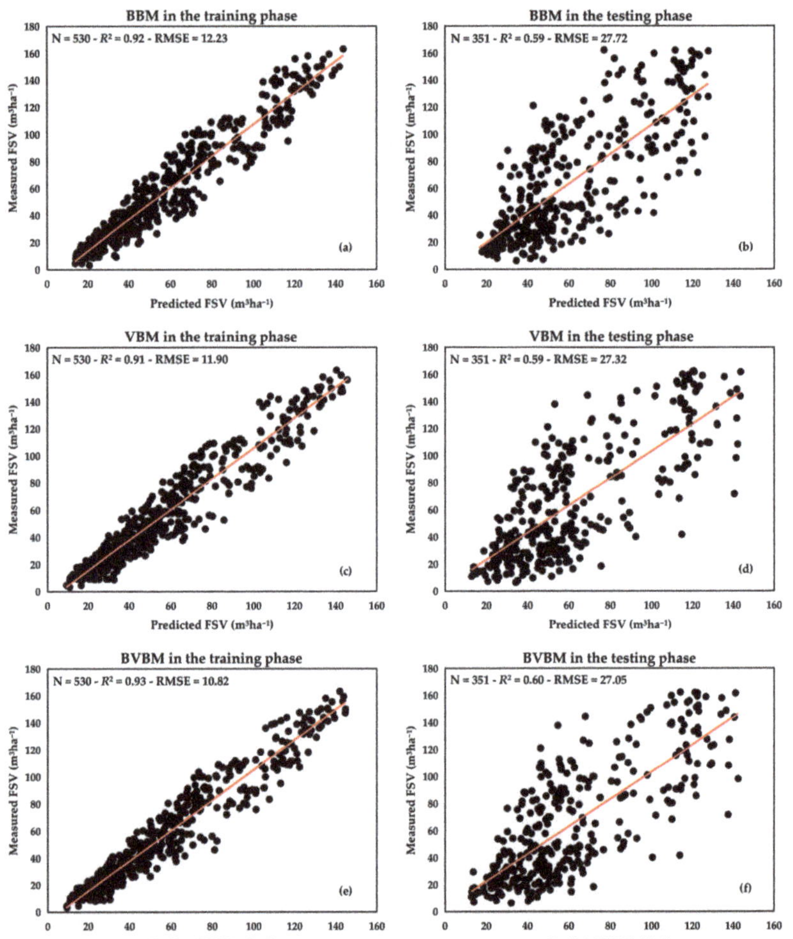

Figure 8. Comparison of the measured FSV and predicted FSV by the three models. (**a**), BBM in the training phase. (**b**), BBM in the testing phase. (**c**), VBM in the training phase. (**d**), VBM in the testing phase. (**e**), BVBM in the training phase. (**f**), BVBM in the testing phase.

Table 6. Characterization of FSV predicted by the three models.

Statistical Category	Training Phase (m^3ha^{-1})			Testing Phase (m^3ha^{-1})		
	BBM	VBM	BVBM	BBM	VBM	BVBM
Minimum	13.69	9.27	9.21	16.88	12.75	12.66
Maximum	143.83	145.66	144.76	127.55	143.64	142.48
Median	48.17	47.75	47.81	50.52	52.38	50.51
Mean	56.71	56.62	56.88	60.50	60.68	60.79

3.5. Mapping FSV Distribution of Helan Mountains

Based on the results shown in Figure 8, we have concluded that the BVBM is the best-performing model in this study, and we calculated the FSV of the Helan Mountains by the BVBM combined with the forest distribution pattern. Figure 9 is the final FSV map, the minimum value of the unit area FSV of the Helan Mountains is 9.63 m^3ha^{-1} and the maximum value is 143.96 m^3ha^{-1}. The total amount of FSV in the Helan Mountains was estimated to be 1,062,727.25 m^3. According to the FSV data released by the Helan Mountains National Nature Reserve in Ningxia Province (http://www.hlsbhq.com/, accessed on 22 January 2023), the total FSV of the Helan Mountains is 1,320,721.7 m^3. Therefore, the accuracy of the BVBM to predict the FSV in the Helan Mountains reached 80.46%.

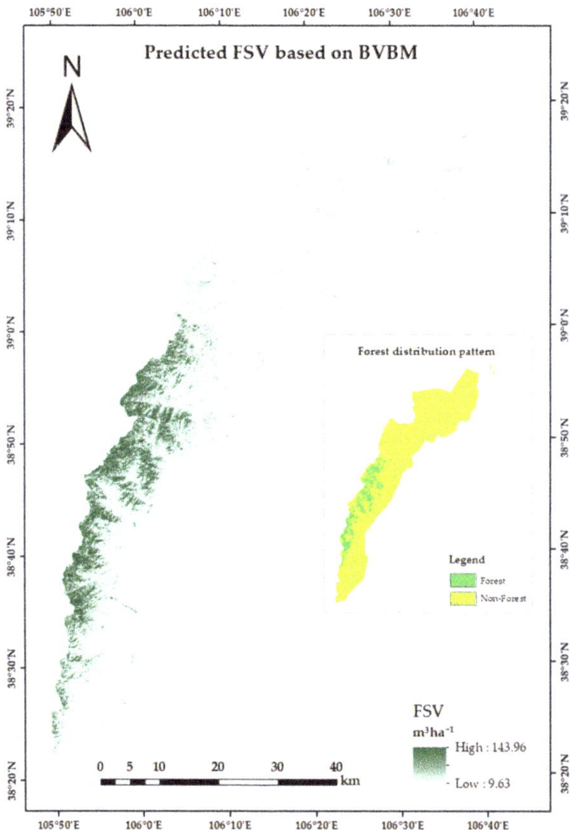

Figure 9. Spatial distribution of the predicted FSV, and forest distribution of the Helan Mountains.

4. Discussion

The carbon sequestration capacity of montane forest ecosystems is very significant and of prime importance in the global carbon cycle. Due to their geographical location and climatic characteristics, montane forests are an integral part of the entire terrestrial forest ecosystem [35,39]. The Helan Mountains are highly representative of montane forest ecosystems, their FSV estimation has a very high reference value for studies across similar landscapes. However, as a result of the inaccessibility and complex spatial heterogeneity of montane forest ecosystems, it is often a daunting task to obtain a sufficient number and sufficiently representative ground samples to estimate FSV in large-scale areas. Although remote sensing images have made it easier, issues related to low-value overestimation and high-value underestimation still occur [15,17]. However, as more and more red-edge bands in Sentinel-2 data are applied, the accurate estimation of vegetation parameters has been greatly improved [1,2]. For example, based on the red-edge band of Sentinel-2, Liu et al. [27] developed several new vegetation indices to estimate the photosynthetic and non-photosynthetic fractional vegetation cover of alpine grasslands on the Qinghai-Tibetan Plateau. Despite exhibiting a more sensitive response at low vegetation coverage, their study found that compared with traditional vegetation indices, the novel vegetation indices can effectively alleviate the high vegetation saturation problem at low vegetation coverage. In a related study in Zhejiang Province, China, Fang et al. [2] used the optimal variable selection method of different dominant tree species to estimate FSV. Their selected variables included a variety of vegetation indices, such as SRre, MSRre, CIre, and NDI45 developed based on the Sentinel-2 red-edge bands. Almost all of these variables appear in the final variable selection results, which also prove the potential of the red-edge band in estimating forest parameters.

In exploring the potential of $NDVI_{RE}$ to estimate FSV based on the Sentinel-2 red-edge bands, in the variable importance results of the VBM and BVBM, the $NDVI_{RE}$ ranks first. It is worth mentioning that the introduction of weighting coefficients "α" and "β" played a key role in the successful construction of the $NDVI_{RE}$. The results of this study also indicate that the model's estimation accuracy of FSV is significantly improved due to the addition of the $NDVI_{RE}$. First of all, an estimation accuracy of 80.46% is impressive in the research on FSV estimation. Moreover, according to Table 6, we found that the minimum and maximum values in the estimated results of the VBM and BVBM with the $NDVI_{RE}$ involvement are superior to those in the BBM, indicating that the $NDVI_{RE}$ mitigates the issue of light saturation to some extent. In addition, the mean values of FSV predicted by the BVBM in the training phase (56.88 m^3ha^{-1}) and the testing phase (60.79 m^3ha^{-1}) are also very close to the mean values of the training data (56.66 m^3ha^{-1}) and the testing data (63.84 m^3ha^{-1}).

Despite the proven efficiency and robustness of the RF algorithm through numerous studies [8,21,35–38], there is still a limitation observed in its ability to predict the minimum and maximum values of FSV in both the training and testing phases when compared to the actual training and testing data. This limitation results in overestimation of low values and underestimation of high values. Therefore, it would be necessary for future studies to incorporate more machine learning algorithms and innovative machine learning algorithms. From another perspective, deep learning, as a kind of non-parametric machine learning algorithm, is widely applied in forest monitoring. Numerous prior studies have demonstrated the outstanding capability of deep learning algorithms when it comes to target detection and vegetation classification [40–44].

Another paramount limitation of this study is the source of sample plot data which were the most recent. Although "one map" contains a large amount of necessary forest information, using these data to carry out research can no longer meet the current requirements for real-time forest monitoring. In order to resolve this problem in future studies, it is necessary to use unmanned aerial vehicles (UAVs) to obtain enough measured sample plots. Similarly, many studies have proposed UAVs equipped with hyper-spectral and LiDAR sensors to obtain the horizontal and vertical structure information of forests [45–51]. Its efficiency in obtaining forest parameters is unmatched by manual investigation. The accuracy of tree height, DBH, and spectral information extracted using UAVs is very close to manual surveys. Therefore, as an innovative research method, it is recommended to use UAVs to replace manual field survey work to improve research efficiency where high-precision forest estimation results can be obtained.

5. Conclusions

This study has effectively estimated and mapped the distribution of FSV in the Helan Mountains, with a resolution of 30 m. Utilizing the RF algorithm in conjunction with data from Sentinel-2, the study has affirmed the potential of $NDVI_{RE}$ in FSV estimation. Among all modeled variables, the novel vegetation index $NDVI_{RE}$, constructed based on the three red-edge bands of Sentinel-2, contributed the most to predicting FSV. Furthermore, the BVBM performed the best among the three models based on the two variables of the band and vegetation index. Finally, this study would assist policymakers in designing forest conservation and management paradigms that could potentially support the sustainability and carbon sequestration dynamics in the Helan Mountains and other montane forest ecosystems.

Author Contributions: Study design: Y.H. and J.W.; Data curation: T.M.; Investigation, D.P., L.C. and X.N.; Methodology: Y.H. and X.L.; Software, T.M.; Supervision, Y.H.; Writing: T.M.; Writing: review & editing: Y.H. and M.B. All authors have read and agreed to the published version of the manuscript.

Funding: This study was supported by the Key Project of Research and Development of Ningxia, China (2021BEB04061, 2022BEG03050), the National Natural Science Foundation of China (32101524), and the National Natural Science Foundation of Ningxia, China (2021AAC03017).

Data Availability Statement: The data presented in this study are available upon request from the corresponding author. The data are not publicly available due to funder regulations.

Conflicts of Interest: The authors declare no conflict of interest.

References

1. Hu, Y.; Xu, X.; Wu, F.; Sun, Z.; Xia, H.; Meng, Q.; Huang, W.; Zhou, H.; Gao, J.; Li, W.; et al. Estimating Forest Stock Volume in Hunan Province, China, by Integrating In Situ Plot Data, Sentinel-2 Images, and Linear and Machine Learning Regression Models. *Remote Sens.* **2020**, *12*, 186. [CrossRef]
2. Fang, G.; Fang, L.; Yang, L.; Wu, D. Comparison of Variable Selection Methods among Dominant Tree Species in Different Regions on Forest Stock Volume Estimation. *Forests* **2022**, *13*, 787. [CrossRef]
3. Yao, W.; Chi-Hui, G.; Xi-Jie, C.; Li-Qiong, J.; Xiao-Na, G.; Rui-Shan, C.; Mao-Sheng, Z.; Ze-Yu, C.; Hao-Dong, W. Carbon peak and carbon neutrality in China: Goals, implementation path and prospects. *China Geol.* **2021**, *4*, 720–746.
4. Sun, Y.; Liu, S.; Li, L. Grey Correlation Analysis of Transportation Carbon Emissions under the Background of Carbon Peak and Carbon Neutrality. *Energies* **2022**, *15*, 3064. [CrossRef]
5. Sun, L.; Cui, H.; Ge, Q. Will China achieve its 2060 carbon neutral commitment from the provincial perspective? *Adv. Clim. Change Res.* **2022**, *13*, 169–178. [CrossRef]
6. Pugh, T.A.M.; Lindeskog, M.; Smith, B.; Poulter, B.; Arneth, A.; Haverd, V.; Calle, L. Role of forest regrowth in global carbon sink dynamics. *Proc. Natl. Acad. Sci. USA* **2019**, *116*, 4382–4387. [CrossRef] [PubMed]
7. Astola, H.; Häme, T.; Sirro, L.; Molinier, M.; Kilpi, J. Comparison of Sentinel-2 and Landsat 8 imagery for forest variable prediction in boreal region. *Remote Sens. Environ.* **2019**, *223*, 257–273. [CrossRef]
8. Li, C.; Zhou, L.; Xu, W. Estimating Aboveground Biomass Using Sentinel-2 MSI Data and Ensemble Algorithms for Grassland in the Shengjin Lake Wetland, China. *Remote Sens.* **2021**, *13*, 1595. [CrossRef]
9. Kumar, L.; Mutanga, O. Remote Sensing of Above-Ground Biomass. *Remote Sens.* **2017**, *9*, 935. [CrossRef]
10. Lu, L.; Luo, J.; Xin, Y.; Duan, H.; Sun, Z.; Qiu, Y.; Xiao, Q. How can UAV contribute in satellite-based Phragmites australis aboveground biomass estimating? *Int. J. Appl. Earth Obs.* **2022**, *114*, 103024. [CrossRef]
11. Georganos, S.; Grippa, T.; Niang Gadiaga, A.; Linard, C.; Lennert, M.; Vanhuysse, S.; Mboga, N.; Wolff, E.; Kalogirou, S. Geographical random forests: A spatial extension of the random forest algorithm to address spatial heterogeneity in remote sensing and population modelling. *Geocarto Int.* **2021**, *36*, 121–136. [CrossRef]
12. Hu, Y.; Sun, Z. Assessing the Capacities of Different Remote Sensors in Estimating Forest Stock Volume Based on High Precision Sample Plot Positioning and Random Forest Method. *Nat. Environ. Pollut. Technol.* **2022**, *21*, 1113–1123. [CrossRef]

13. Lister, A.J.; Andersen, H.; Frescino, T.; Gatziolis, D.; Healey, S.; Heath, L.S.; Liknes, G.C.; McRoberts, R.; Moisen, G.G.; Nelson, M.; et al. Use of Remote Sensing Data to Improve the Efficiency of National Forest Inventories: A Case Study from the United States National Forest Inventory. *Forests* **2020**, *11*, 1364. [CrossRef]
14. Chave, J.; Davies, S.J.; Phillips, O.L. Ground Data are Essential for Biomass Remote Sensing Missions. *Surv. Geophys.* **2019**, *40*, 863–880. [CrossRef]
15. Ou, G.; Li, C.; Lv, Y.; Wei, A.; Xiong, H.; Xu, H.; Wang, G. Improving Aboveground Biomass Estimation of *Pinus densata* Forests in Yunnan Using Landsat 8 Imagery by Incorporating Age Dummy Variable and Method Comparison. *Remote Sens.* **2019**, *11*, 738. [CrossRef]
16. Zhao, P.; Lu, D.; Wang, G.; Wu, C.; Huang, Y.; Yu, S. Examining Spectral Reflectance Saturation in Landsat Imagery and Corresponding Solutions to Improve Forest Aboveground Biomass Estimation. *Remote Sens.* **2016**, *8*, 469. [CrossRef]
17. Lu, D.; Chen, Q.; Wang, G.; Moran, E.; Batistella, M.; Zhang, M.; Vaglio Laurin, G.; Saah, D. Aboveground Forest Biomass Estimation with Landsat and LiDAR Data and Uncertainty Analysis of the Estimates. *Int. J. For. Res.* **2012**, *2012*, 1–16. [CrossRef]
18. Tilly, N.; Aasen, H.; Bareth, G. Fusion of Plant Height and Vegetation Indices for the Estimation of Barley Biomass. *Remote Sens.* **2015**, *7*, 11449–11480. [CrossRef]
19. Kross, A.; McNairn, H.; Lapen, D.; Sunohara, M.; Champagne, C. Assessment of RapidEye vegetation indices for estimation of leaf area index and biomass in corn and soybean crops. *Int. J. Appl. Earth Obs.* **2015**, *34*, 235–248. [CrossRef]
20. Li, C.; Chimimba, E.G.; Kambombe, O.; Brown, L.A.; Chibarabada, T.P.; Lu, Y.; Anghileri, D.; Ngongondo, C.; Sheffield, J.; Dash, J. Maize Yield Estimation in Intercropped Smallholder Fields Using Satellite Data in Southern Malawi. *Remote Sens.* **2022**, *14*, 2458. [CrossRef]
21. Luo, W.; Kim, H.S.; Zhao, X.; Ryu, D.; Jung, I.; Cho, H.; Harris, N.; Ghosh, S.; Zhang, C.; Liang, J. New forest biomass carbon stock estimates in Northeast Asia based on multisource data. *Glob. Chang. Biol.* **2020**, *26*, 7045–7066. [CrossRef]
22. Shen, X.; Cao, L.; Yang, B.; Xu, Z.; Wang, G. Estimation of Forest Structural Attributes Using Spectral Indices and Point Clouds from UAS-Based Multispectral and RGB Imageries. *Remote Sens.* **2019**, *11*, 800. [CrossRef]
23. Xiao, C.; Li, P.; Feng, Z.; Liu, Y.; Zhang, X. Sentinel-2 red-edge spectral indices (RESI) suitability for mapping rubber boom in Luang Namtha Province, northern Lao PDR. *Int. J. Appl. Earth Obs.* **2020**, *93*, 102176. [CrossRef]
24. Roy, D.; Li, Z.; Zhang, H. Adjustment of Sentinel-2 Multi-Spectral Instrument (MSI) Red-Edge Band Reflectance to Nadir BRDF Adjusted Reflectance (NBAR) and Quantification of Red-Edge Band BRDF Effects. *Remote Sens.* **2017**, *9*, 1325. [CrossRef]
25. Bramich, J.; Bolch, C.J.S.; Fischer, A. Improved red-edge chlorophyll-a detection for Sentinel 2. *Ecol. Indic.* **2021**, *120*, 106876. [CrossRef]
26. Delegido, J.; Verrelst, J.; Alonso, L.; Moreno, J. Evaluation of Sentinel-2 Red-Edge Bands for Empirical Estimation of Green LAI and Chlorophyll Content. *Sensors* **2011**, *11*, 7063–7081. [CrossRef]
27. Liu, J.; Fan, J.; Yang, C.; Xu, F.; Zhang, X. Novel vegetation indices for estimating photosynthetic and non-photosynthetic fractional vegetation cover from Sentinel data. *Int. J. Appl. Earth Obs.* **2022**, *109*, 102793. [CrossRef]
28. Bilal, M.; Nichol, J.E. Evaluation of the NDVI-Based Pixel Selection Criteria of the MODIS C6 Dark Target and Deep Blue Combined Aerosol Product. *IEEE J. Sel. Top. Appl. Earth Obs. Remote. Sens.* **2017**, *10*, 3448–3453. [CrossRef]
29. Asgarian, A.; Amiri, B.J.; Sakieh, Y. Assessing the effect of green cover spatial patterns on urban land surface temperature using landscape metrics approach. *Urban Ecosyst.* **2015**, *18*, 209–222. [CrossRef]
30. Ouma, Y.O.; Noor, K.; Herbert, K. Modelling Reservoir Chlorophyll-a, TSS, and Turbidity Using Sentinel-2A MSI and Landsat-8 OLI Satellite Sensors with Empirical Multivariate Regression. *J. Sens.* **2020**, *2020*, 1–21. [CrossRef]
31. Warren, M.A.; Simis, S.G.H.; Martinez-Vicente, V.; Poser, K.; Bresciani, M.; Alikas, K.; Spyrakos, E.; Giardino, C.; Ansper, A. Assessment of atmospheric correction algorithms for the Sentinel-2A MultiSpectral Imager over coastal and inland waters. *Remote Sens. Environ.* **2019**, *225*, 267–289. [CrossRef]
32. Huang, H.; Roy, D.; Boschetti, L.; Zhang, H.; Yan, L.; Kumar, S.; Gomez-Dans, J.; Li, J. Separability Analysis of Sentinel-2A Multi-Spectral Instrument (MSI) Data for Burned Area Discrimination. *Remote Sens.* **2016**, *8*, 873. [CrossRef]
33. Jin, Y.; Yang, X.; Qiu, J.; Li, J.; Gao, T.; Wu, Q.; Zhao, F.; Ma, H.; Yu, H.; Xu, B. Remote Sensing-Based Biomass Estimation and Its Spatio-Temporal Variations in Temperate Grassland, Northern China. *Remote Sens.* **2014**, *6*, 1496–1513. [CrossRef]
34. Xue, J.; Su, B. Significant Remote Sensing Vegetation Indices: A Review of Developments and Applications. *J. Sens.* **2017**, *2017*, 1–17. [CrossRef]
35. Liu, Z.; Ye, Z.; Xu, X.; Lin, H.; Zhang, T.; Long, J. Mapping Forest Stock Volume Based on Growth Characteristics of Crown Using Multi-Temporal Landsat 8 OLI and ZY-3 Stereo Images in Planted Eucalyptus Forest. *Remote Sens.* **2022**, *14*, 5082. [CrossRef]
36. Wang, L.; Zhou, X.; Zhu, X.; Dong, Z.; Guo, W. Estimation of biomass in wheat using random forest regression algorithm and remote sensing data. *Crop J.* **2016**, *4*, 212–219. [CrossRef]
37. Torre-Tojal, L.; Bastarrika, A.; Boyano, A.; Lopez-Guede, J.M.; Graña, M. Above-ground biomass estimation from LiDAR data using random forest algorithms. *J. Comput. Sci.* **2022**, *58*, 101517. [CrossRef]
38. Li, X.; Ye, Z.; Long, J.; Zheng, H.; Lin, H. Inversion of Coniferous Forest Stock Volume Based on Backscatter and InSAR Coherence Factors of Sentinel-1 Hyper-Temporal Images and Spectral Variables of Landsat 8 OLI. *Remote Sens.* **2022**, *14*, 2754. [CrossRef]
39. Glushkova, M.; Zhiyanski, M.; Nedkov, S.; Yaneva, R.; Stoeva, L. Ecosystem services from mountain forest ecosystems: Conceptual framework, approach and challenges. *Silva Balc.* **2020**, *21*, 47–68. [CrossRef]

40. Zhang, L.; Shao, Z.; Liu, J.; Cheng, Q. Deep Learning Based Retrieval of Forest Aboveground Biomass from Combined LiDAR and Landsat 8 Data. *Remote Sens.* **2019**, *11*, 1459. [CrossRef]
41. Ayhan, B.; Kwan, C.; Budavari, B.; Kwan, L.; Lu, Y.; Perez, D.; Li, J.; Skarlatos, D.; Vlachos, M. Vegetation Detection Using Deep Learning and Conventional Methods. *Remote Sens.* **2020**, *12*, 2502. [CrossRef]
42. Bhatnagar, S.; Gill, L.; Ghosh, B. Drone Image Segmentation Using Machine and Deep Learning for Mapping Raised Bog Vegetation Communities. *Remote Sens.* **2020**, *12*, 2602. [CrossRef]
43. Liu, M.; Fu, B.; Xie, S.; He, H.; Lan, F.; Li, Y.; Lou, P.; Fan, D. Comparison of multi-source satellite images for classifying marsh vegetation using DeepLabV3 Plus deep learning algorithm. *Ecol. Indic.* **2021**, *125*, 107562. [CrossRef]
44. Lees, T.; Tseng, G.; Atzberger, C.; Reece, S.; Dadson, S. Deep Learning for Vegetation Health Forecasting: A Case Study in Kenya. *Remote Sens.* **2022**, *14*, 698. [CrossRef]
45. Qin, H.; Zhou, W.; Yao, Y.; Wang, W. Estimating Aboveground Carbon Stock at the Scale of Individual Trees in Subtropical Forests Using UAV LiDAR and Hyperspectral Data. *Remote Sens.* **2021**, *13*, 4969. [CrossRef]
46. Qin, H.; Zhou, W.; Yao, Y.; Wang, W. Individual tree segmentation and tree species classification in subtropical broadleaf forests using UAV-based LiDAR, hyperspectral, and ultrahigh-resolution RGB data. *Remote Sens. Environ.* **2022**, *280*, 113143. [CrossRef]
47. Dashti, H.; Poley, A.; Glenn, N.F.; Ilangakoon, N.; Spaete, L.; Roberts, D.; Enterkine, J.; Flores, A.N.; Ustin, S.L.; Mitchell, J.J. Regional Scale Dryland Vegetation Classification with an Integrated Lidar-Hyperspectral Approach. *Remote Sens.* **2019**, *11*, 2141. [CrossRef]
48. Zhu, W.; Sun, Z.; Peng, J.; Huang, Y.; Li, J.; Zhang, J.; Yang, B.; Liao, X. Estimating Maize Above-Ground Biomass Using 3D Point Clouds of Multi-Source Unmanned Aerial Vehicle Data at Multi-Spatial Scales. *Remote Sens.* **2019**, *11*, 2678. [CrossRef]
49. Picos, J.; Bastos, G.; Míguez, D.; Alonso, L.; Armesto, J. Individual Tree Detection in a Eucalyptus Plantation Using Unmanned Aerial Vehicle (UAV)-LiDAR. *Remote Sens.* **2020**, *12*, 885. [CrossRef]
50. Santos, A.A.D.; Marcato Junior, J.; Araújo, M.S.; Di Martini, D.R.; Tetila, E.C.; Siqueira, H.L.; Aoki, C.; Eltner, A.; Matsubara, E.T.; Pistori, H.; et al. Assessment of CNN-Based Methods for Individual Tree Detection on Images Captured by RGB Cameras Attached to UAVs. *Sensors* **2019**, *19*, 3595. [CrossRef]
51. Sankey, T.T.; McVay, J.; Swetnam, T.L.; McClaran, M.P.; Heilman, P.; Nichols, M.; Pettorelli, N.; Horning, N.; Pettorelli, N.; Horning, N. UAV hyperspectral and lidar data and their fusion for arid and semi-arid land vegetation monitoring. *Remote Sens. Ecol. Conserv.* **2018**, *4*, 20–33. [CrossRef]

MDPI

Article

Early Yield Forecasting of Maize by Combining Remote Sensing Images and Field Data with Logistic Models

Hongfang Chang [1,2], Jiabing Cai [1,2,*], Baozhong Zhang [1,2], Zheng Wei [1,2] and Di Xu [1,2]

1 State Key Laboratory of Simulation and Regulation of Water Cycle in River Basin, China Institute of Water Resources and Hydropower Research, Beijing 100038, China
2 National Center for Efficient Irrigation Engineering and Technology Research-Beijing, Beijing 100048, China
* Correspondence: caijb@iwhr.com; Tel.: +86-10-68786532

Abstract: Early forecasting of crop yield from field to region is important for stabilizing markets and safeguarding food security. Producing a precise forecasting result with fewer inputs is an ongoing goal for the large-area yield evaluation. We present one approach of yield prediction for maize that was explored by incorporating remote-sensing-derived land surface temperature (LST) and field in-season data into a series of logistic models with only a few parameters. Continuous observation data of maize were utilized to calibrate and validate the corresponding logistic models for regional biomass estimating based on field temperatures (including crop canopy temperature (Tc)) and relative dry/fresh biomass accumulation. The LST maps from MOD11A1 products, which are considered to be matched as Tc in large irrigation districts, were assimilated into the validated models to estimate the biomass accumulation. It was found that the temporal-scale difference between the instantaneous LST and the daily average value of field-measured Tc was eliminated by data normalization method, indicating that the normalized LST could be input directly into the model as an approximation of the normalized Tc. Making one observed biomass in-season as the driving force, the maximum of dry/fresh biomass accumulation (DBA/FBA) at harvest could be estimated. Then, grain yield forecasting could be achieved according to the local harvest index of maize. Silage and grain yields were evaluated reasonably well compared with field observations based on the regional map of LST values obtained in 2017 in Changchun, Jilin Province, China. Here, satisfactory grain and silage yield forecasting was provided by assimilating once measured value of DBA/FBA at the middle growth period (early August) into the model in advance of harvest. Meanwhile, good results were obtained in the application of this approach using field data in 2016 to predict grain yield ahead of harvest in the Jiefangzha sub-irrigation district, Inner Mongolia, China. This study demonstrated that maize yield can be forecasted accurately prior to harvest by assimilating remote-sensing-derived LST and field data into the logistic models at a regional scale considering the spatio-temporal scale extension of ground information and crop dynamic growth in real time.

Keywords: yield forecasting; remote sensing; logistic model; normalization method; crop canopy temperature; maize

Citation: Chang, H.; Cai, J.; Zhang, B.; Wei, Z.; Xu, D. Early Yield Forecasting of Maize by Combining Remote Sensing Images and Field Data with Logistic Models. *Remote Sens.* **2023**, *15*, 1025. https://doi.org/10.3390/rs15041025

Academic Editors: Kenji Omasa, Shan Lu and Jie Wang

Received: 9 January 2023
Revised: 10 February 2023
Accepted: 10 February 2023
Published: 13 February 2023

1. Introduction

Early estimates of crop yield will contribute to addressing the key issues of crop production management, future market output, and deep processing. According to the early prediction of crop yield, farmers can adjust and optimize irrigation decision making in a timely way to maximize yield for enhancing profits, while policymakers also take reasonable measures to deal with potential trade risks in order to safeguard food security and ensure market stability [1]. It should be emphasized that the earlier yield forecasting information is provided, the more effective measures are likely to be undertaken [2].

Plentiful studies have been implemented to predict final yield in the past decades based on different methods including field survey, statistical methods, and crop growth models [3–5]. Field survey can assess yield by capturing the ground truth; nonetheless, it is highly time-consuming and labor-intensive. The core of statistical methods lies in the acquisition of empirical relationships between crop yields and specific related indicators. Despite reasonably accurate results in specific field areas, it is restricted when scaling up this relationship to large areas. Crop growth models can be applied to describe plant dynamic growth. Biomass is a critical biophysical indicator with a close link to yield at harvest. Therefore, one way to predict yield is to acquire biomass estimates via crop growth models and then implement the in-season evaluations of yield on account of their good correlations. Crop growth models include process-based models and experiment regression ones [6]. The former requires many parameters as inputs [7,8], making it difficult to execute the models in data-scarce regions [9] though it is more mechanistic than the latter. In contrast, with only a few parameters, statistical regression models have been developed (e.g., the Richards, Compertz, and Weibull equations) and continue to be widely used to illustrate crop growth dynamics including biomass [10–12], but this method has the limitation of extending model parameters to large areas.

Given that these methods have limitations, data assimilation has been developed to solve these problems in yield forecasting. Intrinsically, data assimilation is used to incorporate observations into the model to obtain the optimal possible estimates [13]. The rapid advancement of satellite images allows for large-scale crop growth monitoring [14,15]. Numerous studies have shown that incorporating remote sensing data into crop models can improve regional yield estimates [16–18]. The research has focused mainly on the assimilation of mechanism models [2,19–21] despite the fact that they require numerous input parameters. By contrast, it is worth investigating the potential of experiment regression models in early-season yield forecasting for large areas via data assimilation when there is a lack of detailed input information.

As one of the most important crops, maize yield forecasting is critical for the development of agriculture and livestock [22] and can serve as an excellent reference for other crop research. For different purposes, farmers can harvest maize as silage or grain yield, and correspondingly, dry biomass accumulation (DBA) and fresh biomass accumulation (FBA) need to be predicted as an important precondition for yield forecasting.

The logistic model is one of the most commonly applied regression models for crop growth processes such as DBA throughout the growing season [23–27]. However, the logistic model fails to fully describe the development process of FBA or leaf area index (LAI) due to the existence of a downtrend process after the milk stage caused by leaf senescence. A revised logistic model proposed by Wang [28] overcame this limitation and performed well when describing LAI changes [29], but it has not yet been used for FBA. In addition, the logistic model was originally developed for individual plants that neglected regional applicability. Elings [30] acquired maize leaf area dynamic growth in various environments using a set of parameters for the data normalization method. This method has been used to establish a normalized logistic model of relative DBA (RDBA) for simulating regional crop growth patterns [4,31].

Indeed, another major obstacle to the application of the logistic model in region is the choice of input parameter, which serves as a vital bridge linking the point-based and regional applications. Furthermore, previous studies have demonstrated the growth curve of maize using the logistic models with air or soil temperatures as input [32–34]. The cropping environment is the most influential factor in plant yield [35,36]. Canopy temperature (Tc) is obviously a better indicator for reflecting crop water message responses to field conditions. Moreover, Tc can be considered as matched to land surface temperature (LST) in large agricultural areas, which can be inversed from remote sensing images [37]. Therefore, integrating LST data into the normalized logistic models can acquire the values of RDBA in combination with the once-measured DBA on a certain date and harvest index (HI), which presents an opportunity to achieve early yield forecasting in regions.

Therefore, an approach was developed to retrieve early prediction of maize yield using logistic models in combination with daily LST images and in-season field observations.

The major objectives are as follows: (1) to calibrate and validate the corresponding logistic models for simulating the maize growth curves including RDBA and relative FBA (RFBA) based on different independent variables including temperatures of air, canopy, and soil at 20 cm or 40 cm in the root zone; (2) to determine the applicability of Tc in crop monitoring as well as the appropriate model parameters; (3) to forecast maize yield in region by HI and biomass maximum including DBA and FBA, which can be acquired by incorporating the normalized LST from remote sensing as an approximation of the normalized Tc into the corresponding optimal models with once-observed DBA or FBA as the driving force; and (4) to test the portability of this approach by producing grain yield maps in other agricultural districts and comparing them to local observations.

2. Materials and Methodology

2.1. Study Areas

The first study region was in Changchun area, Jilin Province (about 2.05 million ha), as shown in Figure 1a,b. This region was characterized by a northern temperate continental monsoon climate, with an average annual rainfall of 520–755 mm [38], of which more than 60% occurs in the summer. The annual average daily temperature is 4.8 °C, and the sunshine duration is approximately 2700 h. The data for model developments were obtained from field experiments of maize growth (May–September) from 2017 to 2019 in an agricultural research station (43°38′39.92′′N, 125°19′7.77′′E, 248 m a.s.l.) near Changchun city, as shown in Figure 1f (about 73 ha). The maize cultivar was Xianyu 335. The predominant soil types are black and meadow soil, and the soil texture is mainly sandy loam soil. The field capacity (Fc) and wilting point (Wp) were measured as 37% and 16% in an 80 cm average of the crop root zone, respectively. Precipitation and soil water content in 2017–2019 were monitored, revealing an optimal soil moisture range for maize growth (Figure 2).

Another study region for this work was the Jiefangzha sub-irrigation district (approximately 0.229 million ha), which is one main component of Hetao irrigation district in Inner Mongolia, China (Figure 1a,c). Maize is one of the major crops in this region. The yearly average daily temperature is 9 °C, with an annual rainfall of approximately 151.3 mm. The field monitoring system was conducted in the Shahaoqu experimental station in 2016, as shown in Figure 1b (40°55′8″N, 107°8′16″E, 1036 m a.s.l.). The values of Fc and Wp are 35% and 15% in the crop root zone, respectively. More information can be found in the report by Bai et al. [39].

2.2. Field Measurements

In Changchun, the five typical maize plots (named H1 to H5, Figure 1f) were equipped with five sets of canopy temperature and meteorology monitoring systems (CTMS) (Figure 1e), which are composed of a stainless-steel stand column, solar panels, and various sensors. This system can synchronously monitor field data at 30 min intervals, including wind speed (014MINI-MetOne, Washington, USA), solar radiation (SP110-Apogee/SQ110-Apogee, Logan, USA), the crop canopy temperature (TPiS 1 T 1252B, Excelitas, Waltham, Germany), air temperature/humidity (HMP60-Vaisala, Vantaa, Finland), soil temperature/moisture (20 cm and 40 cm in the root zone) (SM10D, Beijing, China), and more by the corresponding sensors. It should be emphasized that the canopy temperature at 30 min intervals was computed as the average of multi-point values around the equipment by the rotation measurement of a thermal infrared sensor (TPiS 1 T 1252B, Excelitas, Waltham, Germany) installed at the end of a cantilever perpendicular to the stand column. Sensors were set up at a height of 3 m during the whole growth stage, which could be adjusted as need. Regular weekly maintenance ensured the normal functioning of the equipment. A more detailed description of the system was provided by Cai et al. [37] and Huang et al. [40].

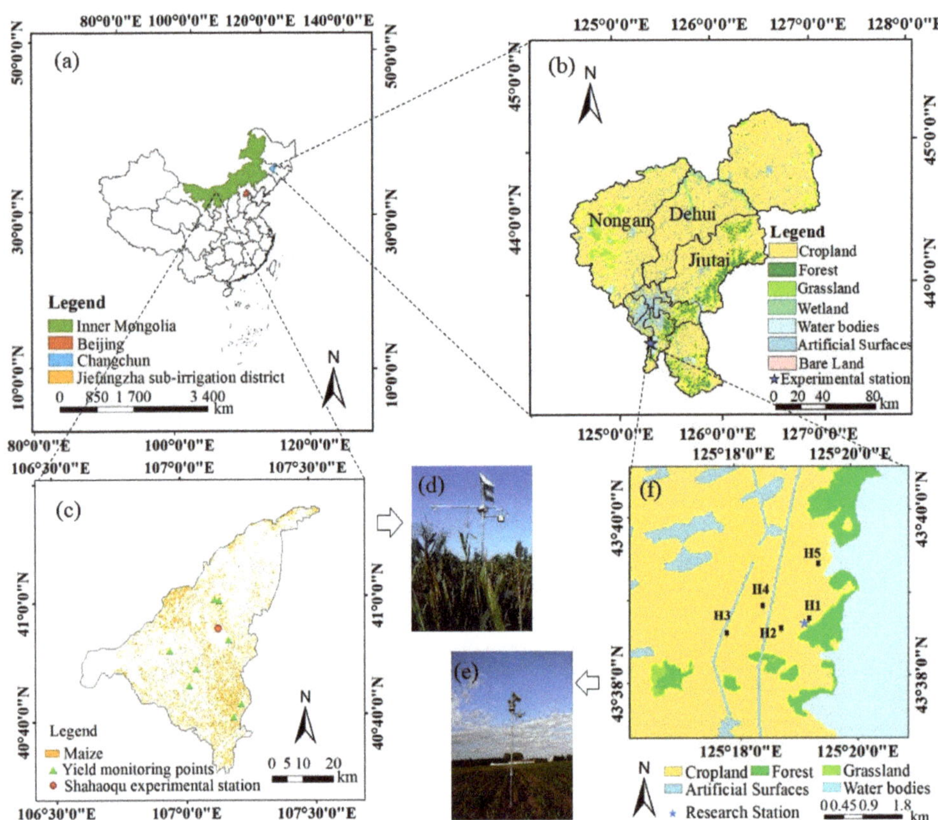

Figure 1. Overview of the two study areas: (**a**) locations in China; (**b**) land-use and -cover map in Changchun from the GlobeLand30 platform (in 2020); (**c**) distributions of experimental station and eight yield monitoring points in Jiefangzha sub-irrigation district; (**d**,**e**) a typical CTMS equipment in Jiefangzha/Changchun; (**f**) locations of the five sets of CTMS equipment (H1 to H5) in Changchun.

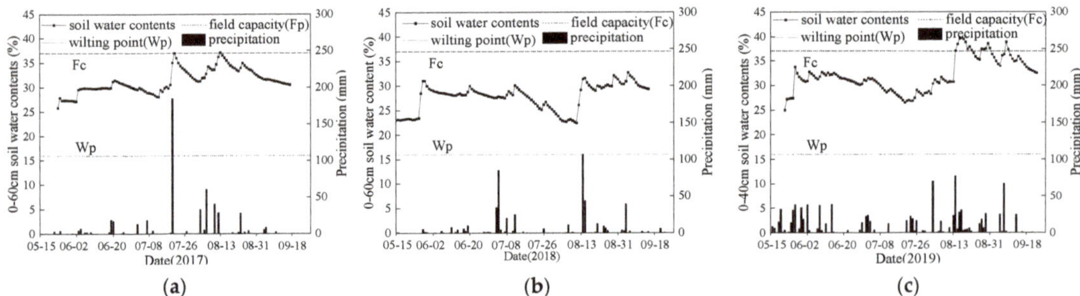

Figure 2. Precipitations and soil water contents changing during maize growing season in Changchun during three years: (**a**) 2017; (**b**) 2018; (**c**) 2019.

Above-ground biomass was sampled every 10–15 days by removing three representative plants from each of monitoring plots. The collected samples were immediately weighted as the amount of FBA. Then, these samples were put in an oven at 105 °C for 30 min to stop the plant life activities and subsequently were dried at 80 °C to a constant weight. (The final weight would be the DBA.) During the harvest, the last sampling of above-ground biomass was recorded, and the grain yield (1 m^2) and planting density were measured in each experimental point at the same time. In conclusion, six samples were collected from each plot per year.

Similarly, a CTMS system was installed in the Shahaoqu experimental station to collect the same data as Changchun, and eight yield monitoring points were erected to measure final grain yield for evaluation in the Jiefangzha sub-irrigation district (Figure 1c). Furthermore, above-ground biomass in the experimental site was sampled and recorded four times as a driving force for yield forecasting. More details can be seen in the report from Bai et al. [39].

2.3. Remote Sensing Data

For the subsequent spatial-scale research in Changchun, the Landsat 8 images (30 m) and MOD11A1 data (1 km) in 2017 were downloaded from the website of USGS (https://earthexplorer.usgs.gov/ accessed on 21 June 2021) with the aim of acquiring the LST data. The available remote sensing images can be seen in Table S1. Daily LST data were derived from the MOD11A1 product (1 km). Due to the cloudiness, there were only four clear Landsat 8 images in 2017 relevant to maize growth that would be used for mapping maize in the research area and verified by the field monitoring data, cropland map from the GlobeLand30 platform, and statistics data.

To obtain the maize distribution in Changchun, some calculations and preprocessing needed to be finished. Firstly, the normalized difference vegetation index (NDVI) and land surface water index (LSWI) were calculated by Equations (1) and (2), which were also followed for four Landsat 8 images (Operational Land Imager, OLI) [41,42]. The Landsat 8 image on 25 September, which is the best one just in the maize growth stage, was chosen as a sample to classify the region of interest (ROI). The classification thresholds could be determined by combining the features of vegetation in different stages and the changes in NDVI and LSWI of ROI samples. The process contained the following four steps (Figure S1): (1) to distinguish the vegetation and the others according to the NDVI values on 25 September (Julian day 268), which belonged to the optimum discrimination period of vegetation in four images; (2) to obtain the thresholds by analyzing the NDVI values of forest and crop samples on 5 June (Julian day 156), when the crops were in the early growth stage, and the NDVI of the forest should be obviously higher; (3) to employ the LSWI of rice samples in ROI on 5 June (Julian day 156) to determine the corresponding thresholds considering the rice was irrigated during this stage, and the LSWI values of rice should be higher; and (4) to determine the relative thresholds between the maize and the other vegetations by analyzing the LSWI ranges of them in ROI on 5 June (Julian day 156) and 25 September (Julian day 268). Based on these thresholds, the distribution of maize + rice could be obtained according to the classification rules of decision tree classification.

$$\text{NDVI} = (\rho_{NIR} - \rho_{RED})/(\rho_{NIR} + \rho_{RED}) \tag{1}$$

$$\text{LSWI} = (\rho_{NIR} - \rho_{SWIR})/(\rho_{NIR} + \rho_{SWIR}) \tag{2}$$

where ρ_{NIR} (band5), ρ_{RED} (band4), and ρ_{SWIR} (band6) are the reflectivity of near-infrared, red, and shortwave infrared band, respectively.

Based on the decision tree classification mentioned above, the combination pattern for maize + rice in Changchun was obtained (Figure 3a). As the dominant field crops are maize and rice here, it was assumed that these results are mostly cropland. In order to evaluate the classification precision, the land-cover data with a spatial resolution of 30 m in 2020 (Figure 3b) were downloaded from the platform of GlobeLand30 (http://www.globallandcover.com/home.html accessed on 3 October 2021), published by the Ministry of Natural Resources of China. Meanwhile, the statistical data of crop areas were obtained from the Statistic Bureau of Jilin Province (http://tjj.jl.gov.cn/tjsj/ accessed on 3 October 2021) for further evaluating the classification precision.

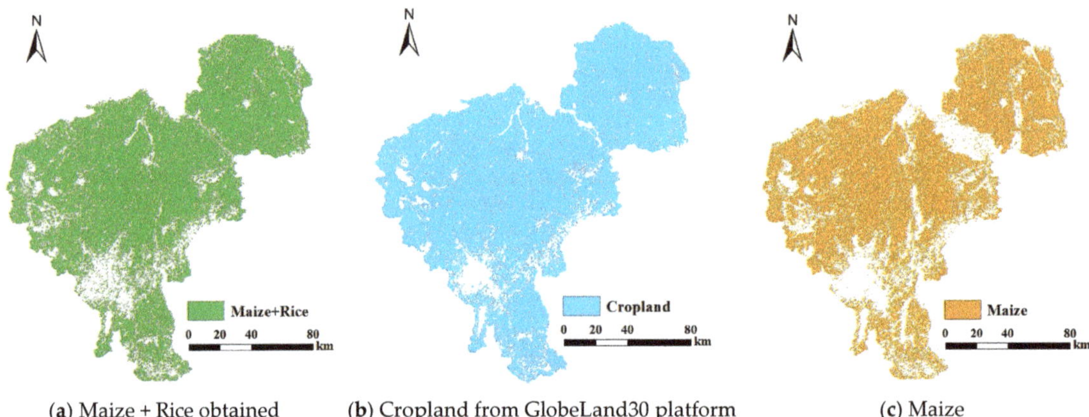

(**a**) Maize + Rice obtained (**b**) Cropland from GlobeLand30 platform (**c**) Maize

Figure 3. Crop patterns and cropland in Changchun area: (**a**) maize and rice obtained by decision tree classification; (**b**) cropland data in 2020 from the GlobeLand30 platform; (**c**) maize mapping by decision tree classification.

Using the confusion matrix, the value of the producer's accuracy (ratio of the number of estimated correct pixels to reference pixels) about cropland was 82.18%, determined through comparing Figure 3a,b. This value proved that the classification method established above was available and appropriate. It should be emphasized that the validation method used here was restricted by the accuracy of the GlobeLand30 product. Meanwhile, this product is not offered for every year, which will affect the accuracy evaluation of crop mapping. The maize mapping could be obtained by the same method (Figure 3c), which took up 87% of cropland in Figure 3b. The ratio was consistent with the value from statistics data (84%).

For the Jiefangzha sub-irrigation district, the spatial distribution of maize was derived from the report from Bai et al. [39], as shown in Figure 1c. Fortunately, the images from Landsat 8 (30 m) and MOD11A1 (1 km) can support the enhanced spatial and temporal adaptive reflectance fusion model (ESTARFM) algorithm [43] to improve the spatial accuracy of LST. Therefore, the fused LST was used in this region. Details of the extraction process can be found in research by Huang et al. [40].

2.4. Logistic Models

2.4.1. Logistic Model

The logistic model depicts a sigmoidal curve [44] that increases gradually at first, more rapidly in the middle, and then slowly at the end before leveling off at a maximum value [45,46], such as the growth curve of DBA [47]. The model equation is as follows:

$$y_D = a/(1 + b\exp(-kt))\ t = \sum_{i=1}^{n}(t_i - 10) \quad (3)$$

where y_D is the dependent growth parameter (DBA, kg ha^{-1}); a denotes the uppermost asymptote, implying the theoretical upper limit of DBA growth; b and k are model parameters; t is the effective accumulated temperature after emergence in the present study (hereinafter referred to as the effective accumulated temperature, °C). Notably, t_{air}, t_{canopy}, t_{20}, and t_{40} (°C), represent the effective accumulated temperature of the air, canopy, and soil at 20 cm or 40 cm of the root zone, respectively. t_i is the day i value (from crop emergence) of daily average temperature of the air, canopy, or soil at 20 cm or 40 cm depth of the crop root zone. (It is calculated as 30 °C when t_i exceeds 30 °C, and it will be calculated as 10 °C if the value is less than 10 °C [48].) n is the total number of days from crop emergence to harvest.

2.4.2. Normalized Logistic Model (N-Logistic Model)

The logistic model was originally developed for individual applications that may be inappropriate for spatial prediction of crop growth. The normalization method transforms the raw data into an interval of 0 to 1, which can eliminate the dimensional differences between plots. Therefore, the RDBA and relative effective accumulated temperature (T) were used to set up the growth model, lower data dispersion (from different plots), and form a regional model. Here, the logistic model with normalization is called N-logistic model, which has the same form as the logistic model but with different parameters:

$$Y_D = A/(1 + B\exp(-KT)) \; Y_D = y_D/y_{Dm} \; T = t/t_m \tag{4}$$

where Y_D is the RDBA, which is the ratio of y_D (DBA in the maize growing season) to y_{Dm} (DBA at harvest); A is the upper most asymptote implying the upper limit of RDBA; and B and K are model parameters; t is the same as in Equation (3); t_m is equal to the value of t at harvest; T is relative effective accumulative temperature (T_{20}, T_{40}, T_{canopy}, and T_{air} mean the values in soil at 20 cm and 40 cm under surface, crop canopy, and air, respectively), which is the ratio of t to t_m. Theoretically, the value of Y_D equals A when T ($0 \le T \le 1$) reaches 1. Therefore, the value of A represents the theoretical upper limit of the RDBA.

2.4.3. Revised Logistic Model (R-Logistic Model)

Like the simulation of LAI, the R-logistic model [28] was employed to verify the FBA growth pattern:

$$y_F = c/(1 + \exp(gt^2 + et + f)) \tag{5}$$

where y_F is the above-ground FBA (kg ha^{-1}); c, g (>0), e, and f are the model parameters; t is the same as in Equation (3). When $t = 0$, $y_F = c/(1 + \exp(f))$ (the above-ground FBA in maize emergence); when $t = (-e/2g)$, the value of $(gt^2 + et + f)$ reaches a minimum, and the value of y_F reaches a maximum; when $t > (-e/2g)$, the value of y_F begins to decline. These situations are consistent with the actual growth curve of FBA.

2.4.4. Normalized Revised Logistic Model (NR-Logistic Model)

Similar to the logistic model, the R-logistic model was initially developed for individual plants. To scale up the simulation from a single plot to a region, the NR-logistic model was developed by the normalization method mentioned above. It takes the same form as the R-logistic model but with key parameters:

$$Y_F = C/(1 + \exp(GT^2 + ET + F)) \; Y_F = y_F/y_{Fm} \tag{6}$$

where Y_F represents the relative fresh biomass accumulation (RFBA); C, G (>0), E, and F are parameters; T ($0 \le T \le 1$) is the same as in Equation (4); y_{Fm} represents the maximum FBA (g m^{-2}). When $T = (-E/2G)$, the value of $(GT^2 + ET + F)$ reaches a minimum, and Y_F reaches a maximum; when $(-E/2G) < T < 1$, the value of Y_F declines as the value of T increases.

2.5. Yield Forecasting

When the model was validated to obtain suitable parameters, namely the RDBA and y_{Dm}, the RFBA and y_{Fm} could be simulated by Equations (4) and (6) with field monitoring data once at least, respectively, when combined with the map of the independent variable (LST) inversed from remote sensing images. Grain yield (Y) will be forecasted by the DBA in harvest period and HI as follows:

$$Y = \text{HI} \times y_{Dm} \tag{7}$$

where Y is grain yield; HI is the weight of a harvested product as a percentage of the total plant weight of a crop; and y_{Dm} is the DBA at harvest time as mentioned above. The values of maize HI in some subareas in Changchun [49] and the measured ones in the experimental station (Table S2) were used to obtain the HI map (Figure S2) in Changchun by the kriging interpolation method.

A flow chart of this approach using canopy temperature to forecast yield is shown in Figure 4. Firstly, the logistic model/R-logistic model was used to prove the possibility to simulate DBA/FBA based on Tc. Secondly, the logistic model/R-Logistic model was changed to N-logistic model/NR-logistic model by the data normalization method for regional biomass (RDBA/RFBA) estimation. Finally, the grain/silage yield can be forecasted using remote-sensing-derived LST and once-measured biomass (DBA/FBA) as inputs to the N-logistic model/NR-logistic model after obtaining the HI value. In Section 3, the grain/silage yield in Changchun would be forecasted using measured DBA on three dates (7/16, 8/10, and 8/31 in 2017), and the grain yield in Jiefangzha would be forecasted by once-measured DBA on four dates (7/6, 7/21, 8/4, and 8/26 in 2016). The independent variable for the N-logistic model and NR-logistic model, i.e., T_{canopy}, was a scale factor for yield forecasting from point to area through the LST map from remote sensing images (T_{LST}) in a large irrigation district.

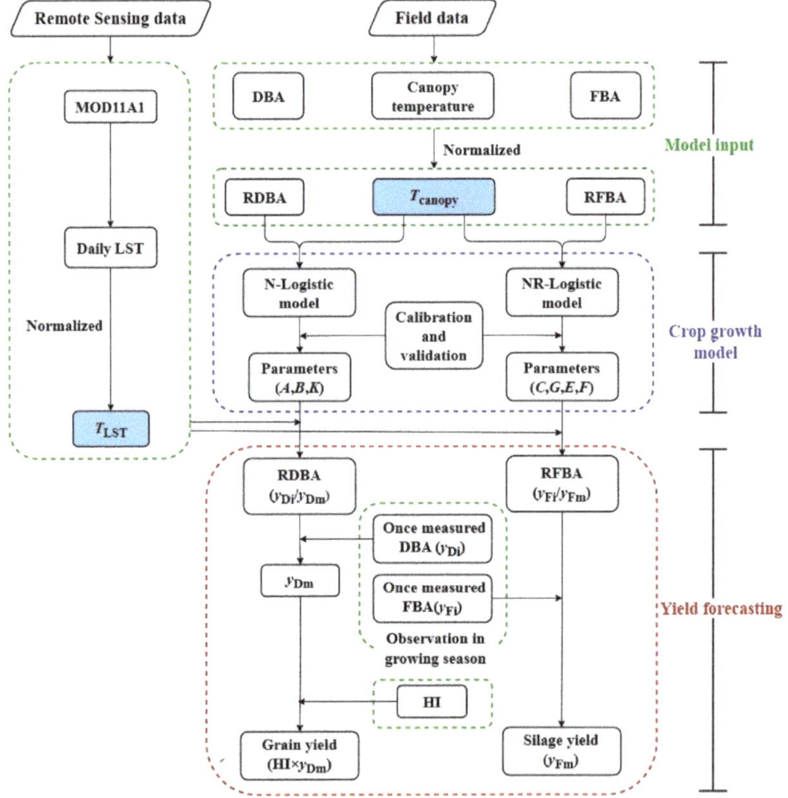

Figure 4. Schematic of the approach for yield forecasting using crop canopy temperature. Notes: DBA is dry biomass accumulation, kg ha^{-1}; FBA is fresh biomass accumulation, kg ha^{-1}; RDBA is relative DBA; RFBA is relative FBA; T_{canopy} represents relative effective accumulated temperature in canopy; LST is land surface temperature,°C; T_{LST} represents relative effective accumulative temperature calculated by LST; y_{Di} is DBA in the maize growing season, kg ha^{-1}; y_{Dm} is DBA at harvest, kg ha^{-1}; y_{Fi} is the above-ground FBA in the maize growing season, kg ha^{-1}; y_{Fm} represents the maximum FBA, kg ha^{-1}; HI is harvest index.

2.6. Statistical Evaluation

The index of model agreement (*d*), root mean square error (*RMSE*), relative error (*RE*), the coefficient of determination (R^2), and the coefficient of variation (*CV*) were used to evaluate the models. Model accuracy increased as the values of *d* and R^2 approached 1.0, and the values of *RMSE* and *RE* decreased. Origin Pro 9.1 software was used to calculate and fit the data to the model. Statistical analyses were performed in Microsoft Excel 2013. The calculation formulas are listed in Equations (S1)–(S5).

3. Results

3.1. Evaluating the Values of LST from MOD11A in Changchun

The regional map of LST can be obtained by remote sensing images products—MOD11A1. Afterwards, the grain yield in an area might be estimated through the validated model and retrieval values. Due to cloud cover, there were some incomplete data in MOD11A1 images from 25 May to 21 September 2017. The kriging interpolation method was used to fill the gaps in data.

It is necessary to obtain high-quality input data for precise estimates of crop yield. Therefore, the accuracy of the MOD11A1-LST retrievals was evaluated by comparison against the Tc observed in situ by the CTMS system in experimental fields. Figure 5 (Figure S3) shows the linear regressions between the LST (time of passing territory: 11:30 a.m.) and Tc (measured at 11:30 a.m.) in same pixel (sample number is 58). The R^2 values here ranged from 0.714 to 0.828, which were a little lower than the results of [40] in Hetao irrigation district of Inner Mongolia Autonomous Region. Those values of LST fused from the Landsat 8 images (30 m) and MOD11A1 data (1 km) might have more precision in large irrigation districts. Regardless, the R^2 values near 0.8 indicate that the retrieved LST directly is reliable.

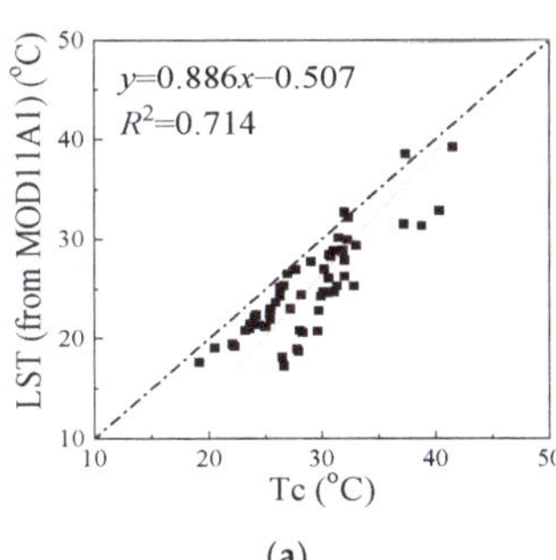

(a)

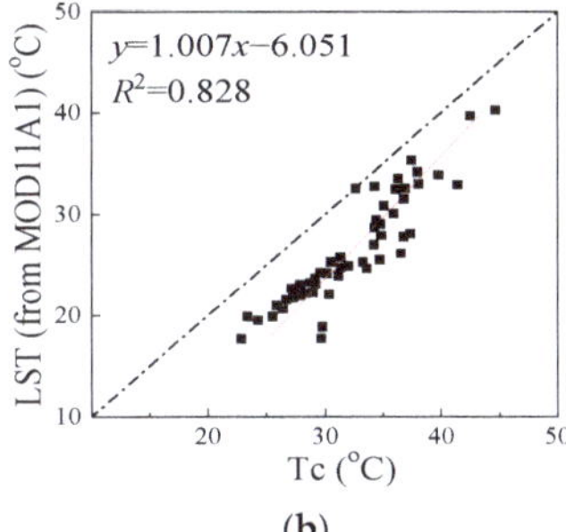

(b)

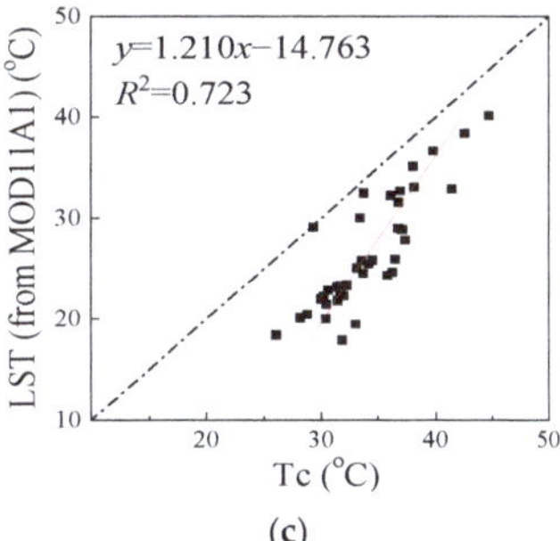

(c)

Figure 5. Regressions between the LST from MOD11A1 product and the observed Tc in field in 2017 (sample number = 58, only at local satellite transit time). (**a**) H1; (**b**) H2; (**c**) H3.

As earlier stated, independent variables (t_{canopy} and T_{canopy}) in corresponding logistic models are calculated by the daily average value of Tc. However, the retrieved LSTs from MOD11A1 represent instantaneous values in time of passing territory. There is a need to verify the feasibility of instantaneous LST values replacing daily average ones to determine T_{canopy}, when Equation (4) or Equation (6) is used in area.

Here, the relative effective accumulated temperature, calculated by instantaneous LST values of MOD11A1 (T_{LST}) at 11:30 a.m. and daily average values observed from the CTMS system, were compared during maize growth period (Figures Figure 6 and S4). The linear regression results of points indicated the strong agreement between instantaneous and daily average values to obtain T_{canopy} ($R^2 > 0.987$). The high consistency ($RMSE < 0.05$) means that the normalized LSTs from MOD11A1 can be used directly as independent variable in models as a robust approximation for the normalized Tc (daily average values). The result highlights that the normalization method in Equations (4) and (6) can eliminate the temporal-scale difference between measured daily average value and the instantaneous value inversed from remote sensing images of crop canopy temperature.

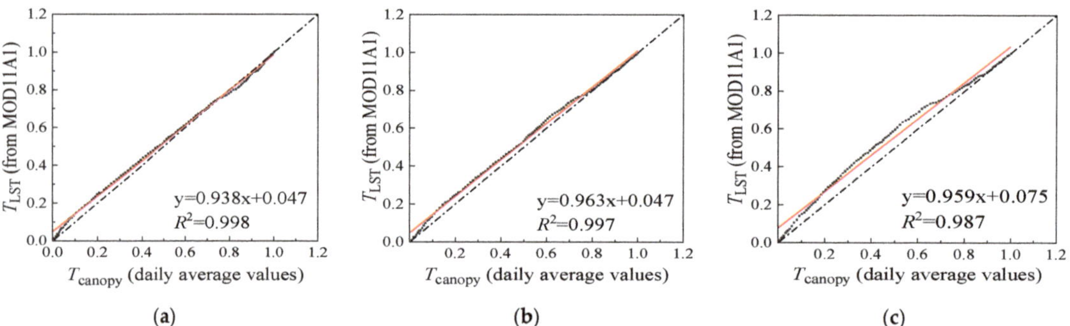

Figure 6. Regressions between the T_{LST} calculated by the remote sensing instantaneous values at 11:30 a.m. (interpolation results) and daily average values (T_{canopy}) observed from the CTMS system in 2017 (sample number = 116, with Supplemented Data). (**a**) H1; (**b**) H2; (**c**) H3.

To verify its accuracy in spatial scale, the LSTs from MOD11A1 over two days (coupled with the time of passing territory of Landsat 8) were resampled to 30 m spatial resolution. These values were used to compare with the inversed LSTs from Landsat 8 using the inversion method in the reference of [40]. The values map of *RE* between two kinds of LST are mostly between −10% and 10% (Figure 7). Such high accuracy indicated that the MOD11A1-LST was reliable to be used to simulate maize growth and estimate the forthcoming yield.

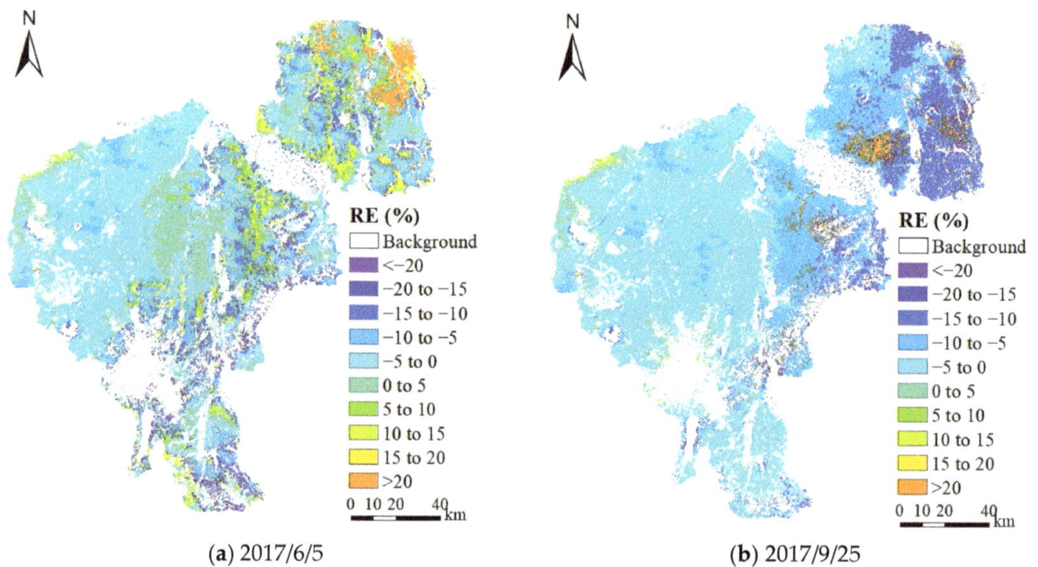

Figure 7. Maps of *RE* values of LST (30 m) between Landsat 8 and MOD11A1 resample products.

3.2. Grain Yield Forecasting in Changchun

3.2.1. Calibration Results Based on the Logistic Model of DBA

Achieving high-quality estimates of DBA is necessary to predict crop yield. Using the logistic model, all of the DBA changes were simulated based on the field observations from

2017–2019. The simulations with four kinds of effective accumulated temperatures ran well, with R^2 average values exceeding 0.95 for five plots (Figure 8a). Figure 8b shows the results of DBA simulating in 2017 of five plots by using t_{canopy} as model input. Each curve is extremely consistent with the measured values, revealing that it is feasible to realize crop growth monitoring by utilizing the Tc.

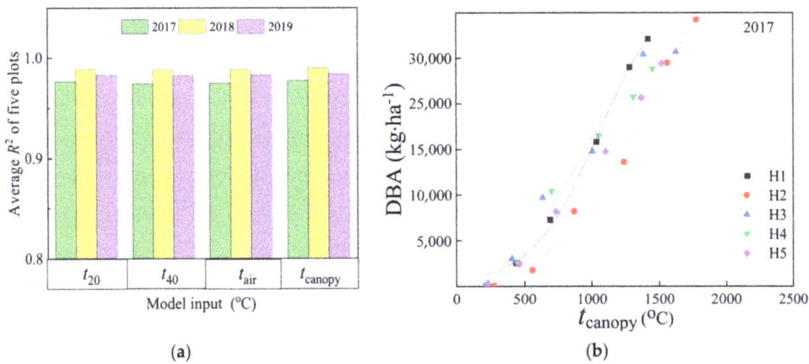

Figure 8. The performance of the DBA simulation results based on the logistic models. (**a**) Average values of R^2 of DBA simulating at five plots based on the logistic model with four kinds of effective accumulated temperature in 2017, 2018, and 2019; (**b**) DBA simulating in five plots based on the logistic model with effective accumulated canopy temperature (t_{canopy}) in 2017.

However, model parameters (*a*, *b*, and *k*) calibrated in different plots represent an obvious discrepancy, as shown in Figure 9, which provides the *CV* values for each parameter among five plots in different years. The *CV* values of *a* and *b* fluctuate significantly more than the *k* value. In addition, the t_{canopy}-based coefficients present as more stable due to lower *CV* values. Apparently, it is still hard to select universal model coefficients that best-simulate regional DBA owing to the existing variation in different plots and years.

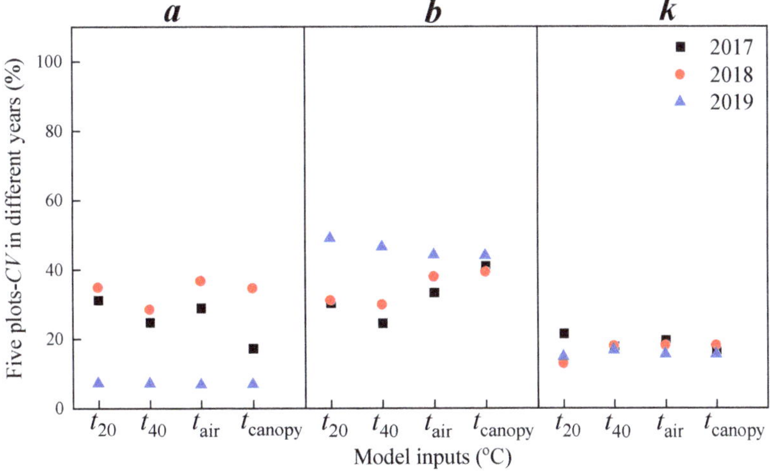

Figure 9. The *CV* values for each logistic model parameter (*a*, *b*, *k*) with four inputs (t_{20}, t_{40}, t_{air}, t_{canopy}) among five plots in 2017–2019.

3.2.2. Calibration Results Based on the N-Logistic Model of RDBA

To address this issue, the N-logistic model was employed with different relative effective accumulated temperatures for the raw data from all the plots in 2017, 2018, and 2019 separately. The simulations for 2018 show the RDBA changes with T_{20}, T_{40}, T_{air}, and T_{canopy} (Figure 10), in which high values of R^2 (>0.98) suggest that it is feasible to simulate crop growth with good accuracy in regions when the data are normalized.

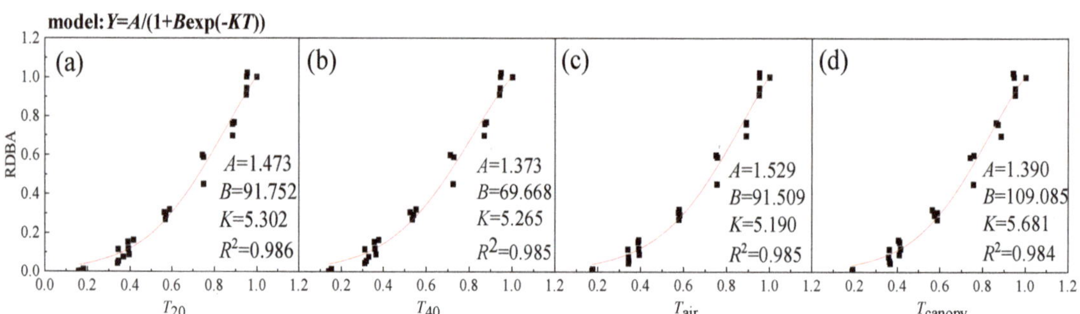

Figure 10. Simulations of RDBA based on the N-logistic model with four inputs from all plots in 2018: (**a**) T_{20}; (**b**) T_{40}; (**c**) T_{air}; (**d**) T_{canopy}.

The calibrated results of the N-logistic model parameters with T_{20}, T_{40}, T_{air}, and T_{canopy} in 2017–2019 presented the inter-annual differences of the model parameters, in which the *CV* values of *A* and *K* were relatively lower (Table 1). The results for T_{canopy} performed better than other variables with lower *CV* values, suggesting that it is a good indicator to include.

Table 1. Calibration results and inter-annual differences of the N-logistic model parameters with T_{20}, T_{40}, T_{air}, and T_{canopy} in 2017–2019.

	A				*B*				*K*			
Year	T_{20}	T_{40}	T_{air}	T_{canopy}	T_{20}	T_{40}	T_{air}	T_{canopy}	T_{20}	T_{40}	T_{air}	T_{canopy}
2017	1.244	1.193	1.248	1.163	33.090	27.140	31.940	41.131	4.863	4.884	4.825	5.465
2018	1.473	1.373	1.529	1.390	91.752	69.668	91.509	109.085	5.302	5.265	5.190	5.681
2019	1.041	1.010	1.056	1.056	43.966	38.916	46.139	58.246	5.905	6.091	5.830	6.111
CV	0.173	0.152	0.186	0.142	0.555	0.485	0.550	0.509	0.098	0.114	0.096	0.057

3.2.3. Validation Results Based on the N-Logistic Model of RDBA

To account for the inter-annual gap in the parameters of the N-logistic models, the calibrated models above were validated by the field observations of the other two years to identify the ideal set of regional parameters. A summary of the statistical characters of the N-logistic models with T_{20}, T_{40}, T_{air}, and T_{canopy} is presented in Table 2.

For the calibrated models in 2017, the measured and predicted values were in better agreement in 2018 than in 2019 because of high values for *d* and R^2 and low values of *RMSE* in 2018. The validation results for 2018 models were better in 2017 than in 2019. Likewise, the validation results in 2017 were better than in 2018 for the calibrated models in 2019. A comparison of all of the validation results showed that the statistical characters performed best in the calibrated models in 2019, with lower *RMSE* and *RE* and higher *d* and R^2.

For the simulations based on T_{canopy}, there were no large differences compared with T_{20}, T_{40}, and T_{air} (Table 2), suggesting that it is feasible to simulate RDBA during the growing season.

Table 2. Validation results of the N-logistic model of RDBA with T_{20}, T_{40}, T_{air}, and T_{canopy} between the simulated and observed data from five plots in 2017–2019.

Calibrated Model	Independent Variable	*RMSE*	*d*	R^2	*RE* (%)	*RMSE*	*d*	R^2	*RE* (%)
		Validation by field data of 2018				Validation by field data of 2019			
In 2017	T_{20}	0.094	0.984	0.978	6.8	0.168	0.942	0.937	4.7
	T_{40}	0.093	0.984	0.979	6.8	0.168	0.942	0.939	5.0
	T_{air}	0.098	0.982	0.976	7.3	0.170	0.941	0.946	5.7
	T_{canopy}	0.101	0.981	0.971	7.3	0.169	0.942	0.947	6.0
		Validation by field data of 2017				Validation by field data of 2019			
In 2018	T_{20}	0.099	0.983	0.951	−5.4	0.114	0.974	0.907	−3.7
	T_{40}	0.098	0.983	0.952	−5.3	0.111	0.974	0.909	−3.4
	T_{air}	0.104	0.981	0.948	−6.0	0.107	0.977	0.917	−3.1
	T_{canopy}	0.112	0.978	0.938	−6.2	0.103	0.978	0.921	−2.8
		Validation by field data of 2017				Validation by field data of 2018			
In 2019	T_{20}	0.068	0.991	0.969	−1.3	0.096	0.994	0.963	4.8
	T_{40}	0.069	0.991	0.968	−1.4	0.094	0.994	0.963	4.6
	T_{air}	0.071	0.990	0.969	−2.3	0.090	0.995	0.965	4.2
	T_{canopy}	0.079	0.988	0.963	−2.8	0.085	0.996	0.968	3.7

3.2.4. Grain Yield Forecasting in Area by MOD11A1-LST Values

Based on the results above, the validated N-logistic model for 2019 in Table 1 was used to simulate the pattern of RDBA in Changchun with T_{canopy}, which was supposed to equal the T_{LST} derived by the daily LST from MOD11A1. The y_{Dm} (DBA at harvest) was ascertained by incorporating at least once-measured DBA (y_D) in field through the growing season, and final grain yield could be forecasted then by the HI map. Here, the data in 2017 were used as an example to calculate and simulate to compare due to the limitation of research conditions and field observations.

Figure 11 demonstrates the spatial grain yield forecasted based on the field monitoring DBA of three different days in the growth period. Assimilating the DBA observation on 16 July into the model, the forecasted final grain yield was approximately 9750–10,500 kg ha^{-1} in most regions (Figure 11a). However, these values were 10,500–11,250 kg ha^{-1} and 12,000 kg ha^{-1} while assimilating the DBA observed on 10 August and 31 August (Figure 11b,c), respectively. As expected, assimilating closer to the harvest date enables the yield prediction to reach greater values, coinciding with the trend in crop growing.

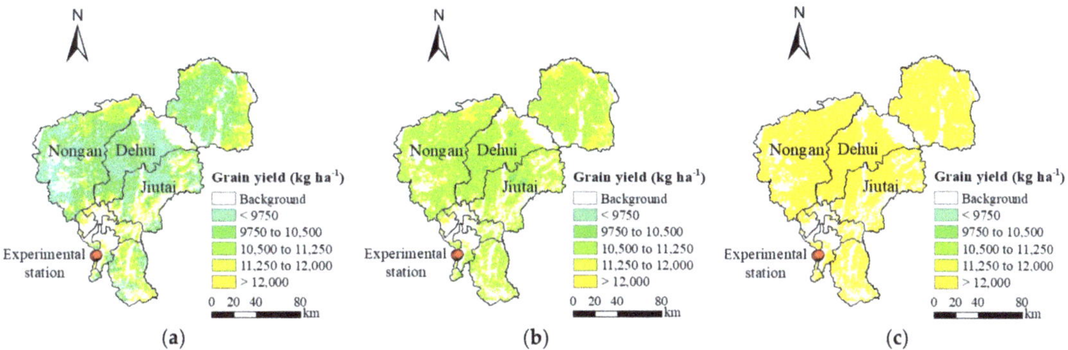

Figure 11. Forecasting results of grain yield using the N-logistic model calibrated in 2019 based on the field observations in three different days: (**a**) 2017/7/16; (**b**) 2017/8/10; (**c**) 2017/8/31.

These predicted results were compared with the measured maize grain yield from the reference of An [49], which provided an average value of grain yield of 11,364.3 kg ha^{-1}

in three subareas of Dehui, Jiutai, and Nongan (Table 3). Simulated average values in the same regions in Figure 11 were 10,126.2 kg ha^{-1}, 10,885.35 kg ha^{-1}, and 13,492.8 kg ha^{-1}, with corresponding *RE* values of −10.89%, −4.21%, and 18.73%, respectively. In addition, the simulated results were 10,778.57 kg ha^{-1}, 10,976.90 kg ha^{-1}, and 13,501.05 kg ha^{-1} based on the DBA values in three days at the experimental site, respectively. When such values were compared with the field yield observations (12,442.74 kg ha^{-1} on average here), then the values of *RE* were −13.38%, −11.78%, and 8.51% (Table 3), correspondingly. The yield prediction results confirm that the forecasting would be more precise along with the acquisition date of once-measured DBA closer to the harvest date.

Table 3. Comparisons between the forecasted grain yields based on the field observations in different days and the measurements [1] in three subareas and experimental station.

	Observation Date of Model Simulation Based on	Measured Data in Experimental Station (kg ha^{-1}) [2]	Forecasting Results (kg ha^{-1})	*RE* (%)	Measured Data in Three Subareas (kg ha^{-1}) [3]	Forecasting Results (kg ha^{-1})	*RE* (%)
Grain yield	198 (2017/7/16)	12,442.74	10,778.57	−13.38	11,364.30	10,126.20	−10.89
	223 (2017/8/10)		10,976.90	−11.78		10,885.35	−4.21
	244 (2017/8/31)		13,501.05	8.51		13,492.80	18.73

[1] Sample numbers in experimental station and three subareas are 13 and 30, respectively. [2] The data were measured on 2017/9/17. [3] The data were measured on 2017/10/1.

It is significant for the accuracy of estimated results to choose the sampling time of DBA when the N-logistic model is used to estimate the grain yield in regions. In theory, the closer sampling time is to harvest time, the more accurate the yield estimate is. However, from the viewpoint of practical application, it is preferable to estimate grain yield early in the growth stage so as to promptly adjust the irrigation and agronomic management according to the estimated results.

3.3. Silage Yield Forecasting in Changchun

3.3.1. Calibration Results Based on the R-Logistic Model of FBA

Utilizing Equation (5), the FBA growth patterns were simulated and calibrated by the observed field values from 2017–2019. Similarly, the FBA simulations of each plot in three years based on the R-logistic model presented a high R^2 ($R^2 > 0.95$), indicating that the R-logistic model (previously applied to LAI growth) was capable of simulating the FBA patterns (Figure 12a). The FBA curves simulated in 2017 with the R-logistic model based on t_{canopy} are shown in Figure 12b. It is apparent that the curves of FBA included an exponential increase at the beginning of growth, followed by a bell-shaped pattern around the peak period, and then a decline toward physiological maturity (similar to LAI). Among them, the disparity in H2 performance could be attributed to sampling error. The maximum silage yield occurs at the peak of this curve, indicating that this may be an ideal harvest period if only the silage yield is considered.

However, the parameters of the calibrated model varied across years and plots. Figure 13 depicts the *CV* values for each model parameter among five plots in different years using the computation method consistent with Figure 9. Apparently, the *CV* values of all coefficients in R-logistic models appear to be higher than those in Figure 9. The reason for this result may be that the occurring time of maximum FBA (silage yield) is harder to pin down since farmers usually harvest silage maize in advance of full maturity.

In brief, the calibrated model by ontogenetic growth data struggles to explain regional maize growth because of the variation in model parameters between years and plots. Therefore, it is necessary to determine a set of universal model parameters for depicting maize growth in large areas.

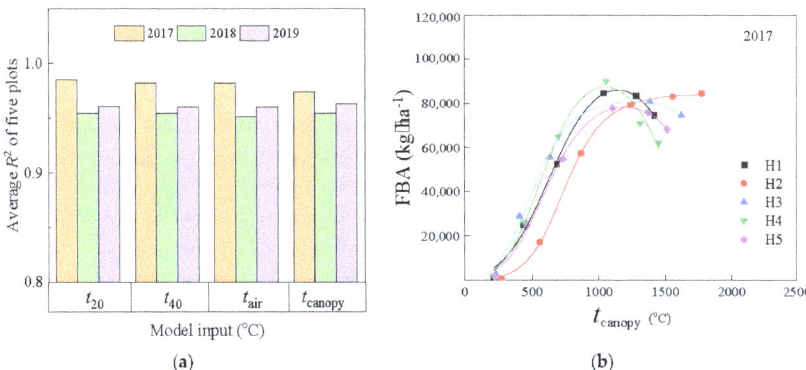

Figure 12. The performance of the FBA simulation results based on the R-logistic models. (**a**) Average R^2 values of FBA simulating at five plots based on the R-logistic model with four kinds of effective accumulated temperature in 2017–2019; (**b**) FBA simulating in five plots based on the R-logistic model with effective accumulated canopy temperature (t_{canopy}) in 2017.

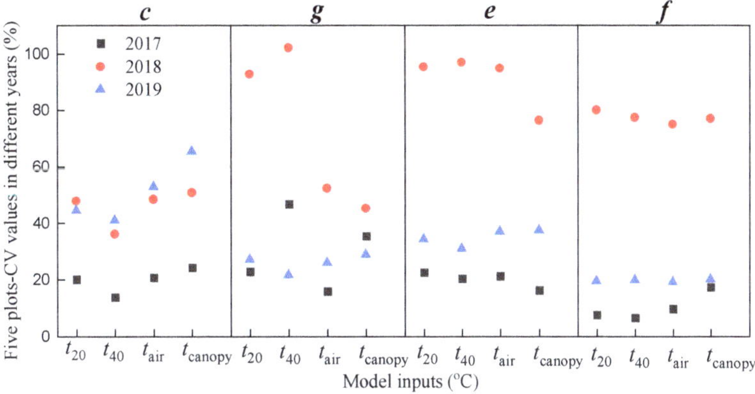

Figure 13. CV values for each R-logistic model parameter (c, g, e, f) with four inputs (t_{20}, t_{40}, t_{air}, t_{canopy}) among five plots in 2017–2019.

3.3.2. Calibration Results Based on the NR-Logistic Model of RFBA

The NR-logistic model was used to simulate the RFBA and to address the issues of regional application. All of the RFBA simulations were based on T_{20}, T_{40}, T_{air}, and T_{canopy} for the raw data from 2017, 2018, and 2019 separately. During the results for 2017, the RFBA growth curve climbed to a peak and subsequently declined as the relative effective accumulated temperature increased (Figure 14). The values of R^2 (>0.94) imply that it is acceptable to simulate RFBA in the research area with the model calibrated by the relative effective accumulated temperature. Meanwhile, no significant differences in R^2 were found for models calibrated with T_{20}, T_{40}, T_{air}, and T_{canopy}.

The RFBA and different relative effective accumulated temperatures in the five plots were used to calibrate the NR-logistic model each year. The calibration results for the model parameters with T_{20}, T_{40}, T_{air}, and T_{canopy} are displayed in Table 4. The CV values in the model parameters G, E, and F were relatively lower than C, indicating different interannual variations in different parameters. The comparison of the parameters derived by different independent variables shows that the yearly gap of T_{canopy} was lower, and the CV values of C, G, E, and F were 0.235, 0.047, −0.045, and 0.105, respectively.

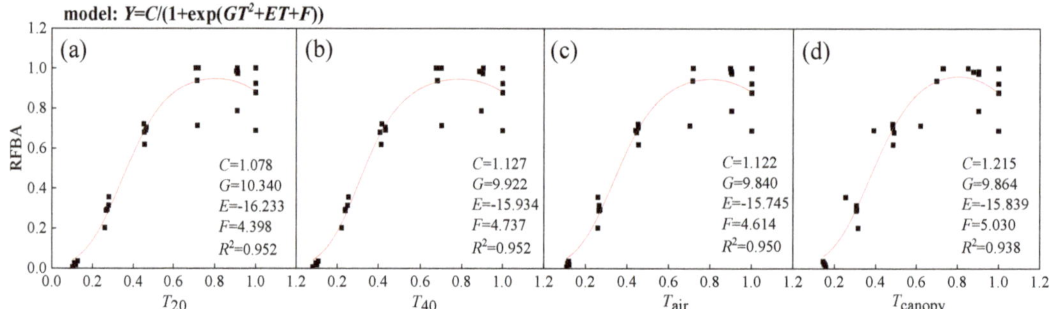

Figure 14. Simulations of RFBA based on the NR-logistic model with four relative effective accumulated temperatures from all plots in 2017: (**a**) T_{20}; (**b**) T_{40}; (**c**) T_{air}; (**d**) T_{canopy}.

Table 4. Calibration results and inter-annual differences of NR-logistic model parameters with T_{20}, T_{40}, T_{air}, and T_{canopy} in 2017–2019.

	Independent Variable	**2017**	**2018**	**2019**	***CV***
C	T_{20}	1.127	2.098	1.270	0.350
	T_{40}	1.078	1.703	1.200	0.250
	T_{air}	1.122	2.086	1.276	0.346
	T_{canopy}	1.215	1.848	1.306	0.235
G	T_{20}	9.922	9.299	10.335	0.053
	T_{40}	10.340	9.713	10.820	0.054
	T_{air}	9.840	8.962	10.119	0.063
	T_{canopy}	9.864	10.375	10.845	0.047
E	T_{20}	−15.934	−14.554	−16.214	−0.057
	T_{40}	−16.233	−14.855	−16.653	−0.059
	T_{air}	−15.745	−14.04	−16.006	−0.070
	T_{canopy}	−15.839	−16.240	−17.276	−0.045
F	T_{20}	4.737	5.797	5.141	0.102
	T_{40}	4.398	5.341	4.911	0.097
	T_{air}	4.614	5.603	5.152	0.097
	T_{canopy}	5.030	6.212	5.774	0.105

3.3.3. Validation Results Based on the NR-Logistic Model of RFBA

Each NR-logistic model was validated by field observations of the other two years in order to test its performance in providing estimates of RFBA. The agreement between the measured and predicted values of the RFBA was evaluated via the statistical characters of *RMSE*, *RE*, R^2, and *d* (Table 5). The calibrated model for 2017 was better validated in 2019 than in 2018, with lower values of *RMSE* and *RE* and higher values of *d* and R^2. For the calibrated model in 2018, there were no differences between the validations in 2017 and 2019. However, the validation results for 2017 were better than in 2018 when using the calibrated model in 2019.

There were no extreme variations in the validated results of the calibrated model with T_{20}, T_{40}, T_{air}, and T_{canopy} each year. The calibrated model in 2019 showed the optimal simulation precision for RFBA (compared to the other two years) even though it is somewhat poorer than the homologous model of RDBA in Table 2.

Using the NR-Logistic model calibrated in 2019 with T_{20}, T_{40}, T_{air}, and T_{canopy}, a scatter plot of the predicted and measured values of RFBA in 2017 and 2018 was added to evaluate the model (Figure 15). The excellent agreement between them can be verified by the high R^2 values ($R^2 > 0.92$). Additionally, the R^2 values were very close among the results for T_{20}, T_{40}, T_{air}, and T_{canopy}. With respect to the results from 2017 (Figure 15a–d), as observed, the fitting data were evenly distributed on both sides of the 1:1 line, indicating a strong concordance between the measured and predicted RFBA. As for the results from 2018 (Figure 15e–h), the fitting data were somewhat over the 1:1 line, which showed that the RFBA was overestimated slightly. To summarize, the NR-logistic model calibrated in 2019 can predict the RFBA better in 2017 than in 2018.

Table 5. Validation results of NR-logistic model of RFBA with T_{20}, T_{40}, T_{air}, and T_{canopy} between the simulated and observed data from five plots in different years.

Calibrated Model	Independent Variable	*RMSE*	*d*	R^2	*RE* (%)	*RMSE*	*d*	R^2	*RE* (%)
		Validation by field data of 2018				Validation by field data of 2019			
In 2017	T_{20}	0.135	0.953	0.902	13.5	0.085	0.986	0.950	3.4
	T_{40}	0.139	0.950	0.899	14.0	0.088	0.985	0.951	5.3
	T_{air}	0.139	0.949	0.898	13.9	0.086	0.985	0.955	6.1
	T_{canopy}	0.130	0.957	0.907	12.8	0.087	0.985	0.954	5.9
		Validation by field data of 2017				Validation by field data of 2019			
In 2018	T_{20}	0.121	0.972	0.916	−9.9	0.111	0.976	0.936	−7.8
	T_{40}	0.123	0.971	0.914	−10.1	0.106	0.984	0.940	−6.6
	T_{air}	0.121	0.971	0.915	−9.9	0.099	0.980	0.946	−5.4
	T_{canopy}	0.120	0.972	0.915	−9.6	0.096	0.987	0.947	−4.5
		Validation by field data of 2017				Validation by field data of 2018			
In 2019	T_{20}	0.079	0.988	0.951	−0.9	0.118	0.974	0.920	11.7
	T_{40}	0.082	0.987	0.948	−1.8	0.115	0.976	0.920	11.0
	T_{air}	0.084	0.986	0.946	−2.4	0.114	0.976	0.916	10.1
	T_{canopy}	0.091	0.984	0.936	−2.4	0.110	0.979	0.918	9.3

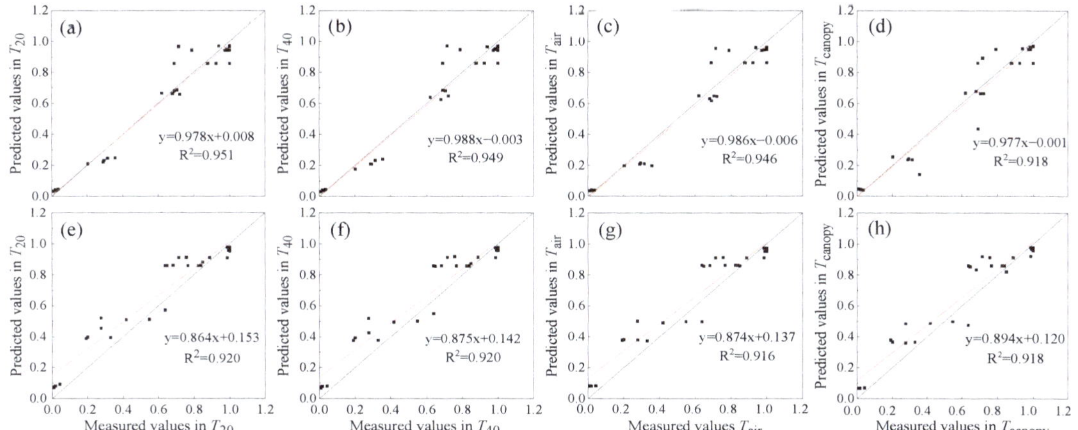

Figure 15. Regressions between the predicted and measured values of RFBA in 2017–2018 using the NR-logistic models calibrated in 2019 with four inputs: (**a**) T_{20}, (**b**) T_{40}, (**c**) T_{air}, (**d**) T_{canopy}, in 2017; (**e**) T_{20}, (**f**) T_{40}, (**g**) T_{air}, (**h**) T_{canopy}, in 2018.

The validated results with T_{20}, T_{40}, T_{air}, and T_{canopy} and the model parameters (Table 4) can be used to assess the RFBA. The selection of the model independent variable may depend only on the way to monitor temperature in situ. The model application will be more convenient if the temperature can be collected easily. However, it is important to highlight that the Tc is a good factor for scale expansion.

3.3.4. Silage Yield (Maximum FBA) Forecasting in Area by MOD11A1-LST Values

The shapes of the FBA and RFBA curves depicted above suggest that the silage yield should be near the peak of the curve to maximize profits. Therefore, the areal silage yield could be simulated by the validated Equation (6) in 2019 with T_{LST} from MOD11A1 as a substitute for T_{canopy} in combination with the field FBA observations on three different days at different growth stages (Figure 16). The differences in spatial distribution were captured. Similarly, the predicted FBA increased and got closer to the maximum output when the field observation date was near harvest.

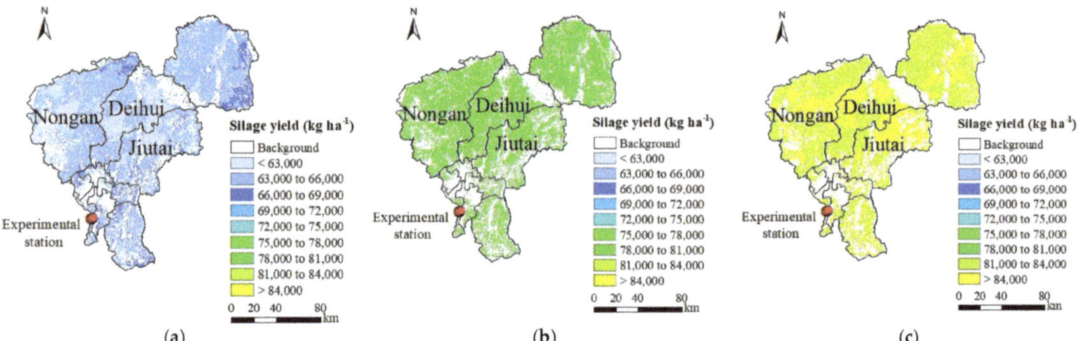

Figure 16. Predicting silage yield values using the NR-logistic model calibrated in 2019 based on the field observations in three different days: (**a**) 2017/7/16; (**b**) 2017/8/10; (**c**) 2017/8/31.

According to the summarized data in Table 6, the model simulation showed only a slight difference on the observation day (31 August) in the experimental station. This is not surprising given that the measured silage yield was recorded on the same date. The results on 10 August still produced a satisfying accuracy, which indicates that it might be an appropriate harvest date for silage yield in view of various factors. There were no measurements of silage yield in the three subareas, so the comparison could not be analyzed and displayed between observations and predictions in the region.

Table 6. Comparisons between the forecasted silage yields (maximum FBAs) based on the field observation [1] in different days and experimental station.

	Observation Date of Model Simulation Based on	Measured Data in Experimental Station (kg ha^{-1}) [2]	Forecasting Results (kg ha^{-1})	*RE* (%)
Silage yield (maximum FBA)	198 (2017/7/16)	84,605.70	65,187.70	−22.95
	223 (2017/8/10)		79,447.25	−6.10
	244 (2017/8/31)		83,715.78	−1.05

[1] Sample numbers in experimental station are 13 and 30. [2] The data were measured on 2017/8/31.

3.4. Verification in Jiefangzha Sub-Irrigation District

As mentioned in Section 2.3, an LST map (30 m) of Jiefangzha showed the results fused from Landsat 8 and MOD11A1 images by the ESTARFM algorithm. Scatters of the fused LSTs and Landsat-LSTs were evenly distributed on both sides of the 1:1 line, showing the fused LSTs were relatively reliable (Figure 17a). The fused LSTs were also compared with the Tc recorded at 11:30 a.m., in which the values of R^2 (0.547), *RMSE* (3.96°C), and *d* (0.79) indicate good consistency between the observed values and the fused ones (Figure 17b). The grain yield forecasting map was constructed by employing the N-Logistic model described previously with the fused LSTs (Figure 18).

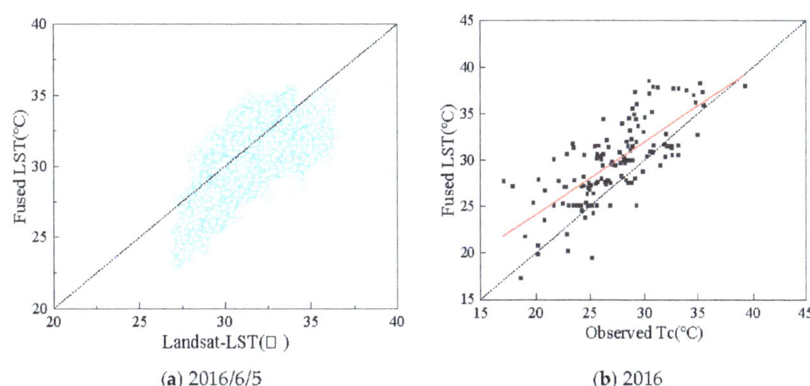

(a) 2016/6/5 (b) 2016

Figure 17. Regressions for accuracy evaluation of the fused LST: (**a**) the fused LST vs. the inversed values from Landsat 8; (**b**) the fused LST vs. the observed Tc in experimental station in 2016.

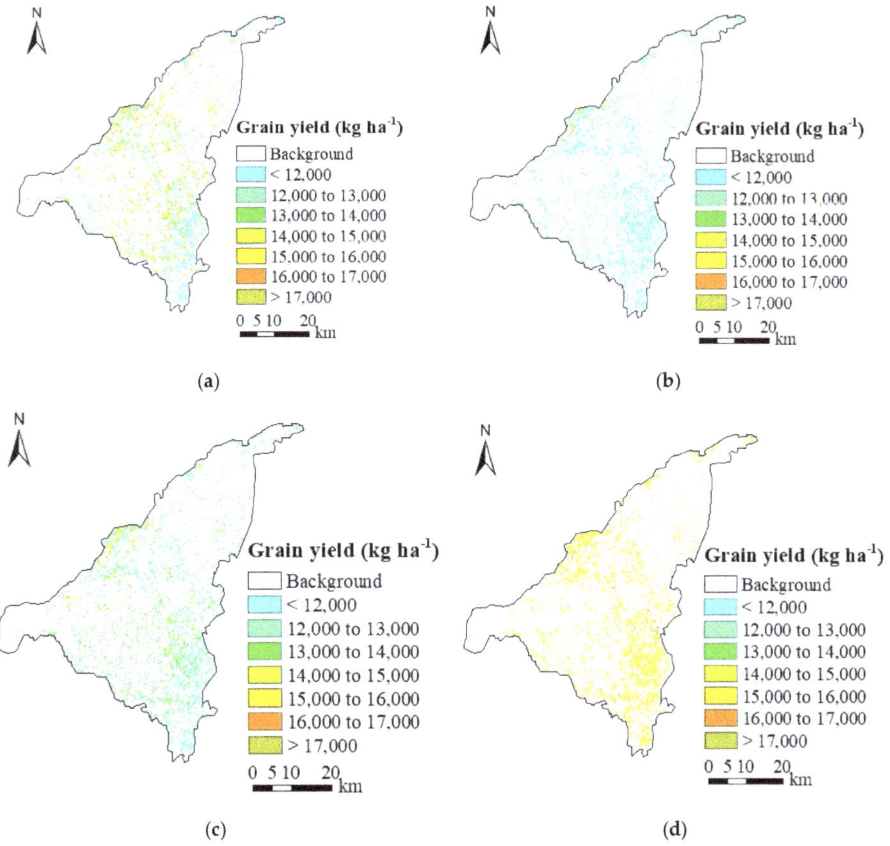

Figure 18. Forecasting results of grain yield in the Jiefangzha sub-irrigation district using the 2019 calibrated model based on the field observations in four different days: (**a**) 2016/7/4; (**b**)2016/7/21; (**c**) 2016/8/4; (**d**) 2016/8/26.

Figure 18 shows the spatial grain yield estimates retrieved using DBA measured on four acquisition dates that include (a) 4 July, (b) 21 July, (c) 4 August, and (d) 26 August in 2016 based on the model calibrated in 2019. The predicted yield showed a significant upward trend as the acquisition date of DBA approached harvest. Meanwhile, the comparisons between the predicted and measured yield in situ were conducted with statistical parameters as evaluation metrics of accuracy (Table 7). The predicted yields coincide with the measured ones with *RE* values, ranging from −16.14% to 9.84%, which confirms that assimilating once-measured data into the model can produce a great estimate of grain yield. Comparatively speaking, DBA assimilation closer to the harvest date allows for a more accurate prediction of yield except for the results based on the data on 21 July, which may be caused by the irrigation measures or the model parameters. Assimilating DBA measured on 4 July also provides a reliable result, albeit with a slightly lower R^2. Such a result proves the feasibility of the approach for early yield forecasting in other large areas though the optimal date of sampling still needs to be explored and determined further.

Table 7. Comparisons between the forecasted grain yields in the Jiefangzha sub-irrigation district based on field observation in different days and the measurements [1].

Observation Date of Model Simulation Based on	*RMSE* (kg ha^{-1})	R^2	*RE* (%)	*d*
186 (2016/7/4)	933	0.63	3.52	0.86
203 (2016/7/21)	2334	0.77	−16.14	0.56
217 (2016/8/4)	1520	0.83	9.84	0.70
239 (2016/8/26)	888	0.88	5.01	0.85

[1.] The data were measured on 2016//9/15.

4. Discussion

According to results reported above, our data on maize biomass show that the Tc should be included as a valuable proxy of other independent variables in models because the Tc can be recognized as LST in large regions covered by maize that can be derived from remote sensing data [50,51].

This study demonstrated the benefits of integrating remote sensing LST into crop growth models in combination with once-observed values (DBA or FBA) to enhance yield prediction. The introduction of remote sensing LST to the logistic models offers effective information about regional crop status, overcoming the limitation of model in regional application. Moreover, the ground crop truth was profitably considered by using once-measured observations as the drive of the yield forecasting model. Another point worth emphasizing is that the subsequent consequences, after undertaking necessary agronomic measures based on the yield forecasting results, can be assessed with just one observation as input again because the effect on the crop can be reflected by LST and measured biomass. All in all, LST as a key factor for characterizing field drought [52] provides an intuitive basis to determine the water shortage of crops in large regions. This offers a solid theoretical basis for the synergistic prediction of future crop drought and yield evaluation in combination using remote sensing technology.

In addition, to obtain a more robust estimate of maize yield, data fusion technology may be recommended to improve the spatio-temporal resolution of LST [40]. Unfortunately, this method is unavailable due to the scarce Landsat images caused by the cloud cover on most days in maize growing season in Changchun. Our attempt in the Jiefangzha sub-irrigation district indicated that the data fusion technology improved the performance of spatial variation of grain yield forecasting under available conditions of suitable Landsat images. It follows that high-resolution satellite imagery such as Sentinel-2 might be a further exploration tool for improving yield forecasting precision.

The values of t_m in models of RDBA and RFBA are key input data, which can be obtained for different hydrological years by analysis of the local yearly rainfall. When the hydrological year of the current growing season is estimated, the N-logistic model and the NR-logistic model can be utilized with the corresponding value of t_m. To obtain the values of y_{Dm} and y_{Fm}, the data of DBA and FBA should be collected in field at least once during the maize growing period. Furthermore, the growth patterns of DBA and FBA can be simulated when the models are set up and calibrated. HI, an empirical value describing the relationship between DBA and grain yield, offers an opportunity to evaluate grain yield. By combining the growth curves for DBA and FBA with the future market demands, maize can be flexibly harvested as silage or grain during the growing season.

Apparently, the early yield forecasting accuracy varies depending on the growth stages, which may be caused by many factors. Firstly, the retrieval of t_m is pivotal, as mentioned above. Secondly, the quality of remote sensing images obtained over time varies slightly, which can introduce some uncertainty into the yield forecasting results. Thirdly, the date of acquiring DBA or FBA is particularly critical for yield forecasting accuracy. As shown in Tables 3 and 7, the optimal predicting date is clearly different in different areas. In Changchun, the optimal date at which the DBA provides more accurate yield forecasting is approximately 38 days (10 August) ahead of harvest. As for the Jiefangzha sub-irrigation district, the results on 26 August produced a better prediction of yield, relatively speaking. Furthermore, the latter is better than the former in statistical parameters. This phenomenon may be caused by a variety of factors. For instance, a rare downpour amounting to more than 160 mm of rain occurred in Changchun on 21 July 2017, which may destroy plants and influence yield forecasting results. In addition, the area of Changchun is larger than the Jiefangzha sub-irrigation district, which may have an impact on the results. Taking the three results into account (Tables 3, 6 and 7), it is suggested to forecast yield by using the field data (DBA or FBA) measured at the middle growth period (early August). Furthermore, it is necessary to determine the optimal date by taking into consideration multiple factors.

Another noteworthy point about the results presented here is that the field soil water content was sufficient or at least was not in water deficit in the research area. The FBA, RFBA, and silage yield may be impacted by the crop and soil water conditions. Future efforts to prove model generality should include examination of changes in the model parameters for different levels of plant and field soil moisture.

5. Conclusions

An approach for yield forecasting from plot level to large scale was developed by incorporating remote sensing LST of a measured biological indicator (DBA and FBA) into corresponding logistic models. The main conclusions are as follows:

(1) The model of 2019 based on T_{canopy} performed better result than others. Crop canopy temperature can be used as input parameter in logistic models to simulate DBA and FBA. It is thus a potentially valuable index to facilitate model development in regions.

(2) The normalization method can eliminate the difference in temporal scale between measured daily average values of Tc and instantaneous remote sensing LSTs. Therefore, the normalized LST retrieved from MOD11A1 can be used directly as an independent variable in models to simulate crop biomass for yield forecasting in areas.

(3) The yield forecasting accuracy is reliable in regions with this approach. Satisfactory grain and silage yield forecasting in Changchun were provided by assimilating DBA or FBA measured on 10 August ahead of harvest with *RE* values of −4.21% and −6.1%, respectively.

(4) The application in the Jiefangzha sub-irrigation district demonstrated that it is possible to apply this approach to predict yield in other regions. These simulation results hold broad potential to provide a real-time reference in maize growing stages for farmers and the grain futures market to make decisions.

Supplementary Materials: The following supporting information can be downloaded at: https://www.mdpi.com/article/10.3390/rs15041025/s1, Figure S1: Rules of decision tree classification for Landsat8 images based on the values of NDVI and LSWI in ROI in 2017 (No.268 (2017/9/25) and No.156 (2017/6/5)); Figure S2: Map of HI obtained by the kriging interpolation method in Changchun area; Figure S3: Regressions between the LST from MOD11A1 product and the observed Tc in field in 2017. (**a**) H4; (**b**) H5; Figure S4: Regressions between the T_{LST} calculated by the remote sensing instantaneous LST values at 11:30 a.m. (interpolation results) and daily average values observed (T_{canopy}) from the CTMS system in 2017. (**a**) H4; (**b**) H5; Table S1: Data list of remote sensing images in 2017; Table S2: Values of harvest index (HI) of maize collected or measured in Changchun and its surrounding areas; Equations (S1)–(S5): Supplementary calculation formula of Section 2.6.

Author Contributions: Conceptualization, J.C.; data curation, H.C.; formal analysis, J.C.; funding acquisition, B.Z.; investigation, H.C.; methodology, H.C.; project administration, Z.W.; resources, J.C.; supervision, D.X.; validation, H.C.; writing—original draft, H.C.; writing—review and editing, J.C. All authors have read and agreed to the published version of the manuscript.

Funding: This research was funded by the National Key Research Program (grant number NK2022180403), the Project of National Natural Science Foundation of China (grant numbers 52130906, 51979286), and the Institute-City Cooperation Program (grant number HBAT02242202010-CG).

Data Availability Statement: The data that support the findings of this work are available from the corresponding author upon reasonable request.

Acknowledgments: The authors would like to thank the anonymous reviewers for their long-term guidance and constructive comments. The authors are grateful to the Ministry of Natural Resources of China for providing the land-cover map in Changchun and the Statistic Bureau of Jilin Province for providing the statistical data of crop areas. The authors also acknowledge the United States Geological Survey (USGS) for offering the images of Landsat 8 and MOD11A1 products.

Conflicts of Interest: The authors declare no conflict of interest.

Abbreviations

LST	Land surface temperature, °C
Tc	Canopy temperature, °C
DBA	Dry biomass accumulation, kg ha^{-1}
FBA	Fresh biomass accumulation, kg ha^{-1}
LAI	Leaf area index
RDBA	Relative dry biomass accumulation
HI	Harvest index
RFBA	Relative fresh biomass accumulation
Fc	Field capacity
Wp	Wilting point
CTMS	Canopy temperature and meteorology monitoring systems
NDVI	Normalized difference vegetation index
LSWI	Land surface water index
ROI	Region of interest
p_{NIR}	Reflectivity of near-infrared band
p_{RED}	Reflectivity of red band
p_{SWIR}	Reflectivity of shortwave infrared band
y_D	Dependent growth parameter
t	Effective accumulated temperature after emergence, °C
t_i	Mean daily temperature in the air, canopy, or soil at 20 cm or 40 cm in the root zone, °C
a	The theoretical upper limit of growth of dry biomass accumulation
b,k	Parameters of the logistic model
t_{air}	Effective accumulative air temperature, °C
t_{canopy}	Effective accumulative canopy temperature, °C
t_{20}	Effective accumulative soil temperature at 20 cm in root zone, °C
t_{40}	Effective accumulative soil temperature at 40 cm in root zone, °C

T	Relative effective accumulated temperature
Y_D	Relative dry biomass accumulation
y_{Dm}	Dry biomass accumulation at harvest, kg ha^{-1}
t_m	Effective accumulative temperature at harvest, °C
A	The upper limit of relative dry biomass accumulation
B, K	Parameters of the normalized logistic model
T_{20}	Relative effective accumulative soil temperature at 20 cm in root zone
T_{40}	Relative effective accumulative soil temperature at 40 cm in root zone
T_{canopy}	Relative effective accumulative canopy temperature
T_{air}	Relative effective accumulative air temperature
y_F	Above-ground fresh biomass accumulation, kg ha^{-1}
c, g, e, f	Parameters of the revised logistic model
Y_F	Relative fresh biomass accumulation
y_{Fm}	Maximum relative fresh biomass accumulation
C, G, E, F	Parameters of the normalized revised logistic model
Y	Grain yield, kg ha^{-1}
T_{LST}	The relative effective accumulative canopy temperature calculated by the remote sensing instantaneous values of MOD11A1
d	Index of agreement
$RMSE$	Root mean square error
RE	Relative error
R^2	Coefficient of determination
CV	Coefficient of variation

References

1. Chen, Y.; Tao, F.L. Potential of remote sensing data-crop model assimilation and seasonal weather forecasts for early-season crop yield forecasting over a large area. *Field Crop. Res.* **2022**, *276*, 108398. [CrossRef]
2. Ziliani, M.G.; Altaf, M.U.; Aragon, B.; Houborg, R.; Franz, T.E.; Lu, Y.; Sheffield, J.; Hoteit, I.; McCabe, M.F. Early season prediction of within-field crop yield variability by assimilating CubeSat data into a crop model. *Agric. For. Meteorol.* **2022**, *313*, 108736. [CrossRef]
3. Basso, B.; Liu, L. Chapter Four—Seasonal crop yield forecast: Methods, applications, and accuracies. *Adv. Agron.* **2019**, *154*, 201–255. [CrossRef]
4. Liu, Y.Q.; Song, W. Modelling crop yield, water consumption, and water use efficiency for sustainable agroecosystem management. *J. Clean. Prod.* **2020**, *253*, 119940. [CrossRef]
5. Paudel, D.; Boogaard, H.; Wit, A.D.; Janssen, S.; Osinga, S.; Pylianidis, C.; Athanasiadis, I.N. Machine learning for large-scale crop yield forecasting. *Agric. Syst.* **2021**, *187*, 103016. [CrossRef]
6. Hoogenboom, G. Contribution of agrometeorology to the simulation of crop production and its applications. *Agric. For. Meteorol.* **2000**, *103*, 137–157. [CrossRef]
7. Liu, S.; Yang, J.Y.; Drury, C.F.; Liu, H.L.; Reynolds, W.D. Simulating maize (*Zea mays* L.) growth and yield, soil nitrogen concentration, and soil water content for a long-term cropping experiment in Ontario, Canada. *Can. J. Soil Sci.* **2014**, *94*, 435–452. [CrossRef]
8. Mubeen, M.; Ahmad, A.; Wajid, A.; Khalip, T.; Hammad, H.M.; Sultana, S.R.; Ahmad, S.; Fahad, S.; Nasim, W. Application of CSM-CERES-Maize model in optimizing irrigated conditions. *Outlook Agric.* **2016**, *45*, 173–184. [CrossRef]
9. Wu, W.; Chen, J.L.; Liu, H.B.; Garcia, A.G.; Hoogenboom, G. Parameterizing soil and weather inputs for crop simulation models using the VEMAP database. *Agric. Ecosyst. Environ.* **2010**, *135*, 111–118. [CrossRef]
10. Birch, C.P.D. A new generalized Logistic sigmoid growth equation compared with the Richards growth equation. *Ann. Bot.* **1999**, *83*, 713–723. [CrossRef]
11. West, G.B.; Brown, J.H.; Enquist, B.J. A general model for ontogenetic growth. *Nature* **2001**, *413*, 628–631. [CrossRef]
12. Bontemps, J.; Duplat, P. A non-asymptotic sigmoid growth curve for top height growth in forest stands. *Forestry* **2012**, *85*, 353–368. [CrossRef]
13. Lewis, J.M.; Lakshmivarahan, S.; Dhall, S. *Dynamic Data Assimilation: A Least Squares Approach*; Cambridge University Press: Cambridge, UK, 2006; Available online: http://www.gbv.de/dms/goettingen/508439248.pdf (accessed on 16 March 2022).
14. Schwalbert, R.A.; Amado, T.J.C.; Nieto, L.; Varela, S.; Corassa, G.M.; Horbe, T.A.N.; Rice, C.W.; Peralta, N.R.; Ciampitti, I.A. Forecasting maize yield at field scale based on high-resolution satellite imagery. *Biosyst. Eng.* **2018**, *171*, 179–192. [CrossRef]
15. Sharifi, A. Yield prediction with machine learning algorithms and satellite images. *J. Sci. Food Agric.* **2021**, *101*, 891–896. [CrossRef]
16. Veloso, A.; Mermoz, S.; Bouvet, A.; Toan, T.L.; Planells, M.; Dejoux, J.-F.; Ceschia, E. Understanding the temporal behavior of crops using Sentinel-1 and Sentinel-2-like data for agricultural applications. *Remote Sens. Environ.* **2017**, *199*, 415–426. [CrossRef]
17. Gaso, D.V.; Wit, A.D.; Berger, A.G.; Kooistra, L. Predicting within-field soybean yield variability by coupling Sentinel-2 leaf area index with a crop growth model. *Agric. For. Meteorol.* **2021**, *308–309*, 108553. [CrossRef]

18. Jin, N.; Tao, B.; Ren, W.; He, L.; Zhang, D.Y.; Wang, D.H.; Yu, Q. Assimilating remote sensing data into a crop model improves winter wheat yield estimation based on regional irrigation data. *Agric. Water Manag.* **2022**, *266*, 107583. [CrossRef]
19. Liu, Z.C.; Wang, C.; Bi, R.T.; Zhu, H.F.; He, P.; Jing, Y.D.; Yang, W.D. Winter wheat yield estimation based on assimilated Sentinel-2 images with the CERES-Wheat model. *J. Integr. Agric.* **2021**, *20*, 1958–1968. [CrossRef]
20. Zhuo, W.; Huang, J.X.; Xiao, X.M.; Huang, H.; Bajgain, R.; Wu, X.C.; Gao, X.R.; Wang, J.; Li, X.C.; Wagle, P. Assimilating remote sensing-based VPM GPP into the WOFOST model for improving regional winter wheat yield estimation. *Eur. J. Agron.* **2022**, *139*, 126556. [CrossRef]
21. Junior, I.M.F.; Vianna, M.D.S.; Marin, F.R. Assimilating leaf area index data into a sugarcane process-based crop model for improving yield estimation. *Eur. J. Agron.* **2022**, *136*, 126501. [CrossRef]
22. Klopfenstein, T.J.; Erickson, G.E.; Berger, L.L. Maize is a critically important source of food, feed, energy and forage in the USA. *Field Crop. Res.* **2013**, *153*, 5–11. [CrossRef]
23. Yan, D.C.; Zhu, Y.; Wang, S.H.; Cao, W.X. A quantitative knowledge-based model for designing suitable growth dynamics in rice. *Plant Prod. Sci.* **2006**, *9*, 93–105. [CrossRef]
24. Sheehy, J.E.; Mitchel, P.L.; Allen, L.H.; Ferrer, A.B. Mathematical consequences of using various empirical expressions of crop yield as a function of temperature. *Field Crop. Res.* **2006**, *98*, 216–221. [CrossRef]
25. Singer, J.W.; Meek, D.W.; Sauer, T.J.; Prueger, J.H.; Hatfield, J.L. Variability of light interception and radiation use efficiency in maize and soybean. *Field Crop. Res.* **2011**, *121*, 147–152. [CrossRef]
26. Shi, P.J.; Men, X.Y.; Sandhu, H.S.; Chakraborty, A.; Li, B.L.; Ou-Yang, F.; Sun, Y.C.; Ge, F. The "general" ontogenetic growth model is inapplicable to crop growth. *Ecol. Model.* **2013**, *266*, 1–9. [CrossRef]
27. Bakoglu, A.; Celik, S.; Kokten, K.; Kilic, O. Examination of plant length, dry stem and dry leaf weight of bitter vetch [*Vicia ervilia* (L) Willd.] with some non-linear growth models. *Legume Res.* **2016**, *39*, 533–542. [CrossRef]
28. Wang, X.L. How to utilize Logistic model in dynamic simulation of crop dry biomass accumulation. *Chin. J. Agrometeorol.* **1986**, *7*, 14–19. (In Chinese)
29. Liu, Y.H.; Su, L.J.; Wang, Q.J.; Zhang, J.H.; Shan, Y.Y.; Deng, M.J. Chapter Six—Comprehensive and quantitative analysis of growth characteristics of winter wheat in China based on growing degree days. *Adv. Agron.* **2020**, *159*, 237–273. [CrossRef]
30. Elings, A. Estimation of leaf area in tropical maize. *Agron. J.* **2000**, *92*, 436–444. [CrossRef]
31. Yu, Q.; Liu, J.D.; Zhang, Y.Q.; Li, J. Simulation of rice biomass accumulation by an extended Logistic model including influence of meteorological factors. *Int. J. Biometeorol.* **2002**, *46*, 185–191. [CrossRef]
32. Sepaskhah, A.R.; Fahandezh-Saadi, S.; Zand-Parsa, S. Logistic model application for prediction of maize yield under water and nitrogen management. *Agric. Water Manag.* **2011**, *99*, 51–57. [CrossRef]
33. Shabani, A.; Sepaskhah, A.R.; Kamgar-Haghighi, A.A. Estimation of yield and dry matter of rapeseed using Logistic model under water, salinity and deficit irrigation. *Arch. Agron. Soil Sci.* **2014**, *60*, 951–969. [CrossRef]
34. Mahbod, M.; Sepaskhah, A.R.; Zand-Parsa, S. Estimation of yield and dry matter of winter wheat using Logistic model under different irrigation water regimes and nitrogen application rates. *Arch. Agron. Soil Sci.* **2014**, *60*, 1661–1676. [CrossRef]
35. Mayer, F.; Gerin, P.A.; Noo, A.; Foucart, G.; Flammang, J.; Lemaigre, S.; Sinnaeve, G.; Dardenne, P.; Delfosse, P. Assessment of factors influencing the biomethane yield of maize silages. *Bioresour. Technol.* **2014**, *153*, 260–268. [CrossRef] [PubMed]
36. Pede, T.; Mountrakis, G.; Shaw, S.B. Improving corn yield prediction across the US Corn Belt by replacing air temperature with daily MODIS land surface temperature. *Agric. For. Meteorol.* **2019**, *276–277*, 107615. [CrossRef]
37. Cai, J.B.; Xu, D.; Si, N.; Wei, Z. Real-time monitoring system of crop canopy temperature and soil moisture for irrigation decision-making. *T. Chin. Soc. Agric. Mach.* **2015**, *46*, 133–139, (In Chinese with English abstract). [CrossRef]
38. Qi, L.L. Research on Safety of Water Supply in Chang-Ji Economic Circle. Master's Thesis, Jilin University, Changchun, China, 2019. (In Chinese with English abstract).
39. Bai, L.L.; Cai, J.B.; Liu, Y.; Chen, H.; Zhang, B.Z.; Huang, L.X. Responses of field evapotranspiration to the changes of cropping pattern and groundwater depth in large irrigation district of Yellow River basin. *Agric. Water Manag.* **2017**, *188*, 1–11. [CrossRef]
40. Huang, L.X.; Cai, J.B.; Zhang, B.Z.; Chen, H.; Bai, L.L.; Wei, Z.; Peng, Z.G. Estimation of evapotranspiration using the crop canopy temperature at field to regional scales in large irrigation district. *Agric. For. Meteorol.* **2019**, *269–270*, 305–322. [CrossRef]
41. Xiao, X.M.; Zhang, Q.Y.; Braswell, B.; Urbanski, S.; Boles, S.; Wofsy, S.; Moore, B.; Ojima, D. Modeling gross primary production of a deciduous broadleaf forest using satellite images and climate data. *Remote Sens. Environ.* **2004**, *91*, 256–270. [CrossRef]
42. Chandrasekar, K.; Sesha Sai, M.V.R.; Roy, P.S.; Dwevedi, R.S. Land Surface Water Index (LSWI) response to rainfall and NDVI using the MODIS Vegetation Index product. *Int. J. Remote Sens.* **2010**, *31*, 3987–4005. [CrossRef]
43. Zhu, X.L.; Chen, J.; Gao, F.; Chen, X.H.; Masek, J.G. An enhanced spatial and temporal adaptive reflectance fusion model for complex heterogeneous regions. *Remote Sens. Environ.* **2010**, *114*, 2610–2623. [CrossRef]
44. Wu, R.L.; Ma, C.-X.; Chang, M.; Littell, R.C.; Wu, S.S.; Yin, T.M.; Huang, M.R.; Wang, M.X.; Casella, G. A Logistic mixture model for characterizing genetic determinants causing differentiation in growth trajectories. *Genet. Res.* **2002**, *79*, 235–245. [CrossRef] [PubMed]
45. Zhao, J.F.; Guo, J.P.; Mu, J. Exploring the relationships between climatic variables and climate-induced yield of spring maize in Northeast China. *Agric. Ecosyst. Environ.* **2015**, *207*, 79–90. [CrossRef]

46. Ding, D.Y.; Feng, H.; Zhao, Y.; Hill, R.L.; Yan, H.M.; Chen, H.X.; Hou, H.J.; Chu, X.S.; Liu, J.C.; Wang, N.J.; et al. Effects of continuous plastic mulching on crop growth in a winter wheat-summer maize rotation system on the loess plateau of China. *Agric. For. Meteorol.* **2019**, *271*, 385–397. [CrossRef]
47. Meade, K.A.; Cooper, M.; Beavis, W.D. Modeling biomass accumulation in maize kernels. *Field Crop. Res.* **2013**, *151*, 92–100. [CrossRef]
48. Liu, Y.; Li, Y.F.; Li, J.S.; Yan, H.J. Effects of mulched drip irrigation on water and heat conditions in field and maize yield in sub-humid region of Northeast China. *T. Chin. Soc. Agric. Mach.* **2015**, *46*, 93–104, 135, (In Chinese with English abstract). [CrossRef]
49. An, Q. Research on Methods of Maize Yield Estimation by Remote Sensing in Changchun Region. Master's Thesis, Jilin University, Changchun, China, 2018. (In Chinese with English abstract).
50. Kustas, W.; Anderson, M. Advances in thermal infrared remote sensing for land surface modeling. *Agric. For. Meteorol.* **2009**, *149*, 2071–2081. [CrossRef]
51. Vancutsem, C.; Ceccato, P.; Dinku, T.; Connor, S.J. Evaluation of MODIS land surface temperature data to estimate air temperature in different ecosystems over Africa. *Remote Sens. Environ.* **2010**, *114*, 449–465. [CrossRef]
52. Aghakouchak, A.; Farahmand, A.; Melton, F.S.; Teixeira, J.; Anderson, M.C.; Wardlow, B.D.; Hain, C.R. Remote sensing of drought: Progress, challenges and opportunities. *Rev. Geophys.* **2015**, *53*, 452–480. [CrossRef]

Article

UAV-Hyperspectral Imaging to Estimate Species Distribution in Salt Marshes: A Case Study in the Cadiz Bay (SW Spain)

Andrea Celeste Curcio [1,*], Luis Barbero [1] and Gloria Peralta [2]

[1] Department of Earth Sciences, Faculty of Marine and Environmental Sciences, International Campus of Excellence in Marine Science (CEIMAR), University of Cadiz, 11510 Puerto Real, Spain

[2] Department of Biology, Faculty of Marine and Environmental Sciences, International Campus of Excellence in Marine Science (CEIMAR), University of Cadiz, 11510 Puerto Real, Spain

* Correspondence: andrea.curcio@uca.es

Abstract: Salt marshes are one of the most productive ecosystems and provide numerous ecosystem services. However, they are seriously threatened by human activities and sea level rise. One of the main characteristics of this environment is the distribution of specialized plant species. The environmental conditions governing the distribution of this vegetation, as well as its variation over time and space, still need to be better understood. In this way, these ecosystems will be managed and protected more effectively. Low-altitude remote sensing techniques are excellent for rapidly assessing salt marsh vegetation coverage. By applying a high-resolution hyperspectral imaging system onboard a UAV (UAV-HS), this study aims to differentiate between plant species and determine their distribution in salt marshes, using the salt marshes of Cadiz Bay as a case study. Hyperspectral processing techniques were used to find the purest spectral signature of each species. Continuum removal and second derivative transformations of the original spectral signatures highlight species-specific spectral absorption features. Using these methods, it is possible to differentiate salt marsh plant species with adequate precision. The elevation range occupied by these species was also estimated. Two species of *Sarcocornia* spp. were identified on the Cadiz Bay salt marsh, along with a class for *Sporobolus maritimus*. An additional class represents the transition areas from low to medium marsh with different proportions of *Sarcocornia* spp. and *S. maritimus*. *S. maritimus* can be successfully distinguished from soil containing microphytobenthos. The final species distribution map has up to 96% accuracy, with 43.5% of the area occupied by medium marsh species (i.e., *Sarcocornia* spp.) in the 2.30–2.80 m elevation range, a 29% transitional zone covering in 1.91–2.78 m, and 25% covered by *S. maritims* (1.22–2.35 m). Basing a method to assess the vulnerability of the marsh to SLR scenarios on the relationship between elevation and species distribution would allow prioritizing areas for rehabilitation. UAV-HS techniques have the advantage of being easily customizable and easy to execute (e.g., following extreme events or taking regular measurements). The UAV-HS data is expected to improve our understanding of coastal ecosystem responses, as well as increase our capacity to detect small changes in plant species distribution through monitoring.

Keywords: coastal marsh; continuum removal; hyperspectral; spectral signatures; unmanned aerial vehicle (UAV); vegetation species discrimination; second derivative transformation

Citation: Curcio, A.C.; Barbero, L.; Peralta, G. UAV-Hyperspectral Imaging to Estimate Species Distribution in Salt Marshes: A Case Study in the Cadiz Bay (SW Spain). *Remote Sens.* **2023**, *15*, 1419. https://doi.org/10.3390/rs15051419

Academic Editors: Kenji Omasa, Shan Lu and Jie Wang

Received: 27 January 2023
Revised: 25 February 2023
Accepted: 28 February 2023
Published: 2 March 2023

1. Introduction

Salt marshes are ecological transition zones where marine and terrestrial ecosystems interact [1]. These ecosystems are characterized by a unique and highly specific assemblage of plants and animals [2] and high primary production, with the plant species being a crucial component of the system dynamics [3]. They offer numerous recognized ecosystem services; highlights among them are the services of coastal protection and blue carbon sink [4–7].

Tidal salt marsh vegetation is typically halophyte and has to tolerate regular periods of immersion/emergence, salinity and anoxia [8,9]. To adapt to these stresses, these plants have developed unique morphological, anatomical, and physiological characteristics [10,11]. The distribution of the salt marsh plant species follows a typical zonation pattern along the elevation gradient [12,13]. This elevation gradient includes gradients in salinity, redox potential, soil N content, soil clay content, and soil organic matter [2]. However, elevation seems a major determinant for the establishment of all of them.

Unfortunately, increasing human populations have caused an extensive loss, degradation, and fragmentation of coastal ecosystems worldwide [14]. The main anthropogenic pressures on salt marshes include changes in hydrological and salinity regimes, physical deterioration or removal of coastal features, and urbanisation [15–17]. However, the main concern nowadays in any coastal ecosystem is the survival of the particular ecosystem in a climate change scenario [18]. Although there are numerous examples of modelling these responses in the literature [19–22], a major modelling limitation is still the low availability of adequate datasets. Remote sensing (RS) techniques are changing this scenario with the provision of high-resolution spatial data that will support a new generation of computer models [18].

Sea level rise is probably the major threat to tidal salt marshes [18]. Changes in sea level are equivalent to changes in elevation. Therefore, our capacity for monitoring changes in elevation and plant species distribution is going to be key for developing early warning management plans. RS techniques are a straightforward and cost-effective way to extract information since they provide recurring datasets in short time scales at affordable prices. Maps and assessments of coastal habitats have both benefited greatly from the use of RS techniques [1,23]. For example, the loss and degradation of salt marshes have been successfully evaluated by combining long-term LANDSAT imagery and numerical modelling [24]. Sentinel-2 and Landsat archives proved to be useful tools for tracking long-term salt marsh extent dynamics [25]. More recently, deep learning models, powered by Sentinel imagery, have improved the mapping of low and high salt marsh land cover in South Carolina coastal wetlands [26].

Differences in the biophysical properties of salt marsh plants generate spectral differences that can be detected using hyperspectral (HS) data [27–29]. However, airborne or satellite-based HS imaging has a spatial resolution (meter to tens of meters) that is probably not adequate to identify species distribution due to the considerable spatial heterogeneity in salt marshes [25,30]. Previous works on HS images from the EO-1 Hyperion satellite (30 m pixels) concluded that 30 m is insufficient spatial resolution to accurately distinguish between species with spectral similarities, such as *Sporobolus maritimus* (Curtis) P.M.Peterson & Saarela (previously named *Spartina maritima* (Curtis) Fernald) and *Sarcocornia* spp. A.J.Scott [31]. Combinations of Quickbird images (2.4 m resolution in the multispectral mode) with high spectral data from Hyperion (242 narrow bands and 30 m pixel) have been probed to map different salt-marsh species with acceptable accuracies in classification [32]. Pléiades images provide a robust and consistent global identification of the salt marsh zone. However, the application of its multispectral (MS) 2 m spatial resolution images proved to be insufficient for early assessment of the *Spartina anglica* C.E. Hubb. (currently *Sporobolus anglicus* (C.E. Hubb.) P.M.Peterson & Saarela) invasion, mainly due to the small size of the patches [33].

Nowadays, most of the RS techniques have developed integrable sensors into unmanned aerial vehicles (UAVs). For the intertidal zone, UAVs that fly up to 120 m altitude are suitable to identify spatial heterogeneity in microtopography, canopy height or greenness [34–36]. In addition, UAVs offer significant operational flexibility and minimal costs [34,37], allowing flight dates to be tailored. Therefore, UAVs may provide the necessary spatial and temporal resolution for mapping species distribution and their temporal changes. High-resolution RGB cameras integrated into UAVs have been previously employed in salt marsh environments. Farris et al. [38] used UAV-LiDAR to track the salt marsh shoreline, while Yan et al. [39] used UAVs to examine environmental factors influencing the ecological response of *Spartina alterniflora* Loisel. (currently *Sporobolus alterniflorus* (Loisel.) P.M.Peterson & Saarela). UAV-multispectral (UAV-MS) technology has also shown utility in calculating indices of plant diversity and species richness in wetland communities [40]. Villoslada et al. [41] have shown that maps created from UAV-MS images provide useful data for managing plant communities and assessing the effects of climate change on coastal meadows. However, achieving a high-accuracy classification requires the use of a large variety of vegetation indices and the evaluation of the spectral properties of the training samples.

The use of UAV hyperspectral remote sensing (UAV-HS) in salt marshes combines the advantages of high spatial and spectral resolutions to capture the finer scale of spectral and spatial heterogeneity. UAV-HS has previously been used to classify desert steppe species [42], using spectral transformation to enhance species differences in vegetation indices with an overall accuracy of 87%. UAV-HS is able to detect salt stress in croplands and the accuracy performance of this technique improves in conjunction with other techniques [43]. Although several salt marsh vegetation species have undergone field hyperspectral investigation through field spectrometer measurements [28,44], this is likely the first work using a UAV hyperspectral sensor in a salt marsh environment.

Advances in RS technique applications require adequate study cases. Cadiz Bay offers an excellent system for assessing the capacity of UAV-HS in the discrimination of salt marsh vegetation species distribution. This tidal environment is home to the southernmost tidal salt marshes in Europe and is protected by numerous environmental protection figures at local and international levels. Cadiz Bay was designated a Natural Park in 1989 (Bahía de Cádiz Natural Park, PNBC, [45]) and RAMSAR site in 2002 (site no. 1265, [46]). The system is considered an important resting place on the migratory route of birds and is included in the Natura 2000 network (ES0000140, as SCA and SPA). Located between two seas and two tectonic plates, Cadiz Bay is a key place for biodiversity studies [47–50].

This work examines the potential of high spatial and spectral resolution UAV-HS data to accurately identify and differentiate the distribution of the salt marsh vegetation at the level of species. The specific goals are (1) to determine which is the appropriate UAV-HS dataset to map salt marsh vegetation; (2) to assess the separability of salt marsh interest classes; and (3) to estimate the elevation ranges of the detected species. These findings are expected to become a starting point for the early assessment of salt marsh degradation and help in the selection of areas for salt marsh rehabilitation or detection of the establishment and spread of alien species.

2. Materials and Methods

2.1. Site Description

Cadiz Bay hosts the southernmost European coastal wetland, located where the Mediterranean Sea, the Atlantic Ocean, and the continents of Europe and Africa converge (Figure 1). Located on the Atlantic coastline, precipitation, wind, and waves are influenced by large-scale oceanic weather systems that cross the North Atlantic [51]. Cadiz Bay is delimited by the tombolo of the city of Cadiz, with NNW-SSE orientation, and opens towards the Atlantic Sea to the north [52]. The entire bay is formed by two water bodies, the external and the inner bay, connected by tides through a narrow strait [53]. The external basin has depths up to 20 m, whereas the inner basin has a mean depth of about 2 m, and is sheltered from ocean waves [54]. The intertidal system of the bay is composed of natural salt marshes, salinas, mudflats, and an intricate network of tidal creeks [55]. The tidal regime is mesotidal semi-diurnal with a mean spring tidal range of 2.96 m [56].

The distribution of vegetation in the natural salt marshes of Cadiz Bay follows a conventional mid-latitude zonation [57], although the protective walls of the salinas frequently cut off the high marsh. The low marsh is mostly inhabited by *Sporobolus maritimus*, whereas the medium marsh is dominated by *Sarcocornia* spp., primarily *Sarcocornia fruticosa* (L.) A.J.Scott and *Sarcocornia perennis* (Mill.) A.J.Scott, and other halophytic species in lower abundance (Figure 2). Seagrass beds of *Zostera noltei* Hornem. and *Cymodocea nodosa* Asch., as well as patches of *Zostera marina* L., are found at the lower parts of the intertidal zone [53].

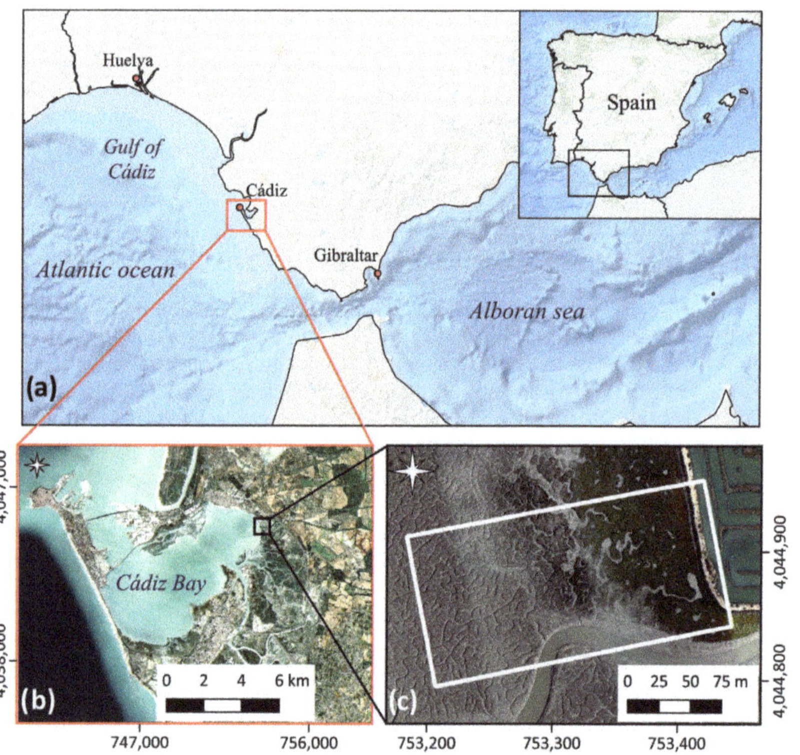

Figure 1. Location of the Cadiz Bay: regional context (**a**) and detail of the study area (**b**,**c**) The white rectangle in picture c represents the flight area for the UAV-HS survey. Coordinates are expressed in ETRS89/UTM zone 29N reference system (EPSG:25829).

Figure 2. Landscapes with dominant vegetation throughout the salt marsh horizons. (**a**) *Sarcocornia* spp. populate the medium salt marsh horizon. (**b**) The vegetation of medium and low horizons overlaps in a narrow fringe here called the transition zone. The abundance of species from the medium and the low horizons can be found in different proportions. (**c**) *Sporobolus maritimus* dominates the low horizon of the salt marsh.

Our study area was selected in an area with a wide and well-developed natural salt marsh zonation, in the north-eastern corner of the inner bay [54].

2.2. UAV and Hyperspectral Sensor

This work was performed with a Matrice 600 hexacopter (DJI) (UAV from now) equipped with a co-aligned VNIR–SWIR hyperspectral (HS from now) system (Headwall Photonics), all property of the University of Cadiz [58].

The HS unit captures continuous information in the 400–2500 nm spectral range, (see [58] for further information). The HS instrument provides VNIR and SWIR data as separate files, but they can be stacked in a single hypercube containing the complete VNIR–SWIR information (see Section 2.4.2).

For data accuracy, the HS system includes an APX-15 GNSS-inertial solution (Trimble Applanix), and the UAV incorporates three built-in GPSs. These GPSs provide an accuracy of ±0.5 m and ±1.5 m in the vertical and horizontal, respectively. However, the post-processing of the APX-15 data increases accuracy to 0.02 m and 0.05 m, respectively.

2.3. UAV-Based Data Acquisition

Flight operations were conducted on 22 October 2021 between 10 am and 12 pm with a low tide of 1.3 m LAT (lowest astronomical tide), covering an area of 4.8 ha (Figure 1c). Clear weather ensured uniform lighting conditions for setting the flight and the sensor. The flight mission was planned with UgCS desktop, version 4.5 (SPH engineering). The flight altitude was set at 120 m and the speed at 5 m/s to ensure radiometric quality. This HS sensor does not require frontal overlap, but a 40% lateral overlap was set to assure the subsequent reconstruction of the orthomosaic. The sensor was calibrated by obtaining a reference spectrum in the 400–1700 nm range from a radiometrically calibrated tarp (3 × 3 m). A Reach RS2+ RTK GNSS antenna (EMLID) was used as a base station (see Section 2.4.1). This antenna allows for obtaining more precise results with accuracies of 4 mm + 1 ppm and 8 mm + 1 ppm, in horizontal and vertical measurements, respectively.

2.4. Hyperspectral Data Processing

This section summarizes the processing to generate the HS products (Figure 3). The processing was performed with ENVI, v. 5.3.6 (L3Harris Geospatial Solutions, Inc., Broomfield, CO, USA -.), whereas QGis, v. 3.26.3 (QGIS Development Team) was used for the visualization and handling of raster deliverables. The projected coordinate system used in this work was the ETRS89, UTM zone 29 N (EPSG:25829).

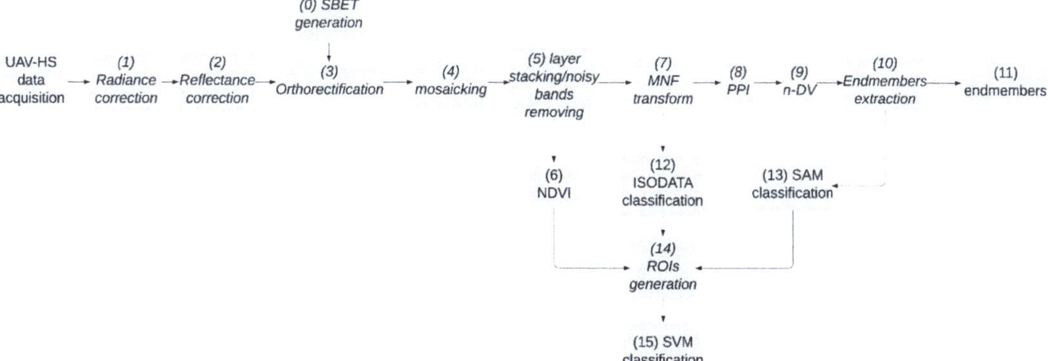

Figure 3. Flowchart showing the hyperspectral processing steps. Rounded boxes indicate data or products, rectangle boxes represent processes. Numbers in brackets refer to the respective explanation in the text. SBET: smoothed best estimate of trajectory; MNF: minimum noise fraction; PPI: pixel purity index; n-DV: n-dimensional visualizer; NDVI: normalized difference vegetation index; SAM: spectral angle mapper; ROIs: regions of interest; SVM: support vector machine. See text for details.

2.4.1. Hyperspectral Pre-Processing

Data from the APX-15 is processed with the POSPac UAV software, v. 8.9 (Trimble Applanix), using the data from the antenna to improve accuracy and create the smoothed best-estimated trajectory (SBET). This file, with root mean square errors (RMSEs) within 0.02–0.05 m, is used for the orthorectification of the hypercubes (Figure 3(0)).

The VNIR and SWIR data cubes are processed separately using SpectralView, v 3.2.0 (Headwall Photonics) as follows: (1) raw data is transformed to radiance (Figure 3(1)) by subtracting the dark reference from the digital numbers (DNs). The dark spectrum is collected in the field by covering the sensor lens after the mission and is considered sensor noise; (2) the reflectance correction is the conversion of radiance to reflectance (Figure 3(2)). This step is used to build a line of best fit between the radiance of the HS sensor and the reflectance measured on the radiometric tarp [59,60]; (3) the orthorectification (Figure 3(3)) is performed by combining a precision DSM and the SBET from step 1. For this operation, a DSM from Curcio et al. [34] was used, with 0.05 m/pixel resolution and 0.01 m mean accuracy; (4) the processed VNIR and SWIR hypercubes are stitched together (mosaicking, Figure 3(4)) into a single final mosaic that is orthorectified and georeferenced.

2.4.2. Hyperspectral Post-Processing

Atmospheric correction is not required when flying at a maximum altitude of 120 m. Once mosaicked, VNIR and SWIR hypercubes are stacked into a single file and the wavelengths associated with water vapour absorbance (i.e., 1350–1460 nm; 1790–1960 nm; 2350–2500 nm [28]) are excluded, resulting in an orthomosaic with 418 exploitable bands (Figure 3(5)).

The Normalized Difference Vegetation Index (NDVI) is used to discriminate the presence of vegetation in the study area (Figure 3(6)) and generate the training set for the final classification step (see Section 2.5). The NDVI map layer is calculated according to Equation (1):

$$NDVI = (NIR - RED)/(NIR + RED) \quad (1)$$

with NIR and RED as the near-infrared and red bands centred at 860 nm and 649 nm, respectively. Very low NDVI values (0.1 and below) are associated with bare soil or water, moderate values (~0.3) correspond to shrub and grassland, and high values (0.6–1) are associated with high-density plants and healthy physiological conditions [61]. Since only shrub and grassland vegetation was considered for this work, the NDVI map layer was masked with a threshold of 0.3. The product of this step is a raster with only vegetation distribution.

2.4.3. Endmember Extraction

An endmember is a pure signature for a class [62] and they are essential for the classification of HS data. Pure signatures from hyperspectral imagery can be found using minimum noise fraction (MNF), the Pixel Purity Index (PPI), and endmember extraction techniques (Figure 3). The MNF maximises the noise-to-signal ratio and reduces dimensionality without sacrificing information [63] (Figure 3(7)). Most of the significant information is contained in the first MNF bands, which are used in successive processing, ignoring the rest of the bands containing only noise [64]. The PPI technique [65] searches for pure spectral signatures by identifying the pixels with the fewest mixed spectral signatures (Figure 3(8)). The PPI image locates the pure pixels of the scene, which will then be used to extract the spectra of the potential endmembers [66]. A region of interest (ROI) is a dataset sample considered important for a particular purpose [67]. In this case, the regions of interest (ROIs) contain the pixels with the pure spectra in the scene and are imported in the n-Dimensional Visualizer scatter plot (n-DV; Figure 3(9)). The n-DV is an ENVI tool for visualising the distribution of pixels in the n-D space (where n is the number of bands), allowing the purest pixels representing the spectral endmembers to be identified and clustered (Figure 3(10,11)). After these steps, the classification procedure can be carried out.

2.5. Classification

The classification process uses unsupervised and supervised algorithms to map the spatial location and abundance of each endmember spectrum. The relevant MNF bands are the input for the ISODATA algorithm (Figure 3(12)), which iteratively clusters the pixels using the least distance approach [68]. The result is a first classified image based on the inherent spectral information of the dataset, with each class represented by a different endmember.

The next level of classification is produced by the spectral angle mapper (SAM) algorithm (Figure 3(13)), which calculates the angle between two spectra to identify their spectral similarity based on a maximum angle threshold [69]. This threshold is set to 0.1 radians to minimize spectral mixing issues. This algorithm uses the endmembers of the unsupervised classification to classify.

The last algorithm selected in this study is the support vector machine (SVM) (Figure 3(15)), selected because of the good results it produces with heterogeneous, complex, and noisy data [70]. The SVM separates the classes using a training set with class samples (i.e., support vectors) [71], with every class represented by an ROI. In this case, ROIs are generated based on three sources of information: (1) the unsupervised classification; (2) the SAM-classified image; (3) the NDVI map (Figure 3(14)).

2.6. Spectral Analysis

To verify that UAV-HS datasets can differentiate vegetation at the species level in salt marshes, the separability of the spectral signatures of the classifications must be tested. This was analysed through spectral transformations. In addition, the usefulness of new spectral indices for the separation of species is explored, using the relevant wavelengths highlighted by the spectral transformations to generate them.

2.6.1. Continuum Removal and Second-Order Derivative

The differences in absorption and reflection spectra between vegetation species can be very small, making classification difficult. Spectral transformation, such as continuum removal (CR) and derivative spectroscopy, have the potential to amplify small differences [28,42,72]. CR is used to normalize the spectra, and sometimes this is enough to highlight differences in absorption and reflection spectra [73]. The second-order derivative method (2nd derivative from now) emphasises the small differences in absorption peaks associated with biochemical properties, allowing for the identification of different species [28,74]. To enhance the signal-to-noise ratio and extract additional hidden spectral features, the 2nd derivative spectrum is filtered using a boxcar average smoothing. All transformed spectra are analysed in four separated wavelength windows: visible region (VIS, 400–700 nm), near-infrared region (NIR, 700–1000 nm), and two regions of short-infrared (SWIR1, 1000–1800 nm; SWIR2, 1800–2350 nm).

2.6.2. Spectral Indices (SI)

New spectral indices (SI) have been constructed from the most outstanding absorption and reflectance features (peaks and valleys of the 2nd derivative, respectively) of the transformed spectrum. These indices can emphasize the distribution of different vegetation species in the salt marsh. Each SI is calculated according to Equation (2) (known as the normalized difference):

$$SI_{B2\text{-}B1} = (B2 - B1)/(B2 + B1) \quad (2)$$

where $SI_{B2\text{-}B1}$ is the calculated spectral index, B1 is the wavelength presenting the absorption feature, and B2 is the wavelength presenting the reflectance feature. This type of equation brings out characteristics not initially visible.

2.7. Validation

2.7.1. Spectral Signatures

The spectral responses of our study area were previously studied in 2014–2015 (FAST project, [75]). In the FAST project, each sampling point was a 1 × 1 m area where five reflectance measurements were made using a field hyperspectral radiometer for the VNIR range (400–1000 nm).

The FAST spectra measurements are utilised here as a reference spectral library to identify species based on their spectral features. All spectra from our classification results are compared to the library using the Spectral Analyst tool from ENVI. The similarities of our classification spectra to those in the library are calculated by providing a similarity score to each spectrum in the library, with the highest score considered the closest match (i.e., the most confident spectral similarity). This analysis considers only the wavelength range available in the FAST library (400–1000 nm).

2.7.2. Classification

Two methods are used to determine the SVM classification accuracy. The first one is the comparison of the classification results with the composition of species observed at 60 randomly sampled points in the study area. To buffer small errors associated with very precise locations, the species from the classification was determined as the prevalent class in a 15 cm diameter buffer area around each point.

The second method is the comparison with random pixels from other sources. In total, seven comparisons are performed from seven sets of random pixels. One set is obtained from the training ROIs. The other six sets are generated from the unsupervised classified image using different sampling methods: (1) two sets are generated with stratified-proportionate samples (SP), with sizes directly related to the size of the classes; (2) two sets of equalized samples (Eq), with fixed size regardless of the class size; and (3) two sets of random samples (R), using 10% and 20% of the total pixels.

For each comparison, the accuracy is determined from (1) the overall accuracy, calculated by counting the correctly classified values and dividing by the total number of values; (2) producer accuracy, which measures the likelihood of correctly classifying a value into each class; (3) user accuracy, which shows the likelihood that a prediction belongs to the correct class. Each probability is determined by dividing the proportion of correct values by the total number of values in a class.

2.8. Elevation of Species Distribution

Once the vegetation classes are confirmed, their elevation distribution is assessed using a DEM with 0.24 m/pixel resolution and a mean accuracy of 0.04 m [34]. The corresponding elevations are extracted for each class using 5 cm sampling grids. After removing outliers, the elevation of each class is characterized using a set of statistical parameters (i.e., minimum, maximum, median, and mode).

3. Results

3.1. Classification Result

For the SVM supervised classification, a total of 15 ROIs were recognized, corresponding to 15 classes, 7 of them associated with vegetation classes. However, only four of these classes are within the salt marsh horizons considered in this work. Regarding the other three vegetation classes, one has been associated with macrophyte debris deposited in the uppermost zone of the salt marsh by an extreme high tide event, and the other two with the typical vegetation of the saline wall. These last two classes are outside the scope of this work and, therefore, will not be taken into consideration. Nevertheless, it is interesting that they can be distinguished from salt marsh species.

Of the 4.8 ha of surveyed salt marsh, the spatial mapping of the endmembers estimates a vegetation cover of 14.7% of the area. These species are distributed parallel to the mean sea line and in different elevation ranges.

In this work, the vegetation classes within the salt marsh horizons include four classes of salt marsh species and a fifth class associated with macroalgal debris. From high to low elevations, the distribution of these classes is macroalgal debris deposits first, followed by vegetation 1 and 2 in the mid-horizon, vegetation 3 within the transition zone, and vegetation 4 in the low horizon. The area covered by these classes is 2.9%, 9.2%, 34.3%, 28.9%, and 24.9% of the vegetated area, respectively (Figure 4).

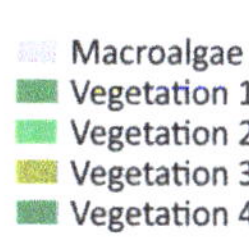

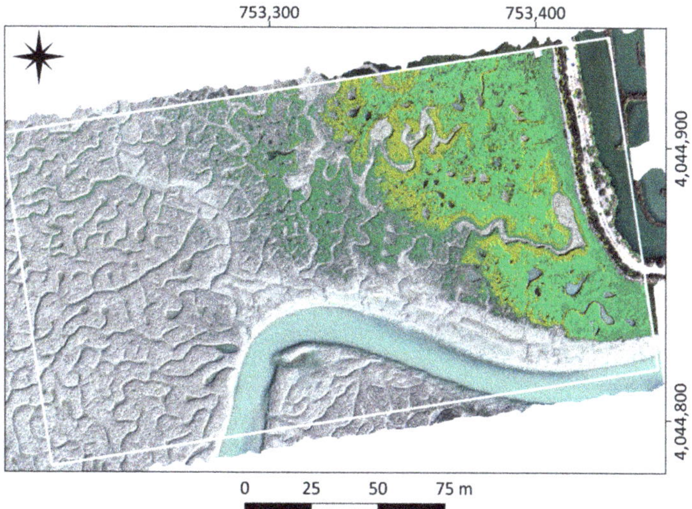

Figure 4. Salt marsh vegetation distribution according to the SVM classification. The white line defines the boundaries of the surveyed area, corresponding to 4.8 ha. The macroalgae class is debris deposited in the uppermost zone of the salt marsh representing less than 3% of the vegetated space. Vegetation 1 and vegetation 2 are distributed along the medium marsh horizon, vegetation 3 corresponds to the transitional zone, and vegetation 4 spreads along the low marsh horizon. The classification is superposed on the orthomosaic obtained by the hyperspectral survey displayed in true colour combination.

3.2. Spectral Analysis

The four marsh vegetation classes show typical plant spectral curves (Figure 5). The peaks of each class are located at the same wavelength (±5 nm), presenting only quantitative differences. The macroalgae class showed clear divergences from this pattern. First, the strong absorption peak in the red region (peak 3 in Figure 5) is less pronounced than in plant classes. It also lacks the absorption peak at 943 nm (peak 6 in Figure 5). There is a significant increase in reflectance from the red-edge region to the SWIR1 region (i.e., 700–1300 nm), and also higher reflectance in entire the SWIR region (1000–2350 nm) (Figure 5, Table 1).

Table 1. Wavelengths of the absorption peaks detected in the spectral signature of the marsh vegetation classes of Cadiz Bay. The numbers in the column headers correspond to the peaks indicated in Figure 5. Units: nm.

Class\Peak	(1)	(2)	(3)	(4)	(5)	(6)	(7)	(8)	(9)	(10)
Macroalgae	447	500	669	776	864	-	997	1182	1523	2025
Vegetation 1	443	492	669	776	860	943	991	1206	1523	2025
Vegetation 2	447	487	669	776	860	943	991	1194	1523	2025
Vegetation 3	447	487	669	776	864	943	997	1188	1523	2025
Vegetation 4	443	483	669	776	864	943	997	1182	1523	2025

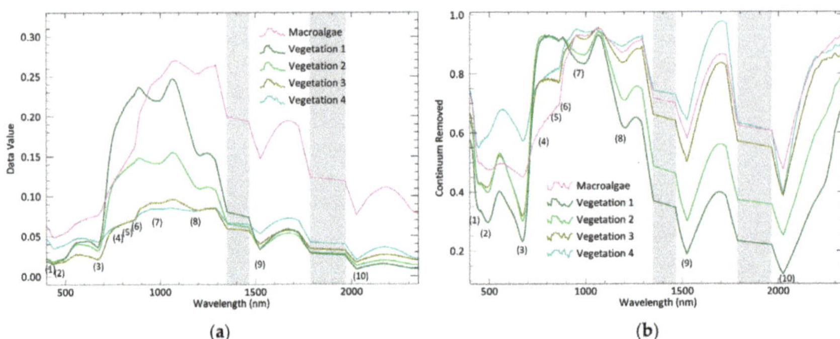

Figure 5. Spectral profiles (**a**) and the corresponding continuum removal transformations (**b**) for the salt marsh vegetation classes identified in Cadiz Bay. The grey areas highlight the water vapour absorbance regions (1350–1460 nm and 1790–1960 nm) excluded from the hyperspectral processing. The numbers in brackets indicate the absorbance peaks of the spectral signatures. Note that minimum values are absorption peaks and maximum values are reflectance peaks.

The 2nd derivative transformation accentuated small differences not previously observed in the reflectance curves (Figures A1–A4). These effects are more pronounced in the SWIR1 region, and the variations between 2013 and 2329 nm (SWIR2 region) are particularly notable.

The spectral indices (SI) were constructed with the significant absorbance peaks at 1057, 1110, 1152, 1182, 1260, and 1331 nm, generating $SI_{1152-1110}$, $SI_{1331-1260}$, $SI_{1182-1057}$, and $SI_{1523-1290}$. These indices may show differences that can be attributed to biophysical differences in vegetation (Figure 6).

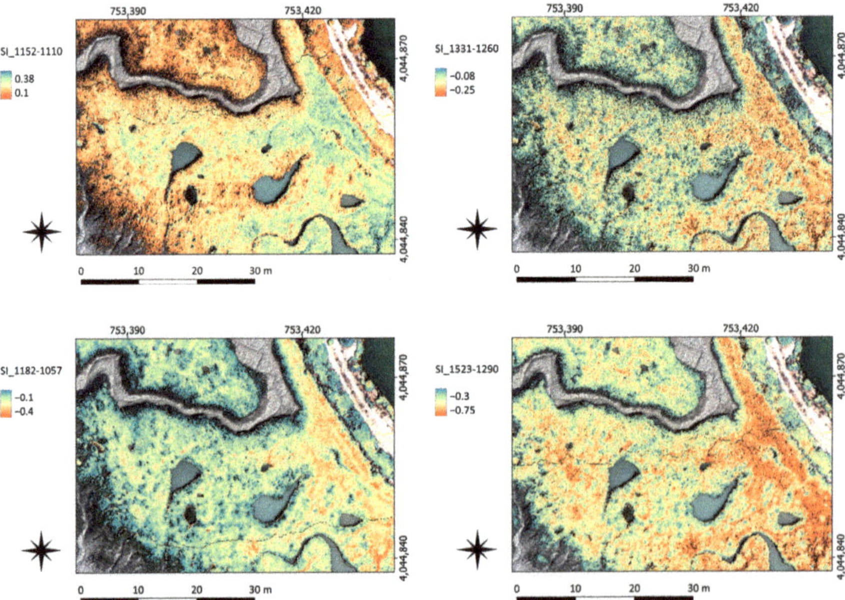

Figure 6. Distribution of spectral indices (SI) values in the salt marsh of Cadiz Bay. The results display only a detail of the study area, and the corresponding SI is indicated in the legend. Wavelengths used for creating SI are suitable to show the variations in canopy cover. Thresholds are adjusted for each index to better enhance differences in canopy cover.

3.3. Validation

According to the scores obtained, vegetation 1 and 2 classes may represent *Sarcocornia* spp., while vegetation 4 class may correspond to Sporobolus maritimus. Vegetation 3 class has been attributed to areas of overlap of different proportions of these species (i.e., the transition zone).

The categories identified at the field reference points were *Sarcocornia*, *Salicornia*, and transition zones only. However, distinguishing between *Sarcocornia* and *Salicornia* species in the field is very challenging and misidentifications can be expected [76]. The accuracy of the classification was estimated to be 46% but increased to 73% when considering that *Sarcocornia* and *Salicornia* could be mistaken for each other in the field. When comparing the classification to the sets of random pixels, the accuracy ranges from 92 to 96% (Table 2).

Table 2. Overall accuracy of the HS image classification. The accuracy is estimated by comparison with field measurements, random samples generated from ROIs, and other groups of random samples (Eq250, Eq500, R10, R20, Sp10, Sp20). Eq250 and Eq500 are the equalized samples groups using 250 and 500 pixels respectively for each class as reference; R10 and R20 are the random samples groups using 10% and 20 % of total pixels as reference respectively; Sp10 and Sp20 are the stratified-proportionate samples groups using 10% and 20 % of total pixels as reference, respectively.

Estimation Method	Accuracy
Eq250	92.5%
Eq500	92.4%
Field measurement	46–73%
ROIs	95.9%
R10	95.6%
R20	95.7%
Sp10	95.9%
Sp20	95.8%

3.4. Elevation Range of Identified Species

Vegetation 1 class (identified as a short *Sarcocornia* spp. or a *Salicornia* spp.) is included within the elevation range of the vegetation 2 class (identified as *S. perennis*) (Table 3). The two classes spread across the same elevation range, but the mode of their elevation range is different, with 2.67 m vs 2.79 m for vegetation 1 and 2, respectively. The transitional class (vegetation 3) extended from 1.91 m to 2.78 m, narrowing to 2.26–2.58 m for Q1–Q3. The *S. maritimus* (vegetation 4) covers a range between 1.22 m and 2.35 m (1.76–2.10 m for Q1–Q3). Macroalgae have a bimodal distribution, with two accumulation zones located at 2.49–2.86 m and 3.35–3.84 m (Figure A5).

Table 3. Estimated elevation range for each salt marsh vegetation class identified in Cadiz Bay. *S. perennis*: *Sarcocornia perennis*; *S. maritimus*: *Sporobolus maritimus*. Figure A5 shows the frequency distribution of the extracted values for each class; Q: quantile.

Class	Specie	0.05 Q	0.25 Q	0.5 Q	0.75 Q	0.95 Q	Mode
Macroalgae	Macroalgae	1.13	2.54	2.75	3.43	3.77	2.81
Vegetation 1	*Salicornia* spp. o *Sarcocornia* spp.	2.34	2.53	2.62	2.69	2.75	2.67
Vegetation 2	*S. perennis*	2.30	2.56	2.67	2.74	2.80	2.79
Vegetation 3	Transitional	1.91	2.26	2.42	2.58	2.78	2.00
Vegetation 4	*S. maritimus*	1.22	1.76	1.94	2.10	2.35	1.68

4. Discussion

The analyses of hyperspectral datasets from low and medium salt marsh areas in Cadiz Bay are adequate to identify vegetation distribution at the species level. Four plant classes distributed along the horizons of the salt marsh and one class of macrophyte debris were recognised. Three of the plant classes have been associated with monospecific vegetation

(*Sarcocornia* spp., *S. perennis*, and *S. maritimus*), while a fourth class represents the mix of species typical of the convergence of distributions (i.e., transition zone). The results from this study are expected to be extrapolated to other mid-latitude tidal marshes since the low and medium tidal marshes of these latitudes usually present similar structural traits [75].

The performance of the SAM classification method is limited in areas with several species due to the mix of spectra [77]. However, this problem can be minimised by using the two supervised classification methods (SAM and SVM) in succession after performing hyperspectral processing procedures. The SVM supervised classification reached up to 98% accuracy, demonstrating its effectiveness in mapping land cover. When looking at the accuracy of individual classes, features such as water bodies and types of soil perfectly match the reference data. Focusing on vegetation classes, the highest accuracy is achieved by vegetation 1, vegetation 4 and macroalgae. The accuracy is lower, although still significant for the remaining classes (Tables A1–A7). The spectral signature of the transition zone (vegetation 3 class), which is a variable mixture of plant species and, therefore, a variable mix of spectra, makes this class the lowest in accuracy (63.16%). Since the training dataset used to produce the map (i.e., ROIs) determines the quality of the algorithm classification, a single ROI for transition zones is considered insufficient to accurately represent the variability associated with different levels of species mix. Still, the overall accuracy provided by our UAV-HS approach is higher than previous classification attempts. Rasel et al. [31] obtained a 43% overall accuracy from space-borne hyperspectral data with 30 m spatial resolution. Rajakumari et al. [28] combined satellite multispectral data and ground spectrometric measurements, achieving 65.8% to 73.55% accuracy for the vegetation and land cover spectral signatures, respectively.

Among the endmembers extracted from Cadiz Bay salt marshes, four exhibit the typical plant pattern, while one shows the macroalgae response with distinct peaks and slopes of the reflectance spectrum. Chlorophyll a (Chl-a), found in both plants and macroalga, determines a typical absorption peak at 669 nm. However, this feature is smoothed in the spectral signature of macroalgae (Figure 5), maybe due to its yellowish-brown colour related to the fucoxanthin content [78]. The class of macroalgae in this study corresponds to debris deposited by an extreme flood event at an elevated position far removed from ordinary tidal cycles. This material dries and decomposes, resulting in high reflectance values in the NIR–SWIR, a typical region where water often attenuates spectral signatures [79]. The reflectance curves of the four remaining classes do not differ much from each other, the only significant variations being in the peak intensity. However, the continuum removal transformation enhances different responses in the 740–864 region, with opposite slopes in the curve between 882 and 949 nm (Figure 7). These findings lead to the acceptance of vegetation 1 and 2 as the spectral signatures of *Sarcocornia* species in the medium marsh horizon, and vegetation 3 and 4 as the signatures of species distributed in the transition and the low marsh, respectively.

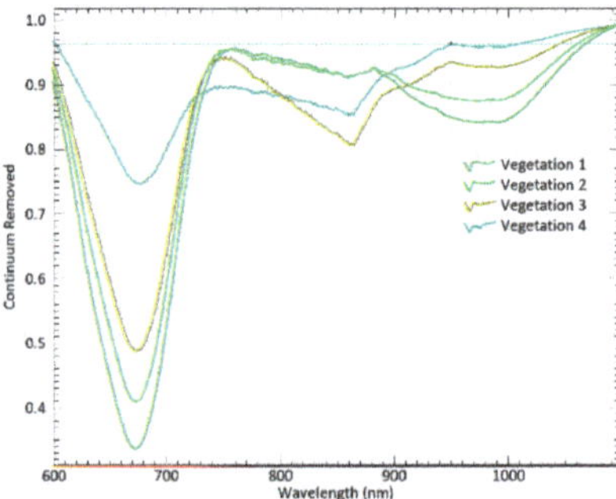

Figure 7. Results of the continuum removal transformation for the discriminated vegetation classes in the salt marsh of Cadiz Bay. Note that vegetation 1 and 2 have the opposite slope to vegetation 3 and 4 in the 882–949 nm region.

Different species of *Sarcocornia* dominate the medium marsh in Cadiz Bay [55,57]. However, because they are morphologically similar to *Salicornia* species, it is very challenging to distinguish them in the field when they coexist, and misidentification can occur [76]. As revealed by the CR spectra, vegetation 1 and 2 differ almost only in the intensity of the peaks, demonstrating optical similarities. However, the 2nd derivative analysis reveals that two groups of halophytes are spectrally distinct (Figures A1–A4), with these spectral differences resulting from biochemical variations between salt marsh species [28]. The pigment content, canopy structure, leaf area and leaf structure all have an impact on the visible region of the reflectance spectrum of plant canopies [42]. The 2nd derivative of our reflectance curves shows differences in the blue and red regions at wavelengths associated with the chlorophyll and xanthophyll peaks [74]. However, our work shows that the largest 2nd derivative peaks are in the SWIR region, and many of them coincide with water absorption wavelengths. Water absorption bands are present at 900 and 967 nm (the water band index [80]), in the 1150–1260 nm region [81] and in the 1450–1940 nm region [82]. Salinity in soils and vegetation is also detectable in the SWIR region [83,84], and Kumar et al. [85] proposed a SWIR-based vegetation index to detect changes in vegetation cover from satellites. All these previous works support our conclusion that SWIR, with its highest spectral variability, is a suitable region to discriminate plant species from salt marshes. The SI established here can reveal differences in the canopy cover (Figure 6), proving that UAV-HS is able to detect variations in canopy cover at the species level. The great advantages of UAVs are the high spatial and temporal resolutions of their products, as well as greater flexibility and lower cost when compared to satellite products. This allows, for example, data collection immediately after an extreme event and then periodically afterwards, providing key data to assess the ability of dynamic systems, such as salt marshes, to return to previous states (or resilience).

The horizon of the low salt marsh in Cadiz Bay is dominated by *S. maritimus* [55]. Its shoot structure and density allow the soil to be exposed, resulting in a mixed spectrum of soil and plant responses that is very similar to the spectral signature of soil with microphytobenthos. This problem may result in misinterpretations when using low spatial resolution sensors [1]. The higher spatial resolution (5 cm/pixel) offered by UAVs not only prevents this issue, but also reduces the occurrence of mixed spectral signatures due to the reduced pixel size. The comparison of the spectral responses of the *S. maritimus* class (vegetation 4) with the soil classes (Figure 8) shows that the influence of the soil is inevitable. However, *S. maritimus* habitats and soil with microphytobenthos can be distinguished by CR and 2nd derivative transformations in the red-edge and SWIR2 regions (Figure 9). This demonstrates that these two habitats can be distinguished in UAV-HS datasets, allowing for more precise mapping of *S. maritimus* and microphytobenthos soil and minimizing overestimation/underestimation issues for these categories.

The zonation of salt marsh plant species depends on elevation, tidal regime, and the gradient of environmental variables, such as salinity, redox potential, soil N, clay, and organic matter content, as well as interspecific relationships [2,86–88]. According to Redondo-Gomez et al. [89], in SW Spain, *S. perennis* subsp. *perennis* occupies from 2.26 to 2.84 m LAT, and *S. perennis* subsp. *alpini* from 2.84 to 3.65 m LAT. Our results agree with these findings, showing that *Sarcocornia* spp. inhabit the salt marsh horizon between 2.30 m and 2.80 m LAT. Previous studies have described *S. fruticose* and *S. perennis* as dominant species in the salt marshes of Cadiz Bay [55]. Unfortunately, the spectral library available in our study area (FAST project, [75]) does not specify which *Sarcocornia* taxa were measured. However, due to differences in the SWIR region, our results suggest that two *Sarcocornia* taxa coexist in the medium salt marsh horizon. Differences in this part of the spectrum have previously been related to differences in the salinity [84], suggesting that soil salinity may be playing a role in the zonation of Cadiz Bay tidal marshes. Although the elevation ranges for vegetation 1 and 2 overlap, suggesting a similar ecological niche, their different mode (2.67 m for vegetation 1 vs 2.79 m for vegetation 2, Table 3) indicate a shift in the optimum range of environmental conditions between the two groups, supporting the existence of two species. Both histograms are left skewed, indicating that these species can populate lower elevations despite performing better at higher positions. As a result of our findings, some resilience is expected in these habitats under sea level rise scenarios.

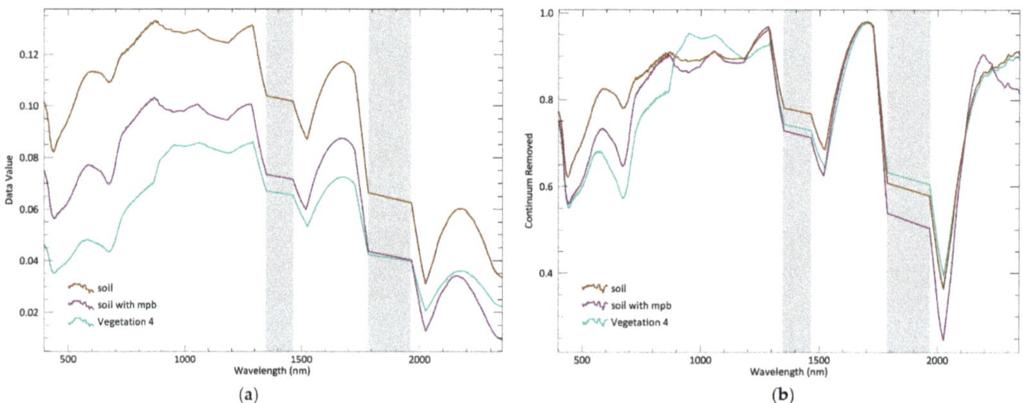

Figure 8. Comparison of the spectral responses of the *S. maritimus* class (vegetation 4) and soil classes: reflectance curve (**a**) and continuum removed spectra (**b**). The water vapour absorbance regions (1350–1460 nm and 1790–1960 nm) excluded from the hyperspectral processing are shown in grey.

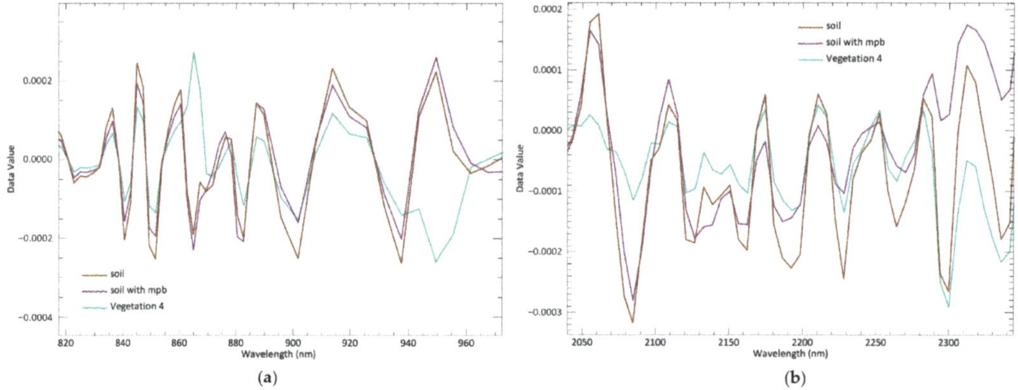

Figure 9. Comparisons of 2nd derivative transformation for the *S. maritimus* (vegetation 4) and soil classes in the red-edge region (**a**) and SWIR2 (**b**).

Regarding the accumulations of macroalgae, the decomposition of these accumulations of organic matter alters the availability of oxygen and the redox potential in the sediments, which could have negative consequences for multiple trophic levels if their incidence increases significantly [90,91]. Understanding the local carbon cycle and the dynamic of the system also requires mapping where macroalgae are deposited [78,91]. In Cadiz Bay, *Sarcocornia* spp. and *S. maritimus* overlap in a narrow area here called the transition zone (vegetation 3). This class has problems with accuracy mainly because of the wide variety of spectral responses due to the different proportions of *Sarcocornia* spp. and *S. maritimus*. Future studies may include more classes for the transition zone, but they will need careful spectral analysis to investigate the spectral response associated with specific proportions of the dominant species.

5. Conclusions

This study demonstrates the potential of UAV-HS technology to identify and map the distribution of plant species in salt marshes, using canopy reflectance information. Salt marsh plant species have very similar spectral shapes. However, hyperspectral technology is capable of detecting spectral differences associated with the water content and salinity of salt marsh plant tissues. The continuum removal and 2nd derivative transformations can detect hidden spectral features in reflectance curves, which can separate plant species with satisfactory accuracy. The classification map obtained through a supervised process reached up to 98% accuracy. The availability of an accurate DEM allows for the estimation of the preferred elevation range for each specie from the distribution of the corresponding classes. The overlap of species distribution generates mixes of spectra with a large variability associated with different species proportions. Future research may reduce these uncertainties but will require an increase in the number of associated classes.

Vegetation distribution is a key indicator in determining the health of salt marshes. The ability to monitor changes in these distributions will improve our understanding of salt marsh dynamics, our modelling capacity to assess responses to sea level rise, and help stakeholders manage these complicated, vulnerable, and valuable ecosystems. UAV-HS data can be used to evaluate salt marsh vulnerability and strengthen conservation efforts by defining critical areas for conservation and examining pressures on crucial ecosystem services, such as blue carbon.

Author Contributions: Conceptualization, A.C.C., L.B. and G.P.; methodology, A.C.C. and L.B.; software, A.C.C.; formal analysis, A.C.C.; investigation, A.C.C.; resources, A.C.C., L.B. and G.P.; data curation, A.C.C.; writing—original draft preparation, A.C.C.; writing—review and editing, A.C.C., L.B. and G.P.; visualization, A.C.C.; supervision, L.B. and G.P.; project administration, A.C.C., L.B. and G.P. All authors have read and agreed to the published version of the manuscript.

Funding: This research received no external funding.

Data Availability Statement: The data presented in this study are available on request from the corresponding author.

Acknowledgments: The authors want to thank all the members of the drone service of the University of Cádiz, which provided all the UAV systems used to carry out the research for this study. The drones service of the University of Cádiz was equipped through the "State Program for Knowledge Generation and Scientific and Technological Strengthening of the R + D + I System State, Subprogram for Research Infrastructures and Scientific-Technical Equipment in the framework of the State Plan for Scientific and Technical Research and Innovation 2017–2020", co-financed by 80% FEDER project ref. EQC2018-004446-P. The authors acknowledge the Program of Promotion and Impulse of the activity of Research and Transfer of the University of Cadiz for the productivity associated with the work. This work is part of the iBESBLUE research project (PID2021-123597OB-I00) funded by MCIN/AEI/10.13039/501100011033/FEDER, EU. Reviewers and editors are acknowledged. All authors have approved each acknowledgment.

Conflicts of Interest: The authors declare no conflict of interest.

Appendix A

Appendix A.1. 2nd Derivative Analysis

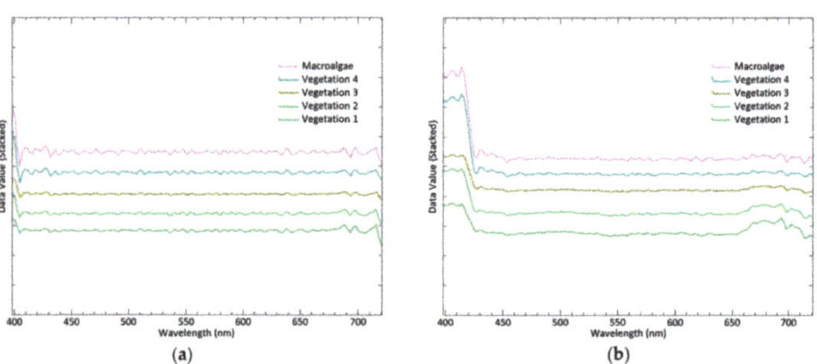

Figure A1. The focus is on the VIS region of the electromagnetic spectrum. The main peaks can be identified in the 2nd derivative spectrum of the five studied spectral signatures (**a**). The boxcar average smoothing filter applied on the 2nd derivative spectrum highlights other important peaks for the studied signatures (**b**). Significant peaks are present at wavelengths where pigments influence the spectral response of vegetation: 427,472, 487, 512, 547, 576, 638, 676, 689, and 698 nm.

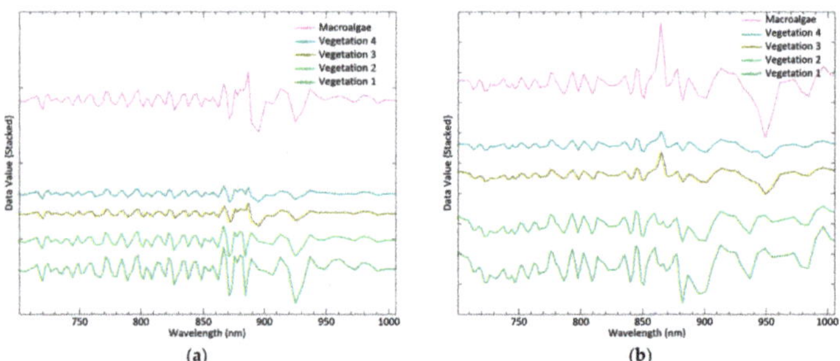

Figure A2. The focus is on the NIR region of the electromagnetic spectrum. The main peaks can be identified in the 2nd derivative spectrum of the five studied spectral signatures (**a**). The boxcar average smoothing filter applied on the 2nd derivative spectrum highlights other important peaks for the studied signatures (**b**). In the NIR region, other important absorbance peaks can be identified at 725, 749, 771, 798, 822, 867, 880, 913, 937, 949, 961, and 997 nm.

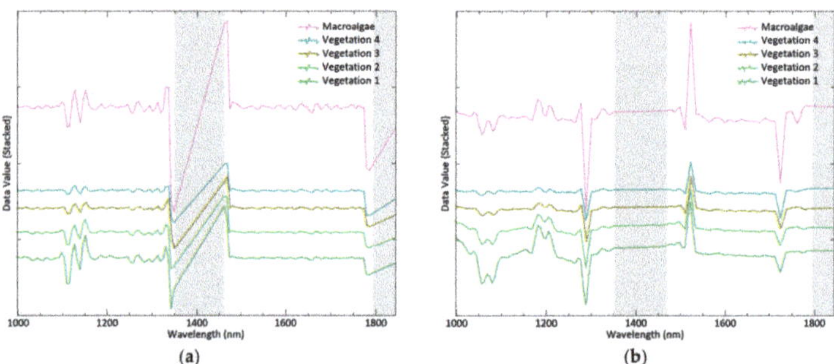

Figure A3. The focus is on the SWIR1 region of the electromagnetic spectrum. The main peaks can be identified in the 2nd derivative spectrum of the five studied spectral signatures (**a**). The boxcar average smoothing filter applied on the 2nd derivative spectrum highlights other important peaks for the studied signatures (**b**). Significant absorption peaks are present at 1039, 1098, 1128, 1152, 1188, 1206, 1331, 1499, 1523, 1594, 1636, 1672, 1690, and 1774 nm. In grey are the water vapour absorbance regions (1350–1460 nm and 1790–1960 nm) excluded from the hyperspectral processing.

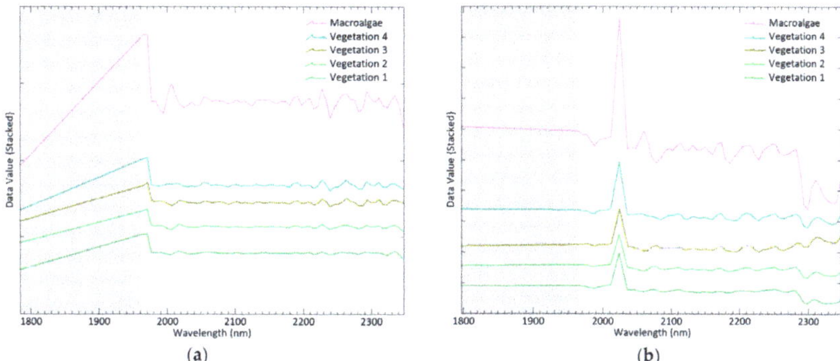

Figure A4. The focus is on the SWIR2 region of the electromagnetic spectrum. The main peaks can be identified in the 2nd derivative spectrum of the five studied spectral signatures (**a**). The boxcar average smoothing filter applied on the 2nd derivative spectrum highlights other important peaks for the studied signatures (**b**). The SWIR2 region presents absorbance peaks at 1971, 2007, 2025, 2054, 2114, 2192, 2228, 2264, 2293, and 2335 nm. In grey is one of the water vapour absorbance regions (1790–1960 nm) excluded from the hyperspectral processing.

Appendix A.2. Accuracy

Table A1. Report for the accuracy of the comparison of classification results to Eq250, the equalized samples groups using 250 pixels for each class as reference. The table summarizes the producer accuracy and user accuracy in percentage and pixels for each class.

Class	Class Name	Prod. Acc. (Percent)	User Acc. (Percent)	Prod. Acc. (Pixels)	User Acc. (Pixels)
ROI #1	Vegetation 1	98.00	98.00	245/250	245/250
ROI #2	Vegetation 2	74.80	87.38	187/250	187/214
ROI #3	Vegetation 3	66.80	81.07	167/250	167/206
ROI #4	Ponds with mpb	100.00	99.60	250/250	250/251
ROI #5	Vegetation 5	85.20	78.89	213/250	213/270
ROI #6	Vegetation 6	88.00	81.48	220/250	220/270
ROI #7	Soil	99.60	89.89	249/250	249/277
ROI #8	Soil with mph	98.40	94.62	246/250	246/260
ROI #9	Tidal channel	100.00	100.00	250/250	250/250
ROI #10	Saline	100.00	100.00	250/250	250/250
ROI #11	Ponds without water	90.40	97.84	226/250	226/231
ROI #12	Shallow water	99.60	100.00	249/250	249/249
ROI #13	Dry soil	100.00	100.00	250/250	250/250
ROI #14	Macroalgae	98.40	100.00	246/250	246/246
ROI #15	Vegetation 4	88.00	79.71	220/250	220/276

Table A2. Report for the accuracy of the comparison of classification results to Eq500, the equalized samples groups using 500 pixels for each class as reference. The table summarizes the producer accuracy and user accuracy in percentage and pixels for each class.

Class	Class Name	Prod. Acc. (Percent)	User Acc. (Percent)	Prod. Acc. (Pixels)	User Acc. (Pixels)
ROI #1	Vegetation 1	98.13	98.13	419/427	419/427
ROI #2	Vegetation 2	74.00	85.87	316/427	316/368
ROI #3	Vegetation 3	66.74	78.51	285/427	285/363
ROI #4	Ponds with mpb	100.00	100.00	427/427	427/427
ROI #5	Vegetation 5	84.31	77.25	360/427	360/466
ROI #6	Vegetation 6	84.78	81.35	362/427	362/445

Table A2. *Cont.*

Class	Class Name	Prod. Acc. (Percent)	User Acc. (Percent)	Prod. Acc. (Pixels)	User Acc. (Pixels)
ROI #7	Soil	99.77	90.83	426/427	426/469
ROI #8	Soil with mph	98.83	96.57	422/427	422/437
ROI #9	Tidal channel	100.00	100.00	427/427	427/427
ROI #10	Saline	100.00	100.00	427/427	427/427
ROI #11	Ponds without water	90.87	99.49	388/427	388/390
ROI #12	Shallow water	100.00	100.00	427/427	427/427
ROI #13	Dry soil	100.00	100.00	427/427	427/427
ROI #14	Macroalgae	98.83	99.29	422/427	422/425
ROI #15	Vegetation 4	89.93	80.00	384/427	384/480

Table A3. Report for the accuracy of the comparison of classification results to random samples generated from ROIs. The table summarizes the producer accuracy and user accuracy in percentage and pixels for each class.

Class	Class Name	Prod. Acc. (Percent)	User Acc. (Percent)	Prod. Acc. (Pixels)	User Acc. (Pixels)
ROI #1	Vegetation 1	97.88	97.13	508/519	508/523
ROI #2	Vegetation 2	73.80	80.25	386/523	386/481
ROI #3	Vegetation 3	66.61	72.40	417/626	417/576
ROI #4	Ponds with mpb	100.00	99.07	427/427	427/431
ROI #5	Vegetation 5	82.86	85.66	986/1190	986/1151
ROI #6	Vegetation 6	86.01	82.09	793/922	793/966
ROI #7	Soil	99.20	98.63	4734/4772	4734/4800
ROI #8	Soil with mph	98.29	97.87	1838/1870	1838/1878
ROI #9	Tidal channel	100.00	100.00	5997/5997	5997/5997
ROI #10	Saline	100.00	99.95	1915/1915	1915/1916
ROI #11	Ponds without water	90.67	93.71	447/493	447/477
ROI #12	Shallow water	99.87	100.00	767/768	767/767
ROI #13	Dry soil	100.00	100.00	508/508	508/508
ROI #14	Macroalgae	98.56	99.39	821/833	821/826
ROI #15	Vegetation 4	89.75	83.13	744/829	744/895

Table A4. Report for the accuracy of the comparison of classification results to R10, the random samples group using 10% of total pixels as reference. The table summarizes the producer accuracy and user accuracy in percentage and pixels for each class.

Class	Class Name	Prod. Acc. (Percent)	User Acc. (Percent)	Prod. Acc. (Pixels)	User Acc. (Pixels)
ROI #1	Vegetation 1	98.39	96.83	61/62	61/63
ROI #2	Vegetation 2	67.44	78.38	29/43	29/37
ROI #3	Vegetation 3	67.92	63.16	36/53	36/57
ROI #4	Ponds with mpb	100.00	100.00	40/40	40/40
ROI #5	Vegetation 5	78.38	83.65	87/111	87/104
ROI #6	Vegetation 6	83.91	79.35	73/87	73/92
ROI #7	Soil	98.82	97.66	501/507	501/513
ROI #8	Soil with mph	100.00	98.32	176/176	176/179
ROI #9	Tidal channel	100.00	100.00	601/601	601/601
ROI #10	Saline	100.00	100.00	188/188	188/188
ROI #11	Ponds without water	79.25	93.33	42/53	42/45
ROI #12	Shallow water	98.90	100.00	90/91	90/90
ROI #13	Dry soil	100.00	100.00	47/47	47/47
ROI #14	Macroalgae	98.55	100.00	68/69	68/68
ROI #15	Vegetation 4	90.11	86.32	82/91	82/95

Table A5. Report for the accuracy of the comparison of classification results to R20, the random samples group using 20% of total pixels as reference. The table summarizes the producer accuracy and user accuracy in percentage and pixels for each class.

Class	Class Name	Prod. Acc. (Percent)	User Acc. (Percent)	Prod. Acc. (Pixels)	User Acc. (Pixels)
ROI #1	Vegetation 1	99.14	97.46	115/116	115/118
ROI #2	Vegetation 2	65.98	76.19	64/97	64/84
ROI #3	Vegetation 3	65.04	74.77	80/123	80/107
ROI #4	Ponds with mpb	100.00	96.43	81/81	81/84
ROI #5	Vegetation 5	83.40	84.45	201/241	201/238
ROI #6	Vegetation 6	88.18	84.04	179/203	179/213
ROI #7	Soil	99.04	98.21	933/942	933/950
ROI #8	Soil with mph	96.49	98.89	357/370	357/361
ROI #9	Tidal channel	100.00	100.00	1162/1162	1162/1162
ROI #10	Saline	100.00	99.75	398/398	398/399
ROI #11	Ponds without water	90.82	89.90	89/98	89/99
ROI #12	Shallow water	100.00	100.00	170/170	170/170
ROI #13	Dry soil	100.00	100.00	77/77	77/77
ROI #14	Macroalgae	97.31	99.45	181/186	181/182
ROI #15	Vegetation 4	91.95	82.47	160/174	160/194

Table A6. Report for the accuracy of the comparison of classification results to Sp10, the stratified-proportionate samples group using 10% of total pixels as reference. The table summarizes the producer accuracy and user accuracy in percentage and pixels for each class.

Class	Class Name	Prod. Acc. (Percent)	User Acc. (Percent)	Prod. Acc. (Pixels)	User Acc. (Pixels)
ROI #1	Vegetation 1	92.31	97.96	48/52	48/49
ROI #2	Vegetation 2	69.23	76.60	36/52	36/47
ROI #3	Vegetation 3	71.43	66.18	45/63	45/68
ROI #4	Ponds with mpb	100.00	100.00	43/43	43/43
ROI #5	Vegetation 5	80.67	87.27	96/119	96/110
ROI #6	Vegetation 6	88.04	81.82	81/92	81/99
ROI #7	Soil	99.58	98.75	475/477	475/481
ROI #8	Soil with mph	98.40	98.40	184/187	184/187
ROI #9	Tidal channel	100.00	100.00	600/600	600/600
ROI #10	Saline	100.00	100.00	192/192	192/192
ROI #11	Ponds without water	89.80	95.65	44/49	44/46
ROI #12	Shallow water	100.00	100.00	77/77	77/77
ROI #13	Dry soil	100.00	100.00	51/51	51/51
ROI #14	Macroalgae	98.80	98.80	82/83	82/83
ROI #15	Vegetation 4	90.36	86.21	75/83	75/87

Table A7. Report for the accuracy of the comparison of classification results to Sp20, the stratified-proportionate samples group using 20% of total pixels as reference. The table summarizes the producer accuracy and user accuracy in percentage and pixels for each class.

Class	Class Name	Prod. Acc. (Percent)	User Acc. (Percent)	Prod. Acc. (Pixels)	User Acc. (Pixels)
ROI #1	Vegetation 1	98.08	97.14	102/104	102/105
ROI #2	Vegetation 2	77.14	80.20	81/105	81/101
ROI #3	Vegetation 3	65.60	70.09	82/125	82/117
ROI #4	Ponds with mpb	100.00	97.70	85/85	85/87
ROI #5	Vegetation 5	81.93	86.67	195/238	195/225
ROI #6	Vegetation 6	87.50	82.14	161/184	161/196
ROI #7	Soil	99.06	98.23	945/954	945/962
ROI #8	Soil with mph	97.59	97.86	365/374	365/373
ROI #9	Tidal channel	100.00	100.00	1199/1199	1199/1199
ROI #10	Saline	100.00	100.00	383/383	383/383

Table A7. *Cont.*

Class	Class Name	Prod. Acc. (Percent)	User Acc. (Percent)	Prod. Acc. (Pixels)	User Acc. (Pixels)
ROI #11	Ponds without water	85.86	90.43	85/99	85/94
ROI #12	Shallow water	100.00	100.00	154/154	154/154
ROI #13	Dry soil	100.00	100.00	102/102	102/102
ROI #14	Macroalgae	98.80	100.00	165/167	165/165
ROI #15	Vegetation 4	89.76	84.66	149/166	149/176

Appendix A.3. Histograms

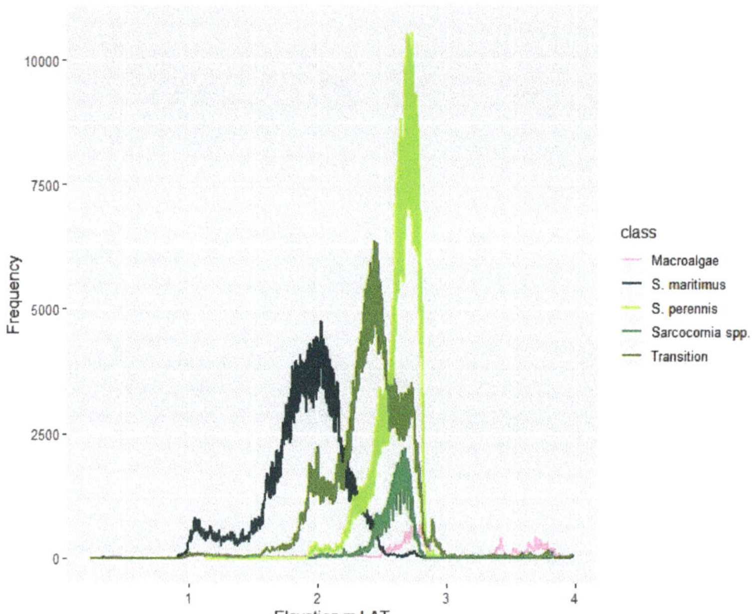

Figure A5. Distribution of the plant species present in the Cádiz Bay with elevation. The elevation refers to LAT.

References

1. Belluco, E.; Camuffo, M.; Ferrari, S.; Modenese, L.; Silvestri, S.; Marani, A.; Marani, M. Mapping Salt-Marsh Vegetation by Multispectral and Hyperspectral Remote Sensing. *Remote. Sens. Environ.* **2006**, *105*, 54–67. [CrossRef]
2. Veldhuis, E.R.; Schrama, M.; Staal, M.; Elzenga, J.T.M. Plant Stress-Tolerance Traits Predict Salt Marsh Vegetation Patterning. *Front. Mar. Sci.* **2019**, *5*, 501. [CrossRef]
3. Adam, P. *Saltmarsh Ecology*; Cambridge University Press: Cambridge, UK, 1990.
4. Duarte, C.M.; Losada, I.J.; Hendriks, I.E.; Mazarrasa, I.; Marbá, N. The Role of Coastal Plant Communities for Climate Change Mitigation and Adaptation. *Nat. Clim. Chang.* **2013**, *3*, 961–968. [CrossRef]
5. McLeod, E.; Chmura, G.L.; Bouillon, S.; Salm, R.; Björk, M.; Duarte, C.M.; Lovelock, C.E.; Schlesinger, W.H.; Silliman, B.R. A Blueprint for Blue Carbon: Toward an Improved Understanding of the Role of Vegetated Coastal Habitats in Sequestering CO_2. *Front. Ecol. Environ.* **2011**, *9*, 552–560. [CrossRef] [PubMed]
6. Möller, I.; Kudella, M.; Rupprecht, F.; Spencer, T.; Paul, M.; van Wesenbeeck, B.K.; Wolters, G.; Jensen, K.; Bouma, T.J.; Miranda-Lange, M.; et al. Wave Attenuation over Coastal Salt Marshes under Storm Surge Conditions. *Nat. Geosci.* **2014**, *7*, 727–731. [CrossRef]
7. Wang, F.; Sanders, C.J.; Santos, I.R.; Tang, J.; Schuerch, M.; Kirwan, M.L.; Kopp, R.E.; Zhu, K.; Li, X.; Yuan, J.; et al. Global Blue Carbon Accumulation in Tidal Wetlands Increases with Climate Change. *Natl. Sci. Rev.* **2021**, *8*, nwaa296. [CrossRef] [PubMed]
8. Davy, A.J.; Brown, M.J.H.; Mossman, H.L.; Grant, A. Colonization of a Newly Developing Salt Marsh: Disentangling Independent Effects of Elevation and Redox Potential on Halophytes. *J. Ecol.* **2011**, *99*, 1350–1357. [CrossRef]

9. Janousek, C.N.; Folger, C.L. Variation in Tidal Wetland Plant Diversity and Composition within and among Coastal Estuaries: Assessing the Relative Importance of Environmental Gradients. *J. Veg. Sci.* **2014**, *25*, 534–545. [CrossRef]
10. Al Hassan, M.; Chaura, J.; López-Gresa, M.P.; Borsai, O.; Daniso, E.; Donat-Torres, M.P.; Mayoral, O.; Vicente, O.; Boscaiu, M. Native-Invasive Plants vs. Halophytes in Mediterranean Salt Marshes: Stress Tolerance Mechanisms in Two Related Species. *Front. Plant Sci.* **2016**, *7*, 473. [CrossRef]
11. Minden, V.; Andratschke, S.; Spalke, J.; Timmermann, H.; Kleyer, M. Plant Trait-Environment Relationships in Salt Marshes: Deviations from Predictions by Ecological Concepts. *Perspect. Plant Ecol. Evol. Syst.* **2012**, *14*, 183–192. [CrossRef]
12. Eleuterius, L.N.; Eleuterius, C.K. Tide Levels and Salt Marsh Zonation. *Bull. Mar. Sci.* **1979**, *29*, 394–400.
13. Roozen, A.J.M.; Westhoff, V. A Study on Long-Term Salt-Marsh Succession Using Permanent Plots. *Vegetatio* **1985**, *61*, 23–32. [CrossRef]
14. Murray, N.J.; Phinn, S.R.; DeWitt, M.; Ferrari, R.; Johnston, R.; Lyons, M.B.; Clinton, N.; Thau, D.; Fuller, R.A. The Global Distribution and Trajectory of Tidal Flats. *Nature* **2019**, *565*, 222–225. [CrossRef] [PubMed]
15. Lowe, M.R.; Peterson, M.S. Effects of Coastal Urbanization on Salt-Marsh Faunal Assemblages in the Northern Gulf of Mexico. *Mar. Coast. Fish.* **2014**, *6*, 89–107. [CrossRef]
16. Schuerch, M.; Spencer, T.; Temmerman, S.; Kirwan, M.L.; Wolff, C.; Lincke, D.; McOwen, C.J.; Pickering, M.D.; Reef, R.; Vafeidis, A.T.; et al. Future Response of Global Coastal Wetlands to Sea-Level Rise. *Nature* **2018**, *561*, 231–234. [CrossRef] [PubMed]
17. Xin, P.; Wilson, A.; Shen, C.; Ge, Z.; Moffett, K.B.; Santos, I.R.; Chen, X.; Xu, X.; Yau, Y.Y.Y.; Moore, W.; et al. Surface Water and Groundwater Interactions in Salt Marshes and Their Impact on Plant Ecology and Coastal Biogeochemistry. *Rev. Geophys.* **2022**, *60*, e2021RG000740. [CrossRef]
18. Fagherazzi, S.; Mariotti, G.; Leonardi, N.; Canestrelli, A.; Nardin, W.; Kearney, W.S. Salt Marsh Dynamics in a Period of Accelerated Sea Level Rise. *J. Geophys. Res. Earth Surf.* **2020**, *125*, e2019JF005200. [CrossRef]
19. Alizad, K.; Hagen, S.C.; Medeiros, S.C.; Bilskie, M.V.; Morris, J.T.; Balthis, L.; Buckel, C.A. Dynamic Responses and Implications to Coastal Wetlands and the Surrounding Regions under Sea Level Rise. *PLoS ONE* **2018**, *13*, e0205176. [CrossRef]
20. Kirwan, M.L.; Mudd, S.M. Response of Salt-Marsh Carbon Accumulation to Climate Change. *Nature* **2012**, *489*, 550–553. [CrossRef]
21. Narayan, S.; Beck, M.W.; Wilson, P.; Thomas, C.J.; Guerrero, A.; Shepard, C.C.; Reguero, B.G.; Franco, G.; Ingram, J.C.; Trespalacios, D. The Value of Coastal Wetlands for Flood Damage Reduction in the Northeastern USA. *Sci. Rep.* **2017**, *7*, 9463. [CrossRef]
22. Watson, E.B.; Oczkowski, A.J.; Wigand, C.; Hanson, A.R.; Davey, E.W.; Crosby, S.C.; Johnson, R.L.; Andrews, H.M. Nutrient Enrichment and Precipitation Changes Do Not Enhance Resiliency of Salt Marshes to Sea Level Rise in the Northeastern U.S. *Clim. Chang.* **2014**, *125*, 501–509. [CrossRef]
23. Silvestri, S.; Marani, M. Salt-Marsh Vegetation and Morphology: Basic Physiology, Modelling and Remote Sensing Observations. In *Ecogeomorphology of Tidal Marshes*; American Geophysical Union: Washington, DC, USA, 2013; Volume 59, pp. 5–25.
24. Lopes, C.L.; Mendes, R.; Caçador, I.; Dias, J.M. Assessing Salt Marsh Loss and Degradation by Combining Long-Term LANDSAT Imagery and Numerical Modelling. *Land Degrad. Dev.* **2021**, *32*, 4534–4545. [CrossRef]
25. Blount, T.R.; Carrasco, A.R.; Cristina, S.; Silvestri, S. Exploring Open-Source Multispectral Satellite Remote Sensing as a Tool to Map Long-Term Evolution of Salt Marsh Shorelines. *Estuar. Coast. Shelf Sci.* **2022**, *266*, 107664. [CrossRef]
26. Li, H.; Wang, C.; Cui, Y.; Hodgson, M. Mapping Salt Marsh along Coastal South Carolina Using U-Net. *ISPRS J. Photogramm. Remote Sens.* **2021**, *179*, 121–132. [CrossRef]
27. Artigas, F.J.; Yang, J. Spectral Discrimination of Marsh Vegetation Types in the New Jersey Meadowlands, USA. *Wetlands* **2006**, *26*, 271–277. [CrossRef]
28. Rajakumari, S.; Mahesh, R.; Saranjith, K.J.; Ramesh, R. Building Spectral Catalogue for Salt Marsh Vegetation, Hyperspectral and Multispectral Remote Sensing. *Reg. Stud. Mar. Sci.* **2022**, *53*, 102435. [CrossRef]
29. Zhuo, W.; Shi, R.; Wu, N.; Zhang, C.; Tian, B. Spectral Response and the Retrieval of Canopy Chlorophyll Content under Interspecific Competition in Wetlands—Case Study of Wetlands in the Yangtze River Estuary. *Earth Sci. Inform.* **2021**, *14*, 1467–1486. [CrossRef]
30. Du, Y.; Wang, J.; Liu, Z.; Yu, H.; Li, Z.; Cheng, H. Evaluation on Spaceborne Multispectral Images, Airborne Hyperspectral, and LiDAR Data for Extracting Spatial Distribution and Estimating Aboveground Biomass of Wetland Vegetation Suaeda Salsa. *IEEE J. Sel. Top. Appl. Earth Obs. Remote Sens.* **2019**, *12*, 200–209. [CrossRef]
31. Rasel, S.M.M.; Chang, H.-C.; Ralph, T.; Saintilan, N. Endmember Identification from EO-1 Hyperion L1_R Hyperspectral Data to Build Saltmarsh Spectral Library in Hunter Wetland, NSW, Australia. In Proceedings of the Remote Sensing for Agriculture, Ecosystems, and Hydrology XVII, Toulouse, France, 22–24 September 2015; Neale, C.M.U., Maltese, A., Eds.; SPIE: Bellinghame, WA, USA, 2015; Volume 9637.
32. Kumar, L.; Sinha, P. Mapping Salt-Marsh Land-Cover Vegetation Using High-Spatial and Hyperspectral Satellite Data to Assist Wetland Inventory. *GIsci. Remote Sens.* **2014**, *51*, 483–497. [CrossRef]
33. Proença, B.; Frappart, F.; Lubac, B.; Marieu, V.; Ygorra, B.; Bombrun, L.; Michalet, R.; Sottolichio, A. Potential of High-Resolution Pléiades Imagery to Monitor Salt Marsh Evolution After Spartina Invasion. *Remote Sens.* **2019**, *11*, 968. [CrossRef]
34. Curcio, A.C.; Peralta, G.; Aranda, M.; Barbero, L. Evaluating the Performance of High Spatial Resolution UAV-Photogrammetry and UAV-LiDAR for Salt Marshes: The Cádiz Bay Study Case. *Remote Sens.* **2022**, *14*, 3582. [CrossRef]

35. Pinton, D.; Canestrelli, A.; Wilkinson, B.; Ifju, P.; Ortega, A. Estimating Ground Elevation and Vegetation Characteristics in Coastal Salt Marshes Using Uav-Based Lidar and Digital Aerial Photogrammetry. *Remote Sens.* **2021**, *13*, 4506. [CrossRef]
36. van Iersel, W.; Straatsma, M.; Addink, E.; Middelkoop, H. Monitoring Height and Greenness of Non-Woody Floodplain Vegetation with UAV Time Series. *ISPRS J. Photogramm. Remote Sens.* **2018**, *141*, 112–123. [CrossRef]
37. Doughty, C.L.; Cavanaugh, K.C. Mapping Coastal Wetland Biomass from High Resolution Unmanned Aerial Vehicle (UAV) Imagery. *Remote Sens.* **2019**, *11*, 540. [CrossRef]
38. Farris, A.S.; Defne, Z.; Ganju, N.K. Identifying Salt Marsh Shorelines from Remotely Sensed Elevation Data and Imagery. *Remote Sens.* **2019**, *11*, 1795. [CrossRef]
39. Yan, D.; Li, J.; Yao, X.; Luan, Z. Integrating UAV Data for Assessing the Ecological Response of Spartina Alterniflora towards Inundation and Salinity Gradients in Coastal Wetland. *Sci. Total Environ.* **2022**, *814*, 152631. [CrossRef] [PubMed]
40. Zhu, H.; Huang, Y.; Li, Y.; Yu, F.; Zhang, G.; Fan, L.; Zhou, J.; Li, Z.; Yuan, M. Predicting Plant Diversity in Beach Wetland Downstream of Xiaolangdi Reservoir with UAV and Satellite Multispectral Images. *Sci. Total Environ.* **2022**, *819*, 153059. [CrossRef]
41. Villoslada, M.; Bergamo, T.F.; Ward, R.D.; Burnside, N.G.; Joyce, C.B.; Bunce, R.G.H.; Sepp, K. Fine Scale Plant Community Assessment in Coastal Meadows Using UAV Based Multispectral Data. *Ecol. Indic.* **2020**, *111*, 105979. [CrossRef]
42. Yang, H.; Du, J. Classification of Desert Steppe Species Based on Unmanned Aerial Vehicle Hyperspectral Remote Sensing and Continuum Removal Vegetation Indices. *Optik* **2021**, *247*, 167877. [CrossRef]
43. Ivushkin, K.; Bartholomeus, H.; Bregt, A.K.; Pulatov, A.; Franceschini, M.H.D.; Kramer, H.; van Loo, E.N.; Jaramillo Roman, V.; Finkers, R. UAV Based Soil Salinity Assessment of Cropland. *Geoderma* **2019**, *338*, 502–512. [CrossRef]
44. Zhou, Q.; Wang, J.; Tian, L.; Feng, L.; Li, J.; Xing, Q. Remotely Sensed Water Turbidity Dynamics and Its Potential Driving Factors in Wuhan, an Urbanizing City of China. *J. Hydrol.* **2021**, *593*, 125893. [CrossRef]
45. *Consejería de Medioambiente Parque Natural de La Bahía de Cádiz*; CMA: Sevilla, Spain, 2006.
46. Ramsar. Available online: https://www.ramsar.org/es (accessed on 23 November 2022).
47. Araújo, C.V.M.; Diz, F.R.; Laiz, I.; Lubián, L.M.; Blasco, J.; Moreno-Garrido, I. Sediment Integrative Assessment of the Bay of Cádiz (Spain): An Ecotoxicological and Chemical Approach. *Environ. Int.* **2009**, *35*, 831–841. [CrossRef]
48. Coll, M.; Carreras, M.; Ciércoles, C.; Cornax, M.J.; Gorelli, G.; Morote, E.; Sáez, R. Assessing Fishing and Marine Biodiversity Changes Using Fishers' Perceptions: The Spanish Mediterranean and Gulf of Cadiz Case Study. *PLoS ONE* **2014**, *9*, e85670. [CrossRef]
49. Pérez Latorre, A.V.; Galán de Mera, A.; Deil, U.; Cabezudo, B. Fitogeografía y Vegetación Del Sector Aljíbico (Cádiz-Málaga, España). *Acta Bot. Malacit.* **1996**, *21*, 241–267. [CrossRef]
50. Rueda, J.L.; González-García, E.; Marina, P.; Oporto, T.; Rittierott, C.; López-González, N.; Farias, C.; Moreira, J.; López, E.; Megina, C.; et al. Biodiversity and Geodiversity in the Mud Volcano Field of the Spanish Margin (Gulf of Cádiz). In Proceedings of the 7° Simpósio Sobre a Margem Ibérica Atlântica—MIA, Lisboa, Portugal, 16–20 December 2012. [CrossRef]
51. Plomaritis, T.A.; Benavente, J.; Laiz, I.; del Río, L. Variability in Storm Climate along the Gulf of Cadiz: The Role of Large Scale Atmospheric Forcing and Implications to Coastal Hazards. *Clim. Dyn.* **2015**, *45*, 2499–2514. [CrossRef]
52. González, M.; Álvarez-Gómez, J.A.; Aniel-Quiroga, Í.; Otero, L.; Olabarrieta, M.; Omira, R.; Luceño, A.; Jelinek, R.; Krausmann, E.; Birkman, J.; et al. Probabilistic Tsunami Hazard Assessment in Meso and Macro Tidal Areas. Application to the Cádiz Bay, Spain. *Front. Earth Sci.* **2021**, *9*, 591383. [CrossRef]
53. Peralta, G.; Godoy, O.; Egea, L.G.; de los Santos, C.B.; Jiménez-Ramos, R.; Lara, M.; Brun, F.G.; Hernández, I.; Olivé, I.; Vergara, J.J.; et al. The Morphometric Acclimation to Depth Explains the Long-Term Resilience of the Seagrass Cymodocea Nodosa in a Shallow Tidal Lagoon. *J. Environ. Manag.* **2021**, *299*, 113452. [CrossRef] [PubMed]
54. Jiménez-Arias, J.L.; Morris, E.; Rubio-de-Inglés, M.J.; Peralta, G.; García-Robledo, E.; Corzo, A.; Papaspyrou, S. Tidal Elevation Is the Key Factor Modulating Burial Rates and Composition of Organic Matter in a Coastal Wetland with Multiple Habitats. *Sci. Total Environ.* **2020**, *724*. [CrossRef] [PubMed]
55. Life Blue Natura 2017 "Caracterización de La Marisma Mareal de La Bahía de Cádiz. Proyecto LIFE 14 CCM/ES/000957 "Blue Natura Andalucía" Expte. 2016/000225/M. Available online: https://life-bluenatura.eu/wp-content/uploads/2017/01/Deliverable_B.Cadiz_cartography.pdf (accessed on 15 October 2022).
56. del Río, L.; Plomaritis, T.A.; Benavente, J.; Valladares, M.; Ribera, P. Establishing Storm Thresholds for the Spanish Gulf of Cádiz Coast. *Geomorphology* **2012**, *143–144*, 13–23. [CrossRef]
57. García de Lomas, J.; García, C.M.; Álvarez, Ó. Vegetación de Las Marismas de Aletas-Cetina (Puerto Real). Identificación de Hábitats de Interés Comunitario y Estimaciones Preliminares de Posibles Efectos de Su Inundación. *Rev. De La Soc. Gaditana De Hist. Nat.* **2006**, *5*, 9–38.
58. Vehículos Aéreos—Servicio de Drones (Uca.Es). Available online: https://dron.uca.es/vehiculos-aereos/ (accessed on 2 November 2022).
59. Barreto, M.A.P.; Johansen, K.; Angel, Y.; McCabe, M.F. Radiometric Assessment of a UAV-Based Push-Broom Hyperspectral Camera. *Sensors* **2019**, *19*, 4699. [CrossRef] [PubMed]
60. Wang, C.; Myint, S.W. A Simplified Empirical Line Method of Radiometric Calibration for Small Unmanned Aircraft Systems-Based Remote Sensing. *IEEE J. Sel. Top. Appl. Earth Obs. Remote Sens.* **2015**, *8*, 1876–1885. [CrossRef]
61. Gandhi, G.M.; Parthiban, S.; Thummalu, N.; Christy, A. Ndvi: Vegetation Change Detection Using Remote Sensing and Gis—A Case Study of Vellore District. *Procedia Comput. Sci.* **2015**, *57*, 1199–1210. [CrossRef]

62. Schwengerdt, R.A. *Remote Sensing: Models and Methods for Image Processing*, 2nd ed.; Academic Press: San Diego, CA, USA, 1997.
63. Bakos, K.L.; Gamba, P. Hierarchical Hybrid Decision Tree Fusion of Multiple Hyperspectral Data Processing Chains. *IEEE Trans. Geosci. Remote Sens.* **2011**, *49*, 388–394. [CrossRef]
64. ENVI—Environment for Visualizing Images. Available online: https://www.l3harrisgeospatial.com/docs/using_envi_Home.html (accessed on 12 December 2022).
65. Boardman, J.W.; Kruse, F.A.; Green, R.O. Mapping Target Signatures via Partial Unmixing of AVIRIS Data. In Proceedings of the Summaries of the Fifth Annual JPL Airborne Earth Science Workshop, Pasadena, CA, USA, 23–26 January 1995; Volume 1.
66. Chang, C.I.; Plaza, A. A Fast Iterative Algorithm for Implementation of Pixel Purity Index. *IEEE Geosci. Remote Sens. Lett.* **2006**, *3*, 63–67. [CrossRef]
67. Jan, M.M.; Zainal, N.; Jamaludin, S. Region of Interest-Based Image Retrieval Techniques: A Review. *IAES Int. J. Artif. Intell.* **2020**, *9*, 520–528. [CrossRef]
68. Abbas, A.W.; Minallh, N.; Ahmad, N.; Abid, S.A.R.; Khan, M.A.A. K-Means and ISODATA Clustering Algorithms for Landcover Classification Using Remote Sensing. *Sindh Univ. Res. J. (Sci. Ser.)* **2016**, *48*, 315–318.
69. Liu, X.; Yang, C. A Kernel Spectral Angle Mapper Algorithm for Remote Sensing Image Classification. In Proceedings of the 6th International Congress on Image and Signal Processing (CISP 2013), Hangzhou, China, 16–18 December 2013; IEEE: Hangzhou, China, 2013; pp. 814–848. [CrossRef]
70. Melgani, F.; Bruzzone, L. Classification of Hyperspectral Remote Sensing Images with Support Vector Machines. *IEEE Trans. Geosci. Remote Sens.* **2004**, *42*, 1778–1790. [CrossRef]
71. Hsu, C.-W.; Chang, C.-C.; Lin, C.-J. *A Practical Guide to Support Vector Classification*; Technical Report; Department of Computer Science and Information Engineering, University of National Taiwan: Taipei, China, 2003; pp. 1–12. Available online: http://www.csie.ntu.edu.tw/~cjlin/papers/guide/guide.pdf (accessed on 12 December 2022).
72. Schmidt, K.S.; Skidmore, A.K. Spectral Discrimination of Vegetation Types in a Coastal Wetland. *Remote Sens. Environ.* **2003**, *85*, 92–108. [CrossRef]
73. Clark, R.N.; Roush, T.L. Reflectance Spectroscopy: Quantitative Analysis Techniques for Remote Sensing Applications. *J. Geophys. Res.* **1984**, *89*, 6329–6340. [CrossRef]
74. Torres-Pérez, J.L.; Guild, L.S.; Armstrong, R.A. Hyperspectral Distinction of Two Caribbean Shallow-Water Corals Based on Their Pigments and Corresponding Reflectance. *Remote Sens.* **2012**, *4*, 3813–3832. [CrossRef]
75. de Vries, M.; van der Wal, D.; Möller, I.; van Wesenbeeck, B.; Peralta, G.; Stanica, A. Earth Observation and the Coastal Zone: From Global Images to Local Information. FP7 FAST Project Syntesis Report. *Zenodo* **2018**. [CrossRef]
76. Steffen, S.; Ball, P.; Mucina, L.; Kadereit, G. Phylogeny, Biogeography and Ecological Diversification of *Sarcocornia* (Salicornioideae, Amaranthaceae). *Ann. Bot.* **2015**, *115*, 353–368. [CrossRef] [PubMed]
77. Didore, V.A.; Vaidya, R.S.; Nalawade, D.B.; Kale, K. v Classification of EO-1 Hyperion Data Using Supervised Minimum Distance Algorithm and Spectral Angle Mapper. *J. Emerg. Technol. Innov. Res.* **2021**, *8*, 148–152.
78. Oppelt, N.; Shulze, F.; Bartsch, I.; Doernhoefer, K.; Eisenhardt, I. Hyperspectral Classification Approaches for Intertidal Macroalgae Habitat Mapping: A Case Study in Heligoland. *Opt. Eng.* **2012**, *51*, 111703. [CrossRef]
79. Liu, G.; Tang, P.; Zhan-qing, C.; Wang, T.; Xu, J. A Study on Effect of Water Background on Canopy Spectral of Wetland Aquatic Plant. *Guang Pu Xue Yu Guang Pu Fen Xi* **2015**, *35*, 2970–2976.
80. Claudio, H.C.; Cheng, Y.; Fuentes, D.A.; Gamon, J.A.; Luo, H.; Oechel, W.; Qiu, H.L.; Rahman, A.F.; Sims, D.A. Monitoring Drought Effects on Vegetation Water Content and Fluxes in Chaparral with the 970 Nm Water Band Index. *Remote Sens. Environ.* **2006**, *103*, 304–311. [CrossRef]
81. Sims, D.A.; Gamon, J.A. Estimation of Vegetation Water Content and Photosynthetic Tissue Area from Spectral Reflectance: A Comparison of Indices Based on Liquid Water and Chlorophyll Absorption Features. *Remote Sens. Environ.* **2003**, *84*, 526–537. [CrossRef]
82. Wang, J.; Xu, R.; Yang, S.; Wang, J.; Xu, R.; Yang, S.; Xu, R. Estimation of Plant Water Content by Spectral Absorption Features Centered at 1,450 Nm and 1,940 Nm Regions. *Environ. Monit. Assess.* **2008**, *157*, 459–469. [CrossRef]
83. Bannari, A.; El-Battay, A.; Bannari, R.; Rhinane, H. Sentinel-MSI VNIR and SWIR Bands Sensitivity Analysis for Soil Salinity Discrimination in an Arid Landscape. *Remote Sens.* **2018**, *10*, 855. [CrossRef]
84. Lugassi, R.; Goldshleger, N.; Chudnovsky, A. Studying Vegetation Salinity: From the Field View to a Satellite-Based Perspective. *Remote Sens.* **2017**, *9*, 122. [CrossRef]
85. Kumar, S.; Arya, S.; Jain, K. A SWIR-Based Vegetation Index for Change Detection in Land Cover Using Multi-Temporal Landsat Satellite Dataset. *Int. J. Inf. Technol.* **2022**, *14*, 2035–2048. [CrossRef]
86. Feng, Y.; Sun, T.; Zhu, M.S.; Qi, M.; Yang, W.; Shao, D.D. Salt Marsh Vegetation Distribution Patterns along Groundwater Table and Salinity Gradients in Yellow River Estuary under the Influence of Land Reclamation. *Ecol. Indic.* **2018**, *92*, 82–90. [CrossRef]
87. Hladik, C.; Schalles, J.; Alber, M. Salt Marsh Elevation and Habitat Mapping Using Hyperspectral and LIDAR Data. *Remote Sens. Environ.* **2013**, *139*, 318–330. [CrossRef]
88. Gabriel, J.R.; Reid, J.; Wang, L.; Mozdzer, T.J.; Whigham, D.F.; Megonigal, J.P.; Langley, J.A. Interspecific Competition Is Prevalent and Stabilizes Plant Production in a Brackish Marsh Facing Sea Level Rise. *Estuaries Coasts* **2022**, *45*, 1646–1655. [CrossRef]
89. Redondo-Gómez, S.; Castillo, J.M.; Luque, C.J.; Luque, T.; Figueroa, M.E.; Davy, A.J. Fundamental Niche Differentiation in Subspecies of Sarcocornia Perennis on a Salt Marsh Elevational Gradient. *Mar. Ecol. Prog. Ser.* **2007**, *347*, 15–20. [CrossRef]

90. García-Robledo, E.; Corzo, A.; García De Lomas, J.; van Bergeijk, S.A. Biogeochemical Effects of Macroalgal Decomposition on Intertidal Microbenthos: A Microcosm Experiment. *Mar. Ecol. Prog. Ser.* **2008**, *356*, 139–151. [CrossRef]
91. Newton, C.; Thornber, C. Ecological Impacts of Macroalgal Blooms on Salt Marsh Communities. *Estuaries Coasts* **2013**, *36*, 365–376. [CrossRef]

Article

Using a Vegetation Index as a Proxy for Reliability in Surface Reflectance Time Series Reconstruction (RTSR)

Pieter Kempeneers *, Martin Claverie and Raphaël d'Andrimont

European Commission, Joint Research Centre (JRC), 21027 Ispra, Italy
* Correspondence: pieter.kempeneers@ec.europa.eu

Abstract: Time series of optical remote sensing data are instrumental for monitoring vegetation dynamics, but are hampered by missing or noisy observations due to varying atmospheric conditions. Reconstruction methods have been proposed, most of which focus on time series of a single vegetation index. Under the assumption that relatively high vegetation index values can be considered as trustworthy, a successful approach is to adjust the smoothed value to the upper envelope of the time series. However, this assumption does not hold for surface reflectance in general. Clouds and cloud shadows result in, respectively, high and low values in the visible and near infrared part of the electromagnetic spectrum. A novel spectral Reflectance Time Series Reconstruction (RTSR) method is proposed. Smoothed values of surface reflectance values are adjusted to approach the trustworthy observations, using a vegetation index as a proxy for reliability. The Savitzky–Golay filter was used as the smoothing algorithm here, but different filters can be used as well. The RTSR was evaluated on 100 sites in Europe, with a focus on agriculture fields. Its potential was shown using different criteria, including smoothness and the ability to retain trustworthy observations in the original time series with RMSE values in the order of 0.01 to 0.03 in terms of surface reflectance.

Keywords: time series; reconstruction algorithm; smoothing; optical remote sensing

Citation: Kempeneers, P.; Claverie, M.; d'Andrimont, R. Using a Vegetation Index as a Proxy for Reliability in Surface Reflectance Time Series Reconstruction (RTSR). *Remote Sens.* **2023**, *15*, 2303. https://doi.org/10.3390/rs15092303

Academic Editor: Lenio Soares Galvao

Received: 19 January 2023
Revised: 20 April 2023
Accepted: 24 April 2023
Published: 27 April 2023

1. Introduction

Optical remote sensing data become available with an increased temporal and spatial resolution, in particular since the Copernicus program launched the Sentinel-2A and Sentinel-2B satellites in 2015 and 2017 [1]. This has opened new opportunities for monitoring vegetation dynamics up to the local scale. Inspecting changes of the spectral signature of vegetation over time can be used to, e.g., monitor crop growth [2], forest degradation [3], and grassland use intensity [4].

However, optical remote sensing is highly affected by atmospheric perturbations, which cause both gaps and noise in the time series. In [5–7], the problem of gaps is addressed by data assimilation from multiple sensors with similar characteristics. Gaps can also be filled by combining data from sensors with different characteristics such as optical and synthetic aperture radar (SAR) sensors [8]. More recently, machine learning models have been introduced to fill gaps by predicting the missing values, e.g., from SAR data [9]. In [10], Landsat and Sentinel-2 surface reflectance have been fused using deep learning techniques. A more simple approach is to fill gaps in the time series via interpolation of clear observations [11]. This avoids issues with data harmonization and co-registration between different sensors [12]. Noise in the time series are due to undetected clouds or other variations in atmospheric conditions such as aerosol concentration. A subsequent noise reduction filter is therefore often applied [13,14]. One of the difficulties in designing such a filter is the trade-off between the ability to reduce unwanted noise and the retention of the relevant changes. Low-pass or smoothing filters are effective at reducing noise in time series with high-frequency fluctuations, but can also affect seasonal vegetation changes. Edges in the vegetation signature risk to be blurred by such filters. These edges are useful features for agriculture applications that study crop harvest practices and rely on, e.g., the identification of the start and end of the growing season [15,16]. Furthermore, for the detection of disruptive events such as land cover change, fire, and floods, the design of the smoothing filter requires special care [13].

Within the context of time-series reconstruction, vegetation indices such as the normalized difference vegetation index (NDVI [17]) have the interesting property that they are usually depressed in cloudy or poor atmospheric conditions [18]. In [19], NDVI was analyzed under different observation conditions. The authors demonstrated that the NDVI values for all land cover types increase with lower aerosol concentrations, near-nadir viewing and high solar illumination. It was therefore concluded that "the best possible pixel value for a particular location is achieved by choosing the highest pixel value from multi temporal data". High NDVI values can therefore be assumed to be more trustworthy than low NDVI values. This has been successfully used to reconstruct time series of NDVI by approaching the upper NDVI envelope via an iterative process [20–22]. The maximum NDVI composite is also based on this property. It creates a single noise-free image from a series of images by selecting the acquisition with the maximum NDVI value [19,23,24]. An overview of techniques to reduce noise of NDVI time series is provided in [25].

For some applications, a multi-variate analysis is preferred over a single index. On the use of time series for detecting land disturbance in [26], it was found that the combined use of spectral bands was better than using a single spectral band or index. Furthermore, in the context of analysis-ready data (ARD [27]), the reconstruction of surface reflectance time series is important. A fill-and-fit approach was followed in [28], where missing pixels are filled with a clear observation in the same or a temporally close image. The selection of the clear observation is based on a similarity measure. A subsequent fitting, based on a (linear or non-linear) harmonic model, then reconstructs the time series. In [29], both NDVI and surface reflectance time series from Sentinel-2 were reconstructed using the penalized least-square regression based on the discrete cosine transform (DCT-PLS). Surface reflectance was also reconstructed in [30], by incorporating the upper envelopes of the time-series vegetation index as constraint conditions. The authors reconstructed surface-reflectance time series for MODIS [30] and advanced very high resolution radiometer (AVHRR) data [31]. Recently, a dynamic temporal smoothing (DTS [32]) method was proposed that can also be applied to surface reflectance values. Although the authors focus in their paper on the two-band enhanced vegetation index (EVI2 [33]), the code presented in [32] can be applied to time series of spectral reflectance. However, the DTS algorithm presented in [33] involves an adjustment of the smoothed value of the vegetation index under study (i.e., EVI2) to the upper envelope of the time series. The assumption that relatively high values correspond to trustworthy observations does not hold for surface reflectance in general. On the contrary, the reflectance value in the visible and near infrared part of the electromagnetic spectrum typically increases for cloud covered pixels.

In this study, a new surface Reflectance Time Series Reconstruction (RTSR) method is proposed for vegetation monitoring. It adjusts the smoothed values, similar to existing reconstruction methods that act on vegetation indices. It hereby decouples the vegetation index as a proxy for reliability from the time series of the reflectance values to be reconstructed. Unlike existing methods that let the smoothed time series approach the upper envelope, the surface-reflectance times series here approaches the trustworthy observations. The remaining sections are structured as follows. In Section 2, the materials are covered with a description of the test sites and remote-sensing time series. The RTSR method is described in Section 3 as well the as the evaluation procedure. Results and the evaluation are presented in Section 4. A discussion with limitations of the proposed method is described in Section 5. Conclusions are drawn in Section 6.

2. Material

2.1. Test Sites

A total of 100 sites were selected to test the reconstruction method (see Figure 1), following a similar approach as in AI4Boundaries [34]. In this parallel research project on mapping crop field boundaries, a random stratified sampling method was designed to extract image chips from various landscapes. The sample was drawn from six European countries for which public parcel data are available: Austria, Spain, France, Netherlands, Slovenia, and Sweden. Each test site covers 256 by 256 pixels, corresponding to 2560 by 2560 m. This is still sufficiently large to contain some contextual information for visual interpretation and can be processed on a computer with relatively low memory constraints (2 GB). The majority of pixels was vegetated land, the land cover in focus for this study. Both agricultural land (57% of all pixels) and forest (27%) were represented. Agricultural land included non-irrigated arable land (27%), pastures (11%), and a mosaic of small cultivated land parcels with different cultivation types (13%). Non-irrigated arable land is defined as cultivated land parcels under rainfed agricultural use for annually harvested non-permanent crops, normally under a crop rotation system, including fallow lands within such crop rotation [35]. Forested pixels were either broad-leaved forest (12%), coniferous forest (13%), or a mixture of both (2%). The remaining non-vegetated pixels included artificial surface and water bodies.

Figure 1. Test sample: 100 sites of 2560 by 2560 m distributed over six European countries (Austria, Spain, France, Netherlands, Slovenia, and Sweden).

2.2. Remote Sensing Time Series

Sentinel-2 products with Level-2A bottom of atmosphere (BOA) reflectance data were downloaded from the Copernicus Open Access Hub [36]. The proposed RTSR method is applicable to the complete set of spectral bands, but only five bands have been selected here in order not to overload the section on results and the figures. The spectral bands B2, B3, B4, and B8 were selected based on their superior spatial resolution (10 m). These bands only cover the visible to near infrared part of the electromagnetic spectrum. They were therefore complemented with spectral band B12, to also cover the short-wave infrared. The proposed method can apply to other data sources, if a vegetation index can be calculated and the assumption that relatively high vegetation index values can be considered as trustworthy holds. In practice, this will be for most optical remote-sensing imagery with a red and near infrared band. Here, the NDVI was calculated using the red (B4) and near infrared (B8) bands:

$$\mathrm{NDVI} = \frac{\mathrm{B8} - \mathrm{B4}}{\mathrm{B4} + \mathrm{B8}} \tag{1}$$

The ground sampling distance (GSD) of the bands in the visual and near infrared is 10 m, whereas the GSD of band B12 in the short wave infrared is 20 m. All selected bands were resampled to 10 m based on the nearest neighbor to obtain a regular gridded data cube. In addition to the spectral bands, Sentinel-2 products are delivered with a scene classification (SCL) band, which is an output of the Sen2Cor atmospheric correction processor [37]. It distinguishes 11 classes including information on clouds and cloud shadows. The products downloaded covered all acquisitions from beginning of December 2018 to the end of January 2020 with a cloud cover of less than 70%. A subset of eight acquisitions for one of the test sites in France under various atmospheric conditions is shown in Figure 2.

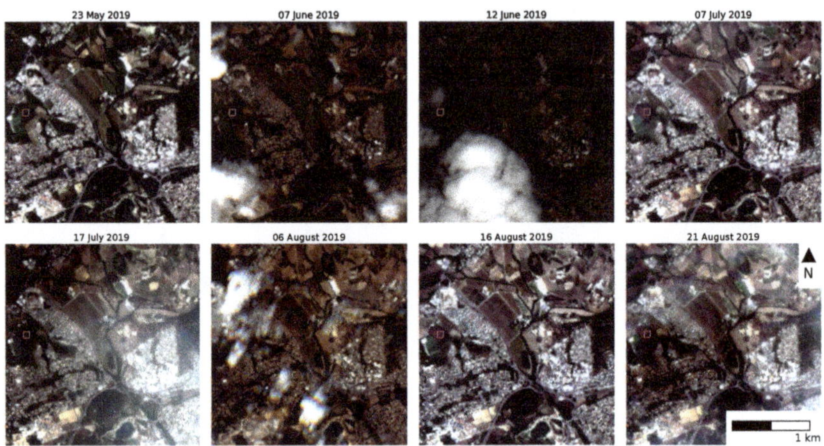

Figure 2. True color images (bands B4, B3, B2) for a subset of the time series acquired with Sentinel-2 under various atmospheric conditions (selected acquisition dates: 23 May 2019, 7 June 2019, 12 June 2019, 7 July 2019, 17 July 2019, 6 August 2019, 16 August 2019, 21 August 2019). The red square area corresponds to an agriculture field near Montpellier, France (43°39′53.64″N, 3°54′47.88″E).

3. Methods

3.1. Masking

The masking of outliers in the time series is a common step in most reconstruction methods [20–22]. The risk of not masking outliers is that contaminated reflectance values propagate to the reconstructed time series and result in residual noise. In the context of this study, masking was based on the SCL band. Pixels identified as dark (SCL = 2), vegetated (SCL = 4), not-vegetated (SCL = 5), water (SCL = 6), and unclassified (SCL = 7) were considered as clear. The remaining classes were masked as "not clear": no data (SCL = 0), saturated or defective (SCL = 1), cloud shadows (SCL = 3), clouds (SCL = 8–9), thin cirrus (10), and snow or ice (SCL = 11). To mitigate some of the omission errors in the SCL band at the cloud edges, a distance-based buffer of five pixels (50 m) was added to the masked pixels. Larger buffer sizes of 100–300 m are commonly used, as proposed in [38]. Here, a relatively small buffer was found to be effective (e.g., masking the contaminated observation corresponding to the acquisition on 12 June 2019), without removing a large amount of usable imagery [39].

Part of time series (April to September 2019) of the reflectance in the red (band B4) and near infrared (band B8) as well as the NDVI for a pixel in an agricultural field (red square in Figure 2) is shown in Figure 3. The eight vertical lines correspond to the eight acquisition dates that have been selected in Figure 2. The masked observations are represented as empty dots in Figure 3. Some of them can easily be identified as outliers, based on the time series in Figure 3. For instance, the observation corresponding to 6 August 2019 (vertical line 6) has a higher reflectance value than expected from the time series (see Figure 3d). This observation was indeed identified by Sen2Cor as cloudy (with medium probability: SCL = 8). Furthermore, masked were the observations acquired on 7 July 2019, 17 July 2019 (thin cirrus: SCL = 10) and 21 August 2019 (SCL = 8).

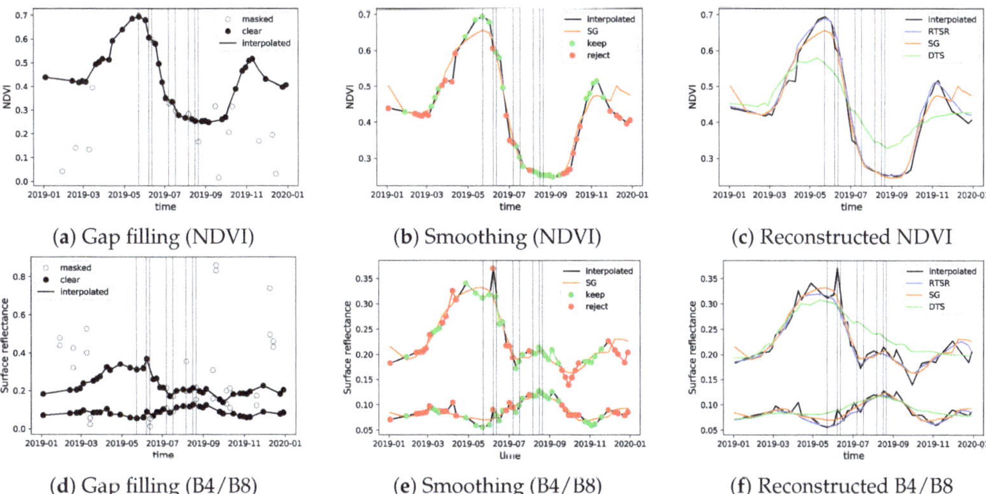

Figure 3. Reconstruction of NDVI (top) and spectral bands B4 and B8 (bottom). Vertical lines correspond to selected acquisition dates in Figure 2. Masked pixels are interpolated to fill gaps in the time series (left column). The iterative reconstruction process (middle column) checks whether observations are trustworty (green) or not (orange) based on the NDVI value. The NDVI values in green that are above the orange long-term change trend curve (SG) are retained for the next iteration. Those in red that are below will be replaced by the corresponding values on SG curve. Reconstructed time series (right column) for the proposed RTSR method (in blue) compared to SG (orange) and the dynamic temporal smoothing (DTS, in green).

3.2. Adaptive Smoothing

The proposed reconstruction algorithm requires complete time series with contiguous observations in the temporal domain for each spectral band. Gaps due to missing or masked observations are filled with linearly interpolated values of clear observations. In case values are masked near the beginning and end of the time series, the nearest clear observation is extrapolated.

Even after interpolating the masked values, the reflectance values $R_{\lambda,i}$ (with λ representing the wavelength and i the acquisition time) can still be noisy. For instance, the acquisition of 7 June 2019 corresponds to a "clear" observation (vertical line 2 in Figure 3). However, the reflectance in bands B4 and B8 is higher than expected, based on the previous and next clear observations (see Figure 3d). The isolated high reflectance can be interpreted as an outlier. This is also confirmed by the corresponding NDVI value, N_i (see Figure 3a). Similar to the time series of the interpolated spectral reflectance, $R_{\lambda,i}$, the time series for the interpolated NDVI value, N_i, also shows an outlier for the acquisition from 7 June 2019. The isolated drop in NDVI is not compatible with a gradual change of vegetation and can be regarded as noisy. By inspecting Figure 3a, it is shown that the image scene indeed contains some cloud cover that could explain the noise.

Several algorithms for smoothing time series have been proposed in the literature, including the Whittaker smoother [40], the Savitzky–Golay filter [41], and its modifications [20,42]. Because the proposed RTSR algorithm was inspired by the work in [20], the same smoothing filter (Savitzky–Golay) was chosen here. It is important to note that other smoothing algorithms can easily be adopted in the proposed reconstruction method. Similar to the approach in [20] to reconstruct the NDVI time series, a first Savitzky–Golay filter, referred to as SG^1, creates the long-term change trend (represented by the orange lines in Figure 3). It has a relatively wide half-width of the smoothing window (7). This value is based on our own experiments and on values found in the literature [20]. Likewise, the parameter for the polynomial degree was set to 2 for SG^1 (suggested between 2 and 4 in [20]). Larger values are supposed to better follow higher frequencies in the time series.

As illustrated in Figure 3b, the long-term change trend of the NDVI, $SG^1(N_i)$, underestimates the NDVI values for the clear observations in May. Based on the assumption that relatively high NDVI values correspond to trustworthy observations, the upper NDVI envelope is expected to better reflect the dynamic changes of interest in the NDVI temporal profile [20–22]. The acquisitions with a calculated NDVI above the long-term change trend (green dots) are therefore assumed to be more reliable than those with a calculated NDVI value below the long-term change trend (orange dots). Following this reasoning, the original spectral reflectance values $R_{\lambda,i}$ observed in May can be considered as reliable. Unlike the NDVI, however, the reflectance values are lower than the long-term change trend $SG^1(R_{\lambda,i})$ (indicated in orange in Figure 3e). The assumption that clouds or poor atmospheric conditions depress NDVI values does not hold for the individual spectral bands. For instance, reflectance values in red and near infrared typically increase in case of cloud cover, but decrease in case of cloud shadow. Neither the upper nor the lower envelope reflect the dynamic changes of interest in the spectral reflectance temporal profile that would be observed in clear conditions. A different algorithmic approach for spectral reflectance was therefore developed, which is listed in Algorithm 1.

The reconstructed reflectance time series $\hat{R}_{\lambda,i}$ is initialized as the long-term change trend $SG^1(R_{\lambda,i})$. If, for a given acquisition i, the interpolated NDVI value N_i (calculated from the interpolated reflectance $R_{\lambda,i}$) shows a higher value than the current NDVI estimate $\hat{N}_i$ (calculated from the estimated reflectance $\hat{R}_{\lambda,i}$), then the corresponding interpolated reflectance value $R_{\lambda,i}$ is assumed to be reliable and replaces the current estimate $\hat{R}_{\lambda,i}$ (green dots in Figure 3e). Else, the current estimate $\hat{R}_{\lambda,i}$ is retained in favor of the interpolated reflectance value $R_{\lambda,i}$ that is assumed to be noisy for the given acquisition i (e.g., orange dot in Figure 3e). By comparing the green and orange dots in both Figure 3b,e, it is shown that the decision which value to select (smoothed or interpolated version) is driven by the NDVI time series.

The algorithm proceeds by smoothing the retained spectral values with a second version of the SG filter, referred to as SG^2. The smoothed values are then used for the next iteration. The reconstructed spectral reflectance time series $\hat{R}_{\lambda,i}$ after the final iteration is shown in Figure 3c,f in blue. Different stopping criteria for number of iterations can be implemented. In [20], a fitting-effect index was proposed. Here, the number of iterations was experimentally fixed to five. More iterations did not further improve results. As a comparison, the reconstructed time series based on the dynamic temporal smoothing (DTS [33]) method is also shown (in green).

Algorithm 1 Reflectance reconstruction algorithm

```
R̂_{λ,i} ← SG¹(R_{λ,i})        ▷ initialize reconstructed reflectance with long-term change trend
while Stopping criterion not met (e.g., 5 iterations) do
    if N̂_i < N_i then                                        ▷ observation is clear
        R̂_{λ,i} ← R_{λ,i}                        ▷ update reconstructed reflectance
    end if
    R̂_{λ,i} ← SG²(R̂_{λ,i})                      ▷ smoothen reconstructed reflectance
end while
```

3.3. Evaluation Procedure

Different metrics were defined to evaluate the proposed reconstruction method, as requirements of a successful approach are also diverse. Noise should be reduced while changes related to vegetation dynamics of interest should be retained. A first metric was defined to quantify the smoothness, the time-series smoothness index (TSI) as proposed in [43]. It is calculated as the average absolute difference between the reconstructed reflectance value $\hat{R}_{\lambda,i}$ at acquisition time i and the mean of the corresponding values before and after that acquisition. The difference is then averaged over all T acquisitions in the time series for each pixel (p):

$$\mathrm{TSI}_{\lambda}(p) = \frac{1}{T}\sum_{i=1}^{T}|\hat{R}_{\lambda,i}(p) - \frac{\hat{R}_{\lambda,i-1}(p) + \hat{R}_{\lambda,i-1}(p)}{2}| \tag{2}$$

Low TSI values indicate smooth time series and high TSI values indicate noisy time series for a specific spectral band λ. A similar evaluation method was conducted in [42,44]. Reconstruction methods with a high level of smoothing are expected to perform well on the TSI metric. However, they risk concealing the temporal changes of interest. Clear observations with low noise content should be represented in the reconstructed time series. A second evaluation metric was therefore based on the error between the reconstructed time series $\hat{R}_{\lambda,i}$ and the original observation $R_{\lambda,i}$ in clear conditions ($i = i_{\mathrm{clear}}$). For each test site, i_{clear} was selected as the acquisition with the maximum number of clear pixels. The reconstructed time series $\hat{R}_{\lambda,i}$ was then compared to the observed time series $R_{\lambda,i}$ at acquisition time i_{clear} (see top of Figure 4). Scatterplots were created per spectral band based on all clear pixels p in a test site (N in total). Pixels masked as not clear were not taken into account. In addition, the root mean squared error (RMSE^0) was calculated as:

$$\mathrm{RMSE}^0_{\lambda} = \sqrt{\frac{\sum_{p=1}^{N}(\hat{R}_{\lambda,i_{\mathrm{clear}}}(p) - R_{\lambda,i_{\mathrm{clear}}}(p))^2}{N}} \tag{3}$$

Next, the most clear observation at acquisition time i_{clear} was masked, resulting in the metric RMSE^1 (see center of Figure 4). This prevented the reconstruction algorithm using it as a trustworthy observation. Notice the extra gap that was introduced at acquisition time i_{clear} in addition to the existing gaps due to the already masked pixels. Similarly, RMSE^3 was calculated by introducing three consecutive gaps in the time series, i.e., masking the most clear and its two surrounding observations (see bottom Figure 4).

A final evaluation metric, $\mathrm{RMSE}^{\mathrm{MVC}}$, was defined based on the maximum NDVI composite over the four-month period 1 May 2019–31 August 2019. This is illustrated in Figure 5. With an average of 41 available observations, the maximum NDVI value composite of the original time series is expected to minimize problems common to single-date remote sensing studies, such as cloud contamination, atmospheric attenuation, surface directional reflectance, and view and illumination geometry [19]. Reconstructed time series that retain high-quality pixel observations should have a similar maximum NDVI composite and result in low values of $\mathrm{RMSE}^{\mathrm{MVC}}$. On the other hand, higher values of $\mathrm{RMSE}^{\mathrm{MVC}}$ are expected for reconstruction methods that alter high-quality observations.

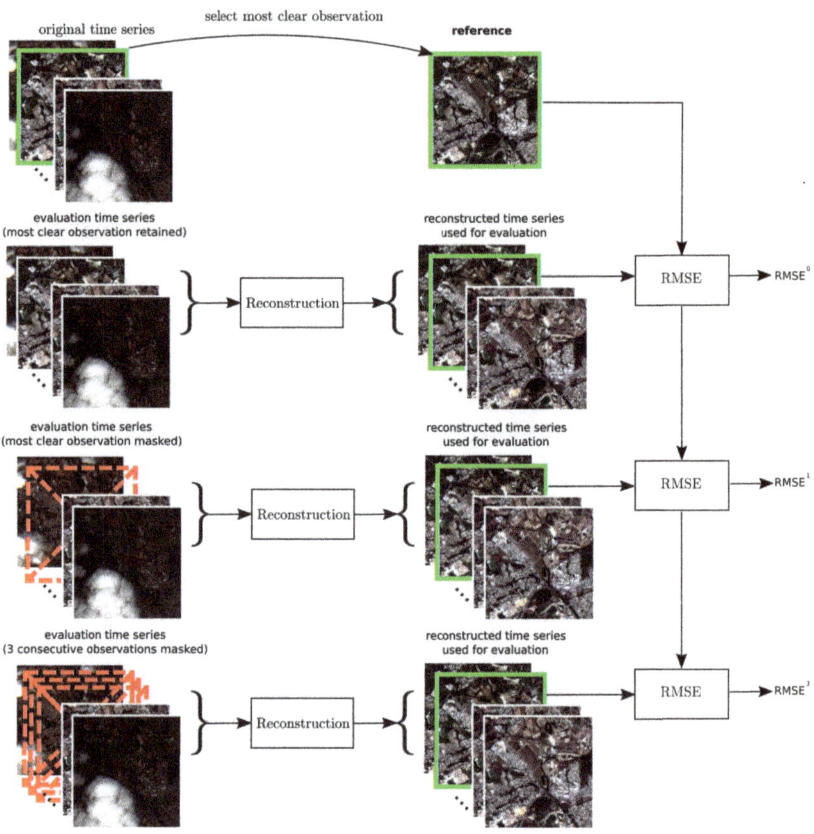

Figure 4. Metrics $RMSE^0$, $RMSE^1$, and $RMSE^3$ to evaluate the reconstruction method on the ability to retain high-quality pixels using the most clear observation as a reference.

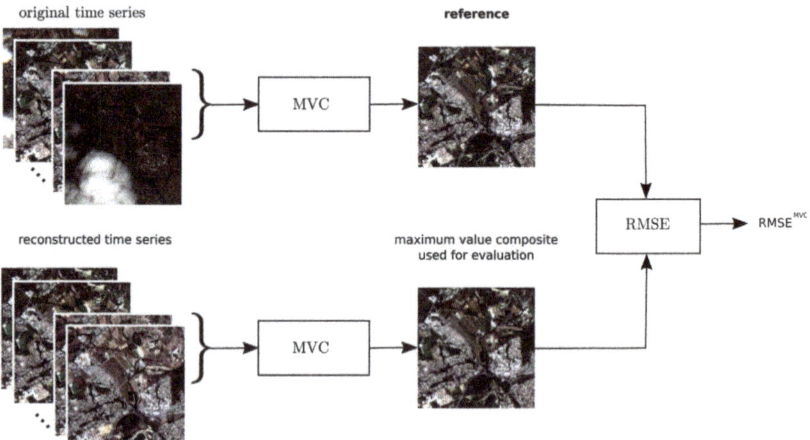

Figure 5. Evaluation metric based on the maximum NDVI value composite over the four-month period 1 May 2019–31 August 2019.

4. Results

The RTSR reconstruction algorithm was implemented in Python using the open source pyjeo [45] library. Processing was performed on the Big Data Analytics Platform (BDAP [46], the in-house storage and computing platform of the Joint Research Centre (JRC) of the European Commission (formerly known as the JEODPP). The results confirmed the potential of the adaptive smoothing using a vegetation index as a proxy for reliability of the reflectance values. A more detailed analysis of the test site near Montpellier is presented first, followed by an analysis of the metrics based on all test sites.

4.1. Evaluation of Test Site near Montpellier

In Figure 6, a subset of true color images for the reconstructed time series is shown. The subset corresponds to the eight acquisitions that are also shown in Figure 2 and illustrates qualitatively that cloudy and shadow pixels were reconstructed successfully.

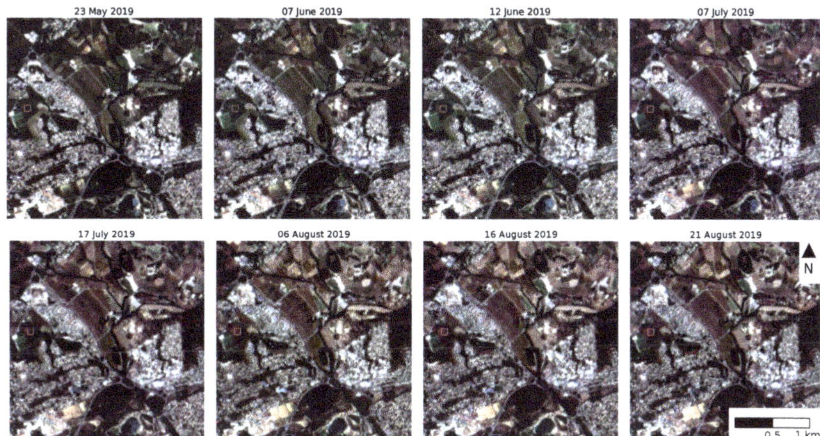

Figure 6. Reconstructed true color images corresponding to the acquisitions in Figure 2.

Scatterplots for the spectral bands under study were then created for the same test site (Figure 7). The pixel values of the observed and the reconstructed time series are compared at acquisition time i_{clear} and visualized based on the kernel density estimate. It is shown that the proposed RTSR method outperforms SG and DTS. The linear regression model indicates a better fit for RTSR, with a strong correlation (R^2). In general, the SG- and DTS-reconstructed surface reflectance values over-estimated the observed surface reflectance value. This is also illustrated for the example pixel indicated by the red square in Figure 7 (except for DTS in B8). The time series in Figure 3f is based on the same pixel. Both SG- and DTS-reconstructed values for the clear observation on 23 May 2019 (first vertical line) indeed over-estimate the observation in B4 (lower graph). For this particular pixel and acquisition time, only SG over-estimated the observation in B8. The DTS-reconstructed surface reflectance in B8 was slightly lower than the clear observation, as can seen in the upper graph in Figure 3f. The RTSR reconstructed value shows little bias in any of the bands.

Furthermore, the RMSE based on the difference between the reconstructed and clear observations on 23 May 2019 was lower for RTSR than for the other two method, except for B8, where it was slightly larger the for this specific test site (0.22 with respect to 0.20 for SG and DTS). However, it will be shown that for the majority of test sites, RTSR has a smaller RMSE for all tested bands.

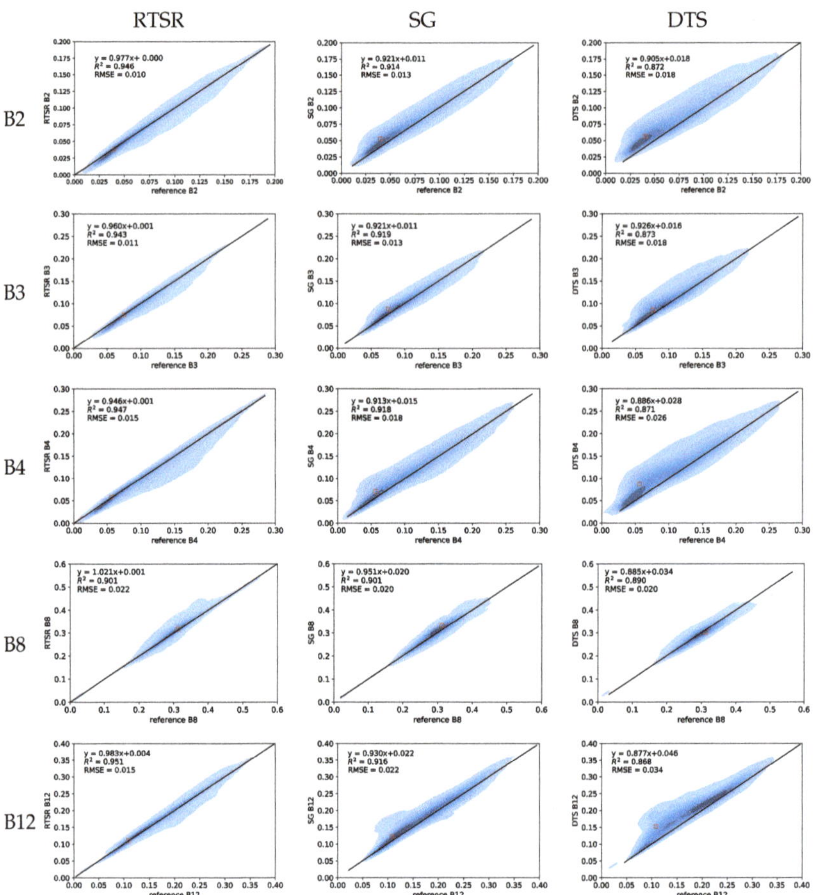

Figure 7. Scatterplot comparing the pixel values of the observed and the reconstructed time series for each spectral bands at acquisition time i_{clear}. The red square corresponds to the pixel that is also indicated in Figures 2 and 6. The RMSE value, slope, intercept, and coefficient of determination (R^2) of the linear regression are also shown.

4.2. Smoothness Index for Aggregated Test Sites

The smoothness for the three methods was evaluated by calculating TSI according to Equation (2). A cumulative frequency distribution representing all pixels aggregated for all test sites is shown in Figure 8. The results show little difference in the smoothness between the three reconstructed methods, with values for the 95 percentile that are almost identical. Most variation was found for spectral band B2, with smoothest results obtained for the non-adaptive Savitzky–Golay filter (SG) followed by the proposed RTSR and DTS. This is not surprising, as the SG-reconstructed time series represents the long-term change trend with a relatively wide half-width of the smoothing window. Both RTSR and DTS constrain the smoothing filter to capture dynamic changes of interest in vegetation.

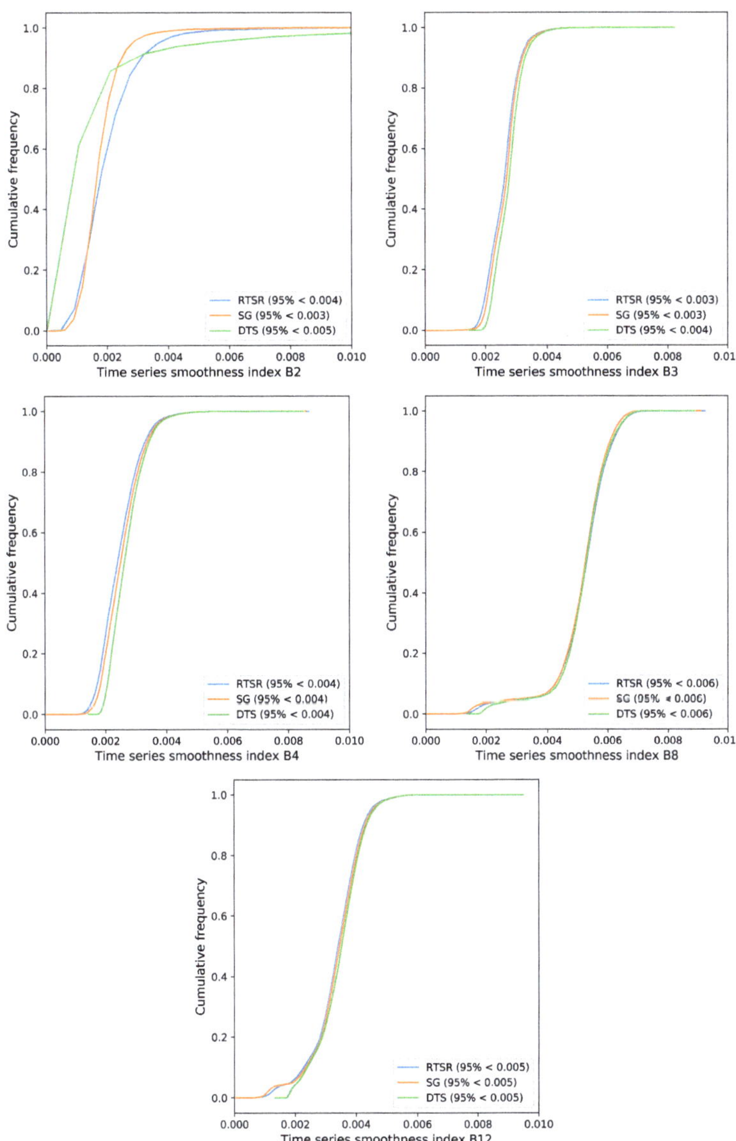

Figure 8. Cumulative frequency distribution for the time-series smoothness index in the visible (B2, B3, B4), near infrared (B8) and short-wave infrared (B12) calculated for three reconstruction methods (SG, DTS, and the proposed RTSR). Small values indicate smoother time series. The 95 percentile is indicated in brackets.

4.3. Statistical Analysis of RMSE Metrics for Aggregated Test Sites

As shown by the box plot in Figure 9a, the RMSE for RTSR was found to be smaller than for SG and DTS for the majority of test sites. This was the case for all spectral bands. The difference was most distinct in the visual to near infrared part of the electomagnetic spectrum. Furthermore, in the short wave infrared (band B12), the minimum as well as the 25 percentile value (Q1) and the median value of the RMSE was minimum over all test sites for the proposed RTSR. However, for some test sites the RMSE was higher for the RTSR than for the SG without adjusting the smoothed value, as expressed by the slightly higher 75 percentile value (Q3) and upper whisker ($Q_3 + 1.5 \times (Q_3 - Q_1)$).

Following the addition of an extra gap in the time series, the three tested reconstruction methods performed more similarly (see Figure 9b). This can be expected given that the gap introduced exactly at acquisition time i_{clear} was also used as a reference. This trustworthy reflectance value in the time series was not available to adjust the smoothed value in the RTSR-reconstructed time series. Only clear observations from nearby acquisition times were able to contribute to the reconstructed time series. When introducing three consecutive gaps ($i_{\text{clear}} - 1, i_{\text{clear}}, i_{\text{clear}} + 1$), the difference between the compared methods was even more reduced, resulting in a comparable RMSE^3 (see Figure 9c).

The results for the maximum-value composite over the four-month period 1 May 2019–31 August 2019 are shown in Figure 9d. The advantage of the proposed RTSR reconstruction method with respect to the other two reference methods is more expressed than in the previous metrics. A potential explanation is that only a single observation was used for the previous metrics. Although the most clear observation was selected, varying atmospheric conditions over the test site result in reflectance values that are not homogeneous. The pixels in this reference image contain some noise, which blurs differences in the evaluation metrics. As motivated in other evaluation studies that obtained reference data from the 10-year average (2011–2020) of NDVI data [25], aggregating data over time reduces noise in the reference data.

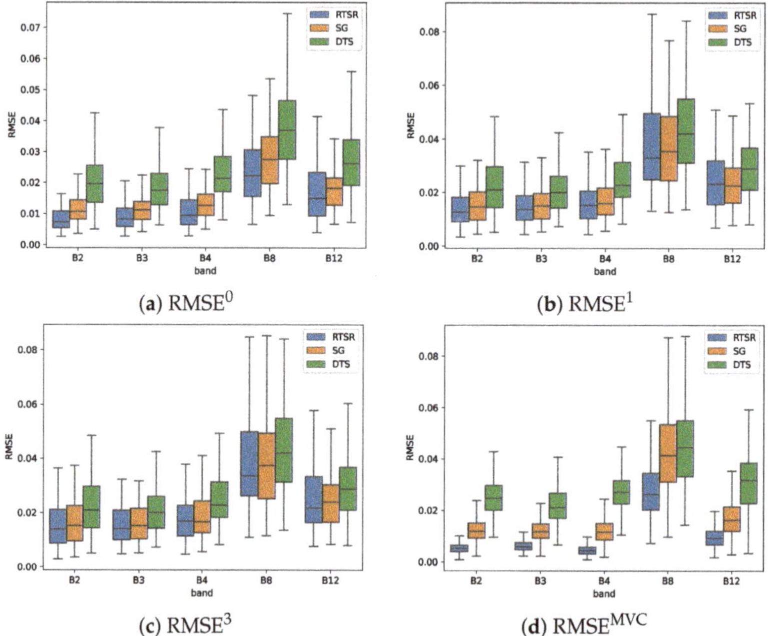

(**a**) RMSE^0 (**b**) RMSE^1

(**c**) RMSE^3 (**d**) RMSE^{MVC}

Figure 9. RMSE based on the difference between the reconstructed image and the original observation for the most clear observation (RMSE^0, RMSE^1, and RMSE^3) and for the maximum NDVI value composite over a four-month period (1 May 2019–31 August 2019).

5. Discussion

5.1. Reference Data and Model Parameters

A major difficulty with the quantitative evaluation of reconstructed time series is the lack of reference data, which must match the test data in temporal, spatial, and spectral dimensions. Several studies have used modeled time series based on the average of multi-year data as a reference [13,25,47,48]. Noise is often added to the modeled reference time series to evaluate the performance of the reconstruction method. This assumes that the modeled data and added noise realistically represent the actual data that will be used for the model that is evaluated. In [25], reference data were obtained from the 10-year average (2011–2020) of NDVI data. The advantage of this is a relatively noise-free reference time series. Taking the average produces a smooth signal as frequent changes related to, e.g., bidirectional effects, crop-harvesting practice, and sudden changes in land cover, are smoothed out. Using averaged data as a reference risks favoring reconstruction algorithms that include a higher level of smoothing. There is a trade-off between reconstructing a smooth noise-free signal on one hand and retaining the original values of high-quality pixels that have been observed in clear conditions on the other. The optimal approach depends on the application. This is also reflected by the tuning parameters that often accompany reconstruction algorithms [20,33]. The idea is that users can then optimize these parameters for their application at hand. However, optimizing these parameters adds to the complexity of a fair comparison of different techniques. The DTS algorithm proposed in [33] contains 13 parameters, wherein the authors optimized these parameters using a reference dataset that captured a range of ecosystems and land use/cover types. They claim the parameters should therefore be fairly robust for applications using vegetation indices such as the EVI2 or NDVI. However, as reflectance values can have a different variance structure or dynamic ranges, different optimal parameters might apply.

In a production environment where products are created at regional and continental scales, the fine-tuning of parameters for individual areas can become tedious and potentially result in unwanted border effects between areas where parameters differ. Therefore, the parameters for the three methods under comparison were fixed. A first parameter of the proposed RTSR is related to the cloud buffer. Due to omission errors of the Sen2Cor cloud detection algorithm, not all cloudy pixels have been identified as such. The observation corresponding to the acquisition on 7 June 2019 (second vertical line in Figure 3) can also be considered as an outlier, but was identified by the SCL band as clear and thus not masked. Cloudy or shadow pixels that are not masked will not be interpolated during the gap-filling step of the method. The corresponding surface-reflectance values will be taken into account and add to the noise in the time series. Because omission errors are more frequent on cloud edges [49], an extra buffer around clouds masks is commonly used [38]. Here, a relatively small buffer of five pixels (50 m) was added. It was able to mask the observation corresponding to the acquisition on 12 June 2019 that was considered by the SCL band as clear (third vertical line in Figure 3). Using a larger buffer provides the potential to mask more pixels. For instance, buffer sizes of 100–300 m were proposed in [38]. However, as suggested in [39], this also removes a large amount of usable imagery. The impact on the reconstructed time series will be reduced, because the observed surface-reflectance value near the cloud edge will most likely have a relatively low calculated NDVI value. The iterative RTSR algorithm will then reject this value and keep the smoothed value.

The parameters of the smoothing filters SG^1 and SG^2, were fixed based on the values found in the literature and experimental results. For the half-width of the smoothing window, values between 4 and 7 were suggested in [20]. Small values result in less smoothing with the risk of over-fitting the data points, whereas larger values risk not picking up important variations in the time series. For the first version of the SG filter, SG^1, a value of 7 was selected with a polynomial degree of 2 (suggested between 2 and 4 in [20]). This creates a relatively smooth signal that is able to represent the slowly varying long-term change trend. A lower filter width and a higher polynomial degree was set for SG^2 (3 and 3 instead of 4 and 2) so that important variations potentially smoothed out with SG^1, can still be picked up by the iterative algorithm. This is illustrated in Figure 3c,f. In contrast to the long-term change trend (in orange), the reconstructed time series (in blue) represents well the original values near the green peak (23 May 2019). The reconstruction based on DTS (in green) neither represents the original values near the maximum nor the minimum greenness. The relatively poor performance of DTS is potentially due its application to surface reflectance data. Although the code in [32] is claimed to be applicable to time series of satellite observations in general, the DTS presented in [33] was applied to EVI2 time series and involved an adjustment to the upper envelope. As for the NDVI, the assumption is that relatively high values correspond to a trustworthy observation. This assumption does not hold for surface reflectance. The DTS-reconstructed time series was obtained with the original code in [32] without this adjustment. Nevertheless, for most test sites the RMSE values of DTS are still in line with the RMSE values obtained in other studies. In [28], a reconstruction algorithm on Sentinel-2 surface reflectance data was evaluated using a metric similar to $RMSE^0$ and obtained RMSE values of 0.0426 (B2), 0.0407 (B3), 0.0405 (B4), and 0.0449 (B8). In [30], reconstructed MODIS Terra MOD09A1 data were evaluated on cloud-free MODIS Aqua MYD09A1 data: 0.0366 (red band), 0.0519 (near infrared band), and 0.0499 (short-wave infrared band).

5.2. Limitations

The relatively low number of test sites (100) and the restricted geographic location (Europe) limit the importance of the evaluation results. Rather than showing the superiority of the proposed RTSR method, the objective of the evaluation was to show the potential of dynamic smoothing where the time series approaches the trustworthy observations. The smoothing method can hereby be adopted from other methods, such as that proposed in DTS. A diverse set evaluation metric was hereby selected. Results of the TSI showed minor differences in smoothness for the three methods under comparison. All three methods were also able to fill gaps in the time series, which is driven by interpolating clear observations near the missing values. The strength of the proposed RTSR is that in addition to producing a smooth reconstructed time series, it is capable of retaining trustworthy observations in the original time series.

A number of limitations of the proposed RTSR method do apply. The time series to be reconstructed must include the spectral bands to calculate the index used as a proxy for the reliability of observations (e.g., the B4 and B8 bands of the Sentinel-2 sensor in the case of NDVI). Another limitation is the assumption that clouds or poor atmospheric conditions depress NDVI values [18]. This holds mostly for vegetated land but not for all land cover types. Clear observations of water bodies, for example, can result in lower NDVI values than cloudy observations. This is not a limitation of the method, but of the proxy used for trustworthy observations. The proxy can be adapted, for example, by distinguishing pixels using a water mask. Instead of NDVI, water surface-reflectance values in the near infrared band can be compared in the iterative algorithm. Clear water pixels usually have lower reflectance values in this part of the electromagnetic spectrum than clouds. When applying the RTSR reconstruction algorithm to create analysis-ready data for other applications beyond vegetation monitoring, the proxy could be adapted in this way. Further evaluation will be needed and is to be part of future research.

The Savitzky–Golay smoothing filter used in the reconstruction algorithm imposes another limitation. The RTSR reconstruction method requires a time series of at least 15 observations, i.e., the full width of the Savitzky–Golay filter, to obtain the long-term change trend curve. A longer time series is needed in practice to improve the results, typically 20 observations or more. In the case of Sentinel-2 acquisitions, this corresponds to a seasonal to annual coverage. The maximum length of the time series is only constrained by the memory resources available, as our implementation reads the entire time series in memory. Because the RTSR algorithm is a pixel-wise operation and requires no spatial contextual information, there is no upper limit in the spatial dimension. The image can be split into smaller tiles that can be run independently in parallel. Tiles can be made as small as needed to fit in the available memory. For instance, the test sites of 256 by 256 pixels and covering 62 acquisitions can be processed with less than 2 GB of memory.

Due to the extent of the smoothing window size, there is an edge effect inherent to the filtering process that impacts the first and last observations of the reconstructed time series. In this study, the acquisition period was extended by one month before and after the period of interest. The respective observations (December 2018 and January 2020) were afterwards removed from the reconstructed time series. This approach is not suited for near real-time applications. The Whittaker smoother has a number of advantages over the SG filter [21,40,50,51]. It employs only past observations and would be a good candidate for the smoothing filter, in particular for near real-time applications. This will be part of future research. Another interesting topic is the effect of the cloud mask on the reconstructed time series and to investigate whether more sophisticated cloud masks can improve results.

6. Conclusions

A spectral reflectance time-series reconstruction (RTSR) method has been proposed to reconstruct optical remote-sensing time-series data that are hampered by missing or noisy observations due to varying atmospheric conditions. This method differs from existing approaches in that it adjusts smoothed values of surface reflectance values to approach those of trustworthy observations, using a vegetation index as a proxy for reliability. The results show that the RTSR method is effective in retaining trustworthy observations in the original time series, with RMSE values in the order of 0.01 to 0.03 in terms of surface reflectance. The method was evaluated on 100 sites in Europe, with a focus on vegetation. The Savitzky–Golay filter was employed herein as the smoothing algorithm, but other filters can also be used. The potential of this method is substantial now that optical remote-sensing data have become publicly available at a finer spatial and temporal resolution. We hope that this study will encourage further research in this area, and that the RTSR method will be applied to create analysis-ready data for other applications beyond vegetation monitoring.

Author Contributions: Conceptualization, P.K., M.C. and R.d.; methodology, P.K.; software, P.K.; evaluation, P.K. and M.C.; writing—original draft preparation, P.K.; review and editing, all authors; All authors have read and agreed to the published version of the manuscript.

Funding: This research received no external funding.

Institutional Review Board Statement: This study was reviewed by the Joint Research Centre editorial review board (JERB).

Informed Consent Statement: Not applicable.

Data Availability Statement: Source code of main image-processing library used for this work (pyjeo) is freely available online https://github.com/ec-jrc/jeolib-pyjeo (accessed on 5 December 2022). Data and results are available on https://jeodpp.jrc.ec.europa.eu/ftp/jrc-opendata/DRLL/RTSR/ (accessed on 19 April 2023).

Conflicts of Interest: The authors declare no conflict of interest.

References

1. Drusch, M.; Del Bello, U.; Carlier, S.; Colin, O.; Fernandez, V.; Gascon, F.; Hoersch, B.; Isola, C.; Laberinti, P.; Martimort, P.; et al. Sentinel-2: ESA's Optical High-Resolution Mission for GMES Operational Services. *Remote Sens. Environ.* **2012**, *120*, 25–36.
2. Ma, C.; Liu, M.; Ding, F.; Li, C.; Cui, Y.; Chen, W.; Wang, Y. Wheat growth monitoring and yield estimation based on remote sensing data assimilation into the SAFY crop growth model. *Sci. Rep.* **2022**, *12*, 5473. [CrossRef]
3. Lambert, J.; Drenou, C.; Denux, J.P.; Balent, G.; Cheret, V. Monitoring forest decline through remote sensing time series analysis. *Gisci. Remote Sens.* **2013**, *50*, 437–457. [CrossRef]
4. Griffiths, P.; Nendel, C.; Pickert, J.; Hostert, P. Towards national-scale characterization of grassland use intensity from integrated Sentinel-2 and Landsat time series. *Remote Sens. Environ.* **2020**, *238*, 111124.
5. Moreno-Martínez, Á.; Izquierdo-Verdiguier, E.; Maneta, M.P.; Camps-Valls, G.; Robinson, N.; Muñoz-Marí, J.; Sedano, F.; Clinton, N.; Running, S.W. Multispectral high resolution sensor fusion for smoothing and gap-filling in the cloud. *Remote Sens. Environ.* **2020**, *247*, 111901. [CrossRef] [PubMed]
6. Kempeneers, P.; Sedano, F.; Piccard, I.; Eerens, H. Data Assimilation of PROBA-V 100 and 300 m. *IEEE J. Sel. Top. Appl. Earth Obs. Remote. Sens.* **2016**, *9*, 3314–3325. [CrossRef]
7. Sedano, F.; Kempeneers, P.; Hurtt, G. A Kalman Filter-Based Method to Generate Continuous Time Series of Medium-Resolution NDVI Images. *Remote Sens.* **2014**, *6*, 12381–12408. [CrossRef]

8. Inglada, J.; Vincent, A.; Arias, M.; Marais-Sicre, C. Improved Early Crop Type Identification By Joint Use of High Temporal Resolution SAR Furthermore, Optical Image Time Series. *Remote Sens.* **2016**, *8*, 362. [CrossRef]
9. Lasko, K. Gap Filling Cloudy Sentinel-2 NDVI and NDWI Pixels with Multi-Frequency Denoised C-Band and L-Band Synthetic Aperture Radar (SAR), Texture, and Shallow Learning Techniques. *Remote Sens.* **2022**, *14*, 4224. [CrossRef]
10. Xiong, S.; Du, S.; Zhang, X.; Ouyang, S.; Cui, W. Fusing Landsat-7, Landsat-8 and Sentinel-2 surface reflectance to generate dense time series images with 10 m spatial resolution. *Int. J. Remote Sens.* **2022**, *43*, 1630–1654. [CrossRef]
11. Inglada, J.; Arias, M.; Tardy, B.; Morin, D.; Valero, S.; Hagolle, O.; Dedieu, G.; Sepulcre, G.; Bontemps, S.; Defourny, P. Benchmarking of algorithms for crop type land-cover maps using Sentinel-2 image time series. In Proceedings of the 2015 IEEE International Geoscience and Remote Sensing Symposium (IGARSS), Milan, Italy, 26–31 July 2015; pp. 3993–3996. [CrossRef]
12. Saunier, S.; Pflug, B.; Lobos, I.M.; Franch, B.; Louis, J.; De Los Reyes, R.; Debaecker, V.; Cadau, E.G.; Boccia, V.; Gascon, F.; et al. Sen2Like: Paving the Way towards Harmonization and Fusion of Optical Data. *Remote Sens.* **2022**, *14*, 3855. [CrossRef]
13. Hird, J.N.; McDermid, G.J. Noise reduction of NDVI time series: An empirical comparison of selected techniques. *Remote Sens. Environ.* **2009**, *113*, 248–258. [CrossRef]
14. Moreno-Martínez, Á.; García-Haro, F.J.; Martínez, B.; Gilabert, M.A. Noise Reduction and Gap Filling of fAPAR Time Series Using an Adapted Local Regression Filter. *Remote Sens.* **2014**, *6*, 8238–8260. [CrossRef]
15. Meroni, M.; d'Andrimont, R.; Vrieling, A.; Fasbender, D.; Lemoine, G.; Rembold, F.; Seguini, L.; Verhegghen, A. Comparing land surface phenology of major European crops as derived from SAR and multispectral data of Sentinel-1 and -2. *Remote Sens. Environ.* **2021**, *253*, 112232. [CrossRef] [PubMed]
16. Karkauskaite, P.; Tagesson, T.; Fensholt, R. Evaluation of the Plant Phenology Index (PPI), NDVI and EVI for Start-of-Season Trend Analysis of the Northern Hemisphere Boreal Zone. *Remote Sens.* **2017**, *9*, 485. [CrossRef]
17. Tucker, C.J. Red and photographic infrared linear combinations for monitoring vegetation. *Remote Sens. Environ.* **1979**, *8*, 127–150. [CrossRef]
18. Goward, S.N.; Markham, B.; Dye, D.G.; Dulaney, W.; Yang, J. Normalized difference vegetation index measurements from the Advanced Very High Resolution Radiometer. *Remote Sens. Environ.* **1991**, *35*, 257–277. [CrossRef]
19. Holben, B.N. Characteristics of maximum-value composite images from temporal AVHRR data. *Int. J. Remote Sens.* **1986**, *7*, 1417–1434. [CrossRef]
20. Chen, J.; Jönsson, P.; Tamura, M.; Gu, Z.; Matsushita, B.; Eklundh, L. A simple method for reconstructing a high-quality NDVI time-series data set based on the Savitzky–Golay filter. *Remote Sens. Environ.* **2004**, *91*, 332–344. [CrossRef]
21. Atzberger, C.; Eilers, P.H. A time series for monitoring vegetation activity and phenology at 10-daily time steps covering large parts of South America. *Int. J. Digit. Earth* **2011**, *4*, 365–386. [CrossRef]
22. Liu, R.; Shang, R.; Liu, Y.; Lu, X. Global evaluation of gap-filling approaches for seasonal NDVI with considering vegetation growth trajectory, protection of key point, noise resistance and curve stability. *Remote Sens. Environ.* **2017**, *189*, 164–179. [CrossRef]
23. Huete, A.; Justice, C.; Van Leeuwen, W. MODIS vegetation index (MOD13). *Algorithm Theor. Basis Doc.* **1999**, *3*, 295–309.
24. Maisongrande, P.; Duchemin, B.; Dedieu, G. VEGETATION/SPOT: An operational mission for the Earth monitoring; presentation of new standard products. *Int. J. Remote Sens.* **2004**, *25*, 9–14. [CrossRef]
25. Li, S.; Xu, L.; Jing, Y.; Yin, H.; Li, X.; Guan, X. High-quality vegetation index product generation: A review of NDVI time series reconstruction techniques. *Int. J. Appl. Earth Obs. Geoinf.* **2021**, *105*, 102640. [CrossRef]
26. Zhu, Z.; Zhang, J.; Yang, Z.; Aljaddani, A.H.; Cohen, W.B.; Qiu, S.; Zhou, C. Continuous monitoring of land disturbance based on Landsat time series. *Remote Sens. Environ.* **2020**, *238*, 111116. . [CrossRef]
27. Dwyer, J.L.; Roy, D.P.; Sauer, B.; Jenkerson, C.B.; Zhang, H.K.; Lymburner, L. Analysis Ready Data: Enabling Analysis of the Landsat Archive. *Remote Sens.* **2018**, *10*, 1363. [CrossRef]
28. Yan, L.; Roy, D.P. Spatially and temporally complete Landsat reflectance time series modelling: The fill-and-fit approach. *Remote Sens. Environ.* **2020**, *241*, 111718. [CrossRef]
29. Yang, K.; Luo, Y.; Li, M.; Zhong, S.; Liu, Q.; Li, X. Reconstruction of Sentinel-2 Image Time Series Using Google Earth Engine. *Remote Sens.* **2022**, *14*, 4395. [CrossRef]
30. Xiao, Z.; Liang, S.; Wang, T.; Liu, Q. Reconstruction of Satellite-Retrieved Land-Surface Reflectance Based on Temporally-Continuous Vegetation Indices. *Remote Sens.* **2015**, *7*, 9844–9864. [CrossRef]
31. Xiao, Z.; Liang, S.; Tian, X.; Jia, K.; Yao, Y.; Jiang, B. Reconstruction of Long-Term Temporally Continuous NDVI and Surface Reflectance From AVHRR Data. *IEEE J. Sel. Top. Appl. Earth Obs. Remote Sens.* **2017**, *10*, 5551–5568. [CrossRef]
32. Dynamic Temporal Smoothing (DTS). Available online: https://github.com/jgrss/satsmooth (accessed on 28 October 2022).
33. Graesser, J.; Stanimirova, R.; Friedl, M.A. Reconstruction of satellite time series with a dynamic smoother. *IEEE J. Sel. Top. Appl. Earth Obs. Remote Sens.* **2022**, *15*, 1803–1813. [CrossRef]
34. d'Andrimont, R.; Claverie, M.; Kempeneers, P.; Muraro, D.; Yordanov, M.; Peressutti, D.; Batič, M.; Waldner, F. AI4Boundaries: An open AI-ready dataset to map field boundaries with Sentinel-2 and aerial photography. *Earth Syst. Sci. Data* **2023**, *15*, 317–329. [CrossRef]
35. Buttner, G.; Feranec, J.; Jaffrain, G.; Mari, L.; Maucha, G.; Soukup, T. The CORINE land cover 2000 project. *EARSeL eProceedings* **2004**, *3*, 331–346.
36. Copernicus Open Access Hub. Available online: https://scihub.copernicus.eu/dhus (accessed on 16 December 2021).

37. Sentinel-2 Level-2A Algorithm Theoretical Basis Document. Available online: https://sentinels.copernicus.eu/documents/247904/446933/Sentinel-2-Level-2A-Algorithm-Theoretical-Basis-Document-ATBD.pdf (accessed on 12 September 2022).
38. Zekoll, V.; Main-Knorn, M.; Alonso, K.; Louis, J.; Frantz, D.; Richter, R.; Pflug, B. Comparison of Masking Algorithms for Sentinel-2 Imagery. *Remote Sens.* **2021**, *13*, 137. [CrossRef]
39. Hughes, M.J.; Kennedy, R. High-Quality Cloud Masking of Landsat 8 Imagery Using Convolutional Neural Networks. *Remote Sens.* **2019**, *11*, 2591. [CrossRef]
40. Whittaker, E.T. On a new method of graduation. *Proc. Edinb. Math. Soc.* **1922**, *41*, 63–75. [CrossRef]
41. Savitzky, A.; Golay, M.J. Smoothing and differentiation of data by simplified least squares procedures. *Anal. Chem.* **1964**, *36*, 1627–1639. [CrossRef]
42. Verger, A.; Baret, F.; Weiss, M. A multisensor fusion approach to improve LAI time series. *Remote Sens. Environ.* **2011**, *115*, 2460–2470. [CrossRef]
43. Weiss, M.; Baret, F.; Garrigues, S.; Lacaze, R. LAI and fAPAR CYCLOPES global products derived from VEGETATION. Part 2: Validation and comparison with MODIS collection 4 products. *Remote Sens. Environ.* **2007**, *110*, 317–331. [CrossRef]
44. Claverie, M.; Ju, J.; Masek, J.G.; Dungan, J.L.; Vermote, E.F.; Roger, J.C.; Skakun, S.V.; Justice, C. The Harmonized Landsat and Sentinel-2 surface reflectance data set. *Remote Sens. Environ.* **2018**, *219*, 145–161. [CrossRef]
45. Kempeneers, P.; Pesek, O.; De Marchi, D.; Soille, P. pyjeo: A Python Package for the Analysis of Geospatial Data. *ISPRS Int. J. Geo-Inf.* **2019**, *8*, 461. [CrossRef]
46. Soille, P.; Burger, A.; De Marchi, D.; Kempeneers, P.; Rodriguez, D.; Syrris, V.; Vasilev, V. A versatile data-intensive computing platform for information retrieval from big geospatial data. *Future Gener. Comput. Syst.* **2018**, *81*, 30–40. [CrossRef]
47. Xu, L.; Li, B.; Yuan, Y.; Gao, X.; Zhang, T. A Temporal-Spatial Iteration Method to Reconstruct NDVI Time Series Datasets. *Remote Sens.* **2015**, *7*, 8906–8924. [CrossRef]
48. Zhou, J.; Jia, L.; Menenti, M.; Gorte, B. On the performance of remote sensing time series reconstruction methods–A spatial comparison. *Remote Sens. Environ.* **2016**, *187*, 367–384. [CrossRef]
49. Skakun, S.; Wevers, J.; Brockmann, C.; Doxani, G.; Aleksandrov, M.; Batič, M.; Frantz, D.; Gascon, F.; Gómez-Chova, L.; Hagolle, O.; et al. Cloud Mask Intercomparison eXercise (CMIX): An evaluation of cloud masking algorithms for Landsat 8 and Sentinel-2. *Remote Sens. Environ.* **2022**, *274*, 112990. [CrossRef]
50. Shao, Y.; Lunetta, R.S.; Wheeler, B.; Iiames, J.S.; Campbell, J.B. An evaluation of time-series smoothing algorithms for land-cover classifications using MODIS-NDVI multi-temporal data. *Remote Sens. Environ.* **2016**, *174*, 258–265. [CrossRef]
51. Schmid, M.; Rath, D.; Diebold, U. Why and How Savitzky–Golay Filters Should Be Replaced. *ACS Meas. Sci. Au* **2022**, *2*, 185–196. [CrossRef]

Article

Optimizing Wheat Yield Prediction Integrating Data from Sentinel-1 and Sentinel-2 with CatBoost Algorithm

Asier Uribeetxebarria *, Ander Castellón and Ana Aizpurua

NEIKER—Basque Institute for Agricultural Research and Development, Basque Research and Technology Alliance (BRTA), Parque Científico y Tecnológico de Bizkaia, P812, Berreaga 1, 48160 Derio, Spain
* Correspondence: auribeetxebarria@neiker.eus; Tel.: +34-607-142-018

Abstract: Accurately estimating wheat yield is crucial for informed decision making in precision agriculture (PA) and improving crop management. In recent years, optical satellite-derived vegetation indices (Vis), such as Sentinel-2 (S2), have become widely used, but the availability of images depends on the weather conditions. For its part, Sentinel-1 (S1) backscatter data are less used in agriculture due to its complicated interpretation and processing, but is not impacted by weather. This study investigates the potential benefits of combining S1 and S2 data and evaluates the performance of the categorical boosting (CatBoost) algorithm in crop yield estimation. The study was conducted utilizing dense yield data from a yield monitor, obtained from 39 wheat (*Triticum* spp. L.) fields. The study analyzed three S2 images corresponding to different crop growth stages (GS) GS30, GS39-49, and GS69-75, and 13 Vis commonly used for wheat yield estimation were calculated for each image. In addition, three S1 images that were temporally close to the S2 images were acquired, and the vertical-vertical (VV) and vertical-horizontal (VH) backscatter were calculated. The performance of the CatBoost algorithm was compared to that of multiple linear regression (MLR), support vector machine (SVM), and random forest (RF) algorithms in crop yield estimation. The results showed that the combination of S1 and S2 data with the CatBoost algorithm produced a yield prediction with a root mean squared error (RMSE) of 0.24 t ha^{-1}, a relative RMSE (rRMSE) 3.46% and an R^2 of 0.95. The result indicates a decrease of 30% in RMSE when compared to using S2 alone. However, when this algorithm was used to estimate the yield of a whole plot, leveraging information from the surrounding plots, the mean absolute error (MAE) was 0.31 t ha^{-1} which means a mean error of 4.38%. Accurate wheat yield estimation with a spatial resolution of 10 m becomes feasible when utilizing satellite data combined with CatBoost.

Keywords: backscatter; gradient boosting; machine learning; NDVI; precision agriculture

Citation: Uribeetxebarria, A.; Castellón, A.; Aizpurua, A. Optimizing Wheat Yield Prediction Integrating Data from Sentinel-1 and Sentinel-2 with CatBoost Algorithm. *Remote Sens.* **2023**, *15*, 1640. https://doi.org/10.3390/rs15061640

Academic Editors: Kenji Omasa, Shan Lu and Jie Wang

Received: 4 February 2023
Revised: 10 March 2023
Accepted: 14 March 2023
Published: 17 March 2023

1. Introduction

Agriculture plays a crucial role in the global economy and, as the world's population continues to grow, the pressure on agricultural production also increases [1]. Historically, the primary method for increasing agricultural production was to expand the cultivated land [2]. This was typically conducted until the early years of the "Green Revolution" (GR), when cereal production tripled while the area devoted to agriculture increased by just 30% [3]. This improvement was driven by heavy public investments in infrastructure and research, as well as the implementation of agricultural promotion policies. The GR was characterized by the widespread use of mechanization, chemical fertilizers, and pesticides, together with genetic improvements in major crops, aspects that played a significant role in yield increases from the 1990s onward [4]. Nitrogen, a key component of fertilizers, is particularly detrimental to the environment when used in excess [5,6]. To address this issue, the European Union has launched the "Farm to Fork" strategy, which aims to reduce the use of pesticides and fertilizers. As crop nutrient requirements are related to production, reliable yield estimates are essential if fertilizer inputs are to be adjusted and losses reduced [7].

Recent studies, such as those conducted by Zambon et al. [8], have demonstrated that technological advances can play a crucial role in achieving sustainable intensification in agriculture. The development of precision agriculture (PA) began in the late 1990s as a strategy for improving the sustainability of agricultural production through the consideration of temporal and spatial variability. The utilization of various sensors, including weather stations [9], multispectral cameras [10], electroconductivity meters [11], and LiDAR [12], is a common practice within the framework of PA. The implementation of PA allows input reduction while maintaining yield levels [13] through the targeted distribution of inputs according to specific crop requirements rather than a uniform application [14]. Despite the availability of PA technologies, adoption among farmers, particularly smallholders, remains low [15]. Partially this phenomenon can be attributed to the economic burden associated with acquiring new technology. Additionally, as technology becomes more sophisticated and data-intensive, farmers may require expert assistance to validate their decisions [16].

Despite the challenges faced by small- and medium-sized farmers to adopt PA techniques, the recent deployment of the Sentinel-2 (S2) satellite constellation by the European Space Agency (ESA) has the potential to enhance their utilization. Specifically, the twin satellites of the S2 series (A and B) were engineered to cater to requirements of the agricultural sector and researchers [17]. These satellites provide high resolution images, with 13 multispectral bands and a rapid revisit rate, all of which are available free of charge through ESA's Copernicus program (https://scihub.copernicus.eu/, accessed on 13 March 2023). The different bands of the sensor allow the calculation of various vegetation indices (VIs), which are related to a range of crop parameters, including crop growth [18], crop classification [19], and soil conditions [20]. For example, Vallentin et al. [21] conducted an analysis utilizing a time series of 13 years to examine the correlation between crop yield and different VIs. Comparison of various satellites led to the conclusion that those of higher resolution, such as the Rapid Eye or S2, performed better when compared to other lower resolution satellite imagery.

VIs have been widely used in agriculture to estimate crop yield because stressed and healthy crops emit energy at different wavelengths. For example, the normalized difference vegetation index (NDVI) is calculated based on the reflectance of vegetation in the red and near-infrared bands of the electromagnetic spectrum. As plants absorb more red light and reflect more near-infrared light as they become more vigorous, the NDVI value increases as the canopy density and biomass increase, and in consequence, the grain yield. Therefore, NDVI can be used as an indicator of plant health and biomass production. Although the use of VIs for this purpose dates to the early 1980s [22], it was not until the 1990s that it became more common [23,24]. With the release of images provided by satellites such as S2 [25], Landsat [26], MODIS [27], and SPOT [28], the use of VIs has exponentially increased. Recent studies, such as that proposed by the authors of [29], have utilized VIs derived from S2 in combination with random forest (RF) to estimate yield within individual plots across multiple wheat fields in England. VIs have also been used to estimate yield across entire countries [30]. Incorporating satellite-derived information into agrometeorological models has been shown to improve their accuracy [31,32]. For example, Vicente-Serrano et al. [33] in Spain combined advanced very high resolution radiometer (AVHRR) and NDVI data as well as drought indices at different time scales to predict wheat yield in advance. In other cases, VIs have been used to estimate yield directly [34]. More recently, publications such as [35,36] have taken a step further by combining machine learning techniques with satellite information to estimate the yield of specific plots using data from other plots.

However, one major limitation of S2 is cloud cover [37], which can restrict the amount of usable data available for certain areas and applications. Additionally, while S2 images have a high spatial resolution, they may not be sufficient for some applications that require very high resolution data as, for example, field work with vineyards or early disease detection. Other impediments include misalignment with other remotely sensed data, such as Landsat 8, the lack of panchromatic and thermal bands, and variations in the spatial resolution of the bands [38].

Sentinel-1 (S1) data are also available for free through the Copernicus program. S1 is a synthetic aperture radar (SAR) designed for radar imaging and can provide data in various modes and polarizations (VV, HH, VH or HV), depending on the emission and acquisition signal mode. S1 operates in the C polarimetric band, which ranges from 5.405 to 5.625 GHz and has a wavelength of 5.6 cm. S1 provides information about objects after being impacted by microwaves (C-band). Importantly, radar data are not affected by atmospheric conditions such as clouds and can also be acquired at night. The spatial resolution of S1 is 10 m, similar to the maximum resolution of S2, and it typically has a revisit period of 6 days [39]. However, the interpretation of the signal from S1 is complex and requires specialized analysis. For example, for a vegetated surface, the C-band signal is a combination of contributions from the soil, canopy, volume scattering within the canopy, and interactions between the soil and vegetation [40]. As a result, its use in agriculture is not as widespread as that of S2.

The computational development and utilization of machine learning techniques have become increasingly important in the field of PA [41]. These technologies allow for the processing and analysis of large amounts of data collected from various sources, including satellite imagery, drones, and Internet of Things (IoT) sensors, to generate accurate and detailed predictions [42]. Different types of machine learning algorithms can be employed in this process, including supervised and unsupervised algorithms. Supervised learning algorithms, such as decision trees, RF, and support vector machines (SVMs), can be used to classify different crops, predict crop yields or detect patterns in crop growth [43–45]. Unsupervised learning algorithms, such as *k-means* and principal component analysis (PCA), can be utilized to identify patterns or delineate site-specific management zones (SSMZs) [46].

Over the past few years, a variety of algorithms have been tested to estimate wheat yield. Tang et al. [47] utilized multiple linear regression (MLR) to estimate yield, with root mean squared error (RMSE) values ranging from 0.54 to 1.02. In the same study, the backpropagation neural network (BPNN) was also tested, obtaining better results with RMSE values ranging from 0.30 to 0.68. Hunt et al. [29] used the RF algorithm to estimate wheat yield in different plots. These results were compared with those obtained from MLR. The RF algorithm consistently obtained superior results for all the considered scenarios. Support vector machine (SVM) is another commonly used algorithm for this purpose. In the study published by Bebbie et al. [25], the coefficient of determination (R^2) value obtained was always greater than 0.80. Meraj et al. [48] compared the ability of SVM and RF to estimate the area of wheat cultivation in large areas of India, obtaining better results with RF. Finally, deep learning algorithms such as the long short-term memory (LSTM) also produced adequate results, with an RMSE of 0.64 t ha^{-1} when estimating wheat grain yield [49]. Srivastava et al. [50] compared the performance of eight different algorithms using a 20-year time series and found that the convolutional neural network (CNN) produced the best results. Finally, Cao et al. [51] compared the performance of MLR, SVM, RF, and XgBoost to estimate winter wheat yield in northern China combining machine learning with a global dynamical atmospheric prediction system.

Recently, in the latter part of the 1990s, a new type of supervised algorithm involving gradient boosting emerged. Gradient boosting is a machine learning technique that aims to enhance the accuracy of predictive models. The method operates by repeatedly training a sequence of base models and assigning increased weights to examples previously misclassified by prior models with the purpose of focusing on the most challenging samples. These algorithms involve the combination of multiple simple models with the goal of creating a robust ensemble model. The first of these algorithms to be developed was the adaptive boosting (AdaBoost) algorithm, published by Yoav Freund and Robert Schapire in [52]. The gradient boosting machine (GBM), proposed by Jerome Friedman in [53], is an extension of AdaBoost, but instead of assigning weights to examples, it utilizes gradient descent to optimize the parameters of the base model. GBM is an iterative algorithm that generates a series of decision trees, with each tree being intended to correct the errors made by the preceding tree. Another gradient boosting algorithm, the extreme gradient boosting (XGBoost) algorithm, was developed by Tianqi Chen in [54] and is optimized for working with large datasets. In 2017, the categorical boosting (CatBoost) algorithm was released by Prokhorenkova et al. [55], which is optimized to handle categorical variables. Currently, CatBoost is considered a powerful algorithm and is widely used owing to its ability to process categorical data and its high capacity to generalize. However, its application in agriculture is not yet widespread.

The challenge of yield estimation in modern agriculture presents numerous opportunities for decision making at both farmer and institutional level, including future action planning, the modulation of input supply according to crop needs, and harvest storage. In this regard, it should be noted that several global-scale works, in addition to satellite and yield information, use weather data [56] and soil information [57] in their yield estimation models. However, it is difficult to have weather and soil information at a sufficient level of detail when making yield estimation at intra-plot level.

Remote sensing technologies also offer new possibilities for improving yield estimation through the use of more advanced algorithms. Taking these considerations into account, the aim of the present study is to conduct a comprehensive analysis of the potential of remote sensing and machine learning techniques for yield estimation. More specifically, the study aims to determine whether the utilization of information obtained from S1 and S2 satellite imagery on different days enhances the accuracy of yield predictions. The study also evaluates the potential benefits of combining S1 and S2 data and, finally, aims to determine the effectiveness of the CatBoost algorithm in comparison to other commonly used methods such as MLR, SVM, and RF.

The analyses are conducted with a practical approach that applies to agriculture. High resolution wheat yield data from 39 plots, obtained with a yield monitor during the 2021 season, are used. Additionally, three cloud-free S2 images representing different phenological stages of wheat are analyzed, from which 13 VIs are calculated. A total of three S1 images, acquired on dates close to those of S2, are also examined for their backscattering values in vertical-vertical (VV) and vertical-horizontal (VH) polarizations.

2. Materials and Methods

2.1. Study Area

This study was carried out with data collected in the 2021 season in 39 wheat (*Triticum aestivum* L.) plots located in the Llanada Alavesa region, situated in the center of the province of Araba/Álava in northern Spain (Figure 1). This region is characterized by agricultural fields growing mainly winter cereals (wheat and barley), potatoes, colza, legumes, forage maize, and sugar beet. Wheat sowing was carried out at a density of 230 kg ha^{-1} with Filón variety seeds between 20 and 30 November 2020. All fields were fertilized with chemical fertilizers, averaging 53 kg ha^{-1} of N, 36 kg ha^{-1} of P, and 102 kg ha^{-1} of K in the growth stage (GS) GS21 that corresponds to tillering [58]. For the top-dressing fertilization, 117 kg ha^{-1} N was applied in the stem elongation phase (GS30).

According to the Köppen classification, the Llanada Alavesa region has a temperate oceanic climate (Cfb) [59] characterized by an average annual air temperature of 11.7 °C. During the summer months, the average temperature reaches 20 °C, while the winter months are relatively mild with an average temperature of 6 °C. Average annual rainfall was 750 mm, with July and August being the driest months with less than 50 mm of precipitation. The study plots were established over two distinct soil types developed on two different lithologies. Thus, soils on the lithology from the Cretaceous geological era are characterized by steep and irregular terrain, are relatively shallow (20–70 cm), and have a high concentration of calcium carbonate ($CaCO_3$) of over 50%. The dominant soil fraction is silt, which has a concentration exceeding 40%. The second type of soil, sourced from Quaternary material, is deeper (over 120 cm), has a lower concentration of $CaCO_3$ (<25%) and a loamier texture, and the stone content is higher [60].

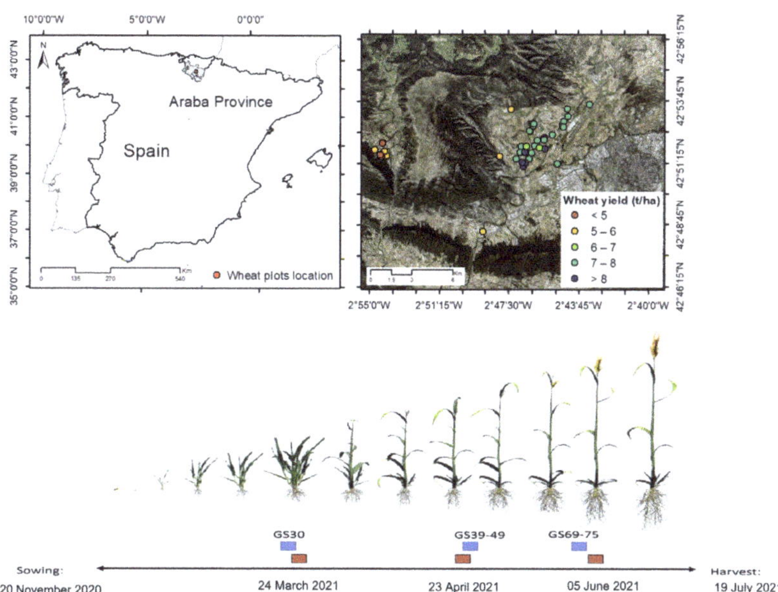

Figure 1. Upper left-hand side of the image shows the general location of the study area within Spain. Upper right-hand side shows detail of the study area with the average yield of each plot (right part). Below, wheat phenological stage and dates when satellite images were acquired. Red squares represent S1 and blue squares represent S2.

The average grain yield of the studied plots (Figure 1) ranged from 4.76 t ha^{-1} for the G32 plot to 8.91 for the G7 plot, with an average value for all plots of 7.01 t ha^{-1}. Plot size ranged from 0.72 to 9.42 ha, with 2.46 ha being the average, representing well Llanada Alaves's plot diversity.

2.2. Sentinel-2 Data and Derived Vegetation Indices

The S2 mission, operated by the European Space Agency (ESA), consists of two twin satellites launched in June 2015 and March 2017. These satellites provide multispectral imagery with 13 bands (https://sentinels.copernicus.eu/web/sentinel/user-guides/sentinel-2-msi/resolutions/spatial, accessed on 13 March 2023). In this study, four spectral bands were utilized: blue (B2, centered at 492.4 nm), green (B3, centered at 559.8 nm), red (B4, centered at 664.6 nm), and near-infrared (B8, centered at 832.8 nm) with a spatial resolution of 10 m. In theory, the combined use of both satellites provides an image of the study area every five days. However, in reality, the availability of cloud-free images is much lower. For this study, three cloud-free images of the study area were selected. The first image (Day 1) was obtained on 24 March 2021, when the crop was in the initial stage of stem elongation (GS30 according to Zadocks [58]). On the second date (Day 2), 23 April 2021, the crop was between GS39 (flag leaf ligule just visible) and GS49 (first awns visible).

The final image, taken on 5 June 2021, (Day 3), depicts the crop between complete anthesis and medium milk stage (GS69-75). The satellite data were downloaded from the Copernicus Open Access Hub (https://scihub.copernicus.eu/, accessed on 13 March 2023) in the form of Level 2A products (Bottom-of-Atmosphere reflectance images), which have undergone atmospheric correction [61]. The tile 30 TWN of satellite S2 fully covered the study area.

In this study, the SNAP software was used to calculate the 13 VIs (Table 1) used for wheat or barley (*Hordeum vulgare* L.) grain yield estimation.

Table 1. Vegetation indices calculated in this study with their formulae according to the S2 bands used.

Vegetation Index	Abbreviation	Formula	Reference
Normalized Difference Vegetation Index	NDVI	(B8 − B4)/(B8 + B4)	[62]
Green Ratio Vegetation Index	GRVI	B8/B3	[63]
Green Normalized Difference Vegetation Index	GNDVI	(B8 − B3)/(B8 + B3)	[64]
Green Difference Vegetation Index	GDVI	B8 − B3	[65]
Enhanced Vegetation Index 2	EVI2	2.4 × ((B8 − B4)/(B8 + B4 + 1))	[66]
Chlorophyll Vegetation Index	CVI	B8 × (B4/(B3 × B3))	[67]
Color Index	CI	(B4 − B2)/B4	[68]
Wide Dynamic Range Vegetation Index	WDRVI	((0.1 × B8) − B4)/((0.1 × B8) + B4)	[69]
Transformed Vegetation Index	TVI	$\sqrt{((B8-B4)/(B8+B4)+0.5)}$	[70]
Soil Adjusted Vegetation Index	SAVI	((B8 − B4)/(B8 + B4 + 0.5)) × (1 + 0.5)	[71]
Simple Ratio 800/670 Ratio Vegetation Index	RVI	B8/B4	[72]
Optimized Soil Adjusted Vegetation Index	OSAVI	(1 + 0.16) × ((B8 − B4)/(B8 + B4 + 0.16))	[73]
Nonlinear Vegetation Index	NLI	((B8 × B8) − B4)/((B8 × B8) + B4)	[74]

2.3. Wheat Grain Yield Acquisition, Preprocessing, and Connection with Sentinel Data

Spatially dense wheat grain yield data were obtained by installing a yield monitor and a GPS receiver on a John Deere T560 harvester. The GPS receiver could receive RX corrections, enabling it to be spatially positioned with an error lower than 15 cm, making it suitable for PA. Yield data were collected between 19 and 25 July 2021. To prepare the yield data for further analysis, they were pre-processed to eliminate anomalous measurements that can greatly affect the results [75]. Firstly, data with incorrect latitude/longitude measurements were removed. In the pre-processing steps, data with moisture concentrations below 8%, or values recorded when the harvester was operating at an inadequate speed, were removed to ensure the accuracy of the data. Afterwards, some steps of the methodology described by Taylor et al. [76] were applied. In the first step, yield values that exceeded or did not reach the established threshold were eliminated. In the next step, data points that were more than 2.5 standard deviations above and below the plot mean were removed. In the following step, the local Moran's I test [77] was applied to eliminate spatial outliers in our case, high yield measures surrounded by low yield measures or vice versa. In addition, to ensure that every S2 pixel was entirely within the study plot, a safety buffer of 15 m was established in each plot to minimize the distortion produced by the edge effect. Pixels located out of the buffer were removed. Data were then interpolated by ordinary kriging to a continuous yield map by selecting the semivariogram that best fit to the yield data for each plot. The most frequently used semi-variograms were exponential, spheric or rational quadratic. The maps were re-sampled to a resolution of 10 × 10 m and aligned with S1 and S2 pixels. Finally, a grid of points was generated in vector format (ESRI shapefile) and the information from the different rasters was transferred to the vector layer using the 'extract values' function in ArcGIS 10.8. This process resulted in a dataset composed by 6219 yield measures.

2.4. Sentinel-1 Data and Retro Dispersion Calculation

S1 ground range detected (GRD) images [78] were used in this study. These images are synthetic aperture radar (SAR) data acquired by the S1 satellite with a resolution of 5 × 20 m and a swath width of 250 km. The interferometric wide (IW) mode of acquisition was used, resulting in the acquisition of two polarization types: VV and VH. The images provide backscatter intensity information and are pre-processed at Level 1, resulting in geolocated, radiometrically calibrated, and terrain-corrected complex data in the slant range. Three images (27 March 2021, 20 April 2021, and 7 June 2021) acquired on days close to those acquired for S2 were downloaded from the Copernicus Open Access Hub (https://scihub.copernicus.eu/, accessed on 13 March 2023). To make the acquired images useful, they were processed using the SNAP software following the procedure outlined in Figure 2. This processing included adjusting the image tile size to the study area and calculating accurate orbits, as the metadata provided with the radar products is not always sufficiently accurate. Precise orbit information was obtained by using the 'apply orbit file' function, which is available a few days after image capture. Other necessary steps included improving image quality through the removal of thermal noise and radiometric artifacts from image edges, image calibration to obtain radiometrically calibrated backscatter images, and the elimination of granular noise caused by backscatter from certain elements (salt and pepper effect). The Lee Sigma filter was used in this process. Geographical coordinates were subsequently added to the images and, in the last step, backscatter values were finally converted to decibels (Figure 2). VV and VH backscatter information were extracted using the same georeferenced grid used to extract the information from S2 VI data. In total, two variables (VV and VH polarization backscatter information) were obtained for each date.

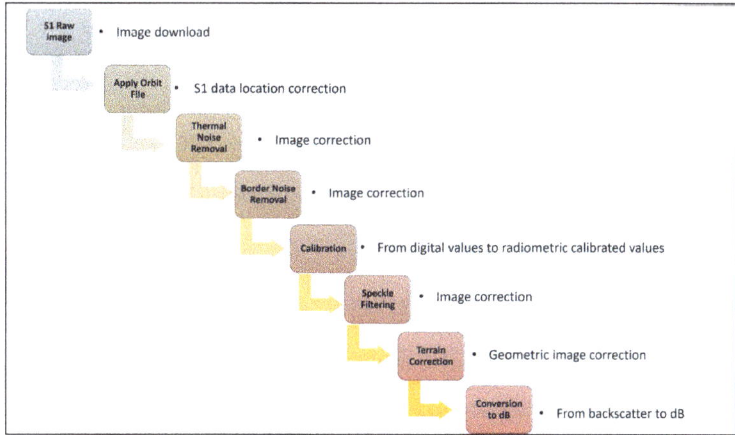

Figure 2. Workflow followed for the pre-processing of S1 images to obtain backscatter information.

2.5. Machine Learning Algorithms

Selecting the appropriate algorithm for a specific task is a crucial step in machine learning. The literature suggests that no single algorithm is the best, and the selection should be based on data characteristics and the desired outcome [79]. Therefore, it is imperative to evaluate the suitability of different algorithms for a given task to obtain optimal results.

In this study, the performance of four supervised machine learning algorithms was evaluated for a specific task: MLR [80], SVM [81], RF [82], and CatBoost [55]. Their election was based on their popularity and versatility in the modern agriculture literature [83].

Hyperparameter optimization for the SVM, RF, and CatBoost algorithms was performed using the GridsearchCV method implemented in the Scikit-learn library [84]. The MLR algorithm does not require hyperparameter optimization.

In this study, the MLR algorithm was implemented with Lasso regularization to reduce the complexity of the model and mitigate the effects of collinearity present between some of the variables. Collinearity is a phenomenon where two or more predictors in a multiple regression are highly correlated and can inflate the regression coefficients [85]. The Lasso function addresses this issue by limiting the sum of the absolute values of the model coefficients.

For its part, SVMs can be used for classification and regression tasks. One of the key advantages of using SVMs is their ability to identify the optimal boundary or decision surface that separates different classes in a dataset. The main idea behind SVMs is to find the best boundary or decision surface that separates different classes in a dataset. This is achieved by maximizing the margin, which is the distance between the boundary and the closest data points from each class [81]. Additionally, SVMs possess the ability to handle high-dimensional and non-linearly separable data by utilizing kernel functions to map the input data into a higher-dimensional space where a linear boundary can be found. This enables SVMs to perform well on complex and non-linear datasets. In this study, the kernel parameter was changed from 'linear' to 'RBF' to achieve this purpose. However, it should be noted that SVMs are less resistant to overfitting than other algorithms. Overfitting is a prevalent issue in machine learning, where a model performs well on the training data but poorly on unseen data. This is due to the margin maximization technique employed by SVMs being susceptible to overfitting [86]. To mitigate this risk, effective optimization of the 'C' hyperparameter is required. A large value of C results in the generation of a complex model, which minimizes training errors but also increases the likelihood of overfitting. Conversely, a small value of C leads to the production of a simpler model, which is less susceptible to overfitting but may not be as effective in fitting the training data [87]. In the present study, various values (1, 10, 100, 1000) of the C hyperparameter were experimentally evaluated to determine the optimal value that strikes a balance between predictive accuracy and model overfitting. Among all the tests carried out, the best results were obtained with C = 10. The gamma parameter was modified to 0.1.

The third algorithm used in the study is an ensemble algorithm that combines multiple decision trees to make predictions and is known as RF. The principle of RF is to construct a large number of decision trees and combine their predictions through methods such as majority voting or averaging [82]. It works by randomly selecting a subset of features to split the decision trees. This approach reduces the overfitting and variance issues commonly associated with single decision tree models. Additionally, RF can handle high-dimensional and correlated features, and can be used for both classification and regression tasks [88]. Moreover, the algorithm provides an estimate of feature importance, which can be useful for feature selection and understanding the underlying relationships in the data. Despite its advantages, RF is sensitive to noise in the dataset and can be computationally expensive for large datasets. Additionally, the algorithm can be sensitive to the number of trees used in the ensemble, requiring proper tuning to achieve optimal performance. Therefore, one of the hyperparameters optimized using the Gridsearch.cv function was the number of trees used in the ensemble, with the best results achieved with 300 trees. In addition, the maximum number of features parameter was determined using the 'auto' function. This function allows to use all features in each split. After tests with different combinations of tree depth (4, 5, 6, 7, 8, 10, 45, 50), the best result was obtained with 45 nodes.

CatBoost is a gradient boosting algorithm for decision trees that is specifically designed to handle datasets with many categorical variables [55]. The algorithm uses the gradient descent to optimize the parameters of the decision trees, which helps to improve the performance of the model [89]. The algorithm works by building and combining multiple decision trees. It uses a subset of the data to build each decision tree and then combines the predictions of all the decision trees to make the final prediction. The algorithm also utilizes a technique called 'permutation feature importance' to determine the importance of variables in the model. This technique is based on measuring the impact of each feature on the model's performance by randomly shuffling the values of a single feature. The feature with the largest impact on the model's performance is considered the most important [55]. Additionally, CatBoost is able to handle missing values in the data without the need for imputation techniques.

The CatBoost configuration that yielded the best results consisted of 18,000 iterations with an early stopping value of 200, which was implemented to prevent overfitting of the algorithm. The depth of the trees was set to six, and the 'MultiRMSE' loss function was selected, with a learning rate of 0.015. The parameter 'leaf_estimation_iteration' was set to 10. As the dataset was not excessively large, it was trained on the computer's CPU, but CatBoost has the option to train on a GPU if needed.

In addition to utilizing supervised algorithms, the present study incorporated the iterative self-organizing data analysis technique (ISODATA) unsupervised algorithm for data classification. This iterative algorithm begins by assigning an arbitrary mean to each class, and subsequently reassigns pixels based on minimizing the Euclidean distance to the mean value of their assigned class. The iteration process continues until either the final iteration is reached or the threshold for the maximum number of pixels changing class is not exceeded.

In this study, a data partitioning strategy was implemented with the purpose of training and validating the algorithms. The strategy entailed the random selection of 70% of the data for training and 30% for testing. This nearly ensures that testing is performed with data from all plots. However, in Section 3.6, the authors deviated from the standard data partitioning strategy and adopted an alternative approach. Except for data belonging to one plot, the rest were utilized for testing while the data from the excluded plot was reserved for testing. Iteratively the same process was performed for all plots. This methodology aimed to evaluate the algorithm's ability to predict the yield of a particular plot without utilizing information from that plot. Algorithms were trained and tested using functions provided by the Scikit-learn library over our datasets. The performance of the regression algorithms was evaluated using R^2, RMSE, and the percentage of mean absolute error (%MAE).

Obtaining an accurate estimated yield map is the first step towards creating a fertilizer prescription map based on yield data in cases where a yield monitor is not available. With this in mind, in Section 3.6, the G15 plot was selected to demonstrate the possibilities offered by the estimated yield map for creating prescription maps. Since prescription maps usually consist of two or three zones, the unsupervised ISODATA algorithm was selected to divide the datasets into two classes. This procedure was applied to the actual yield data and the estimated yield data. The similarity between the classified estimated yield map and the classified real yield map was measured using the 'accuracy' and Kappa Index (KI) metrics, both widely used to assess the performance of classification algorithms.

2.6. *Accuracy Assessment*

2.6.1. Root Mean Squared Error (RMSE)

The RMSE is a commonly used statistic that measures the difference between predicted values and observed values in a regression problem. It is defined as the square root of the mean of the squared differences between the predicted and observed values. A lower RMSE value indicates a better fit of the model to the data. It is widely used in regression problems to evaluate the performance of a model (Equation (1)):

$$\mathrm{RMSE} = \sqrt{\sum_{i=1}^{N} \frac{(Ei - Oi)^2}{n}} \tag{1}$$

where O represents the observed value, E the estimated value, and n represents the number of samples.

2.6.2. Relative RMSE (rRMSE)

The relative RMSE is the ratio of the RMSE to the mean values of field measurements (yield (t ha^{-1})):

$$rRMSE = \frac{\text{RMSE}}{\frac{\sum_{i=i}^{N} Oi}{N}} \quad (2)$$

where O represents the observed value and N represents the number of samples.

2.6.3. Coefficient of Determination (R^2)

The coefficient of determination is a statistical metric used in the context of predictive modeling. The primary goal is to quantify the proportion of variance in the dependent variable that is predictable from the independent variable(s) in a statistical model. It is calculated as the ratio of the explained variation to the total variation of the dependent variable [90]. Equation (3) shows the R^2 formula:

$$R^2 = 1 - \frac{\sigma_r^2}{\sigma^2} \quad (3)$$

where σ_r^2 is the sum of the squared differences between the predicted values (from the model) and the actual values, and σ^2 is the sum of the squared differences between the actual values and the mean of the actual values.

2.6.4. Percentage of Mean Absolute Error (%MAE)

This is a statistical metric that quantifies the magnitude of the difference between two continuous variables. It is commonly used to evaluate the accuracy of a predictive model by comparing the predicted values to the actual values of the dataset. It is calculated as the average of the absolute differences between the predicted and actual values. Its mathematical formulation is represented in Equation (4):

$$\%MAE = \left(\frac{\frac{1}{n} \sum_{i=1}^{n} |y_i - x_i|}{P} \right) \times 100 \quad (4)$$

where y_i is the value of the prediction, x_i represents the observed value, n the total number of observations, and P the mean observed yield of each plot.

2.6.5. Accuracy

The accuracy error metric is a metric to evaluate the performance of a model with categorical data. Accuracy is calculated as the ratio of the number of correct predictions made by the model to the total number of predictions. The accuracy was expressed as a percentage, with values closer to 100% indicating a higher degree of accuracy:

$$\text{Accuracy} = \left(\frac{Cp}{Tp} \right) \times 100 \quad (5)$$

where Cp are correct predictions and Tp are total predictions.

2.6.6. Kappa Index (KI)

The Kappa index (KI) is a measure of accuracy when comparing actual and predicted yield maps. KI is a widely used statistical metric that quantifies the agreement between two categorical classifications, considering the possibility of agreement by chance. The KI was calculated using the formula:

$$KI = \frac{(Oa - Ea)}{(1 - Ea)} \quad (6)$$

where *Oa* is the observed agreement and *Ea* is the expected agreement.

3. Results

3.1. Relationship between Vegetation Indices and Wheat Yield Using Sentinel-2 Imagery

Figure 3 shows correlation matrices for the three different dates of VIs derived from S2 and yield. A high degree of collinearity among the different VIs was observed on the three dates, with Day 2 (GS39-49) showing the strongest correlation between indices with r values above 0.9. In addition to the negative correlation, when compared to the other VI results, the CI index obtained lower r values, ranging from −0.59 to −0.76 (Figure 3, Day 2). Correlations between the VIs of Day 1 (GS30) were slightly lower but remained above 0.8. Furthermore, the correlation between different VIs for Day 3 is not homogeneous (as indicated by the broader color palette of the matrix), and r values varied from 0.97 to 0.58. Given the high degree of collinearity observed among the VIs, some measures were taken to address this issue during the implementation of the different algorithms. One such measure employed was the Lasso correction of the MLR method.

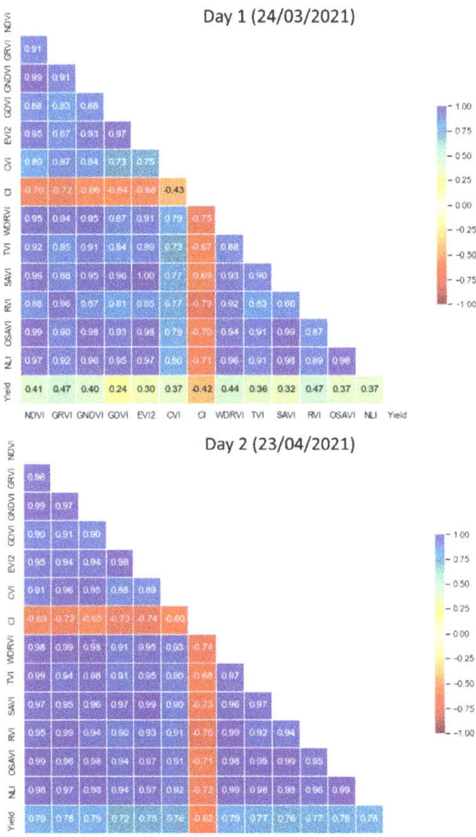

Figure 3. *Cont.*

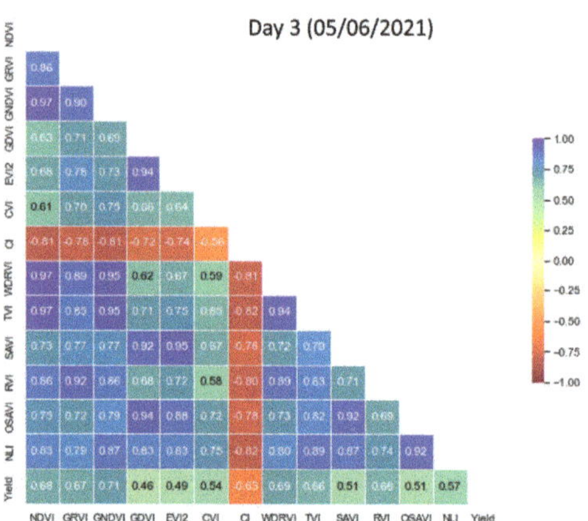

Figure 3. Correlation matrix between the different VIs of the three days (Days 1–3). The last column shows the correlation with the wheat grain yield.

In terms of the relationship between the different VIs and yield, the highest values were measured for Day 2 (GS39-49). Except for the negative correlation of CI (−0.65), the values ranged from 0.81 for GRVI to 0.73 for GDVI. In comparison, for Day 3 (GS69-75), the correlations ranged from 0.72 for GNDVI to 0.54 for GDVI. Although the increase in correlation is not significant (−0.66 compared to −0.65), CI was the only VI that increased the correlation. The lowest correlations were found with the VIs of Day 1 (Figure 3), with values ranging from 0.51 (RVI and GRVI) to 0.26 for GDVI. The correlation of CI was inverse (−0.45). Overall, the highest value was obtained with GRVI for all three dates, whereas GDVI exhibited the lowest values.

3.2. Exploring the Impact of Date Selection on Wheat Yield Prediction Using VIs Derived from Sentinel-2

In this study, the effect of adding different VIs derived from S2 corresponding to the three dates and its combination on the prediction of wheat grain yield was investigated using four different algorithms: CatBoost, SVM, RF, and MLR. All results (RMSE and R^2) (Figure 4) were obtained from the testing dataset. A consistent pattern was observed for all dates, with the best results obtained using CatBoost and the worst using MLR. When using the data from a single day, the R^2 and RMSE values varied greatly depending on the date. The worst results were always obtained when using VIs from Day 1. Thus, RMSE oscillated between 1.20 for CatBoost and 1.45 for MLR while R^2 ranged between 0.45 and 0.33. In contrast, the best results for a single day were obtained with Day 2 and CatBoost, reducing the RMSE to 0.56 and increasing the R^2 to 0.74.

When considering the predictive ability of the model using two different dates, the performance was better than when using each day separately. The R^2 of CatBoost ranged between 0.81 for the Day 1–2 dataset and 0.82 for the Day 2–3 dataset (Figure 4), while the R^2 value of MLR ranged between 0.65 for the Day 1–2 dataset and 0.69 for the Day 2–3 dataset. This result suggests that the best predictions were obtained with the dates corresponding to GS39-49 and GS69-75.

Nonetheless, the results indicate that all algorithms obtained the best results when they were trained with a dataset composed of the three dates (corresponding to GS30, GS39-49, and GS69-75 phenological stages). The R^2 values ranged from 0.859 for the CatBoost algorithm to 0.77 for MLR, while RMSE ranged from 0.32 for CatBoost to 0.50 for MLR.

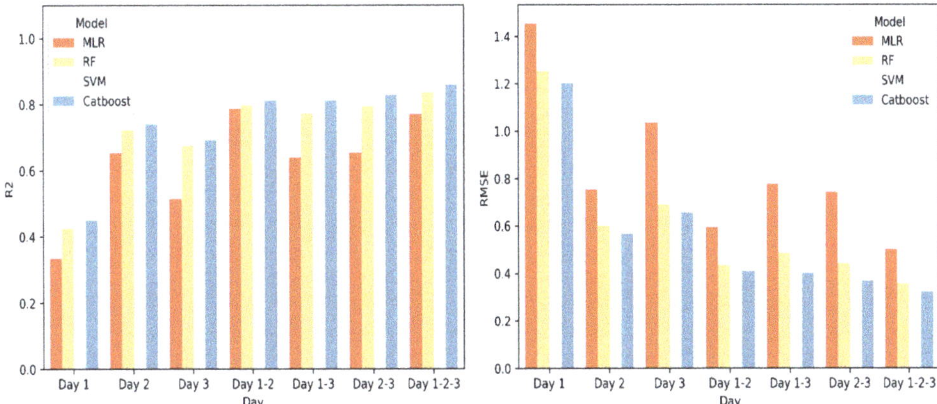

Figure 4. R^2 and RMSE of the four tested algorithms (MLR, Multiple Linear Model; RF, Random Forest; SVM, Support Vector Machine; CatBoost) when trained with VIs derived from S2 corresponding to different dates. It also shows accuracy metrics of the combination of different days.

3.3. Exploring the Impact of Date Selection on Wheat Yield Prediction Using Backscatter Information Derived from Sentinel-1

In this study, the feasibility of using backscatter information obtained from S1 at various dates to train and test machine learning models was evaluated. The results, represented in terms of R^2 and RMSE, obtained during the testing process are presented in Figure 5.

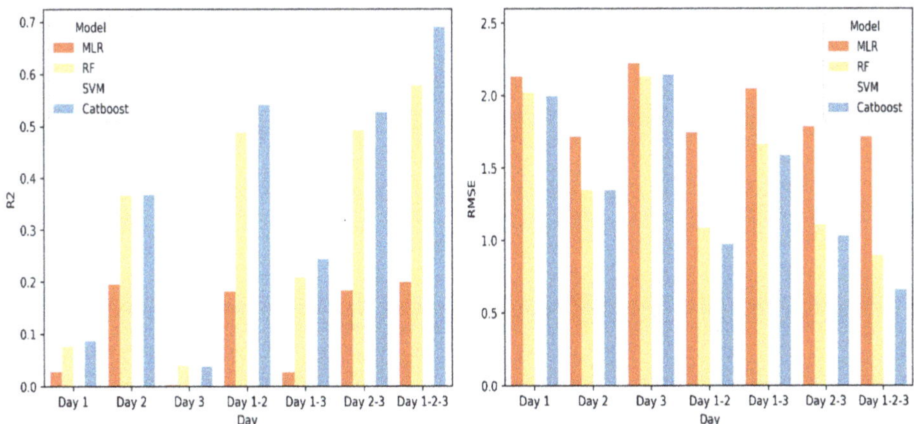

Figure 5. R^2 and RMSE of the four algorithms (MLR, Multiple Linear Model; RF, Random Forest; SVM, Support Vector Machine; CatBoost) when trained with VV and VH polarization backscatter information derived from S1 corresponding to three different dates. It also shows their combined use.

The pattern observed with S2 is repeated with the S1 data, where the best results were obtained using CatBoost and the worst using MLR. In the case of employing single days, the results showed notable variations depending on the selected day. For example, the R^2 value for Day 2 was 0.36, while for Day 3, it decreased to 0.08 when using CatBoost.

For the S1 data, the combination of multiple dates improved the results compared to a single date. The highest R^2 values were obtained when using information from the three days (Days 1–3). Among the algorithms tested, CatBoost showed the best results with an R^2 of 0.69, while the lowest R^2 value of 0.20 was obtained with the MLR model (Figure 5). The RF and SVM models showed similar results, with the latter showing a slightly better performance.

It is noteworthy that combining data from multiple dates did not always result in better performance compared to using data from a single date. For example, the RMSE for Day 2 was 1.34, while the combination of Days 1–3 was 1.59 with the CatBoost algorithm.

Additionally, the greatest differences in the RMSE and R^2 were observed between the algorithms that can analyze non-linear relationships (RF, SVM, and CatBoost) and the one that only analyzes linear relationships (MLR) when compared to the information of S2. In all cases, the non-linear algorithms showed better results (Figure 5).

3.4. Comparison of Machine Learning Algorithms for Estimating Wheat Yield Using Multisource Data

The results presented in the previous section indicate that the best results were consistently obtained using the information from Day 1-2-3. Having determined the optimal date combination, the next objective was to determine which algorithm achieved the best results for it. For this purpose, the RMSE and rRMSE were used. To capture the variability of each algorithm more accurately, the authors trained and validated each algorithm 10 times using different partitions of three datasets (S1, S2, and S1S2), resulting in 30 RMSE and rRMSE values for each algorithm (Table 2).

Table 2. Mean values of RMSE, SD and rRMSE of the four algorithms (MLR, Multiple Linear Model; RF, Random Forest; SVM, Support Vector Machine; CatBoost). Three different datasets were employed: S1 using only data from S1, S2 using data only from S2 and S1S2 using data from S1 and S2.

Algorithm	n *	Mean RMSE (t ha^{-1})	SD	rRMSE (%)
MLR	30	1.1	0.77	15.25
RF	30	0.69	0.35	9.78
SVM	30	0.62	0.34	8.92
CatBoost	30	0.41	0.29	5.91

* Each algorithm was trained and tested with ten different partitions of each dataset (S1, S2 and S1S2).

Table 2 shows the statistics associated to the prediction error obtained after running each algorithm 10 times with each of the three datasets (S1, S2 and S12). CatBoost produced the lowest error with an RMSE of 0.41 t ha-1 and a mean rRMSE of 5.91%. The SD of the RMSE for CatBoost was 0.29, the lowest among the four models. CatBoost not only produced results that were closest to the actual data, but also had less variability in the results compared to the other algorithms. RF and SVM performed similarly, with an average RMSE of 0.69 and 0.62 t ha^{-1}, respectively. The values of rRMSE were 9.78% and 8.92% (Table 2). The SD for both was nearly the same, 0.35 for RF and 0.34 for SVM. Finally, MLR produces the highest mean RMSE of 1.1 t ha^{-1}, with a mean rRMSE of 15.25% and an SD of 0.77.

After determining that CatBoost was the algorithm with the lowest RMSE and rRMSE among the four evaluated algorithms, the subsequent step involved evaluating the performance of CatBoost with each dataset (S1, S2, and S1S2). To this end, CatBoost was trained and tested with each of the three datasets 10 times with different partitions of data to train and test. The results presented in Figure 6 show that the RMSE varied depending on the dataset used for yield estimation. The use of the S1S2 dataset produced the lowest error, with a mean RMSE of 0.24 t ha^{-1}, which is an rRMSE of 3.46%. The RMSE values ranged between 0.22 and 0.26 t ha^{-1}. The mean RMSE obtained with S2 was 0.34 t ha^{-1} and the rRMSE was 4.86%. RMSE values ranged from 0.30 to 0.37 t ha^{-1} (Figure 6). Finally, the highest RMSE values were obtained when using only S1 data, with a mean RMSE of 0.79 t ha^{-1} and values ranging from 0.55 to 0.83 t ha^{-1}. The rRMSE for the S1 dataset was 11.34%. Therefore, the use of combined S1 and S2 (S1S2) data reduced the error by 30%

compared to using S2 data alone. Figure 7 presents the comparison of the predicted values versus the real values using CatBoost with S1S2. The R^2 value was 0.95.

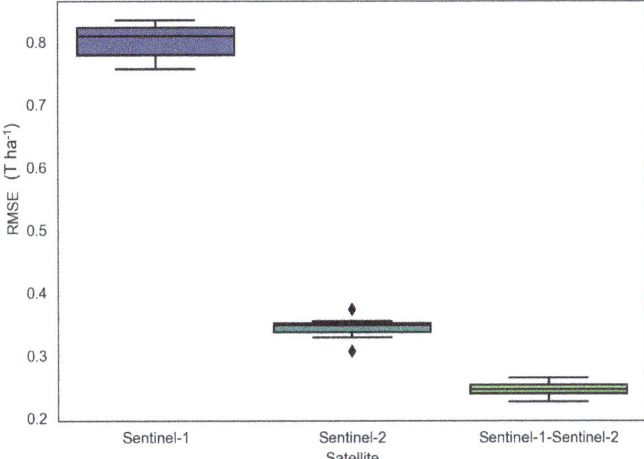

Figure 6. RMSE obtained using the CatBoost algorithm with data from S1 (Sentinel-1), S2 (Sentinel-2), and the combination of both (Sentinel-1 and Sentinel-2).

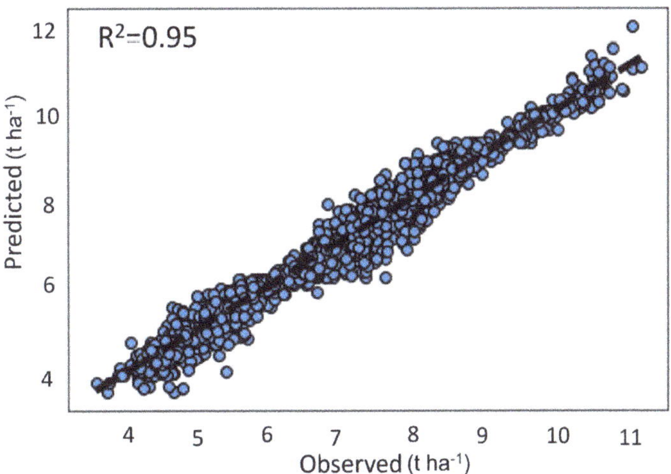

Figure 7. Linear regression between observed and predicted wheat grain yield for the test dataset obtained using the CatBoost algorithm and the S1S2 dataset.

3.5. Contribution of the Variables to the Defintive Algorithm

Figure 8 shows the 10 variables that made the greatest contribution to the CatBoost model, explaining 43.05% of the total variability. Of the 45 variables used (13 VIs and two backscatter variables for each day), the VV polarization variable (VV_Day2) derived from S1 and corresponding to April 20 (Day 2; GS39-49) contributed most to the model, with 6.69% of the explained variability. The second highest contributor was the GRVI_Day2 variable, which explained 5.47% of the variability. This variable, derived from S2, corresponds to April 23. The VH_Day1 variable, as shown in Figure 8, explained 2.99% of the total variability.

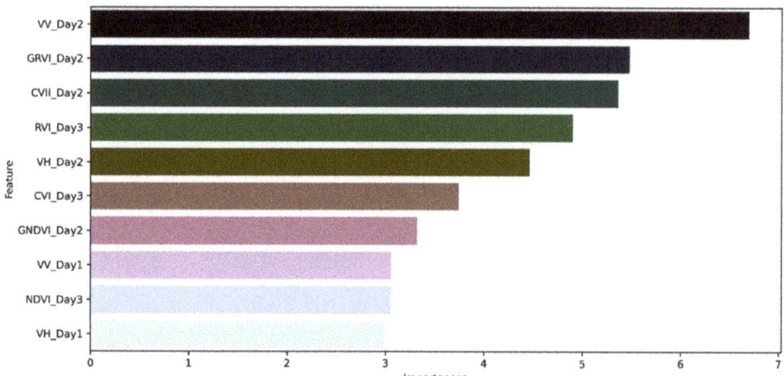

Figure 8. The 10 variables from S1 and S2 that most contributed to the model.

The analysis of the variables derived from S2 revealed a predominance of those obtained on Day 2 (April 20, GS39-49). However, there was also a representation of those from Day 3 (June 5, GS69-75), such as RVI. It is notable that the CVI variable is the only VI represented on two different days. With respect to the variables derived from S1, those corresponding to Day 2 explained more variability. However, in contrast to those derived from S2, in the case of S1 Day 1 (GS30) variables explained more variability than Day 3 (GS69-75) variables. Although the acquisition date is deemed more pertinent, polarization holds significance due to the greater explanatory power of the VV variables compared to the VH variables.

3.6. The Ability of CatBoost to Predict Yield of Entire Plots Using Data from Other Plots

In this section, the study aimed to evaluate the ability of CatBoost to predict the yield of an entire plot using information from other plots. Figure 9 shows that the mean %MAE was 4.38, which is below the acceptable error of 10%. However, plots G1 and G20 exceeded the 10% MAE threshold (Figure 9). To visually represent the difference between the actual and predicted yield values, G15 was selected.

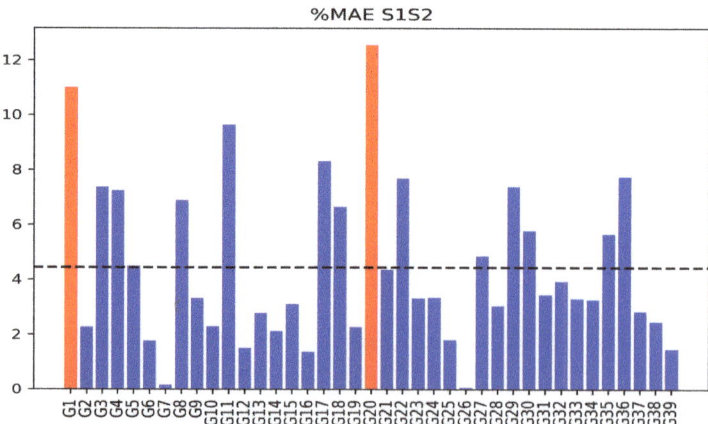

Figure 9. %MAE of the 39 plots when yield was predicted using the information from the rest of the plots. Those plots where %MAE is higher than 10% are shown in red. The dashed black line represents the mean %MAE.

Each dataset (measured and estimated yield data) was classified into two different classes using the ISODATA algorithm, which automatically set the optimal threshold for classification. The threshold for the measured data was set at 5.17 t ha^{-1}, while for the estimated data, it was set at 5.23 t ha^{-1}. To compare the agreement between the two classified maps, the accuracy and KI metrics were used. The accuracy was found to be 91.4%, while the KI was 0.77 (Figure 10). The accuracy and KI metrics show that the two classified maps are similar, indicating that the estimated map has retained the spatial variability of the original data. For G15 plot, the model predicted an average yield error of 0.190 t ha^{-1}, which is less than the maximum established error.

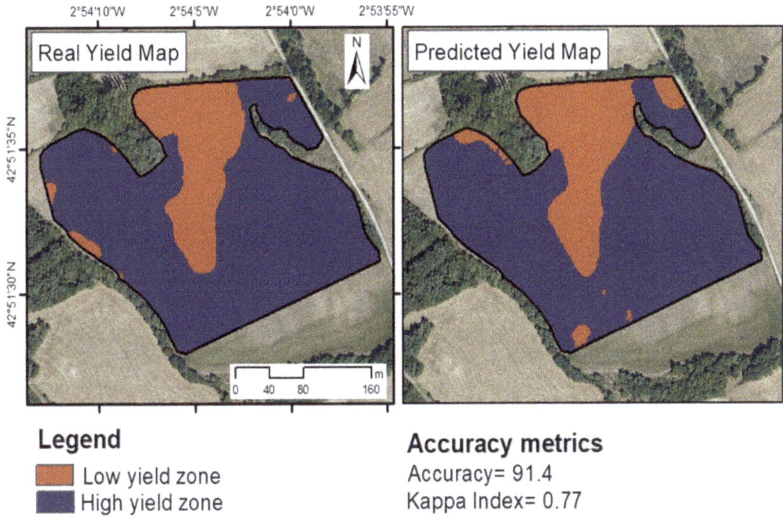

Figure 10. On the left, the classified wheat yield map of plot G15 (6.97 ha). On the right, classified wheat yield map based on the yield data estimated using the CatBoost algorithm with the S1S2 dataset for Days 1–3. The areas with low production are depicted in red, whereas those with high production are shown in blue. The accuracy and KI metrics were used to compare the two maps.

4. Discussion

4.1. Inclusion of Sentinel-1 and Sentinel-2 in the Yield Estimation Model

In this study, an analysis was conducted to examine the impact of incorporating multiple variables derived from S2 bands (VIs) and S1 backscatter information with VV and VH polarization obtained from various dates on yield prediction. The results revealed a consistent pattern in which the most favorable outcomes were consistently achieved when utilizing data from all three specified dates that corresponded to the GS30, GS39-49, and GS69-75 phenological stages.

In this study, the results obtained from VIs were consistent with those reported in prior research by Hunt et al. [29], since the inclusion of data from various dates improved model accuracy. In their study, the RF model was tested using VIs obtained from December to July, and the best results were obtained when using the VIs from all months together. According to the literature, the best grain yield estimation results are typically obtained after the end of the stem elongation phase (>GS39) [91,92], with the strongest relationship occurring during the anthesis or milky grain phase [93]. However, the analysis of VI information using data from only one day revealed that the optimal results were obtained using data corresponding to Day 2 (GS39-49) (Figure 3), which corresponds to the period from the end of stem elongation until the first awns' visible growth stage (24 April). This correlation was slightly higher than that achieved with data from Day 3 when wheat is between complete anthesis and medium milk phase GS69-75 (5 June). Despite the moderate to high collinearity among the VIs on the three dates (Figure 3), the results presented in Section 3.1 suggest that it is beneficial to use all the indices and multiple dates to obtain the best results. Furthermore, it is evident that the use of any model is superior to the use of only one index when predicting wheat yield.

Hunt et al. [29] found that the greatest improvement in the model occurred between December and April for wheat fields in the UK, with the improvement thereafter being less significant. In this study, the mean correlation coefficient between VI and yield on Day 1 (GS30) was 0.36, while on Day 2 (GS39-49), it increased to 0.78 (Figure 3). Additionally, other authors such as Segarra et al. [35] have reported that the best results (R^2 = 0.89) were obtained with the leaf area index (LAI) corresponding to the stem elongation/heading stage, and the results with VIs were similar (R^2 = 0.88). This is not surprising since LAI and some VIs are related [69]. Correlation between grain yield and VIs and LAI at this phase is logical since the phases encompassing stem elongation to ear growth phases are crucial in the vegetative growth of wheat [94] and greatly determine the final grain yield. The models demonstrated a high degree of efficiency in their ability to estimate yield at the end of April (GS39-49). Although it may be late to make decisions that improve yield in rainfed conditions, it could be useful for the planning of future fertilizer decisions within the framework of precision fertilization.

Analysis of the S1 backscatter information revealed that the best results were obtained using data from Day 2, corresponding to 20 April. However, in contrast to the results obtained with S2, the data from Day 1 explained more variability than Day 3 data (as seen in Figure 5). Previous research has reported a positive correlation between wheat yield and the backscattering coefficient from S1 [95]. This correlation can be attributed to the fact that backscattering is sensitive to changes in crop growth, biomass, and soil water content, all crucial factors in determining wheat yield [96]. In the early growth stages, stronger correlations were reported when backscatter information was used [96]. During these stages, the crop is more sensitive to variations in water and nutrient availability [97], and variations in backscattering can indicate crop health and potential yield. Furthermore, the correlation between the backscattering coefficient and wheat yield is more robust in areas where wheat is grown in monoculture. This is because the crop canopy in monoculture is more homogenous, and the backscattering signal can be more directly linked to crop growth and yield.

For the three S1 images, the VV polarization was found to contribute more to the model, in contrast to the results reported by Mandal et al. [98] who found higher correlations with VH polarization. The reason behind this is that VH polarization is more sensitive to changes in surface roughness, which is an indicator of crop growth, whereas VV polarization captures better changes in soil water content and soil moisture [99]. This seems to indicate that soil water content in the crop early stages affects the final yield in a relevant way. It is noteworthy that the correlation between backscattering and wheat yield is not simple, thus it is understandable that a higher R^2 value was obtained when using S2 data than S1 (Figures 4 and 5).

4.2. Reasons Why the Combination of Information from Sentienel-1 and Sentinel-2 Enhances the Yield Estimation Model

Previous studies, such as those published by Mercier et al. [100], have utilized data from S1 to predict the phenological stage of wheat. Other investigations have employed the combined information from S1 and S2 for the same purpose [101]. For example, Chaucha et al. [102] used the combined data from both satellites to determine wheat lodging in specific plots. Thus, there are previous studies in which the combined information from S1 and S2 has been utilized to estimate properties that can impact wheat yield or monitor crop development. However, to date, no studies have been identified in the literature that employ the combined information from both satellites to directly estimate wheat yield.

The findings of this study indicate that the utilization of data from both satellites improves the RMSE when compared to results obtained using only data from S2 (Figure 6). Establishing a relationship between wheat grain yield and S1 backscatter is not straightforward as the correlation is not linear, as shown by the performance of MLR (Figure 5). The backscatter is associated with crop canopy and soil roughness, which is related to crop development, LAI, biomass, and grain yield [103]. On the other hand, VIs derived from S2 data are relatively simple to calculate, are not computationally intensive, and are usually related to the biophysical properties of crops, such as greenness and health [104]. However, multicollinearity is a problem when using multiple VIs (Figure 3), as it reduces model accuracy [105]. The analysis of variable contribution showed that, among the top ten most representative variables, variables from both sources of information were present (Figure 8). Despite the unexpected nature of this finding, the variable that demonstrated the greatest contribution in the model was VV_Day2. This is particularly surprising because VH polarization is usually more sensitive to crop changes than VV [98]. By using data from both S1 and S2 satellite sources together, a more comprehensive understanding of the crop can be obtained, which can lead to more accurate wheat yield predictions. This study demonstrates the potential of using combined S1 and S2 data for crop monitoring and yield prediction and highlights the importance of considering multiple data sources for more accurate crop assessment.

4.3. Algorithm Analysis

The results obtained through the utilization of RF, SVM, and CatBoost algorithms surpassed those obtained through the utilization of MLR in all scenarios. The greatest error measured with RMSE was observed when the model was trained with S1 data, as depicted in Figure 5. The reason for this is that the connection between backscatter and yield is not linear, and MLR is not able to handle non-linear relationships. Although the relationship between VIs and wheat yield is primarily linear, it possesses enough non-linearity for other algorithms to yield superior results [106]. The capacity to handle non-linear relationships is a key advantage of some algorithms (SVM, RF, CatBoost), as it enables the analysis of complex multivariate relationships between different types of data, which is not feasible with MLR. The results obtained through the utilization of RF and SVM are comparable, with those obtained using the SVM model being slightly superior, which is in contrast to those reported by other authors [35,107] in the field of wheat yield prediction. Although RF generally outperforms SVM, in some areas of PA such as disease detection, SVM has performed better than RF [108]. However, in this study, the best results were achieved using the CatBoost algorithm, which is a member of the boosting algorithm family. The algorithms belonging to this family have produced inconsistent outcomes within the domain of PA. For example, Bebie et al. [25] reported the worst results when the boosting regression (BR) algorithm was used, while Heremans et al. [108] obtained the best outputs with the same algorithms. CatBoost, like Xtreme Gradient Boosting (XGBoost), is a gradient boosting algorithm that belongs to the next generation of boosting algorithms, and XGBoost has been used successfully in PA to predict monthly NDVI evolution [109]. However, the use of this group of algorithms is not as prevalent in PA as RF or SVM. As an example, the Scopus database revealed a limited number of articles, only seven, that employ CatBoost within any field of PA. In contrast, it is widely utilized in other areas such as industries, finance, healthcare, and online advertising.

Although in this case it has not been used because all the variables are quantitative, one of the main advantages of CatBoost over other algorithms is its ability to handle categorical variables because it can automatically deal with them without any additional pre-processing, such as 'one hot encoding' reducing considerably matrix dimensions. Moreover, CatBoost is specifically engineered to handle large datasets, as it facilitates training on graphics processing units (GPUs), thereby significantly decreasing computation time. In terms of performance, CatBoost has been shown to have high performance and generalization ability, outperforming other algorithms such as RF and the generalized regression neural network (GRNN) algorithm [110]. Additionally, CatBoost has a built-in mechanism for handling overfitting, which can be a problem with other algorithms like deep neural networks (DNNs) [111] and missing values. Finally, CatBoost also has a built-in feature importance mechanism that allows users to understand the importance of each feature in the dataset.

4.4. *CatBoost Algorithm as a Tool for Processing Heterogeneous Data in Precision Agriculture*

Use of the CatBoost algorithm in PA can provide significant advantages in terms of scaling up results. This algorithm is based on gradient boosting and is specifically designed to handle both numerical and categorical variables. This characteristic makes it suitable for PA, where a large amount of heterogeneous data are generated.

Compared to traditional machine learning algorithms such as RF, CatBoost has demonstrated improved performance in terms of accuracy and speed. The algorithm utilizes decision trees as weak learners and combines them in an iterative manner to make a strong prediction model. This results in a model that can generalize well to new data and is able to handle large amounts of data more efficiently than traditional algorithms.

In PA, the use of remote sensing data is increasingly common. This technology allows the acquisition of information on the physical, chemical, and biological characteristics of crops. Integration of the CatBoost algorithm with remote sensing data can provide valuable insights into crop growth. Another advantage of CatBoost is its ability to operate effectively even in the presence of missing records in a database. This is a common challenge faced when utilizing information from multiple sensors, as failures of individual sensors can occur at any point in time. The application of techniques to address such situations is not ideal, as it involves the addition of estimated information, which does not enhance the model. Furthermore, CatBoost data does not require scaling, leading to reduced time and effort in data preprocessing.

4.5. *Potential of S1 Backscatter and VIs for Precise Yield Mapping in Rainfed Areas Using the CatBoost Algorithm*

VIs have been widely utilized in PA for various purposes such as yield estimation, SSMZ delimitation, and water stress estimation. For its part, S1 backscatter information has been used for crop classification or for measuring land transformation changes. However, its use for yield estimation is not common. As previously mentioned, its relationship with growth is not direct, but it has been associated with key factors such as soil moisture, roughness or crop height. Therefore, it is imperative to conduct new studies to understand the underlying relationship between wheat yield and the S1 backscatter signal.

This study represents a preliminary step towards the goal of modulating fertilizer application according to crop needs. The underlying theoretical basis of this approach is that in rainfed areas, the fertilizer needs of the crop are generally associated to the potential yield. The high resolution of this study allowed for the estimation of precise yield maps. In this sense and according to Figure 9, the average %MAE was 4.38%, equivalent to an error of 0.31 t ha^{-1}. This level of precision would enable farmers to adjust fertilizer rates at the plot level with an acceptable margin of error. Figure 10 takes this approach one step further by comparing the yield maps generated from the yield monitor data with those generated using the proposed methodology. The classification of pixels was found to be consistent between the two maps in 91.4% of cases, suggesting that this approach captures intra-plot yield spatial variability. Therefore, this would enable farmers who do not have a yield monitor installed on their harvesters but have a variable rate fertilizer applicator to create and employ intra-plot prescription maps based on estimated yield maps. In addition, thanks to the auxiliary information source used (VI and backscatter derived from satellites), this methodology can be scalable and applicable to larger areas. The results, however, were obtained using satellite images acquired between Day 1 (GS30) and Day 3 (GS69-75), with the latter date being too late to increase yield by fertilizing. Considering this, the authors believe that future works should be directed at studying the combined capability of CatBoost with remote sensing data at early phenological stages of the crop to vary the fertilization strategy during the growing cycle.

Finally, it is worth noting that the results presented in this study are promising, but only correspond to one year. Thus, future works should encompass data from several years to verify that the results remain consistent across all campaigns. Furthermore, it would be interesting in future studies to incorporate high resolution climate and soil information in order to better understand the reasons behind yield spatial variability.

5. Conclusions

The models developed to estimate yield using information from S1 and S2 satellites showed better results than the correlation analysis. Among the evaluated models, CatBoost, which is still relatively underutilized in agriculture, provided the best results. Furthermore, using all available images that correspond to the GS30, GS39-49 and GS69-75 wheat phenological phases improved the performance of the models. Additionally, combining images from S1 and S2 substantially improved predictions, providing a level of precision sufficient to consider yield maps for fertilizer adjustment. This is an important aspect because most farmers in the area do not have yield monitors.

Despite its potential, the methodology proposed in this article has some limitations. Operationally, the biggest challenge lies in the clouds that impact the usability of the S2 images. While, theoretically, S2 provides an image every five days, in reality only three images were obtained throughout the whole crop growing cycle which were free of clouds and hence suitable for analysis. Moreover, to effectively train the algorithm, it is imperative to have access to high resolution yield data, such as that provided by yield monitors, although the use of such equipment is not yet widespread.

Combining the backscatter information of S1 with that of S2 resulted in improved outcomes of only using data from S2. However, further research is necessary to gain a better understanding of the relationship between backscattering and crop yield. In addition, this study focused solely on VIs and backscattering as they provide information on crop status. Future research could benefit from incorporating high resolution meteorological and edaphic variables, such as temperature, precipitation, and soil moisture, to better comprehend the factors influencing crop yield.

Author Contributions: A.U. worked in the following: Conceptualization, Methodology, Software, Data Processing, Formal Analysis, Original Draft Preparation, Visualization, Investigation, Interpretation. A.C. worked in the following: Methodology, Data Acquisition, Results Analysis, Resources. A.A. worked in the following: Conceptualization, Methodology, Writing, Reviewing and Editing, Supervision of Parameter Computing, Funding Acquisition, Project Administration. All authors have read and agreed to the published version of the manuscript.

Funding: This work was funded by the AgritechZeha project of the Basque Government, Department of Economic Development, Sustainability and Environment. It also was partially elaborated in the context of the CLIMALERT project SOE3/P4/F0862 UNION EUROPE. So, we want to express our gratitude to Interreg Sudoe Programme which a is part of the European territorial cooperation objective known as Interreg (financed by one of the European structural funds: the European Regional Development Fund (ERDF)).

Data Availability Statement: Data are available in a publicly accessible repository that does not issue DOIs. The raw satellite information data can be found in https://scihub.copernicus.eu/dhus/#/home, accessed on 30 January 2023.

Acknowledgments: The authors would like to thank Javier Alava, a farmer in the GARLAN cooperative, for providing the possibility to carry out the research in his plots and giving us high resolution yield information.

Conflicts of Interest: The authors declare no conflict of interest. The funders had no role in the design of the study, in the collection, analyses, or interpretation of data, in the writing of the manuscript, or in the decision to publish the results.

References

1. Giller, K.E.; Delaune, T.; Silva, J.V.; Descheemaeker, K.; van de Ven, G.; Schut, A.G.; van Wijk, M.; Hammond, J.; Hochman, Z.; Taulya, G.; et al. The future of farming: Who will produce our food? *Food Secur.* **2021**, *13*, 1073–1099. [CrossRef]
2. Pingali, P.L. Green Revolution: Impacts, limits, and the path ahead. *Proc. Natl. Acad. Sci. USA* **2012**, *109*, 12302–12308. [CrossRef]
3. Wik, M.; Pingali, P.; Broca, S. *Global Agricultural Performance: Past Trends and Future Prospects*; World Bank: Washington, DC, USA, 2008.
4. Hazell, P. *Handbook of Agricultural Economics*; Pingali, P., Evenson, R., Eds.; Elsevier: Amsterdam, The Netherlands, 2010; pp. 3469–3530.
5. Randall, G.; Goss, M. Nitrate Losses to SurfaceWater through Subsurface, Tile Drainage. In *Nitrogen in the Environment: Sources, Problems, and Management*; Elsevier: Amsterdam, The Netherlands, 2008.
6. Snyder, C.; Bruulsema, T.; Jensen, T.; Fixen, P. Review of greenhouse gas emissions from crop production systems and fertilizer management effects. *Agric. Ecosyst. Environ.* **2009**, *133*, 247–266. [CrossRef]
7. Ziliani, M.G.; Altaf, M.U.; Aragon, B.; Houborg, R.; Franz, T.E.; Lu, Y.; Sheffield, J.; Hoteit, I.; McCabe, M.F. Early season prediction of within-field crop yield variability by assimilating CubeSat data into a crop model. *Agric. For. Meteorol.* **2022**, *313*, 108736. [CrossRef]
8. Zambon, I.; Cecchini, M.; Egidi, G.; Saporito, M.G.; Colantoni, A. Revolution 4.0: Industry vs. Agriculture in a Future Development for SMEs. *Processes* **2019**, *7*, 36. [CrossRef]
9. Mumtaz, R.; Baig, S.; Fatima, I. Analysis of meteorological variations on wheat yield and its estimation using remotely sensed data. A case study of selected districts of Punjab Province, Pakistan (2001–2014). *Ital. J. Agron.* **2017**, *12*, 897. [CrossRef]
10. Sandonís-Pozo, L.; Llorens, J.; Escolà, A.; Arnó, J.; Pascual, M.; Martínez-Casasnovas, J.A. Satellite multispectral indices to estimate canopy parameters and within-field management zones in super-intensive almond orchards. *Precis. Agric.* **2022**, *23*, 2040–2062. [CrossRef]
11. Uribeetxebarria, A.; Arnó, J.; Escolà, A.; Martínez-Casasnovas, J.A. Apparent electrical conductivity and multivariate analysis of soil properties to assess soil constraints in orchards affected by previous parcelling. *Geoderma* **2018**, *319*, 185–193. [CrossRef]
12. Del-Moral-Martínez, I.; Rosell-Polo, J.R.; Company, J.; Sanz, R.; Escolà, A.; Masip, J.; Martínez-Casasnovas, J.A.; Arnó, J. Mapping Vineyard Leaf Area Using Mobile Terrestrial Laser Scanners: Should Rows be Scanned On-the-Go or Discontinuously Sampled? *Sensors* **2016**, *16*, 119. [CrossRef]
13. Daberkow, S.G.; McBride, W.D. Farm and Operator Characteristics Affecting the Awareness and Adoption of Precision Agriculture Technologies in the US. *Precis. Agric.* **2003**, *4*, 163–177. [CrossRef]
14. Chen, W.; Bell, R.W.; Brennan, R.F.; Bowden, J.W.; Dobermann, A.; Rengel, Z.; Porter, W. Key crop nutrient management issues in the Western Australia grains industry: A review. *Soil Res.* **2009**, *47*, 1–18. [CrossRef]
15. Barnes, A.; Soto, I.; Eory, V.; Beck, B.; Balafoutis, A.; Sánchez, B.; Vangeyte, J.; Fountas, S.; van der Wal, T.; Gómez-Barbero, M. Exploring the adoption of precision agricultural technologies: A cross regional study of EU farmers. *Land Use Policy* **2019**, *80*, 163–174. [CrossRef]
16. Ingram, J. Agronomist–farmer knowledge encounters: An analysis of knowledge exchange in the context of best management practices in England. *Agric. Hum. Values* **2008**, *25*, 405–418. [CrossRef]
17. Segarra, J.; Buchaillot, M.L.; Araus, J.L.; Kefauver, S.C. Remote Sensing for Precision Agriculture: Sentinel-2 Improved Features and Applications. *Agronomy* **2020**, *10*, 641. [CrossRef]
18. Ghosh, P.; Mandal, D.; Bhattacharya, A.; Nanda, M.K.; Bera, S. Assessing crop monitoring potential of sentinel-2 in a spatio-temporal scale. *ISPRS Int. Arch. Photogramm. Remote Sens. Spat. Inf. Sci.* **2018**, *XLII-5*, 227–231. [CrossRef]
19. Yi, Z.; Jia, L.; Chen, Q. Crop Classification Using Multi-Temporal Sentinel-2 Data in the Shiyang River Basin of China. *Remote Sens.* **2020**, *12*, 4052. [CrossRef]
20. Sadeghi, M.; Babaeian, E.; Tuller, M.; Jones, S.B. The optical trapezoid model: A novel approach to remote sensing of soil moisture applied to Sentinel-2 and Landsat-8 observations. *Remote Sens. Environ.* **2017**, *198*, 52–68. [CrossRef]
21. Vallentin, C.; Harfenmeister, K.; Itzerott, S.; Kleinschmit, B.; Conrad, C.; Spengler, D. Suitability of satellite remote sensing data for yield estimation in northeast Germany. *Precis. Agric.* **2022**, *23*, 52–82. [CrossRef]
22. Barnett, T.; Thompson, D. Large-area relation of landsat MSS and NOAA-6 AVHRR spectral data to wheat yields. *Remote Sens. Environ.* **1983**, *4*, 277–290. [CrossRef]
23. Maselli, F.; Conese, C.; Petkov, L.; Gilabert, M.-A. Use of NOAA-AVHRR NDVI data for environmental monitoring and crop forecasting in the Sahel. Preliminary results. *Int. J. Remote Sens.* **1992**, *13*, 2743–2749. [CrossRef]
24. Hamar, D.; Ferencz, C.; Lichtenberger, J.; Tarcsai, G.; Ferencz-Árkos, I. Yield estimation for corn and wheat in the Hungarian Great Plain using Landsat MSS data. *Int. J. Remote Sens.* **1996**, *17*, 1689–1699. [CrossRef]
25. Bebie, M.; Cavalaris, C.; Kyparissis, A. Assessing Durum Wheat Yield through Sentinel-2 Imagery: A Machine Learning Approach. *Remote Sens.* **2022**, *14*, 3880. [CrossRef]
26. Shen, J.; Evans, F.H. The Potential of Landsat NDVI Sequences to Explain Wheat Yield Variation in Fields in Western Australia. *Remote Sens.* **2021**, *13*, 2202. [CrossRef]
27. Trombetta, A.; Iacobellis, V.; Tarantino, E.; Gentile, F. Calibration of the AquaCrop model for winter wheat using MODIS LAI images. *Agric. Water Manag.* **2016**, *164*, 304–316. [CrossRef]

28. Boissard, P.; Guérif, M.; Pointel, J.-G.; Guinot, J.-P. Application of SPOT data to wheat yield estimation. *Adv. Space Res.* **1989**, *9*, 143–154. [CrossRef]
29. Hunt, M.L.; Blackburn, G.A.; Carrasco, L.; Redhead, J.W.; Rowland, C.S. High resolution wheat yield mapping using Sentinel-2. *Remote Sens. Environ.* **2019**, *233*, 111410. [CrossRef]
30. Li, H.; Chen, Z.; Liu, G.; Jiang, Z.; Huang, C. Improving Winter Wheat Yield Estimation from the CERES-Wheat Model to Assimilate Leaf Area Index with Different Assimilation Methods and Spatio-Temporal Scales. *Remote Sens.* **2017**, *9*, 190. [CrossRef]
31. Curnel, Y.; de Wit, A.J.W.; Duveiller, G.; Defourny, P. Potential performances of remotely sensed LAI assimilation in WOFOST model based on an OSS Experiment. *Agric. For. Meteorol.* **2011**, *151*, 1843–1855. [CrossRef]
32. Rodriguez, J.C.; Duchemin, B.; Hadria, R.; Watts, C.; Garatuza, J.; Chehbouni, A.; Khabba, S.; Boulet, G.; Palacios, E.; Lahrouni, A. Wheat yield estimation using remote sensing and the STICS model in the semiarid Yaqui valley, Mexico. *Agronomy* **2004**, *24*, 295–304. [CrossRef]
33. Vicente-Serrano, S.M.; Prats, J.M.C.; Romo, A. Early prediction of crop production using drought indices at different time-scales and remote sensing data: Application in the Ebro Valley (north-east Spain). *Int. J. Remote Sens.* **2006**, *27*, 511–518. [CrossRef]
34. Moriondo, M.; Maselli, F.; Bindi, M. A simple model of regional wheat yield based on NDVI data. *Eur. J. Agron.* **2007**, *26*, 266–274. [CrossRef]
35. Segarra, J.; Araus, J.L.; Kefauver, S.C. Farming and Earth Observation: Sentinel-2 data to estimate within-field wheat grain yield. *Int. J. Appl. Earth Obs. Geoinf.* **2022**, *107*, 102697. [CrossRef]
36. Uribeetxebarria, A.; Castellón, A.; Elorza, I.; Aizpurua, A. Intra-Plot Variable N Fertilization in Winter Wheat through Machine Learning and Farmer Knowledge. *Agronomy* **2022**, *12*, 2276. [CrossRef]
37. Meraner, A.; Ebel, P.; Zhu, X.X.; Schmitt, M. Cloud removal in Sentinel-2 imagery using a deep residual neural network and SAR-optical data fusion. *ISPRS J. Photogramm. Remote Sens.* **2020**, *166*, 333–346. [CrossRef]
38. Phiri, D.; Simwanda, M.; Salekin, S.; Nyirenda, V.R.; Murayama, Y.; Ranagalage, M. Sentinel-2 Data for Land Cover/Use Mapping: A Review. *Remote Sens.* **2020**, *12*, 2291. [CrossRef]
39. Torres, R.; Snoeij, P.; Geudtner, D.; Bibby, D.; Davidson, M.; Attema, E.; Potin, P.; Rommen, B.; Floury, N.; Brown, M.; et al. GMES Sentinel-1 mission. *Remote Sens. Environ.* **2012**, *120*, 9–24. [CrossRef]
40. Ulaby, F.; Moore, R.; Fung, A. *Microwave Remote Sensing Active and Passive-Volume III: From Theory to Applications*; Artech House: Norwood, MA, USA, 1986.
41. Chlingaryan, A.; Sukkarieh, S.; Whelan, B. Machine learning approaches for crop yield prediction and nitrogen status estimation in precision agriculture: A review. *Comput. Electron. Agric.* **2018**, *151*, 61–69. [CrossRef]
42. Mishra, S.; Mishra, D.; Santra, G.H. Applications of Machine Learning Techniques in Agricultural Crop Production: A Review Paper. *Indian J. Sci. Technol.* **2016**, *9*, 1–14. [CrossRef]
43. Shao, Y.; Campbell, J.B.; Taff, G.N.; Zheng, B. An analysis of cropland mask choice and ancillary data for annual corn yield forecasting using MODIS data. *Int. J. Appl. Earth Obs. Geoinf.* **2015**, *38*, 78–87. [CrossRef]
44. Bhosle, K.; Musande, V. Evaluation of Deep Learning CNN Model for Land Use Land Cover Classification and Crop Identification Using Hyperspectral Remote Sensing Images. *J. Indian Soc. Remote Sens.* **2019**, *47*, 1949–1958. [CrossRef]
45. Worrall, G.; Rangarajan, A.; Judge, J. Domain-Guided Machine Learning for Remotely Sensed In-Season Crop Growth Estimation. *Remote Sens.* **2021**, *13*, 4605. [CrossRef]
46. Arno, J.; Martinez-Casasnovas, J.A.; Ribes-Dasi, M.; Rosell, J.R. Clustering of grape yield maps to delineate site-specific management zones. *Span. J. Agric. Res.* **2011**, *9*, 721. [CrossRef]
47. Tang, X.; Liu, H.; Feng, D.; Zhang, W.; Chang, J.; Li, L.; Yang, L. Prediction of field winter wheat yield using fewer parameters at middle growth stage by linear regression and the BP neural network method. *Eur. J. Agron.* **2022**, *141*, 126621. [CrossRef]
48. Meraj, G.; Kanga, S.; Ambadkar, A.; Kumar, P.; Singh, S.K.; Farooq, M.; Johnson, B.A.; Rai, A.; Sahu, N. Assessing the Yield of Wheat Using Satellite Remote Sensing-Based Machine Learning Algorithms and Simulation Modeling. *Remote Sens.* **2022**, *14*, 3005. [CrossRef]
49. Wang, J.; Si, H.; Gao, Z.; Shi, L. Winter Wheat Yield Prediction Using an LSTM Model from MODIS LAI Products. *Agriculture* **2022**, *12*, 1707. [CrossRef]
50. Srivastava, A.K.; Safaei, N.; Khaki, S.; Lopez, G.; Zeng, W.; Ewert, F.; Gaiser, T.; Rahimi, J. Winter wheat yield prediction using convolutional neural networks from environmental and phenological data. *Sci. Rep.* **2022**, *12*, 3215. [CrossRef]
51. Cao, J.; Wang, H.; Li, J.; Tian, Q.; Niyogi, D. Improving the Forecasting of Winter Wheat Yields in Northern China with Machine Learning–Dynamical Hybrid Subseasonal-to-Seasonal Ensemble Prediction. *Remote Sens.* **2022**, *14*, 1707. [CrossRef]
52. Freund, Y.; Schapire, R.E. Experiments with a New Boosting Algorithm. In Proceedings of the 13th International Conference on International Conference on Machine Learning, Bari, Italy, 3–6 July 1996; Morgan Kaufmann Publishers Inc.: San Francisco, CA, USA, 1996; pp. 148–156.
53. Friedman, J.H. Greedy function approximation: A gradient boosting machine. *Ann. Stat.* **2001**, *29*, 1189–1232. [CrossRef]
54. Chen, T.; Guestrin, C. XGBoost: A Scalable Tree Boosting System. In Proceedings of the KDD '16: 22nd ACM SIGKDD International Conference on Knowledge Discovery and Data Mining, San Francisco, CA, USA, 13–17 August 2016; pp. 785–794. [CrossRef]
55. Prokhorenkova, L.; Gusev, G.; Vorobev, A.; Dorogush, A.V.; Gulin, A. CatBoost: Unbiased Boosting with Categorical Features. *arXiv* **2019**, arXiv:1706.09516.

56. Cai, Y.; Guan, K.; Lobell, D.; Potgieter, A.B.; Wang, S.; Peng, J.; Xu, T.; Asseng, S.; Zhang, Y.; You, L.; et al. Integrating satellite and climate data to predict wheat yield in Australia using machine learning approaches. *Agric. For. Meteorol.* **2019**, *274*, 144–159. [CrossRef]
57. Folberth, C.; Skalský, R.; Moltchanova, E.; Balkovič, J.; Azevedo, L.B.; Obersteiner, M.; van der Velde, M. Uncertainty in soil data can outweigh climate impact signals in global crop yield simulations. *Nat. Commun.* **2016**, *7*, 11872. [CrossRef]
58. Zadoks, J.C.; Chang, T.T.; Konzak, C.F. A decimal code for the growth stages of cereals. *Weed Res.* **1974**, *14*, 415–421. [CrossRef]
59. Beck, H.E.; Zimmermann, N.E.; McVicar, T.R.; Vergopolan, N.; Berg, A.; Wood, E.F. Data descriptor: Present and future Köppen-Geiger climate classification maps at 1-km resolution. *Sci. Data* **2018**, *5*, 180–214. [CrossRef]
60. Unamunzaga, O.; Aizpurua, A.; Artetxe, A.; Besga, G.; Castroviejo, L.; Blanco, F.; de la Llera, I.; Ramos, L.; Astola, G. Asistencia Técnica Para la Caracterización Agrológica del Suelo Rústico del Municipio de Vitoria-Gasteiz. Available online: https://docplayer.es/amp/152712108-Asistencia-tecnica-para-la-caracterizacion-agrologica-del-suelo-rustico-del-municipiode-vitoria-gasteiz.html (accessed on 30 January 2021). (In Spanish).
61. Drusch, M.; Del Bello, U.; Carlier, S.; Colin, O.; Fernandez, V.; Gascon, F.; Hoersch, B.; Isola, C.; Laberinti, P.; Martimort, P.; et al. Sentinel-2: ESA's Optical High-Resolution Mission for GMES Operational Services. *Remote Sens. Environ.* **2012**, *120*, 25–36. [CrossRef]
62. Rouse, J.W., Jr.; Haas, R.H.; Deering, D.W.; Schell, J.A.; Harlan, J.C. *Monitoring the Vernal Advancement and Retrogradation (GreenWave Effect) of Natural Vegetation*; NASA/GSFC Type III Final Report 5; NASA: Greenbelt, MD, USA, 1974; p. 371.
63. Gitelson, A.A.; Kaufman, Y.J.; Stark, R.; Rundquist, D. Novel algorithms for remote estimation of vegetation fraction. *Remote Sens. Environ.* **2002**, *80*, 76–87. [CrossRef]
64. Buschmann, C.; Nagel, M. In vivo spectroscopy and internal optics of leaves as basis for remote sensing of vegetation. International Journal of Remote Sensing. *Int. J. Remote Sens.* **1993**, *14*, 711–722. [CrossRef]
65. Tucker, C.; Elgin, J.; McMurtrey, J.; Fan, C. Monitoring corn and soybean crop development with hand-held radiometer spectral data. *Remote Sens. Environ.* **1979**, *8*, 237–248. [CrossRef]
66. Miura, T.; Yoshioka, H.; Fujiwara, K.; Yamamoto, H. Inter-Comparison of ASTER and MODIS Surface Reflectance and Vegetation Index Products for Synergistic Applications to Natural Resource Monitoring. *Sensors* **2008**, *8*, 2480–2499. [CrossRef]
67. Vincini, M.; Frazzi, E.; D'Alessio, P. A broad-band leaf chlorophyll vegetation index at the canopy scale. *Precis. Agric.* **2008**, *9*, 303–319. [CrossRef]
68. Escadafal, R. Remote sensing of soil color: Principles and applications. *Remote Sens. Rev.* **1993**, *7*, 261–279. [CrossRef]
69. Gitelson, A.A. Wide Dynamic Range Vegetation Index for Remote Quantification of Biophysical Characteristics of Vegetation. *J. Plant Physiol.* **2004**, *161*, 165–173. [CrossRef]
70. Bannari, A.; Morin, D.; Bonn, F.; Huete, A.R. A review of vegetation indices. *Remote Sens. Rev.* **1995**, *13*, 95–120. [CrossRef]
71. Huete, A.R. A soil-adjusted vegetation index (SAVI). *Remote Sens. Environ.* **1988**, *25*, 295–309. [CrossRef]
72. Tucker, C.J. Red and photographic infrared linear combinations for monitoring vegetation. *Remote Sens. Environ.* **1979**, *8*, 127–150. [CrossRef]
73. Rondeaux, G.; Steven, M.; Baret, F. Optimization of soil-adjusted vegetation indices. *Remote Sens. Environ.* **1996**, *55*, 95–107. [CrossRef]
74. Goel, N.S.; Qin, W. Influences of canopy architecture on relationships between various vegetation indices and LAI and Fpar: A computer simulation. *Remote Sens. Rev.* **1994**, *10*, 309–347. [CrossRef]
75. García-Escudero, L.A.; Gordaliza, A.; Matrán, C.; Mayo-Iscar, A. A general trimming approach to robust cluster Analysis. *Ann. Stat.* **2008**, *36*, 1324–1345. [CrossRef]
76. Taylor, J.A.; McBratney, A.B.; Whelan, B.M. Establishing Management Classes for Broadacre Agricultural Production. *Agron. J.* **2007**, *99*, 1366–1376. [CrossRef]
77. Zhang, C.; Luo, L.; Xu, W.; Ledwith, V. Use of local Moran's I and GIS to identify pollution hotspots of Pb in urban soils of Galway, Ireland. *Sci. Total. Environ.* **2008**, *398*, 212–221. [CrossRef]
78. European Space Agency (ESA). Sentinel-1 Mission. 2021. Available online: https://www.esa.int/Applications/Observing_the_Earth/Copernicus/Sentinel-1 (accessed on 30 January 2021).
79. Alpaydin, E. Introduction to Machine Learning. 2nd ed. 2010. Available online: https://books.google.nl/books?hl=nl&lr=&id=TtrxCwAAQBAJ&oi=fnd&pg=PR7&dq=introduction+to+machine+learning&ots=T5ejQG_7pZ&sig=0xC_H0agN7mPhYW7oQsWiMVwRnQ#v=onepage&q=introduction-to-machine-learning&f=false (accessed on 30 January 2021).
80. Friedman, J.H.; Hastie, T.; Tibshirani, R. Regularization Paths for Generalized Linear Models via Coordinate Descent. *J. Stat. Softw.* **2010**, *33*, 1–22. [CrossRef] [PubMed]
81. Cortes, C.; Vapnik, V. Support-vector networks. *Mach. Learn.* **1995**, *20*, 273–297. [CrossRef]
82. Breiman, L. Random forests. *Mach. Learn.* **2001**, *45*, 5–32. [CrossRef]
83. Liakos, K.G.; Busato, P.; Moshou, D.; Pearson, S.; Bochtis, D. Machine learning in agriculture: A review. *Sensors* **2018**, *18*, 2674. [CrossRef]
84. Pedregosa, F.; Varoquaux, G.; Gramfort, A.; Michel, V.; Thirion, B.; Grisel, O.; Blondel, M.; Prettenhofer, P.; Weiss, R.; Dubourg, V.; et al. Scikit-Learn: Machine Learning in Python. *J. Mach. Learn. Res.* **2011**, *12*, 2825–2830.
85. Bloniarz, A.; Liu, H.; Zhang, C.-H.; Sekhon, J.S.; Yu, B. Lasso adjustments of treatment effect estimates in randomized experiments. *Proc. Natl. Acad. Sci. USA* **2016**, *113*, 7383–7390. [CrossRef]

86. Rodríguez-Pérez, R.; Bajorath, J. Evolution of Support Vector Machine and Regression Modeling in Chemoinformatics and Drug Discovery. *J. Comput. Mol. Des.* **2022**, *36*, 355–362. [CrossRef]
87. Balfer, J.; Bajorath, J. Systematic Artifacts in Support Vector Regression-Based Compound Potency Prediction Revealed by Statistical and Activity Landscape Analysis. *PLoS ONE* **2015**, *10*, e0119301. [CrossRef]
88. Ho, T.K. Random decision forests. In Proceedings of the 3rd International Conference on Document Analysis and Recognition, Montreal, QC, Canada, 14–16 August 1995; pp. 278–282.
89. Ruder, S. An overview of gradient descent optimization algorithms. *arXiv* **2016**, arXiv:1609.04747.
90. Glantz, S.; Slinker, B. *Primer of Applied Regression and Analysis of Variance*; McGraw-Hill: New York, NY, USA, 1990.
91. Magney, T.S.; Eitel, J.U.; Huggins, D.R.; Vierling, L.A. Proximal NDVI derived phenology improves in-season predictions of wheat quantity and quality. *Agric. For. Meteorol.* **2016**, *217*, 46–60. [CrossRef]
92. Uribeetxebarria, A.; Castellón, A.; Aizpurua, A. A First Approach to Determine If It Is Possible to Delineate In-Season N Fertilization Maps for Wheat Using NDVI Derived from Sentinel-2. *Remote Sens.* **2022**, *14*, 2872. [CrossRef]
93. Babar, M.A.; Reynolds, M.P.; van Ginkel, M.; Klatt, A.R.; Raun, W.R.; Stone, M.L. Spectral Reflectance Indices as a Potential Indirect Selection Criteria for Wheat Yield under Irrigation. *Crop Sci.* **2006**, *46*, 578–588. [CrossRef]
94. Tian, H.; Wang, P.; Tansey, K.; Han, D.; Zhang, J.; Zhang, S.; Li, H. A deep learning framework under attention mechanism for wheat yield estimation using remotely sensed indices in the Guanzhong Plain, PR China. *Int. J. Appl. Earth Obs. Geoinf.* **2021**, *102*, 102375. [CrossRef]
95. Hosseini, M.; McNairn, H. Using multi-polarization C- and L-band synthetic aperture radar to estimate biomass and soil moisture of wheat fields. *Int. J. Appl. Earth Obs. Geoinf.* **2017**, *58*, 50–64. [CrossRef]
96. Ouaadi, N.; Jarlan, L.; Ezzahar, J.; Zribi, M.; Khabba, S.; Bouras, E.; Bousbih, S.; Frison, P.-L. Monitoring of wheat crops using the backscattering coefficient and the interferometric coherence derived from Sentinel-1 in semi-arid areas. *Remote Sens. Environ.* **2020**, *251*, 112050. [CrossRef]
97. Wollmer, A.-C.; Pitann, B.; Mühling, K.H. Grain storage protein concentration and composition of winter wheat (*Triticum aestivum* L.) as affected by waterlogging events during stem elongation or ear emergence. *J. Cereal Sci.* **2018**, *83*, 9–15. [CrossRef]
98. Mandal, D.; Kumar, V.; Ratha, D.; Dey, S.; Bhattacharya, A.; Lopez-Sanchez, J.M.; McNairn, H.; Rao, Y.S. Dual polarimetric radar vegetation index for crop growth monitoring using sentinel-1 SAR data. *Remote Sens. Environ.* **2020**, *247*, 111954. [CrossRef]
99. Bai, X.; He, B.; Li, X.; Zeng, J.; Wang, X.; Wang, Z.; Zeng, Y.; Su, Z. First Assessment of Sentinel-1A Data for Surface Soil Moisture Estimations Using a Coupled Water Cloud Model and Advanced Integral Equation Model over the Tibetan Plateau. *Remote Sens.* **2017**, *9*, 714. [CrossRef]
100. Mercier, A.; Betbeder, J.; Baudry, J.; Le Roux, V.; Spicher, F.; Lacoux, J.; Roger, D.; Hubert-Moy, L. Evaluation of Sentinel-1 & 2 time series for predicting wheat and rapeseed phenological stages. *ISPRS J. Photogramm. Remote Sens.* **2020**, *163*, 231–256. [CrossRef]
101. El Imanni, H.S.; El Harti, A.; Panimboza, J. Investigating Sentinel-1 and Sentinel-2 Data Efficiency in Studying the Temporal Behavior of Wheat Phenological Stages Using Google Earth Engine. *Agriculture* **2022**, *12*, 1605. [CrossRef]
102. Chauhan, S.; Darvishzadeh, R.; Lu, Y.; Boschetti, M.; Nelson, A. Understanding wheat lodging using multi-temporal Sentinel-1 and Sentinel-2 data. *Remote Sens. Environ.* **2020**, *243*, 111804. [CrossRef]
103. Vavlas, N.-C.; Waine, T.W.; Meersmans, J.; Burgess, P.J.; Fontanelli, G.; Richter, G.M. Deriving Wheat Crop Productivity Indicators Using Sentinel-1 Time Series. *Remote Sens.* **2020**, *12*, 2385. [CrossRef]
104. Kamenova, I.; Dimitrov, P. Evaluation of Sentinel-2 vegetation indices for prediction of LAI, fAPAR and fCover of winter wheat in Bulgaria. *Eur. J. Remote Sens.* **2021**, *54* (Suppl. S4), 89–108. [CrossRef]
105. Sohil, F.; Sohali, M.U.; Shabbir, J. An introduction to statistical learning with applications in R: By Gareth James, Dan-iela Witten, Trevor Hastie, and Robert Tibshirani, New York, Springer Science and Business Media, 2013, $41.98, EISBN: 978-1-4614-7137-7. *Stat. Theory Relat. Fields* **2021**, *6*, 87. [CrossRef]
106. Tesfaye, A.A.; Osgood, D.; Aweke, B.G. Combining machine learning, space-time cloud restoration and phenology for farm-level wheat yield prediction. *Artif. Intell. Agric.* **2021**, *5*, 208–222. [CrossRef]
107. Kok, Z.H.; Shariff, A.R.M.; Alfatni, M.S.M.; Khairunniza-Bejo, S. Support Vector Machine in Precision Agriculture: A review. *Comput. Electron. Agric.* **2021**, *191*, 106546. [CrossRef]
108. Heremans, S.; Dong, Q.; Zhang, B.; Bydekerke, L.; Van Orshoven, J. Potential of ensemble tree methods for early-season prediction of winter wheat yield from short time series of remotely sensed normalized difference vegetation index and in situ meteorological data. *J. Appl. Remote Sens.* **2015**, *9*, 097095. [CrossRef]
109. Li, X.; Yuan, W.; Dong, W. A Machine Learning Method for Predicting Vegetation Indices in China. *Remote Sens.* **2021**, *13*, 1147. [CrossRef]
110. Zhang, Y.; Zhao, Z.; Zheng, J. CatBoost: A new approach for estimating daily reference crop evapotranspiration in arid and semi-arid regions of Northern China. *J. Hydrol.* **2020**, *588*, 125087. [CrossRef]
111. Srivastava, N.; Hinton, G.; Krizhevsky, A.; Sutskever, I.; Salakhutdinov, R. Dropout: A Simple Way to Prevent Neural Networks from Overfitting. *J. Mach. Learn. Res.* **2014**, *15*, 1929–1958.

Article

How Sensitive Is Thermal Image-Based Orchard Water Status Estimation to Canopy Extraction Quality?

Livia Katz [1,2,3,4,*], Alon Ben-Gal [4], M. Iggy Litaor [3,5], Amos Naor [3], Aviva Peeters [6,7], Eitan Goldshtein [1], Guy Lidor [1], Ohaliav Keisar [1], Stav Marzuk [1,8], Victor Alchanatis [1] and Yafit Cohen [1]

1 Institute of Agricultural Engineering, Agricultural Research Organization, The Volcani Institute, Rishon-LeZion 7505101, Israel
2 Department of Soil and Water Sciences, The Robert H. Smith Faculty of Agriculture, Food & Environment, The Hebrew University of Jerusalem, Rehovot 7610001, Israel
3 Department of Precision Agriculture, MIGAL Galilee Research Institute, Kiryat Shmona 1101602, Israel
4 Environmental Physics and Irrigation, Agricultural Research Organization, Gilat Research Center, Mobile Post Negev 8528000, Israel
5 Department of Environmental Sciences, Tel Hai College, Upper Galilee, Qiryat Shemona 1220800, Israel
6 TerraVision Lab, Midreshet Ben-Gurion 8499000, Israel
7 School of Architecture, SCE Shamoon College of Engineering, Beer Sheva 8410802, Israel
8 Department of Industrial Engineering and Management, Ben-Gurion University of the Negev, Beer Sheva 8410501, Israel
* Correspondence: livia.katz@mail.huji.ac.il

Abstract: Accurate canopy extraction and temperature calculations are crucial to minimizing inaccuracies in thermal image-based estimation of orchard water status. Currently, no quantitative comparison of canopy extraction methods exists in the context of precision irrigation. The accuracies of four canopy extraction methods were compared, and the effect on water status estimation was explored for these methods: 2-pixel erosion (2PE) where non-canopy pixels were removed by thresholding and morphological erosion; edge detection (ED) where edges were identified and morphologically dilated; vegetation segmentation (VS) using temperature histogram analysis and spatial watershed segmentation; and RGB binary masking (RGB-BM) where a binary canopy layer was statistically extracted from an RGB image for thermal image masking. The field experiments occurred in a four-hectare commercial peach orchard during the primary fruit growth stage (III). The relationship between stem water potential (SWP) and crop water stress index (CWSI) was established in 2018. During 2019, a large dataset of ten thermal infrared and two RGB images was acquired. The canopy extraction methods had different accuracies: on 12 August, the overall accuracy was 83% for the 2PE method, 77% for the ED method, 84% for the VS method, and 90% for the RGB-BM method. Despite the high accuracy of the RGB-BM method, canopy edges and between-row weeds were misidentified as canopy. Canopy temperature and CWSI were calculated using the average of 100% of canopy pixels (CWSI_T100%) and the average of the coolest 33% of canopy pixels (CWSI_T33%). The CWSI_T33% dataset produced similar SWP–CWSI models irrespective of the canopy extraction method used, while the CWSI_T100% yielded different and inferior models. The results highlighted the following: (1) The contribution of the RGB images is not significant for canopy extraction. Canopy pixels can be extracted with high accuracy and reliability solely with thermal images. (2) The T33% approach to canopy temperature calculation is more robust and superior to the simple mean of all canopy pixels. These noteworthy findings are a step forward in implementing thermal imagery in precision irrigation management.

Keywords: canopy temperature; crop water status index; accuracy assessment; peach orchard; stem water potential

Citation: Katz, L.; Ben-Gal, A.; Litaor, M.I.; Naor, A.; Peeters, A.; Goldshtein, E.; Lidor, G.; Keisar, O.; Marzuk, S.; Alchanatis, V.; et al. How Sensitive Is Thermal Image-Based Orchard Water Status Estimation to Canopy Extraction Quality? *Remote Sens.* **2023**, *15*, 1448. https://doi.org/10.3390/rs15051448

Academic Editors: Kenji Omasa, Shan Lu and Jie Wang

Received: 12 January 2023
Revised: 17 February 2023
Accepted: 21 February 2023
Published: 4 March 2023

1. Introduction

Crop water status estimation is significantly affected by canopy temperature [1]. Accurate classification of canopy pixels within an image is central to crop water status estimation. The misclassification of non-canopy pixels, such as soil and mixed pixels, can significantly alter the canopy temperature and crop water status estimation [2] and, thus, may affect orchard irrigation decision making.

1.1. Crop Water Status Estimation for Precision Irrigation Management

Crop water stress index (CWSI) is an indirect measurement of crop water status derived from a thermal image. Absolute canopy temperature is a function of stomata opening and cooling by subsequent crop transpiration and is affected by meteorological factors, including ambient temperature, vapor pressure, wind speed, and radiation [3]. To compare thermal images and eliminate the need to measure all of the meteorological parameters, normalization of canopy temperature via CWSI was proposed as a proxy of crop water status [4,5]:

$$\text{CWSI} = \frac{T_{canopy} - T_{wet}}{T_{dry} - T_{wet}} \tag{1}$$

where T_{canopy} is the temperature of the canopy, T_{wet} is the temperature of a fully transpiring canopy, and T_{dry} is the temperature of a non-transpiring (stressed) canopy. CWSI ranges from zero to one, where higher values indicate higher water stress. The difference between T_{canopy} (°C) and T_{air} (°C) is dependent on vapor pressure deficit (VPD) [4,5]. T_{dry} is typically calculated using an empirical method [6,7], while T_{wet} is determined by employing empirical, theoretical, and statistical methods [6,8] or, for commercial plot scale, by calculating the average temperature of the coolest 5–10% of canopy pixels of each individual thermal image [7–9].

The calculation of T_{canopy} involves two steps. First, canopy pixels need to be extracted from the image and separated from non-canopy pixels, including "mixed pixels" (combinations of canopy, soil, weeds, foreign objects, and shade). The second step is the calculation of canopy temperature. A common approach for calculating T_{canopy} of an area of interest (a whole plot or a management zone) is by using the mean [10] or the median [7] temperature of extracted canopy pixels. Meron et al. [6] proposed using the coldest 33% of canopy pixels for the calculation. Cohen et al. [8] reported an over-estimation of water stress in cotton with the mean of all canopy extracted pixels. When the mean of the coldest 33% was used, water status was better estimated. No such comparison between the approaches used for calculating canopy temperature was found for orchards.

1.2. Approaches of Thermal Image-Based Canopy Extraction

Canopy extraction approaches incorporating thermal imagery include methods that use a single thermal infrared image (1-source) and other methods that use a thermal infrared image and additional remotely sensed images, usually RGB (multi-source). One-source methods include purely threshold-based statistical analysis on the one hand and coupled statistical and spatial analyses on the other hand. Statistical analysis of a temperature histogram to identify canopy pixels within a thermal image has been performed in orchards [11,12] where canopy can be distinguished from soil. In such cases, temperature histograms are characterized by a bimodal distribution, where the canopy and soil pixels are represented by cooler and warmer peaks, respectively. Mixed pixels, which include combinations of canopy, soil, weeds, foreign objects, and shade in a single pixel, are generally composed of a "saddle" area between the two peaks. Depending on the crop architecture, the distance between plants, the degree of complexity, and pixel resolution, there can be significant overlap between mixed pixels, pure-canopy pixels, and pure-soil pixels, creating a challenge in identifying pure-canopy pixels. Additionally, water-stressed trees may have higher canopy temperatures and could be misidentified as mixed or soil pixels [13].

An additional group of 1-source methods incorporates statistical and spatial analyses of a single thermal image. Spatial watershed segmentation has been coupled with binary thresholding to extract pure-canopy pixels in palm trees [14] and in vineyards [15]. Camino et al. [16] incorporated watershed segmentation and quartile histogram analysis in an almond orchard. Superpixel algorithms are used to differentiate meaningful regions of interest in an image [17], such as tree crowns in a forest system [18] and fruit detection in orchards [19]. In peach orchards, one technique involved thresholding to remove non-canopy pixels and then morphological erosion to remove mixed pixels and to extract pure-canopy pixels [20]. A second method used edge detection algorithms followed by morphological dilation to remove mixed pixels [21]. Additional methods include delineation of regions of interest of a single canopy [2] as well as pure edge detection analysis [22]. The incorporation of two types of analyses, statistical and spatial, on thermal images alone has been claimed to improve the quality of canopy extraction in comparison to merely statistical-based analysis [23].

In general, multi-source methods are based on statistical analysis of a visible (RGB) or multispectral image to extract canopy pixels, which is then used as a binary mask that is superimposed on a thermal image. This technique has been implemented in crops including potato [9], mint [24], and grape [25]. Additional feature layers, such as maps of irrigation pipes, can be incorporated to improve canopy extraction [26]. However, poor overlap of RGB and thermal images can cause misidentification of canopy pixels.

Currently, there are many methods of canopy extraction, but, to the best of our knowledge, there is no comprehensive quantitative comparison of canopy extraction methods in the context of orchard water status estimation. Thus, the decision of which canopy extraction method to incorporate and how to calculate canopy temperature may be arbitrary and not based on experimental data. Accurate canopy extraction and temperature calculations are crucial to minimizing inaccuracies in thermal image-based estimation of orchard water status that may directly affect irrigation decisions. This study tested the hypothesis that thermal image-based orchard water status estimation is significantly sensitive to the canopy extraction quality and to the temperature calculation approach. The objective was to determine the sensitivity of thermal image-based orchard water status estimation to canopy extraction methodology and quality. Four canopy extraction methods were evaluated. Three methods followed the 1-source approach (thermal images), incorporating both statistical and spatial analyses: (1) 2-pixel erosion (2PE), where non-canopy pixels were removed by thresholding followed by morphological erosion; (2) edge detection (ED), where edges were identified and then morphologically dilated; and (3) vegetation segmentation (VS) using statistical analysis of the temperature histogram followed by spatial watershed segmentation. A fourth method, denoted RGB-BM, followed the multi-source approach and used an RGB image to statistically extract a binary canopy layer to mask the thermal image. Additionally, two approaches to canopy temperature calculation were assessed by calculating the following: (1) the average of 100% of canopy pixels (T100%), and the average of the coolest 33% of canopy pixels (T33%).

2. Materials and Methods

2.1. Research Area

A field experiment was conducted during the 2019 season in a 4 ha commercial late-harvest peach orchard (Prunus persica cv. 1881) located near Mishmar Hayarden, Israel (33.01°N; 35.60°E) (Figure 1). The elevation of the orchard ranges from 171 to 188 m above sea level, the average slope is 5% to the northwest, and within the orchard, the slope ranges from 0 to 11.3%. The orchard was planted in 2007 with spacing of 2.6 m and 5 m between trees and rows, respectively, and was divided into 22 management cells (MC) of 35 m × 35 m to monitor various orchard parameters, including canopy area and SWP. A precision drip irrigation regime was implemented in the north subplot (MC 1–11), while the south subplot (MC 12–22) was uniformly irrigated. A detailed description of the irrigation design of the entire orchard and the decision-making process in the north subplot using thermal image-based tree water status estimation following the 2PE canopy extraction method is reported in [27]. The experiment was conducted during stage III of fruit development, which is the primary stage of fruit growth and period when most of the annual irrigation is applied.

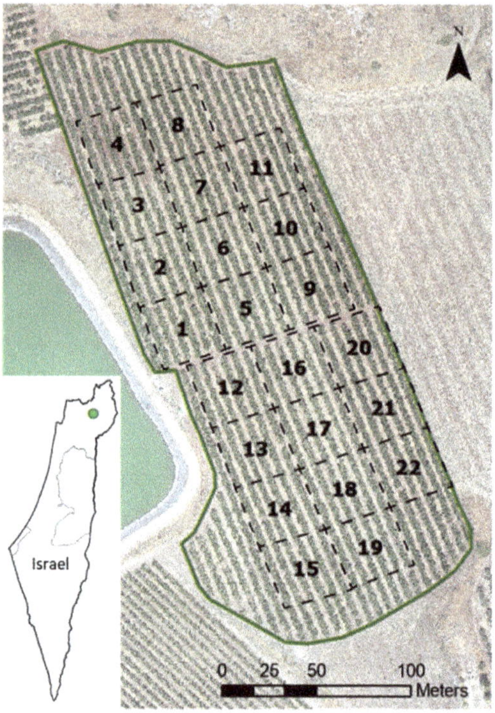

Figure 1. Mishmar Hayarden peach orchard (green line) divided into 22 management cells (MC) (black dashed squares).

The major steps of data acquisition and analysis are presented as a flow chart in Figure 2.

2.2. Image Acquisition

Ten high-resolution thermal images were acquired between 21 July and 26 August 2019. A sensitive (±2 °C) uncooled FLIR SC655 camera (FLIR® Systems, Inc., Billerica, MA, USA) with 640 × 480 resolution was mounted on a six-engine drone (Datamap Group, Bnei Brak, Israel). The flight height for all campaigns was 100 m, and the subsequent ground spatial resolution was approximately 7 cm. All campaigns were conducted midday between 12:30 and 15:15 on cloudless days. Mosaics were created using the ThermCam software (FLIR® Systems, Inc., Billerica, MA, USA) and Pix4D mapper software (Pix4D, Prilly, Switzerland). All of the thermal images were resampled to the average pixel size of the ten images, which was 7.3737 cm.

Two RGB images were acquired on 21 July and 12 August 2019 immediately prior to the respective thermal image campaign using a Phantom 4 Pro V2 (DJI Technology Co., Ltd., Shenzhen, China). The ground spatial resolution was approximately 3 cm.

2.3. Canopy Extraction Methods

Four methods of canopy extraction from thermal images were applied and evaluated, representing a range of techniques found in the literature. In this study, they were executed primarily using the ArcGIS Pro software (ESRI, Redlands, CA, USA).

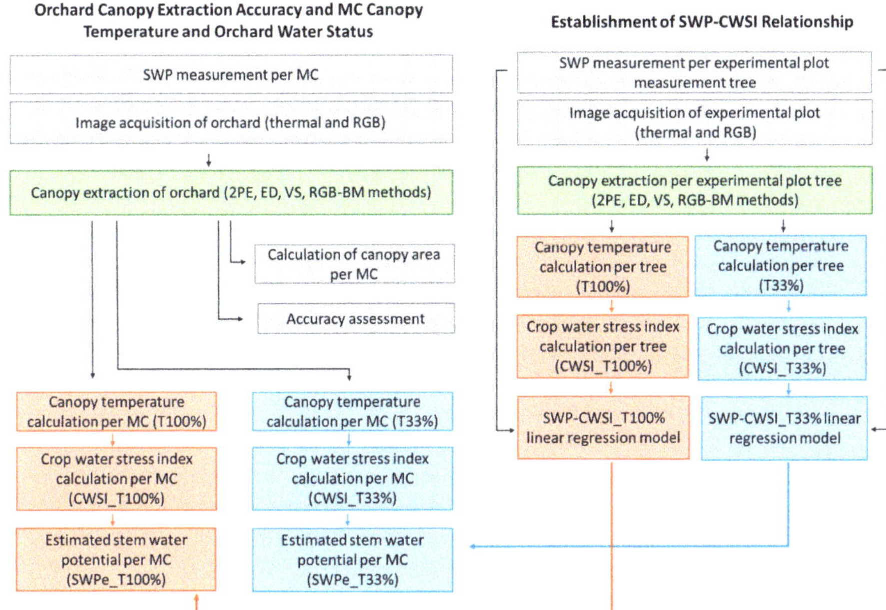

Figure 2. Data acquisition and analysis of orchard canopy extraction accuracy, canopy temperature, and orchard water status using the 2-pixel erosion (2PE), edge detection (ED), vegetation segmentation (VS), and RGB binary masking (RGB-BM) canopy extraction methods (green boxes). Canopy temperature per management cell (MC) was calculated using the average of 100% of canopy pixels (T100%) (orange boxes) and the average of the coolest 33% of canopy pixels (T33%) (blue boxes). Orchard water status was estimated using the crop water stress index (CWSI) and the estimated stem water potential (SWPe). The SWPe was based on a tree-scale stem water potential (SWP) and CWSI relationship established using each canopy extraction method and each canopy temperature calculation approach.

Three methods followed the 1-source approach, incorporating both statistical and spatial analyses, using only thermal images:

(1) 2-pixel erosion (2PE):
   a. Extraction of the coolest two-thirds of temperature pixels from the whole orchard histogram to separate canopy from non-canopy (mixed and soil) pixels [27] (statistical).
   b. Morphological erosion of the two pixels [28] (spatial).

(2) Edge detection (ED) based on [21]:
   a. Image sharpening with high pass filter (spatial).
   b. Determination of edges (statistical).
   c. Morphological expansion of three pixels (spatial).
   d. Thresholding to extract only canopy pixels (statistical).

(3) Vegetation segmentation (VS) based on [29] written in the Matlab R2020a (Mathworks Inc., Matick, MA, USA):
   a. Temperature histogram analysis using the Otsu [30] and full-width-half-maximum [11] algorithms to differentiate between canopy and non-canopy pixels (statistical).
   b. Watershed segmentation to define the basin of each peach tree [14] (spatial).

The temperature values per pixel of the 2PE, ED, and VS methods were retrieved by multiplying the respective final layer of canopy pixels by the original thermal image.

A fourth method followed the multi-source approach, using a thermal and an RGB image:

(4) RGB-based binary masking (RGB-BM):

a. Resampling of the RGB to 7.3737 cm.
b. Georeferencing between the RGB and thermal layers.
c. The excess green index (ExG) (2G-R-B) is calculated per pixel and effectively differentiates between plant and soil pixels [31].
d. Binary thresholding of the ExG layer to separate canopy from non-canopy pixels [30] (statistical).
e. Thermal image masking using the ExG layer (post-binary thresholding) [24] to retrieve the temperature values of each pixel (spatial).

2.4. *Canopy Extraction Quality Evaluation*

The quality of canopy extraction was determined by measuring the canopy area consistency throughout the study as well as assessing the accuracy of each method (ArcGIS Pro 2.9.0 software, ESRI, Redlands, CA, USA).

2.4.1. Canopy Area Consistency

During stage III, vegetative growth of deciduous fruit trees, including peaches, is minimal to non-existent [32]. Hence, canopy area consistency can be used as a measure of extraction quality. The canopy area (m^2) per MC (n = 22) per image was calculated for all four canopy extraction methods. On two dates, 21 July and 12 August, the median values were calculated, and the Student's t-test was used to determine significant differences in the mean canopy area on these dates for each method (JMP statistical software, JMP Inc., Cary, NC, USA). Additionally, the coefficient of variation (CV) was calculated for the canopy area median values per image for the 2PE, ED, and VS methods.

2.4.2. Accuracy Assessment

Accuracy assessment was performed per canopy extraction method on the datasets for 21 July and 12 August. All orchard pixels were reclassified into two categories using the final extraction layer per image as pure-canopy and non-canopy pixels. Synchronization between the final extraction layer (thermal or other) and the RGB ground truth image was verified. One hundred sample points were divided equally between these categories. A different set of 100 sample points was distributed for each of the four methods and two dates. A total of 800 sample points were used in the analysis. Ground truth validation was visually determined per sample point with the original RGB image from each respective date, 21 July and 12 August. The ensuing confusion matrix included the following: sample points that were correctly classified as canopy pixels (true positive—TP); sample points that were classified as canopy but were actually non-canopy pixels (false positive—FP); sample points that were correctly classified as non-canopy pixels (true negative—TN); and sample points that were classified as non-canopy but were actually canopy pixels (false negative—FN). The following parameters were calculated, enabling the evaluation of canopy extraction quality: overall accuracy (Equation (2)); precision (Equation (3)); recall (Equation (4)); and F1-score, which is the harmonic mean of precision and recall [33] (Equation (5)):

$$Overall\ accuracy = (TP + TN)/(TP + TN + FP + FN) \quad (2)$$

$$Precision = TP/(TP + FP) \quad (3)$$

$$Recall = TP/(TP + FN) \quad (4)$$

$$F1 - score = 2 \times (Precision \times Recall)/(Precision + Recall) \quad (5)$$

2.5. Canopy Temperature Calculation

The temperature of all extracted orchard canopy pixels was retrieved, and a histogram was built for each method. Descriptive statistics were calculated for the histograms, including mean, median, standard deviation, minimum, and maximum. The canopy temperature of each MC was estimated using two methods: the average of 100% of canopy pixels (T100%), and the average of the coolest 33% of canopy pixels (T33%). The calculations used the raster [34], rgdal [35], and reshape2 [36] packages in R [37], and the canopy temperature graphs were constructed using the ggplot2 package in R [38].

2.6. Orchard Water Status Estimation

The CWSI was calculated per MC following Equation (1): *Twet* was calculated based on the average of the coolest 5% of canopy pixels of the whole orchard [9]; *Tdry* was calculated empirically as Tair + 2 °C [7,27]; and *Tcanopy* was calculated using the methods described in the previous section for T100% and T33%. The CWSI was denoted CWSI_T100% and CWSI_T33%, respectively. Air temperature values for the day and time of each thermal image campaign were acquired from the nearby Gadot meteorological station (33.03°N; 35.62°E). The CWSI graphs were constructed using the ggplot2 package in R [38].

2.6.1. Establishment of the Relationship between SWP and CWSI

Linear regression models were developed based on a field experiment that took place during the fruit growth stage III of season 2018. Different irrigation levels were applied to the three plots in the orchard to create a range of soil water contents and respective plant water status. A campaign, including stem water potential (SWP) measurements of five trees per plot (n = 15) and thermal imaging, took place on 05 August 2018 during stage III. Plant water status was evaluated by measuring SWP using a Scholander-type pressure chamber (Arimad, MRC Ltd., Holon, Israel). Two shaded leaves were covered with an aluminum foil zip-lock bag 1.5 h before the measurement. Measurements were performed on each measurement tree between the hours of 12:30–15:15, and the results were averaged per tree.

For each canopy extraction method (2PE, ED, VS, and RGB-BM), CWSI_T100% and CWSI_T33% were calculated per measurement tree using the method described in the previous section. Linear regression models were created using the SWP measurements and the CWSI calculations and evaluated using the following parameters: correlation coefficient (R^2), root-mean-square error (RMSE), *p*-value, and the lower and upper 95% confidence intervals of the intercept and slope using the JMP statistical software (JMP Inc., Cary, NC, USA).

2.6.2. Estimated Stem Water Potential

The linear regression models were the basis for calculating the estimated SWP (SWPe). The SWPe was calculated from the CWSI_T100% and CWSI_T33% values per MC for the entire dataset, and its values were denoted SWPe_T100% and SWPe_T33%, respectively. Additionally, SWP was measured on three to four healthy trees of representative canopy size per MC in the north subplot and three trees per MC in the south subplot for each day of data collection. The SWP measurements served as a reference indicating the actual water status of each MC. The specific method and time of measurement is described in Section 2.6.1. The SWPe was subtracted from the measured SWP per MC per day, and descriptive statistics were used to evaluate the datasets: average, standard deviation, maximum, minimum, median, 25% and 75% quartile, mean squared error (MSE), and RMSE. Additionally, each measured SWP and SWPe value was compared to the optimal water status range of stage III, which was defined between −1.17 and −1.43 MPa and based roughly on [39]. Above-range values (>−1.17 MPa) indicated excessive moisture and possible over-irrigation; within-range (optimal) SWP values (between −1.17 and −1.43 MPa) indicated sufficient water status; and below-range values (<−1.43 MPa) specified orchard water stress. The variance of each distribution was calculated.

3. Results

3.1. Evaluation of Canopy Extraction Quality

3.1.1. Canopy Area Consistency

A difference in canopy area was evident between the two RGB-BM images on 21 July and 12 August (Figure 3): the median values were 414 and 474 m^2, respectively. This is a difference of 60 m^2, while slight differences were observed between these dates with the 2PE, ED, and VS methods: 6, 4, and 3 m^2, respectively. Accordingly, a significant difference in canopy area mean (increase) was calculated between the two RGB-BM images ($p < 0.001$, n = 22 MC per date), while no difference was detected for the 2PE, ED, or VS methods ($p > 0.05$, n = 22 MC per date).

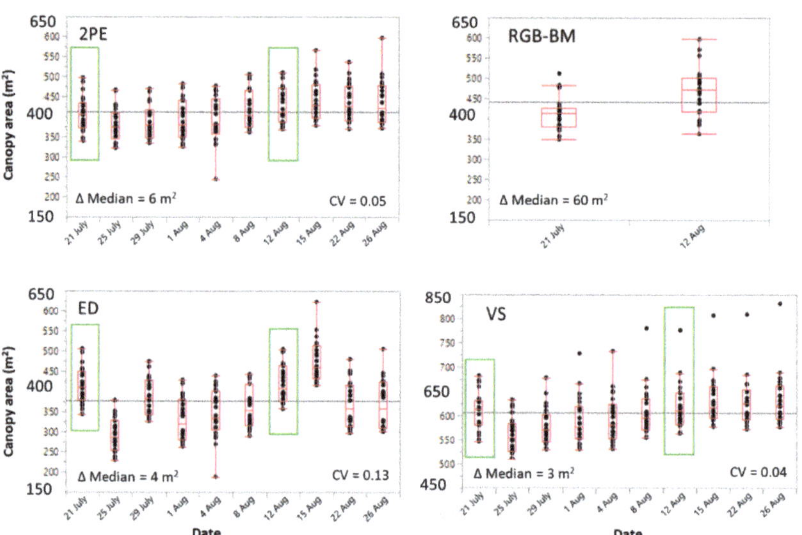

Figure 3. Canopy area (m^2) per management cell (MC) (black dots) and box plot (red) per day of image acquisition (21 July–26 August 2019). The black horizontal line is the grand mean. The green boxes indicate data from 21 July and 12 August of the 2-pixel erosion (2PE), edge detection (ED), and vegetation segmentation (VS) methods. The RGM binary masking (RGB-BM) method was performed only on these dates. Note: the Y-axis range of the VS method is specifically different from the other methods.

The canopy area consistency of the 2PE, ED, and VS methods was evaluated using ten thermal images (Figure 3). The 2PE and VS methods were more consistent compared to the ED method. The median values per date of the 2PE ranged from 374 to 430 m^2 (a difference of 56 m^2), and the values ranged from 555 m^2 to 626 m^2 (a difference of 71 m^2) with the VS method. In contrast, the median values of the ED method ranged from 288 to 460 m^2 (a difference of 172 m^2, 3-fold of the 2PE method), indicating less consistency over time. The coefficient of variation (CV) values of the 2PE, VS, and ED methods were 0.05, 0.04, and 0.13, respectively, highlighting the differences in consistency.

3.1.2. Accuracy Assessment

The differences in canopy area identification accuracy were evident between the canopy extraction methods (Figure 4). The RGB-BM method was found to be the most accurate among the canopy extraction methods, as demonstrated through the overall accuracy and the F1-score values on both 21 July and 12 August. On 21 July, the recall values of all of the methods were high (90–97%), indicating that most of the actual canopy was correctly classified. The precision, or the degree to which the classified map correctly identified canopy, however, varied according to extraction method: the 2PE and RGB-BM methods' precision was fairly high (84%), while the VS method's precision was the lowest (62%). On 12 August, the recall values of the VS and RGB-BM methods were higher than the 2PE and ED methods. The precision of the 2PE method was slightly less than the RGB-BM method (78%), and the ED method's precision was the lowest (68%).

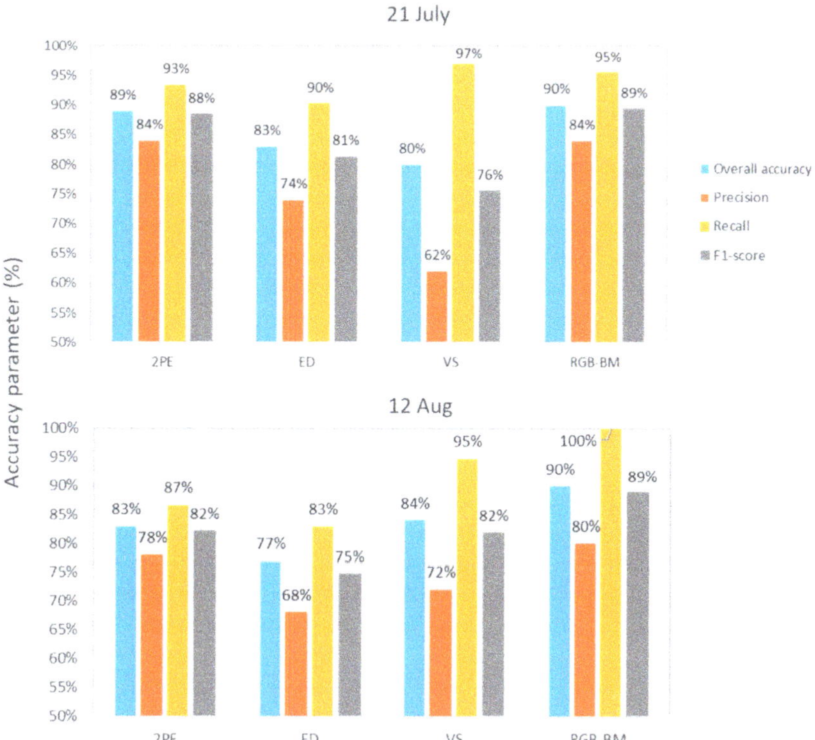

Figure 4. Overall accuracy (blue bars) of canopy/non-canopy classification and precision (red bars), recall (yellow bars), and F1-score (grey bars) parameters of canopy classification as measured with a confusion matrix per date for the 2-pixel erosion (2PE), edge detection (ED), vegetation segmentation (VS), and RGB binary masking (RGB-BM) canopy extraction methods.

3.2. *Canopy Temperature Calculation*

The differences in estimated canopy temperature between the four canopy extraction methods were evident at both the whole orchard and MC scales (Figure 5). The most striking difference between the methods in the whole orchard histograms of canopy temperature was the range of values. The temperature ranges of the extracted canopy pixels using the RGB-BM method was substantially wider (30–64 °C) than that of the 2PE (30–42 °C), ED (30–46 °C), and VS (30–47 °C) methods. The "tail" of the warm pixels of the RGB-BM histogram is the result of non-tree canopy being misclassified as canopy. These warm pixels were located between tree rows that contained grasses and soil but were free of tree canopy material, and they are illustrated in the RGB-BM temperature map of the MC 5 (Figure 5 left image column). The ExG index, which is the basis for the RGB-BM method, seemingly had difficulty differentiating between the different types of plant material. However, this method was noticeably able to detect slight differences between canopy and non-canopy pixels within the tree canopy, in contrast to the 2PE, ED, and VS methods which pixels were all relatively coarse.

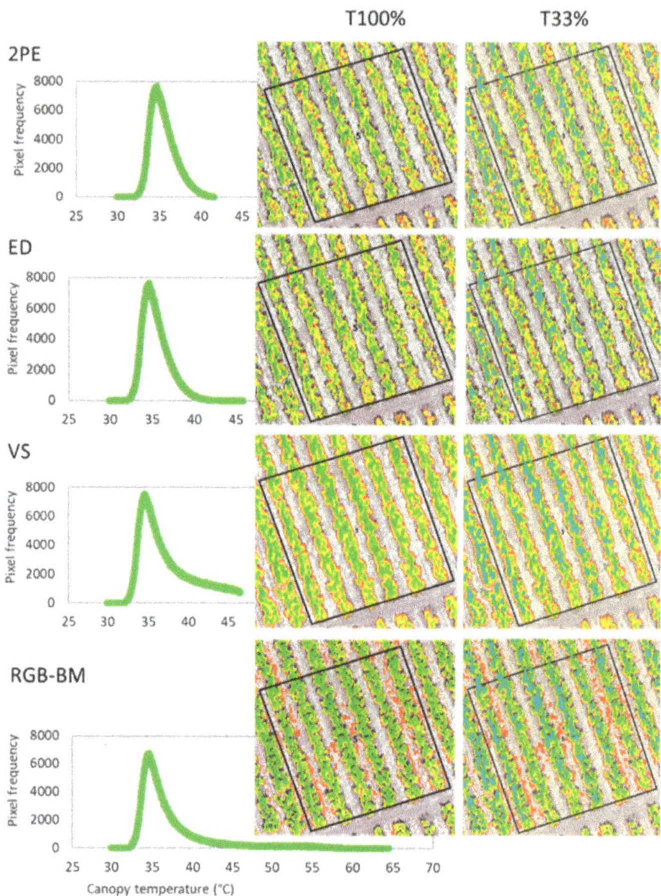

Figure 5. Canopy temperature histogram of the whole orchard for the 2-pixel erosion (2PE), edge detection (ED), vegetation segmentation (VS), and RGB binary masking (RGB-BM) canopy extraction methods on 12 August 2019. Images of all extracted canopy temperature pixels (T100%) of the management cell (MC) 5 (left image column) and the highlighted (turquoise) coolest 33% of canopy temperature pixels (T33%) (right image column) for all canopy extraction methods.

The VS method histogram is characterized by a larger number of pixels between 38 and 47 °C compared to the other methods (Figure 5), indicating that not all mixed pixels have been properly removed. The edges of canopy material (between one and three pixels) are noticeably warmer than other parts of the canopy throughout the orchard. Additionally, many between-row pixels of the MC 4 (not shown) were misidentified as canopy pixels. The MC 4 was defined as stressed and irrigated according to the SWPe value. Over-irrigation supposedly caused waterlogging in specific locations and relatively wet soil in others, and it directly affected the pixel temperature in this MC (Figure 3 outlier).

Visible differences were evident between the spatial patterns of the extracted canopy pixels and the coolest 33% of the canopy pixels for each canopy extraction method (Figure 5); however, relatively small differences were noted in the spatial patterns between the coolest 33% canopy pixels of all extraction methods (Figure 5 right image column). Notable differences were found between the canopy temperatures calculated using the average 100% of canopy pixels (T100%) and the average of the coolest 33% of canopy pixels (T33%) (Figure 6). The average T100% values were higher than those of T33% for each canopy extraction method: 1.28 °C (ED), 1.37 °C (2PE), 2.85 °C (RGB-BM), and 3.02 °C (VS). Additionally, the T100% calculation emphasized the differences between the extraction methods. The RGB-BM and VS methods yielded considerably higher T100% than the 2PE and ED methods. The value of the VS method was, on average, 1.91 °C higher than the 2PE method, while the average differences between the 2PE and ED methods were minimal (0.13 °C). The T33% dataset was characterized by minimal to slight differences between the canopy extraction methods: an average difference of 0.04 °C between the 2PE and ED methods and of 0.26 °C between the VS and 2PE methods.

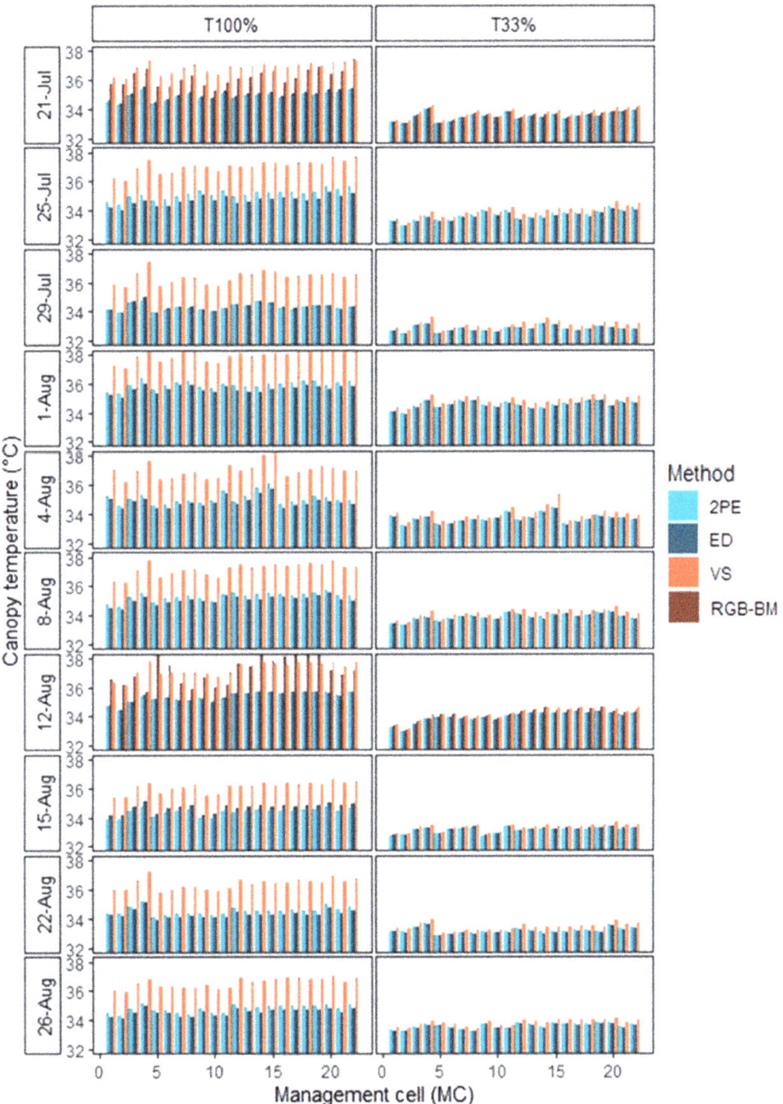

Figure 6. Canopy temperature (°C) calculated by the average 100% (T100%) and by the average of the coolest 33% (T33%) of canopy pixels per management cell (MC) between 21 July and 26 Aug 2019 for the canopy extraction methods: 2-pixel erosion (2PE) (turquoise), edge detection (dark blue), vegetation segmentation (VS) (coral), and RGB binary masking (RGB-BM) (brick red).

3.3. Orchard Water Status Estimation

The CWSI_T100% values were substantially higher than the CWSI_T33% values per MC, per date, and per canopy extraction method, and they mirrored the trends found in the canopy temperature calculated using T100% and T33% (Figure 7). The average difference between the CWSI_100% and CWSI_T33% values for each canopy extraction method was as follows: 0.28 (ED), 0.30 (2PE), 0.67 (RGB-BM), and 0.68 (VS). Within the CWSI_T100% dataset, minimal differences were recorded between the 2PE and ED methods (0.02), while large differences were calculated between the 2PE and VS methods (0.42). In the CWSI_T33% dataset, no difference was found between the 2PE and ED methods, and a difference of 0.03 was calculated between the VS and 2PE methods.

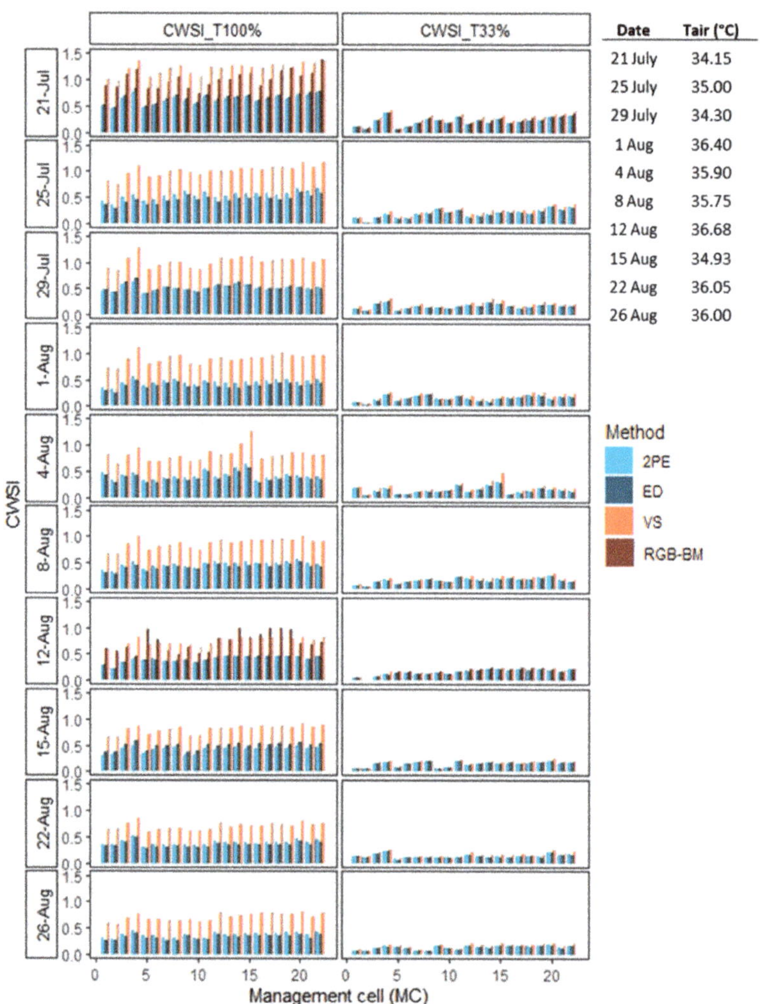

Figure 7. Crop water status index (CWSI) with Tcanopy (°C) calculated using the average 100% (CWSI_T100%) and the average of the coolest 33% (CWSI_T33%) of canopy pixels. Twet = lowest 5% of canopy pixels, and Tdry = Tair + 2 °C. Values per management cell (MC) between 21 July and 26 August 2019 for the canopy extraction methods: 2-pixel erosion (2PE) (turquoise), edge detection (ED) (dark blue), vegetation segmentation (VS) (coral), and RGB binary masking (RGB-BM) (brick red). The table insert shows the air temperature (Tair (°C)) values.

3.3.1. SWP-CWSI Model Comparison

The relationship between the measured SWP and CWSI was modeled for all four canopy extraction methods and the two temperature calculations (Figure 8). The CWSI_T100% values are higher than the CWSI_T33% values per tree as expected. The R^2 is higher and the RMSE is lower for all of the CWSI_T33%-based models in comparison to the CWSI_T100%-based models, regardless of extraction method, which possibly resulted from the higher variability of canopy temperature per tree with the CWSI_T100% calculation. The intercept of the CWSI_T100%-based models is significantly higher than CWSI-T33% for all canopy extraction methods. There is a significant difference in slope between the CWSI_T100%-based and CWSI_T33%-based models for the 2PE and ED methods ($p < 0.0001$), while no significant difference is detected for the VS and RGB-BM methods ($p > 0.05$). The slope signifies the sensitivity of CWSI in relation to the change in measured SWP. Within the CWSI_T100%-based models, the slopes of the 2PE and ED methods are significantly different (steeper) than the VS and RGM-BM methods ($p < 0.0001$) when each model was compared to the other models. No difference is found between the intercepts of these models. Within the CWSI_T33%-based models, no difference is found in the slope or intercept. All eight models are significant ($p < 0.0001$), enabling the estimation of SWP based on these relationships.

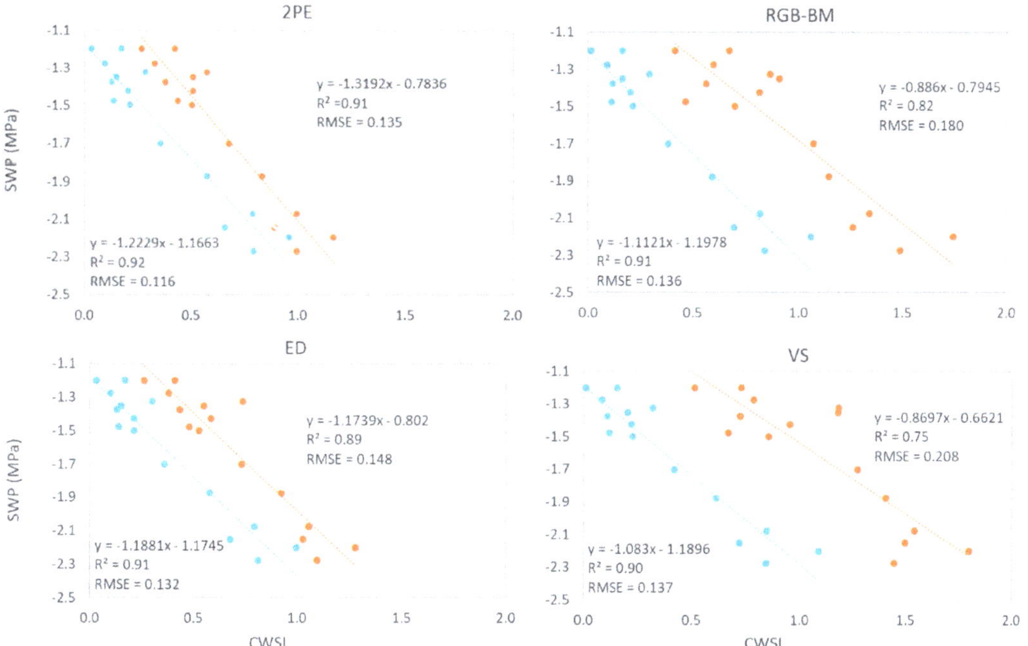

Figure 8. Linear regression model of SWP and CWSI for the 2-pixel erosion (2PE), edge detection (ED), vegetation segmentation (VS), and RGB binary masking (RGB-BM) canopy extraction methods. Crop water status index (CWSI) with Tcanopy (°C) calculated using the average 100% (CWSI_T100%) (red points and lines) and the average of the coolest 33% (CWSI_T33%) (blue points and lines) of canopy pixels. Twet = lowest 5% of canopy pixels, and Tdry = Tair + 2 °C. Each point represents a measurement tree (n = 15).

3.3.2. Estimated Stem Water Potential

The difference between the measured and estimated SWP values was calculated per MC for each canopy extraction and temperature calculation method, highlighting the differences between the datasets (Figure 9). A value of zero indicates no difference between the measured and estimated SWP. Positive values indicate that the estimated SWP is lower (more negative, the MC more stressed) than the measured SWP. Conversely, negative values indicate that the estimated SWP value is higher (less negative, the MC less stressed) than the measured SWP values. The average difference between the measured and estimated SWP (SWPe_T100%) in the RGB-BM dataset is substantially higher in comparison to the other canopy extraction methods and indicates a shift to more positive values, in comparison to the SWPe_T33% values. The MSE and RMSE values reinforce this point and indicate that the SWPe_T100% values of both the RGB-BM and the VS extraction methods are higher than the measured SWP values, indicating that the extraction quality is poorer than the 2PE and ED methods. The average differences between the measured and estimated SWP (SWPe_T33%) for each extraction method are mostly negative and close to zero. The histogram analysis, MSE, and RMSE all indicate that the 2PE, ED, and VS methods are similar to each other, while the RGB-BM is slightly different. These results suggest that theoretical irrigation decisions based on the SWPe_T33% values of the 2PE, ED, and VS methods would yield similar results.

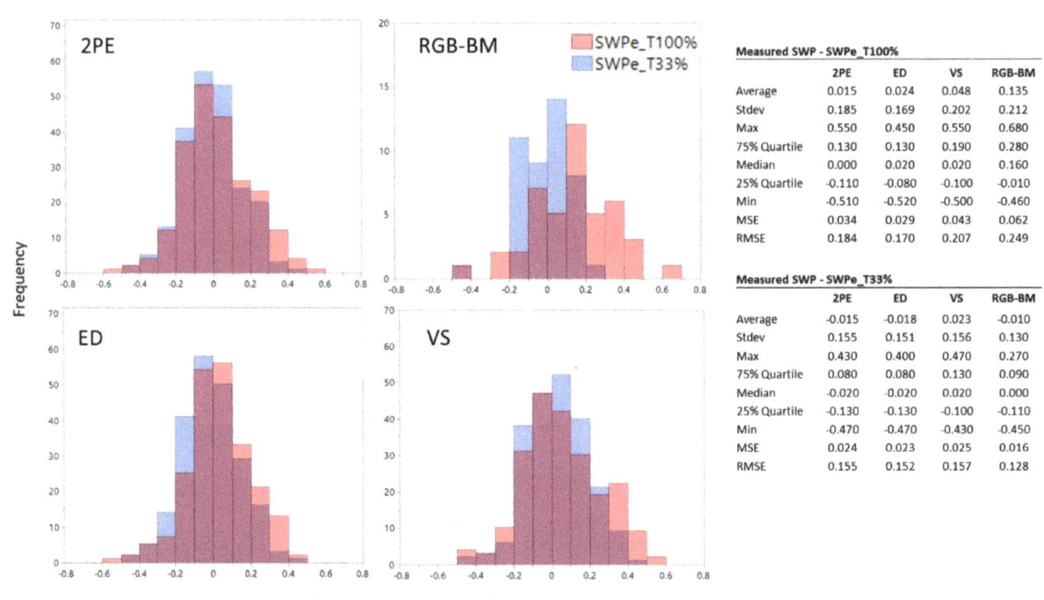

Measured SWP - SWPe_T100%

	2PE	ED	VS	RGB-BM
Average	0.015	0.024	0.048	0.135
Stdev	0.185	0.169	0.202	0.212
Max	0.550	0.450	0.550	0.680
75% Quartile	0.130	0.130	0.190	0.280
Median	0.000	0.020	0.020	0.160
25% Quartile	-0.110	-0.080	-0.100	-0.010
Min	-0.510	-0.520	-0.500	-0.460
MSE	0.034	0.029	0.043	0.062
RMSE	0.184	0.170	0.207	0.249

Measured SWP - SWPe_T33%

	2PE	ED	VS	RGB-BM
Average	-0.015	-0.018	0.023	-0.010
Stdev	0.155	0.151	0.156	0.130
Max	0.430	0.400	0.470	0.270
75% Quartile	0.080	0.080	0.130	0.090
Median	-0.020	-0.020	0.020	0.000
25% Quartile	-0.130	-0.130	-0.100	-0.110
Min	-0.470	-0.470	-0.430	-0.450
MSE	0.024	0.023	0.025	0.016
RMSE	0.155	0.152	0.157	0.128

Figure 9. The histogram of the difference between the measured and estimated stem water potential (SWPe) calculated using the canopy temperature data of the average 100% (SWPe_T100%) (pink bars) and the average of the coolest 33% (SWPe_T33%) (blue bars) of canopy pixels for the 2-pixel erosion (2PE), edge detection (ED), vegetation segmentation (VS), and RGB binary masking (RGB-BM) canopy extraction methods. The frequency refers to the number of management cells (MC). The table insert provides the descriptive statistics of each dataset. Note: the Y-axis range of the RGB-BM method is specifically different from the other methods.

The distribution of the SWPe_T100% and SWPe_T33% values for each canopy extraction method were compared to the defined optimal SWP range for stage III (between −1.17 and −1.43 MPa) (Figure 10). Within the SWPe_T100% dataset, the RGB-BM distribution is noticeably offset to more negative SWP values, and a substantially high percentage of below-range values (75%) were calculated, indicating that the orchard was estimated to be under greater stress in comparison to the 2PE and ED methods. Forty-three percent of the VS method's SWPe values are below the optimum range. The majority of the SWPe values of the 2PE and ED (63%) methods are within the optimal range of orchard water status. The SWPe_T33% dataset is characterized by a higher percentage of values within the optimal range of orchard water status for each canopy extraction method in comparison to the SWPe_T100% dataset. Additionally, the variance of the SWPe_T33% values is substantially smaller in comparison to the SWPe_T100% dataset for each extraction method. A negligible percentage of above-range SWPe values was calculated, indicating that the orchard is theoretically not over-irrigated. The measured SWP distribution is similar to the SWPe_T100% ED method dataset.

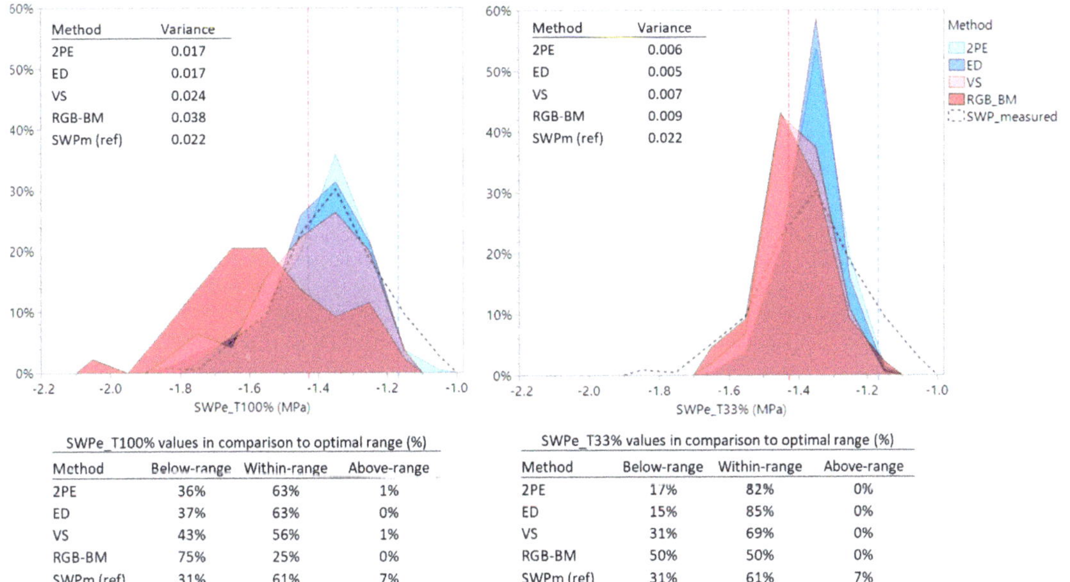

SWPe_T100% values in comparison to optimal range (%)

Method	Below-range	Within-range	Above-range
2PE	36%	63%	1%
ED	37%	63%	0%
VS	43%	56%	1%
RGB-BM	75%	25%	0%
SWPm (ref)	31%	61%	7%

SWPe_T33% values in comparison to optimal range (%)

Method	Below-range	Within-range	Above-range
2PE	17%	82%	0%
ED	15%	85%	0%
VS	31%	69%	0%
RGB-BM	50%	50%	0%
SWPm (ref)	31%	61%	7%

Figure 10. Histogram of percent estimated stem water potential (SWPe) (MPa) calculated using the canopy temperature data of the average 100% (SWPe_T100%) and the average of the coolest 33% (SWPe_T33%) of canopy pixels in comparison to the defined optimal SWP range for stage III: upper (−1.17 Mpa, blue dashed line) and lower (−1.43 Mpa, red dashed line) thresholds. Below-range SWP values indicate orchard stress, while above-range water status values indicate theoretical over-irrigation. The canopy extraction methods tested were 2-pixel erosion (2PE, turquoise polygon), edge detection (ED, blue polygon), vegetation segmentation (VS, pink polygon), and RGB binary masking (RGB-BM, red polygon).

4. Discussion

Canopy extraction that is based purely on the temperature attribute assumes a distinct difference between soil and canopy temperatures. While this is largely true, canopy temperature can be similar to shadowed or wet soil [40], and the temperature of mixed pixels can be similar to canopy suffering from water stress [16]. In comparison, the canopy of RGB images has a different multispectral signature than soil, enabling the use of spectral vegetation indices for canopy classification. Accordingly, the ExG index, a popular index for vegetation identification [41], served as the basis for binary thresholding in this study. RGB images also have higher spatial resolution in comparison to thermal images. These two characteristics led to the assumption that RGB-based canopy extraction would be more accurate than using a single thermal image. This assumption was supported to some extent by this study. Higher accuracy of canopy extraction was obtained by the RGB-based method compared to the thermal-based methods. However, between-row weeds were misclassified as tree canopy with the RGB-based method, leading to an atypical increase in canopy area during stage III. Additionally, inaccuracies in geographical and geometrical fit between the RGB mask and the thermal image are a known drawback with multi-source methods, such as RGB-BM [2], and explain the inclusion of warm canopy edges in the canopy mask.

Between-row weeds and canopy edges are highly affected by surrounding high soil temperatures, therefore leading to the overestimation of canopy temperature and CWSI in this study. Camino et al. [16] also found that warm edge pixels cause significant errors in almond tree canopy temperature and CWSI values. The VS method also included warm temperature pixels on the edges of all trees in the orchard. Similar to the RGB-BM method, the VS canopy temperature and CWSI values were higher in comparison to the 2PE and ED methods. Conversely, the 2PE and ED methods were both able to adequately remove canopy edge pixels by incorporating morphological erosion and edge detection algorithms, respectively. The difference between these two groups of methods, 2PE–ED and VS–RGB-BM, was also evident in the SWP–CWSI linear models calculated using the CWSI_T100% values. The superiority of the 2PE and ED extraction methods over the RGB-BM method implies that the multispectral nature and the high spatial resolution of the RGB images do not obviate the need to incorporate spatial analyses, such as morphological erosion and edge detection algorithms. This suggests that the contribution of the RGB images is not significant for the canopy extraction stage and canopy pixels can be extracted with high accuracy and reliability merely with thermal images. Furthermore, the multi-source approach is slightly more complex and time consuming than the one-source approach, primarily due to the critical georeferencing step. Thus, it is concluded that one-source thermal-based approaches can be preferably used for canopy extraction.

Canopy temperature was estimated in this study using the average of all canopy pixels (T100%) [7,10] and of the coolest 33% canopy pixels (T33%) [6,26,27]. The T100% values were substantially higher than the T33% values for all MCs, dates, and canopy extraction methods. Within-crown temperature variability has been documented for almond trees [16,42] and is partially affected by the inclusion of pixels at the edge of the canopy. The T33% approach is less influenced by canopy temperature heterogeneity [6] and minimizes the effect of mixed pixels. This idea is reinforced in the present study by the similar spatial patterns and canopy temperatures between the canopy extraction methods using the T33% calculation approach. The significant effect on temperature using the T100% approach resulted in a pronounced effect on the CWSI.

Substantial differences were apparent between the extraction methods within the CWSI_T100% dataset (Figure 7). The VS and RGB-BM values reached unexpectedly high values for well-watered peach trees: 0.53–1.37 (VS) and 0.45–1.39 (RGB-BM). Furthermore, the maximum CWSI_T100% values of the 2PE and ED methods were extremely high: 0.77 (2PE) and 0.82 (ED). A CWSI value of one indicates an extremely stressed peach tree with closed stomata. For reference, in one of the experimental plots that formed the basis for the SWP–CWSI models in this study, irrigation was suspended for a total of three weeks prior to the imaging campaign. In this plot, and in stark contrast to the VS method, the CWSI_T100% values of the measurement trees ranged between 0.58 and 0.96. CWSI values higher than one imply that non-canopy pixels are included in the calculation. In contrast, no significant differences were found between the CWSI values that were calculated using the T33% approach. Additionally, and similar to the findings of Cohen et al. (2017) in cotton, the SWP–CWSI models using the T100% approach were inferior in comparison to the T33% approach. Most importantly, the T33% dataset produced similar SWP–CWSI models irrespective of the canopy extraction method used, while the T100% yielded very different models. These results highlight the robustness of the T33% approach and indicate that it is not sensitive to the canopy extraction accuracy.

The robustness of the T33% approach is further emphasized by comparing the SWPe values to the optimal water status range. This optimal range of SWP constitutes the basis for irrigation decision making [27]. Therefore, a comparison of the SWPe distribution to the optimal range can indicate the extent to which a specific canopy extraction method is prone to water stress overestimation and leads to hypothetical over-irrigation as a result.

Within the SWPe_T33%, a large percentage of the estimated SWP values were within range for the 2PE, ED, and VS methods, indicating a theoretical irrigation policy that adequately brings and maintains the MC in the optimal range. Higher percentages of above-range SWP values were calculated with the VS and RGM-BM methods (compared to additional extraction methods), indicating that the orchard was supposedly under a higher degree of stress, necessitating increased irrigation.

The comparison of the SWPe_T100% distribution of values to the optimal water status range further reinforces the fact that the estimated SWP values calculated with the T100% method, and in particular using the RGB-BM canopy extraction method, possibly overestimate orchard water status, hypothetically resulting in more-than-optimal irrigation application with subsequent agronomic and economic consequences [32]. It should be noted that none of the canopy extraction methods or temperature calculation methods sufficiently estimated above-range (less negative) water status or below-range extremely stressed (more negative) values in the SWPe_T33% dataset. This result, rather than indicating the quality of the canopy extraction, signifies a general limitation of water status assessment using thermal images. Thermal-based water status estimation suffers from different types of inaccuracies, including the effect of meteorological conditions and different approaches for determination of T_{wet} and T_{dry} values.

5. Conclusions

The current study explored the sensitivity of thermal image-based orchard water status estimation to canopy extraction quality using four canopy extraction methods, which was previously unaddressed in scientific literature. Three methods used a single thermal image (1-source) (2PE, ED, and VS), while a fourth method incorporated a thermal and an RGB image (multi-source) (RGB-BM). Two approaches to canopy temperature calculation were also evaluated: the average of all canopy pixels (T100%) and the average of the coolest 33% of canopy pixels (T33%). This study found that canopy pixels can be extracted with high accuracy and reliability using only thermal images, primarily using the 2PE and ED methods. The incorporation of an RGB image reduces the overall quality, as between-row weeds and warm canopy edges are misidentified as tree canopy. Additionally, the T33% approach to canopy temperature calculation was found to be robust and not sensitive to canopy extraction accuracy. In comparison, the T100% approach, specifically for the VS and RGB-BM methods, overestimated orchard water stress. These findings indicate that orchard water status is sensitive to canopy extraction quality but is affected to a greater degree by the canopy temperature calculation approach. Future research should explore the relationship between SWP and CWSI on additional days, under different meteorological conditions, and over seasons to strengthen the estimation of orchard water status. Future research should also explore the sensitivity of orchard water status to canopy extraction quality in additional varieties of peach and other fruit trees located in different environments. Such research studies will widen the scope of impact and scale of the main findings from this study, improving irrigation management based on thermal images.

Author Contributions: Conceptualization, L.K. and Y.C.; Methodology, L.K. and Y.C.; Software, E.G.; Validation, L.K.; Formal Analysis, L.K. and S.M.; Investigation, L.K., G.L., O.K. and V.A.; Resources, G.L., O.K. and V.A.; Data Curation, L.K., S.M. and E.G.; Writing—Original Draft Preparation, L.K.; Writing—Review and Editing, L.K., Y.C., A.B.-G., M.I.L., A.P. and E.G.; Visualization, L.K.; Supervision Y.C., A.B.-G. and M.I.L.; Project Administration, L.K., A.N. and V.A.; Funding Acquisition, Y.C., A.B.-G. and V.A. All authors have read and agreed to the published version of the manuscript.

Funding: This research was a part of the "Eugene Kendel" Project for the Development of Precision Drip Irrigation funded via the Ministry of Agriculture and Rural Development in Israel (Grant No. 20-12-0030). The project also received funding from the European Union's Horizon 2020 research and innovation program under Project SHui, grant agreement No 773903.

Data Availability Statement: Data sharing not applicable.

Acknowledgments: The authors would like to thank the grower, Shlomo Cohen, for collaborating and allowing the research to be conducted in his peach orchard; Reshef Elmakais, Tomer Hagai, and Ohad Masad, for field measurements and technical support; and Datamap company for imagery acquisition and pre-processing.

Conflicts of Interest: The authors declare that they have no conflict of interest.

Abbreviations

CWSI	crop water stress index
CWSI_T33%	crop water stress index calculated with the average temperature of the coolest 33% of canopy pixels
CWSI_T100%	crop water stress index calculated with the average temperature of 100% of canopy pixels
ED	edge detection
ExG	excess green index
MC	management cell
RGB-BM	red –green–blue binary masking
SWP	stem water potential (MPa)
SWPe_T33%	estimated stem water potential using the average temperature of the coolest 33% of canopy pixels (MPa)
SWPe_T100%	estimated stem water potential using the average temperature of 100% of canopy pixels (MPa)
T33%	average temperature of the coolest 33% of canopy pixels
T100%	average temperature of 100% of canopy pixels
VS	vegetation segmentation
2PE	2-pixel erosion

References

1. Gonzalez-Dugo, V.; Zarco-Tejada, P.J. Assessing the Impact of Measurement Errors in the Calculation of CWSI for Characterizing the Water Status of Several Crop Species. *Irrig. Sci.* **2022**, 1–13. [CrossRef]
2. Zhou, Z.; Majeed, Y.; Diverres Naranjo, G.; Gambacorta, E. Assessment for Crop Water Stress with Infrared Thermal Imagery in Precision Agriculture: A Review and Future Prospects for Deep Learning Applications. *Comput. Electron. Agric.* **2021**, *182*, 106019. [CrossRef]
3. Jones, H.G. Use of Infrared Thermometry for Estimation of Stomatal Conductance as a Possible Aid to Irrigation Scheduling. *Agric. For. Meteorol.* **1999**, *95*, 139–149. [CrossRef]
4. Idso, S.B.; Jackson, R.D.; Pinter, P.J.; Reginato, R.J.; Hatfield, J.L. Normalizing the Stress-Degree-Day Parameter for Environmental Variability. *Agric. Meteorol.* **1981**, *24*, 45–55. [CrossRef]
5. Jackson, R.D.; Idso, S.B.; Reginato, R.J.; Pinter, P.J. Canopy Temperature as a Crop Water Stress Indicator. *Water Resour. Res.* **1981**, *17*, 1133–1138. [CrossRef]
6. Meron, M.; Tsipris, J.; Orlov, V.; Alchanatis, V.; Cohen, Y. Crop Water Stress Mapping for Site-Specific Irrigation by Thermal Imagery and Artificial Reference Surfaces. *Precis. Agric.* **2010**, *11*, 148–162. [CrossRef]
7. Gonzalez-Dugo, V.; Zarco-Tejada, P.; Nicolás, E.; Nortes, P.A.; Alarcón, J.J.; Intrigliolo, D.S.; Fereres, E. Using High Resolution UAV Thermal Imagery to Assess the Variability in the Water Status of Five Fruit Tree Species within a Commercial Orchard. *Precis. Agric.* **2013**, *14*, 660–678. [CrossRef]
8. Cohen, Y.; Alchanatis, V.; Saranga, Y.; Rosenberg, O.; Sela, E. Mapping Water Status Based on Aerial Thermal Imagery: Comparison of Methodologies for Upscaling from a Single Leaf to Commercial Fields. *Precis. Agric.* **2017**, *18*, 801–822. [CrossRef]
9. Rud, R.; Cohen, Y.; Alchanatis, V.; Levi, A.; Brikman, R.; Shenderey, C.; Heuer, B.; Markovitch, T.; Dar, Z.; Rosen, C.; et al. Crop Water Stress Index Derived from Multi-Year Ground and Aerial Thermal Images as an Indicator of Potato Water Status. *Precis. Agric.* **2014**, *15*, 273–289. [CrossRef]
10. Gonzalez-Dugo, V.; Goldhamer, D.; Zarco-Tejada, P.J.; Fereres, E. Improving the Precision of Irrigation in a Pistachio Farm Using an Unmanned Airborne Thermal System. *Irrig. Sci.* **2015**, *33*, 43–52. [CrossRef]
11. Rud, R.; Cohen, Y.; Alchanatis, V.; Beiersdorf, I.; Klose, R.; Presnov, E.; Levi, A.; Brikman, R.; Agam, N.; Dag, A.; et al. Characterization of Salinity-Induced Efects in Olive Trees Based on Thermal Imagery. In Proceedings of the 10th European Conference on Precision Agriculture; Stafford, J.V., Ed.; Wageningen Academic Publishers: Rishon-LeZion, Israel, 2015; pp. 511–518.
12. Egea, G.; Padilla-Díaz, C.M.; Martinez-Guanter, J.; Fernández, J.E.; Pérez-Ruiz, M. Assessing a Crop Water Stress Index Derived from Aerial Thermal Imaging and Infrared Thermometry in Super-High Density Olive Orchards. *Agric. Water Manag.* **2017**, *187*, 210–221. [CrossRef]

13. Agam, N.; Cohen, Y.; Berni, J.A.J.; Alchanatis, V.; Kool, D.; Dag, A.; Yermiyahu, U.; Ben-Gal, A. An Insight to the Performance of Crop Water Stress Index for Olive Trees. *Agric. Water Manag.* **2013**, *118*, 79–86. [CrossRef]
14. Cohen, Y.; Alchanatis, V.; Prigojin, A.; Levi, A.; Soroker, V.; Cohen, Y. Use of Aerial Thermal Imaging to Estimate Water Status of Palm Trees. *Precis. Agric.* **2012**, *13*, 123–140. [CrossRef]
15. Baluja, J.; Diago, M.P.; Balda, P.; Zorer, R.; Meggio, F.; Morales, F.; Tardaguila, J. Assessment of Vineyard Water Status Variability by Thermal and Multispectral Imagery Using an Unmanned Aerial Vehicle (UAV). *Irrig. Sci.* **2012**, *30*, 511–522. [CrossRef]
16. Camino, C.; Zarco-Tejada, P.J.; Gonzalez-Dugo, V. Effects of Heterogeneity within Tree Crowns on Airborne-Quantified SIF and the CWSI as Indicators of Water Stress in the Context of Precision Agriculture. *Remote Sens.* **2018**, *10*, 604. [CrossRef]
17. Nixon, M.S.; Aguado, A.S. Region-Based Analysis. In *Feature Extraction and Image Processing for Computer Vision*, 4th ed.; Nixon, M.S., Aguado, A.S., Eds.; Academic Press: Cambridge, MA, USA, 2020; pp. 399–432.
18. Zhou, Y.; Wang, L.; Jiang, K.; Xue, L.; An, F.; Chen, B.; Yun, T. Individual Tree Crown Segmentation Based on Aerial Image Using Superpixel and Topological Features. *J. Appl. Remote Sens.* **2020**, *14*, 1. [CrossRef]
19. Maheswari, P.; Raja, P.; Apolo-Apolo, O.E.; Pérez-Ruiz, M. Intelligent Fruit Yield Estimation for Orchards Using Deep Learning Based Semantic Segmentation Techniques—A Review. *Front. Plant Sci.* **2021**, *12*, 684328. [CrossRef]
20. Katz, L.; Ben-Gal, A.; Litaor, M.I.; Naor, A.; Peres, M.; Bahat, I.; Netzer, Y.; Peeters, A.; Alchanatis, V.; Cohen, Y. Spatiotemporal Normalized Ratio Methodology to Evaluate the Impact of Field-Scale Variable Rate Application. *Precis. Agric.* **2022**, *23*, 1125–1152. [CrossRef]
21. Park, S.; Ryu, D.; Fuentes, S.; Chung, H.; Hernández-Montes, E.; O'Connell, M. Adaptive Estimation of Crop Water Stress in Nectarine and Peach Orchards Using High-Resolution Imagery from an Unmanned Aerial Vehicle (UAV). *Remote Sens.* **2017**, *9*, 828. [CrossRef]
22. Bian, J.; Zhang, Z.; Chen, J.; Chen, H.; Cui, C.; Li, X.; Chen, S.; Fu, Q. Simplified Evaluation of Cotton Water Stress Using High Resolution Unmanned Aerial Vehicle Thermal Imagery. *Remote Sens.* **2019**, *11*, 267. [CrossRef]
23. Cohen, Y.; Alchanatis, V.; Meron, M.; Saranga, Y.; Tsipris, J. Estimation of Leaf Water Potential by Thermal Imagery and Spatial Analysis. *J. Exp. Bot.* **2005**, *56*, 1843–1852. [CrossRef] [PubMed]
24. Osroosh, Y.; Khot, L.R.; Peters, R.T. Economical Thermal-RGB Imaging System for Monitoring Agricultural Crops. *Comput. Electron. Agric.* **2018**, *147*, 34–43. [CrossRef]
25. Zhou, Z.; Diverres, G.; Kang, C.; Thapa, S.; Karkee, M.; Zhang, Q.; Keller, M. Ground-Based Thermal Imaging for Assessing Crop Water Status in Grapevines over a Growing Season. *Agronomy* **2022**, *12*, 322. [CrossRef]
26. Bahat, I.; Netzer, Y.; Grünzweig, J.M.; Alchanatis, V.; Peeters, A.; Goldshtein, E.; Ohana-Levi, N.; Ben-Gal, A.; Cohen, Y. In-Season Interactions between Vine Vigor, Water Status and Wine Quality in Terrain-Based Management-Zones in a 'Cabernet Sauvignon' Vineyard. *Remote Sens.* **2021**, *13*, 1636. [CrossRef]
27. Katz, L.; Ben-Gal, A.; Litaor, M.I.; Naor, A.; Peres, M.; Peeters, A.; Alchanatis, V.; Cohen, Y. A Spatiotemporal Decision Support Protocol Based on Thermal Imagery for Variable Rate Drip Irrigation of a Peach Orchard. *Irrig. Sci.* **2022**, *42*, 1118–1126. [CrossRef]
28. Dag, A.; Alchanatis, V.; Zipori, I.; Sprinstin, M.; Cohen, A.; Maravi, T.; Naor, A. Automated Detection of Malfunctions in Drip-Irrigation Systems Using Thermal Remote Sensing in Vineyards and Olive Orchards. In *Precision Agriculture '15*; Wageningen Academic Publishers: Rishon-LeZion, Israel, 2015; pp. 12–23.
29. Kalo, N.; Edan, Y.; Alchanatis, V. Detection of Irrigation Malfunctions Based on Thermal Imaging. In Proceedings of the Precision Agriculture'21; Wageningen Academic Publishers: Noordwijk, The Netherlands, 2021; pp. 2217–2224.
30. Otsu, N. A Threshold Selection Method from Gray-Level Histograms. *IEEE Trans. Syst. Man. Cybern.* **1979**, *9*, 62–66. [CrossRef]
31. Hamuda, E.; Glavin, M.; Jones, E. A Survey of Image Processing Techniques for Plant Extraction and Segmentation in the Field. *Comput. Electron. Agric.* **2016**, *125*, 184–199. [CrossRef]
32. Steduto, P.; Hsiao, T.C.; Fereres, E.; Raes, D. *Crop Yield Response to Water*; Food and Agriculture Organization of the United Nations: Rome, Italy, 2012; Volume 1028.
33. Zhong, L.; Hu, L.; Zhou, H. Deep Learning Based Multi-Temporal Crop Classification. *Remote Sens. Environ.* **2019**, *221*, 430–443. [CrossRef]
34. Hijmans, R.J. Raster: Geographic Data Analysis and Modeling. 2019. Available online: https://rspatial.org/raster (accessed on 10 January 2022).
35. Bivand, R.; Keitt, T.; Rowlingson, B. Rgdal: Bindings for the "Geospatial" Data Abstraction Library. 2019. Available online: http://rgdal.r-forge.r-project.org/ (accessed on 10 January 2022).
36. Wickham, H. Reshaping Data with the {reshape} Package. *J. Stat. Softw.* **2007**, *21*, 1–20. [CrossRef]
37. *R Core Team R: A Language and Environment for Statistical Computing*; R Core Team: Vienna, Austria, 2019.
38. Wickham, H. *Ggplot2: Elegant Graphics for Data Analysis*; Springer-Verlag: New York, NY, USA, 2016; ISBN 978-3-319-24277-4.
39. Shimshowitz, E. The Effect of Irrigation and Crop Load on Crop Yield Anf Fruit Size Distribution in Nectarine Cv. Arctic Mist. Master's Thesis, Tel Hai Academic College, Upper Galilee, Qiryat Shemona, Israel, 2018.
40. Meron, M.; Sprintsin, M.; Tsipris, J.; Alchanatis, V.; Cohen, Y. Foliage Temperature Extraction from Thermal Imagery for Crop Water Stress Determination. *Precis. Agric.* **2013**, *14*, 467–477. [CrossRef]

41. Lee, M.K.; Golzarian, M.R.; Kim, I. A New Color Index for Vegetation Segmentation and Classification. *Precis. Agric.* **2021**, *22*, 179–204. [CrossRef]
42. Gonzalez-Dugo, V.; Zarco-Tejada, P.; Berni, J.A.J.; Suárez, L.; Goldhamer, D.; Fereres, E. Almond Tree Canopy Temperature Reveals Intra-Crown Variability That Is Water Stress-Dependent. *Agric. For. Meteorol.* **2012**, *154–155*, 156–165. [CrossRef]

Review

Remote Sensing Monitoring of Rice and Wheat Canopy Nitrogen: A Review

Jie Zheng [1,2], Xiaoyu Song [1], Guijun Yang [1], Xiaochu Du [2], Xin Mei [2] and Xiaodong Yang [1,3,*]

1 Key Laboratory of Quantitative Remote Sensing in Agriculture of Ministry of Agriculture and Rural Affairs, Information Technology Research Center, Beijing Academy of Agriculture and Forestry Sciences, Beijing 100097, China
2 Faculty of Resources and Environment Science, Hubei University, Wuhan 430062, China
3 Huanan Industrial Technology Research Institute of Zhejiang University, Guangzhou 510700, China
* Correspondence: yangxd@nercita.org.cn

Abstract: Nitrogen(N) is one of the most important elements for crop growth and yield formation. Insufficient or excessive application of N fertilizers can limit crop yield and quality, especially as excessive N fertilizers can damage the environment and proper fertilizer application is essential for agricultural production. Efficient monitoring of crop N content is the basis of precise fertilizer management, and therefore to increase crop yields and improve crop quality. Remote sensing has gradually replaced traditional destructive methods such as field surveys and laboratory testing for crop N diagnosis. With the rapid advancement of remote sensing, a review on crop N monitoring is badly in need of better summary and discussion. The purpose of this study was to identify current research trends and key issues related to N monitoring. It begins with a comprehensive statistical analysis of the literature on remote sensing monitoring of N in rice and wheat over the past 20 years. The study then elucidates the physiological mechanisms and spectral response characteristics of remote sensing monitoring of canopy N. The following section summarizes the techniques and methods applied in remote sensing monitoring of canopy N from three aspects: remote sensing platforms for N monitoring; correlation between remotely sensed data and N status; and the retrieval methods of N status. The influential factors of N retrieval were then discussed with detailed classification. However, there remain challenges and problems that need to be addressed in the future studies, including the fusion of multisource data from different platforms, and the uncertainty of canopy N inversion in the presence of background factors. The newly developed hybrid model integrates the flexibility of machine learning with the mechanism of physical models. It could be problem solving, which has the advantages of processing multi-source data and reducing the interference of confounding factors. It could be the future development direction of crop N inversion with both high precision and universality.

Keywords: rice and wheat; nitrogen remote sensing; quantitative retrieval; research prospect

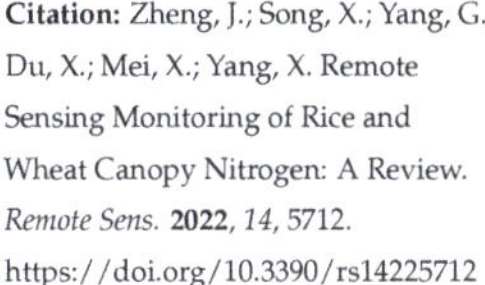

Citation: Zheng, J.; Song, X.; Yang, G.; Du, X.; Mei, X.; Yang, X. Remote Sensing Monitoring of Rice and Wheat Canopy Nitrogen: A Review. *Remote Sens.* **2022**, *14*, 5712. https://doi.org/10.3390/rs14225712

Academic Editors: Kenji Omasa, Shan Lu and Jie Wang

Received: 30 September 2022
Accepted: 6 November 2022
Published: 11 November 2022

Publisher's Note: MDPI stays neutral with regard to jurisdictional claims in published maps and institutional affiliations.

1. Introduction

The effective guarantee of national food security is a key objective for China. Therefore, it has become a need of green agriculture to reduce the amount and increase the efficiency of chemical fertilizers, which could improve the effective supply of agriculture [1–3]. Rice and wheat are the main crops in the world, with a wide distribution and highly suitability. How to achieve high quality and yield is currently a major challenge for agricultural production [4–6]. N made up more than 40% of the mineral elements needed for the growth of rice and wheat [7]. Its content would impact the physiological traits, photosynthesis, and enzyme activities, leading to variations in protein content and grain production [8–10]. Healthy plants have a total N content that ranges from 0.3% to 5% of their dry matter, which directly affects crop production. N deficiency can hinder chlorophyll (Chl) synthesis and reduce the effective number of spikes, thus reducing yields. A surplus of N prevents photosynthetic products from reaching the seeds, delaying maturation [7,11]. Therefore, N content during crop growth has become one of the most important indicators in agricultural production management. Currently it is common to use base fertilizer, or top dressing through experience. It will result in too much or not enough N, which not only inhibits crop growth but also pollutes the environment. How to monitor rice and wheat canopy N timely and accurately, so as to guide variable rate fertilization has become a core research issue [12–15].

Currently the experts in plant protection and agronomy test N manually by observing crop symptoms, plant growth and leaf color. They also use chemical diagnostics, such as plant total N diagnostics and rapid nitrate diagnostics, to detect the N content of each organ. The former has variances in results due to subjective judgments. The latter has good accuracy but destroys the plant and has a time lag, making large-scale application difficult [16]. Therefore, traditional N assessment methods do not facilitate variable rate fertilization, because the information on the timing and extent of crop N abundance and deficiency is not efficiently provided. Remote sensing has enabled the rapid development of multiple scales of application, including satellite, unmanned aerial vehicles (UAVs) and ground. It is the current technology for rapidly acquiring spatial and temporal continuum information on a large range. The spectrum is sensitive to the N response in the Visible–Near Infrared (VIS–NIR). This spectral information can show subtle changes in N content, allowing for more accurate N retrieval [17,18]. Remote sensing monitoring of N in rice and wheat can be non-destructive, effective, and real-time for large-scale studies. It offers significant potential for crop nutrient diagnosis and as a basis of subsequent guidance on fertilizer application [19–21].

Both rice and wheat belong to the gramineous cereal crop in the botanical classification, and both are C3 crops with similar photosynthetic systems [22]. Thus, the absorption of colored light by Chl in both canopies is consistent and the response to the spectrum is similar. N is mainly found in the Chl of the photosynthetic systems and the process of N accumulation in rice and wheat has a high degree of similarity [23,24]. They are both transformed into nutrient bodies at the vegetative growth stages and into reproductive organs at the reproductive growth stages, and there are anisotropic changes in N accumulation in each organ at the growth stage [23,25]. Although the cropping patterns differ, one being dryland and the other paddy, this effect can be attenuated when pre-processing the remote sensing data [26]. Therefore, in remote sensing-based studies, the N transformation processes in rice and wheat are highly consistent, making their remote sensing monitoring systems relatively similar [14,27,28]. In this study, the remote sensing monitoring techniques for both crops are explored in an integrated manner.

To analyze and summarize the current research hotspots and trends in the field of remote sensing of canopy N in rice and wheat, this paper traces the related literature in the Web of Science ™ Core Collection Database. The literature is retrieved and filtered by the topics "nitrogen concentration" or "nitrogen content", "rice" or "wheat" and "remote sensing". An initial collection of 572 apparently relevant records covered the period 2003–2021. An initial screening progress is conducted to exclude literature that is not relevant to the review, such as conference proceedings, patents, etc. However, it still contains some irrelevant literature, and a further screening is necessary. This is followed by a more detailed screening on titles and abstracts to exclude the following: (1) non-targeted research topics (e.g., corn, cotton, grassland, etc.); (2) not directly estimated N (e.g., Chl, protein, etc.); (3) estimated other indicators (e.g., leaf area, plant height, yield, etc.); (4) measured soil N or other trace elements in agricultural fields; and (5) review articles. By carefully reading the titles and abstracts, off-topic papers are obviously manually excluded. Eventually, a total of 174 articles were identified and analyzed in depth. Figure 1 shows the number of studies retrieved from 2003 to 2021 that used remotely sensed data to assess the canopy N status of rice and wheat, reflecting the general trend in research on the application of remote sensing monitoring of canopy N.

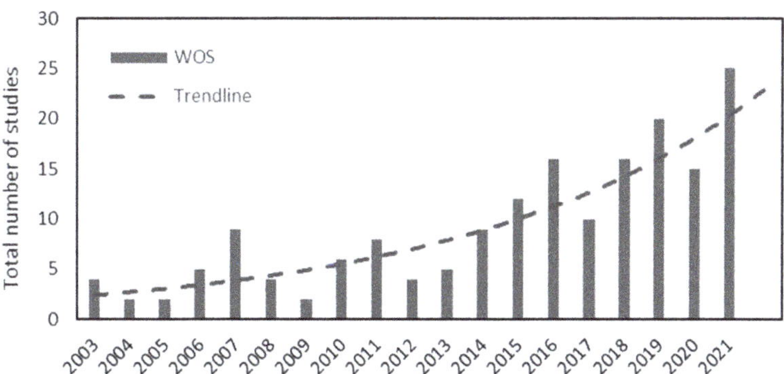

Figure 1. Number of rice and wheat N retrieval research studies per year from 2003 to 2021.

Remote sensing technology has shown great potential for crop N monitoring, providing research ideas from various perspectives. This paper based on the mechanism of remote sensing monitoring of canopy N, respectively summarizes the current techniques and methods from three aspects: remote sensing platforms for canopy N monitoring; correlation between remotely sensed data and N status; and the retrieval methods of N status. Then it discusses the factors affecting the accuracy in remote sensing of canopy N and sketches future areas for research.

2. Mechanisms for Remote Sensing Monitoring of Canopy N

2.1. Physiological Mechanisms of Crop N

Focusing on crop physiological mechanisms, N is closely linked to Chl. N is an important component in the formation of chloroplasts and Rubisco enzymes. Increasing leaf N content will significantly increase leaf chlorophyll content (LCC) and Rubisco enzymes, ultimately leading to a significant increase in crop photosynthetic rate and consequent changes in the external morphology and internal structure of crop leaves [9,29]. Chl can respond to N uptake by crops, and the strength of the relationship directly affects the accuracy of N estimation. However, N is redistributed and reused in the crop during growth. Early in crop development, N is initially concentrated in nutrient bodies such as leaves, and as nutritional and reproductive growth coexist, N starts to be distributed to nutrient bodies and reproductive organs; at the crop maturity stage, N in nutrient bodies is transferred to reproductive organs. This resulted in varying degrees of N and Chl content reduction in the canopy leaves, which changed the correlations between crop N and Chl at various growth stages [25]. In addition to Chl, other mineral deficits, diseases, frost damage, and water stress may also produce leaf yellowing, and using Chl content as a proxy for N is misleading, which limits the ability to estimate N directly from Chl [28]. It has been recommended to utilize leaf protein as a substitute for leaf N content in several studies [30,31], because in contrast to Chl, protein is also a major nitrogenous component in crops and contributes to the varied distribution of N in crop plants. The mechanism of protein–N interactions is under investigation.

2.2. Spectral Response Properties of Canopy N

N affects the spectral reflectance of crops by influencing the Chl content of green crops (Figure 2). Healthy crops' VIS reflectance spectrum is determined by the Chl's absorption effect, which forms a prominent reflectance peak near 550 nm. Multiple reflections in the NIR combine to generate a red edge region of reflectance in the range of 700–780 nm, and the rising slope of the curve reflects the Chl content per unit area to some extent. NIR (780–1350 nm) is closely related to leaf structure and is instructive for exploring whether N is influenced by leaf structure. Under N stress, both the canopy spectral reflectance and the vertical distribution of N will alter.

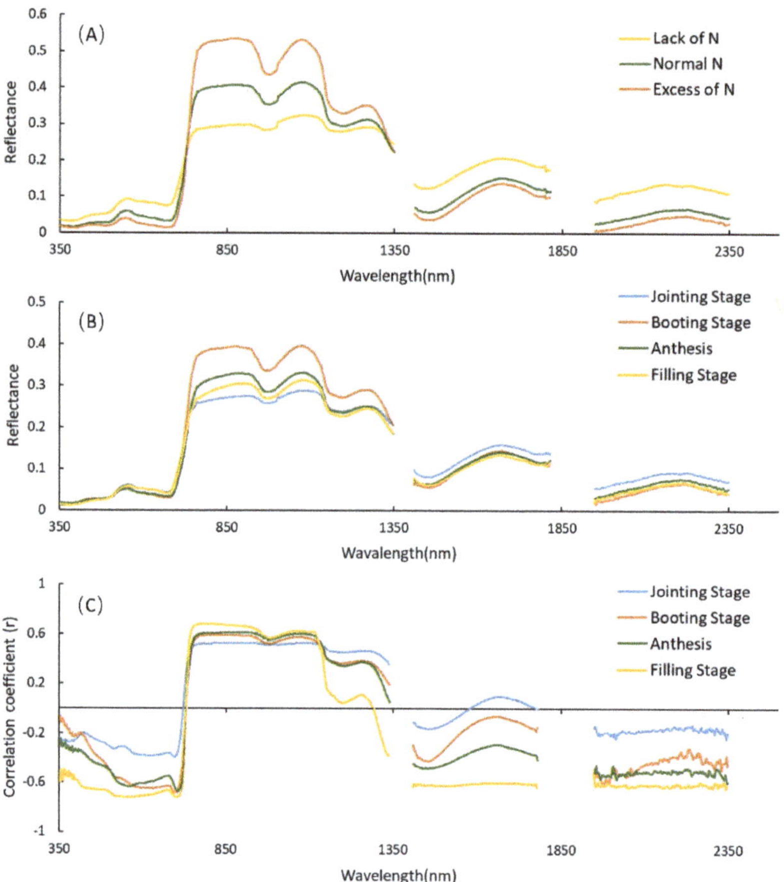

Figure 2. Spectral reflectance properties of the wheat canopy: (**A**) spectral reflectance under N stress; (**B**) spectral reflectance at different growth stages under normal N; and (**C**) correlation between leaf nitrogen concentration (LNC) and spectral reflectance at different growth stages.

During N deficiency, the VIS reflectance of the crop canopy spectrum increased, while the NIR reflectance and red-edge position (REP) decreased; in excess of N, the VIS reflectance decreased, while the NIR reflectance and REP increased (Figure 2) [14,32–34]. As the growth stages proceed, the response of canopy spectral reflectance to crop N status reduced, and the VIS regions also displayed "red shift" and "blue shift" with the development [33,35,36]. It can be found that the analysis of crop N abundance and deficiency using spectral techniques can be specific to a band interval [7,37]. Hyperspectral techniques can even be precise to a specific band [38,39], which provides the possibility of effective identification of crop N deficiency.

The top leaves of the plant under N stress will use the N transferred from the bottom leaves, causing the bottom leaves to yellow and decline prematurely, while the top leaves color changes are not obvious because of being less stressed by N [40,41]. As influenced by the level of soil N supply, the spectral reflectance of leaves at different leaf positions differed erratically in VIS and SWIR, while showing a clear gradient in NIR [42]. Duan et al. [43] suggested that N concentration at different leaf positions decreases from top to bottom at the jointing stage, flowering stages, and filling stage, while the flag leaf stage shows an increasing and then decreasing trend. The vertical distribution of N in the plant is not constant and varies between N conditions, planting densities and growth stages [41,43,44]. Exploring the spectral response properties of the different leaf positions can serve as a foundation for precise N quantification.

3. Techniques for Remote Sensing Monitoring of Canopy N

The continuous development of remote sensing technology on ground-based, UAV-based, and satellite-based platforms provides a wide range of modal options for N status monitoring studies. There is the technical focus in the field of N monitoring, including exploring the correlations between remotely sensed information extracted from multi-source data and N status, while achieving accurate modeling inversion of N status. To summarize the relevant technologies, three main aspects are discussed: remote sensing monitoring platforms; the correlation between remotely sensed data and N status; and the retrieval methods of canopy N status.

3.1. Remote Sensing Platforms for Canopy N Monitoring

3.1.1. Ground-Based Platform

Spectral data collected by ground-based sensors can be divided into non-imaging spectral data and imaging spectral data, with a spectral range primarily in the VIS–NIR, and in several studies involving the SWIR [18,45]. This close-range spectral information has ultra-high spectral resolution and can respond to subtle changes in N. It has been widely used in N monitoring for both leaf and canopy scale. Non-imaging spectral data are point spectral data, and lack spatial information for N estimation at regional scale [46]. Imaging spectral data, on the other hand, combine spatial and spectral features and allow the estimation of canopy parameters from faceted data. However, due to being restricted by data volume and acquisition method, it is generally used for ground study and rarely used directly for diagnosis and applications of large area. Current ground-based spectrometers used commonly include ASD FieldSpec (Analytical Spectral Devices, Boulder, CO, USA), RS-5400 (Spectral Evolution, Haverhill, MA, USA), HR-1024i (Spectra Vista Corporation, Poughkeepsie, NY, USA), SOC710 (Surface Optics Co. Ltd., San Diego, CA, USA), and FISS (Institute of Remote Sensing and Digital Earth, Chinese Academy of Sciences, Beijing, China) [47–50], etc. These instruments provide a stable and high spectral resolution, but some of them are heavy and are generally measured in a backpack or mounted on a tripod in research, with a single angle and way of acquiring data. The advent of handheld instruments such as the RapidSCAN (Holland Scientific Inc., Lincoln, NE, USA), 4300 Handheld FTIR (Agilent Technologies Inc., Santa Clara, CA, USA), and Crop Sense (Beijing Academy of Agriculture and Forestry Sciences, China) [51,52] symbolizes the development of spectral sensors towards lightweight and flexibility. The portable spectrometer can acquire data on a wide range of measurement scales flexibly and efficiently, and it is easy to install on a variety of platforms such as lab benches, black boxes, and rocker arms. In addition, the multi-angle spectral acquisition device [53–55] consists of several moving parts to adjust the observation position and direction, where a goniometer is often used to control the observation of zenith angle changes. The sensor is placed on the goniometer to obtain spectral data in the viewing zenith angles (VZA) ranged from -60° to 60° [53,54]. It is a convenient platform for obtaining multi-angle crop spectral data, which has many applications in the study of the vertical distribution of N in the plant canopy.

3.1.2. UAV-Based Platform

With the development of lighter and smaller sensors and the increased carrying capacity of UAVs, the UAVs carrying sensors for data acquisition have become mainstream platforms in crop N monitoring. It is possible to rapidly acquire ground data with high spatial, temporal, and spectral resolution, facilitating the research at small and medium scales. Compared to airborne platforms (operating at kilometers of altitude), UAVs have the benefit of low cost, low operating altitude, and greater flexibility in terms of data collection arrangements. The sensors currently on UAVs specifically include digital cameras, multispectral/hyperspectral sensors, infrared thermal imagers, chlorophyll fluorescence sensors and LIDAR sensors [56–61]. The hyperspectral imaging spectrometer perfectly combines the advantages of spectroscopic and imaging technology for use in a large area. The payload of UAVs has made lightweight, low-cost sensors a research focus, and airborne hyperspectral imaging spectrometers such as the UHD185-Firefly and Cubert S185 (Cubert GmbH, Ulm, Baden-Württemberg, Germany), and Micro-Hyperspec (Headwall Photonics Inc., Boston, MA, USA) [62–64] are already being used for N monitoring. The PIS112 hyperspectral imaging spectrometer (Beijing Academy of Agriculture and Forestry Sciences, China), GaiaSky-mini hyperspectral imaging camera (Sichuan Dualix Spectral Imaging Technology Co., Ltd., Chengdu, China) [65] and other sensors, as well as the eight-rotor unmanned aircraft system based on RGB and 25-band small multispectral cameras (Zhejiang University, China) [66,67] developed by multiple teams in China, have also been used for agricultural monitoring with good results. Among the image data acquired by the UAV platform, hyperspectral can show subtle changes in crop spectral reflectance features due to its narrow bandwidth and wide continuous spectral range, which is conducive to the fine monitoring of crop N. Most of the multispectral data are small in volume and the spectral range encompasses the N response sensitive VIS–NIR bands. However, the low spectral resolution tends to result in "missing" spectral information while overcoming the high redundancy of hyperspectral information. Studies using multispectral data sources must consider whether the 'missing' spectral information contains sensitive bands and how it can be modeled. UAV remote sensing is not affected by the external environment, such as the atmosphere, and provides better access to high-quality spectral information, making it a common data source for crop N research.

3.1.3. Satellite-Based Platform

Images acquired by satellite platform sensors are characterized by a large range and multiple time phases, making large-scale spatial monitoring possible. Satellite data commonly used for agricultural monitoring include Landsat, Sentinel-2, etc. [68–71]. The small canopy area of rice and wheat requires high temporal and spatial resolution of data in seasonal N management. RapidEye, WorldView-2, etc. have better results in terms of temporal and spatial resolution, and the indices constructed in this way have achieved better results in N monitoring [72–74]. It is of practical significance to monitor crop N in large areas by satellite remote sensing with high spatial and temporal resolution. However, the use of field canopy spectral reflectance to simulate the spectral bands of satellite multispectral sensors, and then transfer the N response models from field experiments to satellite images, inherently results in mistakes due to the low spectral resolution of the data [75,76]. Since the 20th century, China has launched a series of resources satellites, HJ-1A/1B satellites and GaoFen (GF) satellites, etc. [77–81]. The types of sensors and spatial resolution they carry have reached international advanced levels, and their applications in agriculture are gradually spreading. Among them, GaoFen-6 (GF-6) is the first GF satellite for precise agricultural monitoring, for the first time adding green vegetation-sensitive red edge spectral bands and operating in a network with GaoFen-1 (GF-1) to significantly improve the monitoring capability of agricultural resources. GF-6 has been shown to be effective in improving the accuracy of image classification [78–81], and research into the quantitative inversion of crop parameters is still in its infancy. Satellite imagery has a high point and a wide field of view, but the quality has uncontrollable factors. Acquisition is influenced by radiation, aerosols, and weather, increasing the image processing process; analysis is influenced by complex geography and the presence of mixed image elements, increasing the difficulty of N estimation. The effectiveness of crop N estimation models based on satellite imagery still needs to be improved. In addition, as some important reflectance features associated with N can only be measured by hyperspectral sensors, future research could be explored based on satellite hyperspectral imagery.

Remote sensing technologies from ground-based, UAV-based, and satellite-based platforms are maturing and have been widely used in agriculture (Figure 3). The different platforms have their own advantages in remote sensing monitoring, with their mounted sensors having a high degree of overlap in the wavebands sensitive to N (Figure 4). Crop N monitoring based on ground-based and UAV-based platforms has been effective. Because these platforms provide highly accurate data, more valid information can be mined from spectroscopy, graphics, etc. Satellite data, although limited by data volume and accuracy, has become an effective complement to them in expanding the scale of research. Research on multi-sensor and multi-platform joint observations is gradually being developed.

Figure 3. Remote sensing platforms, including ground-based platform, UAV-based platform, and satellite-based platform.

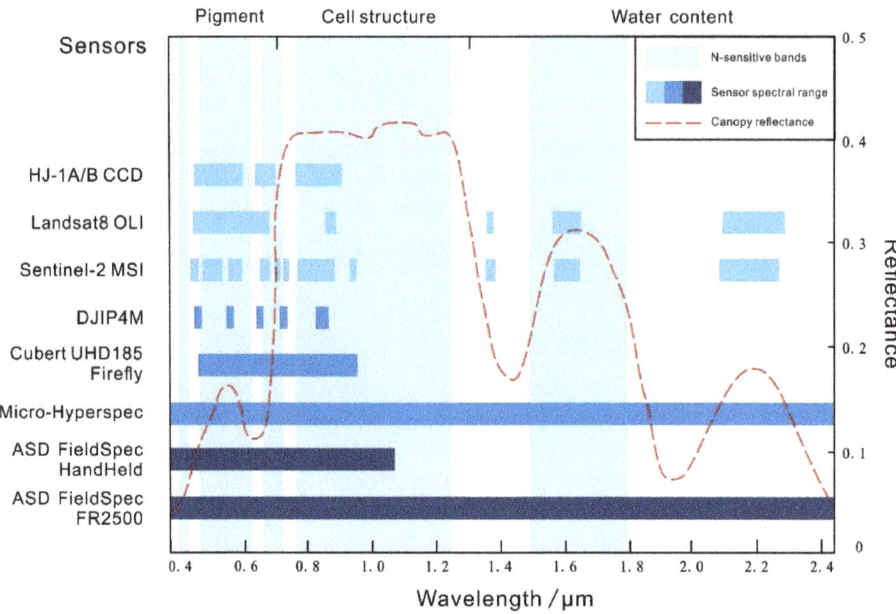

Figure 4. Spectral reflectance properties of crop canopy and spectral range of different platform sensors.

3.2. *Correlations between Remotely Sensed Data and N Status*

Crop N status can be distinguished between two measurement perspectives: one is area-based measure "nitrogen content" (N_{area}, per unit area); another is mass-based measure "nitrogen concentration" (N%, per unit dry matter) [28]. N% can be converted to N_{area} by plant leaf (or other plant organs) dry biomass. In research, N content generally includes leaf/plant nitrogen accumulation per unit soil area (LNA/PNA), etc.; N concentration generally includes LNC, plant nitrogen concentration on a leaf dry weight basis (PNC), canopy N concentration (CNC), canopy N density (CND), etc. [28,82]. On this basis, Nitrogen Nutrition Index (NNI) is defined as the ratio of the actual crop N concentration to the critical N concentration. It is a direct indicator of whether the crop N concentration is at an optimum level [83,84]. In existing studies, the accuracy of remote sensing estimates of N content is significantly higher than that of N concentration, due to the fact that N concentration is more difficult to extract from remote sensing information compared to N content [64,85,86]. However, N concentration is not disturbed by density and is more accurate in responding to crop N status [82]. Throughout the whole growth stage of rice and wheat, N concentration has a narrower range of variation with a decreasing trend, and the change rate decreased before it increased; N content is a product of the combination of N concentration and plant dry biomass, so N content has a wider range of variation with an increasing trend [87,88]. With the support of remote sensing technology, the study directly links N indicators and spectral reflectance.

3.2.1. Sensitive Spectral Extraction

Spectral response under crop N stress varies significantly, and the analysis of original hyperspectral information is an intuitive study in N remote sensing assessment [39,89], but the sensitive bands extracted vary for the same/different crops in different geographical environments [90,91]. As in different studies, the N-sensitive spectral bands of rice include 738 nm, 1362 nm, 1835 nm and 1859 nm [90], and also include NIR (>760 nm), visible (355, 420, 524–534, 583 and 687 nm) and red edge (707 nm) region [39]; the N-sensitive spectral bands of wheat include 440 nm and 610 nm [91], and also include 790.4 nm [92]. Wang et al. [38] explored the best common central bands 822 and 738 nm for LNA estimation in rice and wheat, which can effectively assess the N nutrient status of plants and reasonably reflect the intrinsic N information in different crops. Most studies have extracted several N sensitive bands to estimate crop N by individual spectra or combinations. Although the method is simple, the accuracy is affected by the stability of the spectral information.

Hyperspectral data have hundreds of high-resolution continuous spectral information. When exploring correlations between spectra and N status, using full band data as an input can increase errors and reduce efficiency. Whereas insufficient exploitation and use will lead to data waste, thus losing the significance of high-precision data. Therefore, improving the use efficiency of hyperspectral data still needs to be further explored. Wang et al. [93] divided the spectral data into five groups: blue, green, red, red edge and NIR, and extracted the corresponding N-sensitive bands, in which the red edge (702, 703 and 710 nm) and red edge (706, 733 and 759 nm) correlated with leaf and canopy-scale N status up to 0.92. Yu et al. [94] reduced the spectral data by a discrete wavelet multi-scale decomposition method (DWMD), achieving better results compared to iteratively retaining informative variables, with 16.28–26.23% improvement in coefficient of determination (R^2). Hyperspectral data are very similar between adjacent bands, and their dimensionality reduction can help reduce the complexity of feature extraction. Liu et al. [95] demonstrated that using feature bands extracted by improved adaptive ant colony optimization algorithm as input parameters under the same prediction model can reduce the complexity of the model, while improving the prediction capability. In addition, the autocorrelation matrix (R^2 = 0.86 between N-sensitive bands and N status), the non-negative matrix factorization (R^2 = 0.83), the successive projections algorithm (R^2 = 0.66), and the competitive adaptive reweighted sampling (R^2 = 0.93), etc. are also gradually applied [62,94,96–99]. Improvements to the N-sensitive band extraction method have resulted in the stability of the retrieval N information, but make the processing process more complicated, with a corresponding increase in model performance and computation time [97]. In addition, the bands extracted by these algorithms integrate more reflectance information, which improves stability and anti-interference. This provides a guideline for constructing a unified and generalizable spectral feature extraction method.

3.2.2. Mathematical Transformations of Spectra

The difference in response to N between spectra can be increased by mathematical transformations of spectra, improving its adaptability to inversion N status. The spectral reflectance curve feature can reflect the N change trend, and the first-order derivative of the spectrum indicates the rate of change in the reflectance, which reduces the effect of background information and is widely used in N estimation [63,92,93]. The morphological differences in crop spectra can be described by characteristic parameters such as slope, angle, and rate of change in the curve. Slope and angle are generally calculated based on areas of reflectance that change abruptly, such as peaks and valleys. The slope indicates the rate of reflectance rise/fall, and the angle formed by the sides of the reflection peak and absorption valley indicates the width between the peak and valley [100]. The rate of change is a generalization of the first-order derivative, which can respond not only to the change in reflectance between successive wavelengths, but also to the rate of change in reflectance between any two wavelengths [101]. Based on two integration metrics, normalized area reflectivity curve and reflectivity integration index, Du et al. [58] combined more wavelengths with improved LNC retrieval performance.

The study showed that the red edge region of 700–780 nm is a sensitive band for responding to the growth status of green crops [102,103], and the characteristic parameters of REP, red-edge slope, red-edge peak, red-edge minimum amplitude and red-edge area obtained by mathematical transformation of the red edge band have also become common parameters for N diagnosis in rice and wheat [36,104,105]. The REP based on linear extrapolation method showed better canopy N concentration correlation at larger canopy cover. Since the first-order derivative has "bimodal" characteristics in the two main spectral regions, the conventional REP is not sensitive enough to canopy LNC only using single-peak maximum, and the spectral reflectance data can be fitted to generate continuous REP values to achieve a continuous relationship between REP and N [105]. Li et al. [106] proposed a continuous wavelet transform-based REP extraction technique, wavelet-based red edge position (WREP), which provides a new idea for understanding the spectral variation in red edge region. Guo et al. [107] constructed an algorithm based on the analysis of red edge features, shifting red edge absorption area (sREA), which enables the construction of N absorption models at the regional scale. Due to the possible discontinuity of changes, insignificant amplitude, and large errors in the RE first-order derivative spectra, the estimation of crop N by red edge parameter features is highly dependent on the feature extraction method. It is necessary to select the appropriate method and parameters according to the research needs. When the canopy cover is too high, the response of the red edge region to N becomes gradually slower and then there is saturation, so only using red edge parameters easily causes misjudgment.

The mathematical transformation of spectra can respond to the trend of N changes, but the spectral information is changed by the influence of environmental stress. Deeply mining the bands with more effective information and using them in combination is the key to overcome the external factors and improve the accuracy of the model.

3.2.3. Spectral Indices

The information presented by specific parameter combinations (mainly difference, ratio and normalized values) is more stable and representative, and it has become the first choice for remote sensing inversion of crop parameters. The development of spectroscopy has improved the accuracy of spectral information, and the methods of first-order derivative, continuum removal, wavelet transform and smoothing algorithm (such as Savitzky-Golay smoothing), etc. are applied to the original spectra, which make the spectral calculation more refined and the results more accurate [95,108,109]. Compared with traditional vegetation indices (VIs), the construction parameters of spectral indices are no longer limited to band reflectance, but can also be red edge parameters, other spectral indices, etc. (Table 1) [26,110,111].

For multispectral/hyperspectral remote sensing data, spectral indices are usually constructed by selecting appropriate bands in the visible red region and NIR region. For different varieties of crops, there are significant differences in the slopes of normalized difference vegetation index (NDVI)-based LNC models, and it is difficult to achieve uniform regression analysis across different crops varieties with conventional indices. It has been shown that $NDVI_{(1220, 610)}$ is a good index to estimate LNC in both rice and wheat, with RMSE all less than 13.04% [34], and $NDVI_{(1220, 710)}$ achieves high precision estimation of N status for different varieties of rice [112]. However, the model has not been extended to other regions for validation, and the accuracy and stability of the spectral index remains to be explored. In addition, the use of a multi-band vegetation index for LNC monitoring in rice and wheat is more effective [92,113,114]. Wang et al. [113] used a three-band vegetation index combining NIR, red edge and blue bands to estimate LNC for rice and wheat with R^2 of 0.866 and 0.883, respectively, which were 17.66% (rice) and 7.68% (wheat) more accurate than NDVI, and 40.13% (rice) and 16.18% (wheat) more accurate than RVI. Tan et al. [92] explored the relationship between LNA and parameters such as first-order derivative sum (SD), first-order derivative maximum (D), etc. in VIS and red edge regions, and the new normalized index $(SD_r - SD_b)/(SD_r + SD_b)$, which was constructed by integrating information from multiple bands, was a good fit for wheat LNA (R^2 = 0.935) and was applicable to wheat N inversion in different varieties and regions.

The N estimation model used fixed VIs performed consistently over the same growth stage, but when pooling data from multiple growth stages, accuracy decreased significantly. Canopy structure and background conditions changed as the growth stage progressed, but in all cases, whole leaf pixels showed more stable performance than light and shade leaf pixels [87,115,116]. Traditional spectral indices cannot fully capture the intrinsic relationship between canopy spectra and N in growth stage, and how the predicted intensity of N models varies across growth stages has not been fully explored [117]. The red edge chlorophyll index ($CI_{red\ edge}$) is sensitive to canopy N content and can effectively mitigate the effect of canopy structure on canopy N estimation. Li et al. [118] combined NDVI and $CI_{red\ edge}$ to construct a nitrogen planar domain index (NPDI) with good predictive ability for canopy N uptake in wheat, corn and both combinations. Palka et al. [119] constructed a regression model by combining the canopy chlorophyll content index (CCCI) and the canopy nitrogen index (CNI), and modified the CCCI-CNI to extend N estimation to the end of booting stage for wheat. Quantifying the spectral contribution in the mixed image elements using spectral mixture analysis (SMA) can improve the spectral accuracy, and the spectral indices constructed in this way show superior capability in all growth stages and even in the early evaluation of LNC [116,120]. The selection of sensitive multi-band and multi-index information fused to form a spectral index improves the sensitivity of N inversion, and to some extent overcomes the inconsistency between indices and N for different varieties of crops and different fertility stages.

Based on RGB data, the response of different image indices to N varies widely and has limited response [121]. The RGB three-color channels of UAV images each contain luminance information and are susceptible to interference from lighting conditions. In contrast, in the HSV and Lab color spaces, V and L denote value and luminosity, respectively, and the transformation of RGB to them weakens the influence of luminance information through nonlinear changes, resulting in a more sensitive response to LNC [122]. Nevertheless, RGB is limited by spectral accuracy and still has difficulties in index improvement. In addition to the color information extracted from spectral data, the texture information can be obtained from the local variance function, which reflects the variation relationship between several pixel points and characterizes the canopy structure. Fusing two or more datasets with different feature information can provide a more comprehensive interpretation of the relationship between remote sensing information and N status [123]. Therefore, for RGB data, the "image-spectrum" fusion indices formed by fusing image indices and texture features improves the sensitivity of image data to N features, and the investigation of its ability to diagnose N status is a major development direction at present [60,82]. When extracting texture features through Gray-level Co-occurrence Matrix (GLCM), crop cultivation patterns may cause differences in texture information metrics for N content in different directions, and using texture information calculated along the perpendicular to the row direction to monitor row-grown crops has the best results [85]. In addition, adding depth information to RGB images can break through the limitations of extracting canopy structural features from 2D images [86,124]. Xu et al. [124] fused texture features with 3D structural information and RVI to invert N status with better accuracy and stability, the LNA prediction accuracy of 0.74 during whole growth stages. By adding 2D and 3D structural information, the background and saturation effects can be better reduced, and the N inversion information of the crop canopy can be enhanced.

3.3. Retrieval Methods of Canopy N Status

For crop N status assessment and monitoring, the modeling methods can be divided into statistical analysis, physical analysis, and hybrid methods. Statistical analysis is to obtain high-precision N diagnosis by establishing mathematical relationships between ground truth data and remote sensing spectral information, which can be further classified into two categories: traditional statistical methods and machine learning methods. Physically based methods take the structural information and physicochemical properties of plants and leaves as input parameters, simulate the process of radiation absorption and scattering inside the crop, and form the reflection spectrum and output, so as to establish the correlation between crop parameters and ground reflection spectrum. The models are divided into radiative transfer models (RTM) and geometric optical models, and RTM are often used in research because of the continuous and uniform distribution of rice and wheat in cultivation. Hybrid models have advantages by using multiple categories of models in combination but are currently used rarely in N retrieval (discussed in Section 5).

Table 1. Spectral indices for assessing crop N status.

Index	Formula	Reference
Nitrogen Reflectance Index	$\frac{(R_{800}/R_{550})_{target}}{(R_{800}/R_{500})_{reference}}$	[125]
Red Edge Position: Linear Extrapolation Method	linear extrapolation of two straight lines on the derivative spectral curve (lines formed by 680 nm and 694 nm, and formed by 732 nm and 760 nm)	[126]
Normalized Difference Red Edge	$\frac{R_{790}-R_{720}}{R_{790}+R_{720}}$	[33]
Double-peak Canopy Nitrogen Index	$\frac{(R_{720}-R_{700})/(R_{700}-R_{670})}{R_{720}-R_{670}+0.03}$	[127]
Nitrogen Planar Domain Index	$\frac{Measured\ CI_{red\ edge}-CI_{red\ edge\_}MIN}{CI_{red\ edge\_}MAX-CI_{red\ egde\_}MIN}$	[118]
Water Resistance Nitrogen Index	$\frac{(R_{735}-R_{720})R_{900}}{R(R_{930}-R_{980})(R_{735}+R_{720})_{min}}$	[128]
Canopy Chlorophyll Content Index	$\frac{NDRE-NDRE_{min}}{NDRE_{max}-NDRE_{min}}$	[129]
Modified Chlorophyll Absorption Ratio Index	$\frac{R_{700}-R_{670}-0.2(R_{700}-R_{550})}{R_{700}/R_{670}}$	[130]
MCARI/MTVI2	$\text{MCARI}=\frac{R_{700}-R_{670}-0.2(R_{700}-R_{550})}{R_{700}/R_{670}}$ $\text{MTVI2}=\frac{1.5(1.2(R_{800}-R_{550})-2.5(R_{670}-R_{550}))}{sqrt((2R_{800}+1)^2-(6R_{800}-5sqrt(R_{670}))-0.5)}$	[131]

R_i stands for reflectance at wavelength i nm.

3.3.1. Traditional Statistical Methods

Traditional statistical methods have simple mechanisms, and their inverse accuracy depends more on the rationality of modeling parameters, thus they cannot overcome the influence of environmental and other factors, and there are difficulties in transferring prediction models to other datasets. However, due to the simplicity and convenience of the model and its usability, it is still widely used in the field of crop N monitoring at present. Univariate linear regression and its nonlinear transformation, as the basis of statistics, have achieved good results in exploring whether there is a significant correlation between crop N indicators and individual VIs, and occupy an important position in N status inversion [104,132,133]. Hansen et al. [89] found that applying partial least squares regression (PLSR) to fit N status gave equal and better results than exponential regression, with the R^2 maximum increase of 23%. PLSR was considered a good alternative to univariate statistical models. When multiple growth stages are involved or when there is a lack of phenological information, individual index that do not fully utilize spectral data cannot describe the relationship between N and spectral information, so using multiple linear regression (MLR) is a better choice [134]. MLR can reduce model chance. Whether the parameters involved in modeling exist in response to N and whether there is overfitting between parameters is the key to influencing the multiple regression model. Pearson correlation coefficient is generally used as a measure in the study, and the combination of parameters with larger correlation coefficients is selected to construct MLR models with good interannual scalability [64,69]. PLSR, principal component analysis (PCA), stepwise multiple linear regression (SMLR), and Ridge Regression, etc., play an advantage in dealing with the collinearity problem between parameters. Many co-linear spectral variables were reduced to a few uncorrelated latent variables to avoid overfitting problems [91,123,135]. The selection of a suitable variable dimensionality reduction method based on the quantitative relationship between the number of samples and the dimensionality of variables, combined with multiple regression, significantly improves the inversion accuracy and applicability.

Traditional statistical methods can describe different rates of change between N status and spectral information, providing fitting models for a variety of change conditions. For region-specific datasets, models with good inversion effects can be obtained by comparing different regression methods; for datasets with different growing environments, the best regression models derived from the study will be different [64,89,104,134]. Although the statistical models are not stable enough to overcome environmental problems, the different regression methods in the statistical models are more consistent in principle, have specific mathematical relationships, are easy to understand and apply, and are extremely convenient to use in fixed research areas. Therefore, the traditional statistical methods still dominate the existing N remote sensing monitoring studies.

3.3.2. Machine Learning Methods

Machine learning models are gaining widespread attention for their ability to handle large amounts of input data from multiple platforms and to solve nonlinear tasks. Artificial Neural Network (ANN) and Back Propagation Neural Network (BPNN) are commonly used models for remote sensing estimation of crop N status, which can automatically extract relevant features from data. However, in practical applications, a large training dataset is required, and the number and size of the implied layers, training efficiency, and overfitting are considered. Yang et al. [96] used Gaussian radial basis function as the implied layer of the neural network to avoid the tedious calculation and overfitting phenomenon of BPNN, with structural adaptive features, good generalization ability and fast learning convergence speed, and more stable and reliable application. When constructing neural network models directly, the differences in results for different types of parameters are not obvious, but the accuracy is significantly improved after using PCA for model input parameters [136–139]. The combined use of PCA and machine learning methods shows unique advantages and promising applications.

Support Vector Machine (SVM) is extremely effective for analytically solving high-dimensional data problems. Yao et al. [140] applied traditional regression analysis, ANN and SVM, to compare the prediction accuracy, computational efficiency and complexity level of different methods for inversion of wheat LNC. The results showed that the machine learning models were more accurate, with the SVM method being more stable in dealing with potential confounding factors for most varieties, ecological niches, and growth stages. The kernel function in SVM is the focus of attention, and the multiple-kernel support vector regression (MK-SVR) plays an advantage in estimating N status at different growth stages because it combines the advantages of local kernel function and global kernel function [141]. However, complex optimization algorithms can reduce the computational efficiency of SVM and using a combination of least squares and SVM methods, LS-SVM can solve linear or nonlinear multivariate estimation capability in a relatively fast way, significantly improving the computational efficiency of SVM [17,141,142].

In the presence of weak a priori knowledge, Gaussian Processes Regression (GPR) can perform adaptive nonlinear fitting of complex datasets with flexible probabilistic Bayesian models and simpler parameter optimization applied to crop N status inversion [11,93,143]. Random Forest (RF) integrated with decision trees as the basic unit can rank the importance of variables, reduce redundancy in high-dimensional datasets, and have high stability, with vast application prospects [37,144]. When the entire spectral range of a single band is used as an input variable, the accuracy of regression by RF inversion (R^2 = 0.89) is higher than that of univariate regression with existing VIs; when VIs is used as input features, model accuracy is improved with R^2 of 0.95 [145]. Determining the appropriate input dataset is a key element to exploit the predictive power of the model.

Machine learning methods techniques can be used to reveal the physiological and structural characteristics of plants, and can respond to dynamic differences in physiology due to environmental influences [144]. The study is no longer limited to conventional machine learning models, but improves the models starting from input datasets [138,145], model parameters [96,144], functions and structures [17,96,141], which not only improves the efficiency of data analysis, but also enables higher accuracy N status analysis, making it more efficient and flexible in N monitoring. The input variables have diversified from single spectral information to mathematically transformed spectral data, spectral indices, and texture information, etc. while the machine learning methods are expanding toward efficiency, accuracy, and speed.

3.3.3. Physically Based Methods

RTM use optimization algorithms for inversion to infer the N content of crops from observed spectral data. Depending on the scale of the object of study, it is mainly divided into leaf radiative transfer models and canopy models. At the leaf scale, the PROSPECT model is the most widely used and has been continuously optimized and improved [146–148]; at the canopy scale, the SAIL model is one of the first models applied, mainly for uniformly distributed continuous vegetation surfaces [149]. Combining PROSPECT and SAIL models to invert vegetation physiological and biochemical parameters is a common approach nowadays, mostly around canopy and LCC, water content and leaf area [29,75,150,151]. Most studies derive crop N status through empirical relationships based on significant associations between leaf area or Chl and N. Despite the long-standing stability and reliability of RTM in the inversion of physicochemical parameters, different configurations of the model by users may lead to equally plausible results [152,153], so it is necessary to constrain the model using a priori information. Combining the DSSAT cropping system model CSM and PROSAIL model, complementing the interaction between crop growth stages and the environment for the constraints of the input parameters of the PROSAIL model, plays a unique advantage in the inversion of crop physicochemical parameters, not only with high accuracy, but also with the statistics of physicochemical parameters among different varieties of crops [154].

Yang et al. [155] used N uptake coefficients to equivalently replace the Chl uptake coefficients in the original PROSPECT model, and established the N-PROSPECT model based on the PROSPECT model to directly invert leaf N content. The N-PROSAIL model, established by combining the N-PROSPECT model and the SAIL model, achieves the diagnosis of N status at the leaf and canopy scales, and reduces the model error by setting a priori parameters at different growth stages [156]. The RTM expresses the crop growth process from a physical point of view, which is more stable in the inversion, but has the problem of being time-consuming. Combined with the Lookup Table (LUT) it can reduce the computational demand. Li et al. [157] constructed a multi-LUT for wheat LAI, LND and two spectral indices (MSR and MCARI/MTVI2), which not only reduced the LUT size and improved the computation time, but also had better accuracy of N estimation. On the other hand, since protein is also a major N-containing component in crops, coupling protein specific absorption coefficients into the PROSPECT model to form PROSPECT-PRO, which is combined with the 4SAIL model to form PROSAIL-PRO, can also be used for crop N status diagnosis [31,158]. RTM with strong explanations is better expressed in the inversion, but because the model expression depends on the input of more parameters and complex computational process they are less used in current research. Reducing the complexity of models and complementing the advantages of statistical models, hybrid RTM and machine learning models have become a future research need.

4. Influential Factors on Accuracy of Remote Sensing Monitoring of Canopy N

Spectral reflectance information has been shown to be sensitive to the N content of canopy leaves, but differences in data acquisition, vertical distribution of leaf N, dynamic changes in N during the growth stages, and physiological differences between different plants can all have an impact on the correlation between crop spectral information and canopy N indicators.

4.1. Differences in Data Acquisition Angles

Both portable spectrometers and UAV sensors usually acquire spectral data for crop N monitoring within a range of VZAs, which can vary by 30° and more. The spectral indices in crop N status studies are usually developed from vertical angle data. However, due to the variation in the angle of view of data acquisition caused by different experimental conditions, and the anisotropy of vegetation reflectance, the accuracy and robustness of using these indices directly to estimate the N content are not sufficient. The reflectance in the VIS, red edge and NIR bands decreases gradually from VZAs from −60° to 0°, with relative changes in reflectance ranging from 34.7% (+60°) to 265.5% (−60°), and 81.7% (+60°) to 89.3% (−60°) in the VIS and NIR bands, respectively [53]. Therefore, developing an index that is sensitive to N content and insensitive to VZAs is of great practical importance to adapt to different experimental conditions, improve prediction accuracy and enhance model stability.

Higher viewing angles allow better extraction of crop biochemical information compared to the nadir orientation [159]. The change in view angle significantly affects canopy reflectance, especially in the red and NIR bands, which in turn makes VIs based on these spectral bands sensitive to angle [160]. The introduction of angle-insensitive bands to construct indices, such as the normalized difference red edge (NDRE), the green and blue bands, and the green band Chlorophyll Index (CIgreen), can improve the accuracy of canopy N inversion of different VZAs remote sensing images, significantly expanding the range of suitable viewing angles for determining crop N status by remote sensing, and thus adapting to the differences between different experimental conditions [53,161,162]. For the angular insensitivity index cannot be simply attributed to the effect of a single band; green, blue and NIR bands may have played a joint role in improving the index adaptation. It is difficult to obtain accurate spectral collection perspectives in the applications. A unified N monitoring model under a range of perspectives can help with the flexible application of crop N diagnosis. Like other VIs, Li et al. [53] developed angular insensitivity vegetation index (AIVI) to have the best LNC estimation accuracy at −20° view angle, but at the same time the correlation between AIVI and LNC has high stability at −10° to −40° with R^2 of 0.83. Similarly, floating-position water band index (FWBI) has the highest correlation with LNC at −10° view angle (R^2 = 0.852), also has superior N content estimation accuracy at 0° to 30° (R^2 = 0.835) [161]. The statistics on angular differences show that back-scatter direction has better LNC prediction accuracy than the forward-scatter view angle.

However, the spectral information obtained by whatever VZAs inevitably has information such as soil background, light shading, etc. Different growth stages and different light conditions will change the crop spectral reflectance, which is a common noise in inversion. The study reduces their effects by spectral preprocessing such as first-order differentiation and wavelet transform, suitable vegetation index, and threshold segmentation [27,63,82,93]. The water background of rice is a unique feature that differs from other crops; water has an absorption effect on the NIR band, and when the canopy cover is small, the water depth and turbidity have an isotropic effect on the spectral reflectance of the red-edge region [163]. Therefore, when converting reflectance to vegetation index, this effect can be eliminated or attenuated by calculating between multiple bands. The individual N content was significantly improved in accuracy before and after removal, and the group indicators indicated the total amount per unit area, which was less influenced by background noise and had a smaller enhancement effect [82].

4.2. Vertical Distribution of Leaf N

When the canopy leaves are taken as a whole object, most of the studies are carried out for the upper leaves, ignoring the vertical heterogeneity of the canopy. The LNC within the canopy is not constant from the top to the bottom of the leaf layer, and it varies with growth stages. At jointing stage, flowering stage and filling stage, LNC decreases from top to bottom; and at booting stage, it tends to increase and then decrease [43]. Moreover, it is difficult to effectively identify the information at the bottom of the canopy by acquiring the crop reflectance spectra vertically due to the influence of canopy leaf cover at different growth stages. At present, we have achieved better results in exploring the canopy N content by acquiring reflectance spectrum vertically or from multiple angles to overcome the stratification differences.

The PLS algorithm has better estimation capability for different levels of leaf N status, then becomes an effective tool for early N monitoring [40,164–166]. Huang et al. [164] combine NRI and NPCI to construct a PLSR model which could better retrieve foliage N density in different leaf layers ($R^2 > 0.67$). He et al. [166] also demonstrated that PLSR estimates LNC accuracy better than BPNN and eXtreme gradient boost (XGBoost). However, for the studied spectral information, which must contain information from different leaf layers, it is especially important to determine the contribution of different leaf layers to the spectral reflectance of the canopy. Studies have begun to explore the characteristics of the vertical distribution of N in the canopy of crops at different growth stages, and to develop an effective method for estimating N in each leaf layer or total N in the canopy by determining the correlation between different leaf layers and N status [40,43,44]. Duan et al. [43] used a calibration coefficient to adjust the relationship between the effective layer of remote sensing detection and the whole canopy, and then developed a method for estimating the overall canopy LNC based on GI, mND705 and NDVI. He et al. [44] estimated the canopy top LNC by NDRE, then inputting the results into the LNC vertical distribution model to get the model coefficients; thus the model based on the relative canopy height could obtain LNC in different leaf layers (LNC_{Li}), which was superior because of fewer parameters and higher accuracy. The short plant size of rice and wheat crops and the small vertical distance between different leaf layers can easily mask differences in the spectral response of canopy N status changes [54]. Compared with vertical remote sensing observation, multi-angle observation can reduce the information bias of fixed viewpoints. Using different combinations of VZAs, Wu et al. [54] were able to retrieve LCC in the upper-layer (VZA 10°), middle-layer (VZA 10° and 30°) and bottom-layer (VZA 10°, 30° and 50°) of the plant, respectively. Based on the response of spectral indices to each leaf layer of N status at different VZAs, selecting the best VZAs or combination of VZAs can realize the complementation of canopy spectral information so that the accuracy of crop N monitoring can be more robust and accurate [40,54].

Multi-angle stereoscopic observation can obtain more vertical information about the plant, but when the bottom leaves are too low, the influence of soil background and crop residues, etc., will increase. Therefore, it is necessary to determine the effective depth of crop canopy spectroscopy observation and realize the inversion model of vertical distribution of N content in canopy from "surface" to "three-dimensional". At the same time, the multi-angle measurement will generate a huge amount of data, and how to quickly extract the effective information from it has become an urgent problem to be solved.

4.3. Dynamic Changes in N during the Growth Stages

N content in crops is a long-term accumulation process that changes as growth stages. The correlation between N and spectral information varies at different stages, and an ideal N inversion model needs to overcome the effects of phenological variability and accurately estimate the N content of the crop at different growth stages. Throughout the crop's reproductive stages, temperature levels affect photosynthesis and metabolic processes that are closely related to N assimilation and utilization. Therefore, it is necessary to introduce meteorological information to construct a dynamically changing model.

Crop models such as CERES and APSIM, which are widely used around the world, simulate crop growth processes by inputting meteorological data and field management data, etc., and are important guides for real-time diagnosis of crop N nutrition status under different cropping conditions [167–169]. The construction of growth models relies on numerous experimental parameters, which are data-intensive and cumbersome to process. Cao et al. [170] dynamic obtained relative growing degree days (GDD) based on the physiological development time of crops to participate in modeling, and quantified the model parameters that reduced the effects brought about by different indicators. In addition, the field spectral information obtained based on remote sensing data is accurate, and quantitative analysis of temporal variation between VIs is more conducive to the dynamic monitoring of N status. Double Logistic functions and Gaussian curves fitted to time-series data can effectively describe crop growth and senescence processes [171–173]. By combining effective accumulated temperature and crop growth parameters, such as the spectral index NDRE, a model for monitoring the entire growing period of the winter wheat canopy, constructed with the growing degree-days as a moderating factor, offers the possibility for N estimation throughout the growth stages [174]. Dynamic curves of indices such as NDVI, constructed using accumulated growing degree days (AGDD) as a time driver, provide a reference for N nutrition diagnosis at different periods [172]. Combining multi-temporal VIs with key phenological indicators, the constructed dynamic model has clear biological significance, which not only facilitates crop N monitoring but also significantly enhances the ability of N status early prediction.

4.4. Physiological Differences in Plants

N synthesis is influenced by canopy structure, photosynthesis-related pigments, and water content, etc., and confounding effects are common in canopy reflectance under physiological stress, making N remote sensing monitoring challenging.

Water is the carrier of N transport in the crop and is closely related to the N status of the canopy. Exploring the spectral response to N and water shows that the spectral reflectance of wheat treated under low water conditions increases in both regions of SWIR, with less difference in the NIR region under water differences [27,59]. Thus, indices constructed based on SWIR and NIR reflectance are better able to show differences between N levels in different water treatment environments. The SWIR region contains more water-related information, but sensors that include SWIR are costly so it is relevant to explore water-sensitive bands in the VIS–NIR range. Under the condition of constant N content, the red edge reflectance of crops with different water treatments tends to be the same [128]. Based on the red-edge correlation indices such as NDRE and normalized pigment chlorophyll index (NPCI), the introduction of water-related indices, such as floating-position water band index (FWBI), crop water stress index (CWSI), etc., can significantly improve the interaction between water deficit and N nutrition [128,175–177]. Whether the multi-analysis based on specific VIs or the whole spectrum, studies have been conducted to separate N and water information in the spectrum by reducing spectral mixing effects, thus improving the estimation accuracy of crop N under the influence of water.

Crop growth parameters are not independent of each other and may correlate under different circumstances, such as correlation between N, Chl and LAI when the canopy cover is small. However, when Chl and LAI change driven by other external conditions, there will be errors in estimating N status based on their correlation with N. Research needs indices that are both sensitive to N and resistant to interference from other factors. The combination of two VIs with different sensitivities to Chl and LAI, whose ratios can minimize the effect of LAI and have a better correlation with Chl, such as the joint indices modified chlorophyll absorption ratio index and second modified triangular vegetation index in ratio (MCARI/MTVI2) [131], the red-edge-chlorophyll absorption index and the triangular vegetation index in ratio (RECAI/TVI) [111], the transformed chlorophyll absorption reflectance index and optimized soil-adjusted vegetation index in ratio (TCARI/OSAVI) [177]. The red edge region is influenced by LAI which cannot estimate N well in complex situations. Chen et al. [127] found that there is a double-peaked phenomenon in the first-order derivative spectrum of the red edge region, and that changes in N concentration can be amplified as changes in the relative height of the double-peaked peaks, with which the proposed double-peaked canopy nitrogen index (DCNI) can overcome the influence of LAI. There is a certain similarity between different spectral indices and different physiological parameters, which show hierarchy and aggregation in statistical analyses. Exploring the different relationships that may exist between parameters is therefore important for exploring N status in crops grown under different growing conditions.

5. Challenges and Perspectives

After decades of development, the techniques for remote sensing monitoring of canopy N have made rapid progress and achieved good results, but there are inevitably many difficulties and challenges that need to be addressed.

(1) The development of multi-source data integration from "satellite–airborne–ground" to meet the needs of high-precision monitoring at all scales. In recent years, remote sensing-based crop N monitoring and assessment research has been conducted mainly at the laboratory and field scale, applying small sensor platforms based on ground-based spectral instruments and UAV to acquire data. Research on large farms, counties, cities, or larger regional scales rely on satellite-based multispectral data. Multispectral data are affected by radiation and atmosphere during acquisition, making processing more difficult. In the face of complex geography, the low-resolution images are prone to mixed pixels, making it difficult to achieve accurate estimates of N status. Hyperspectral data are limited at large scales due to their access and data volume, so how to achieve high accuracy monitoring and assessment of crop N over large scale areas is a major challenge currently faced.

Currently, the research on crop N estimation from UAV data is beginning to bear fruit, with good validation in small farm applications. A summary of the spectral data acquisition platforms and their inversion N in existing studies is shown in Table 2. UAV-based research can not only analyze spectral information (sensitive spectra, spectral mathematical transformations, spectral combination calculations, etc.), but also extract image information (texture information, color information, etc.), which shows advantages of high precision due to its high spatial and temporal resolution and the amount of representation information. Satellite-based imaging data are limited in depth, mostly only extract VIs result in low inversion accuracy, yet satellite data is still the most important source of data for large-scale studies. In the existing research situation, the rapid development of sensor technology and remote sensing platforms has extended the scale of research to medium and large farm areas. The results of existing UAV-scale research results are translated to municipal, provincial, national, and even larger scales through algorithms such as multi-scale analysis and reconstruction, and spatiotemporal data fusion, thus enabling N monitoring over large areas. Therefore, the fusion of multiple sources of data from "satellite–airborne–ground" is the basis for large scale applications with inversion accuracy [178]. At present, the spectral resolution of the red edge and NIR bands (N-sensitive bands) in satellite-based hyperspectral sensors is insufficient, and the spatial resolution and revisit period are not advantageous. In this context, the transformation of ground-based research results and the development of high-precision satellite-based hyperspectral sensors deserve even more attention. From the perspective of data acquisition, the complementary advantages of the multiple types of data from "satellite–airborne–ground" could break the limits of geographical scope, and then enable high-precision monitoring and assessment of crop N status at all scales.

(2) Research still has bottlenecks in monitoring crop N in the presence of confounding factors. Under N stress, the spectral properties of vegetation leaves change, and N monitoring is achieved through crop spectral information obtained from ground-based observations by remote sensing technology. However, in practical applications, the inconsistency of crop growth conditions can lead to irregular overall crop deficiencies in water, fertilizer, and cause pests and diseases, all of which could generate yellowing and wilting of crop leaves. The changes in the external structure and intrinsic characteristics of the canopy, result in corresponding changes in the spectral reflectance characteristics [178]. To simply attribute spectral changes to canopy N content would be a misjudgment. Studies are generally set up with variable conditions for different growth stages, locations, field management, species, or plant types to test the stability of the model. However, there is insufficient evidence that the method is effective in overcoming spectral variation due to

physiological differences. The primary way to achieve interpretability and practical application of the model is to start with the principles and isolate the influencing factors [179,180]. When considering only N and another stress factor, overcoming the effects of water can increase R^2 to 0.843 [128]. However, few studies have quantified and differentiated the contribution of leaf biochemical content (including water, diseases, other pigments, etc.) to the spectral band from spectral perspective.

Table 2. Summary of the data platforms, retrieval methods, and research results of the studies cited in the body.

Crop	N Status	Data Platforms	Indices	Retrieval Method	Results	References
Wheat	LNA	ASD FieldSpec Pro	Spectral bands VIs	PLSR SVM RF	R^2 = 0.895 RMSE = 0.903 g/m^2	[109]
Rice	LNC CNC	ASD FieldSpec Pro2500 ASD FieldSpec3	Spectral bands VIs	GPR RF, GPR-RF SVR, GPR=SVR	R^2 > 0.94 NRMSE < 6%	[52]
Wheat	CND	ASD FieldSpec Handheld UAV (UHD 185 Firefly)	VIs	N-PROSAIL	Field: R^2 = 0.83 RMSE = 0.23 UAV: R^2 = 0.74 RMSE = 0.26	[157]
Wheat	N concentration N content	ASD FieldSpec HandHeld	Spectral bands VIs	Statistical analysis	N concentration: R^2 = 0.81 RMSE = 0.72% N content: R^2 = 0.96 RMSE = 0.83 g/m^2	[119]
Wheat	LNA	ASD FieldSpec HandHeld 2 RealSense depth camera D435i	VIs Texture	MLR BP	R^2 = 0.74 RRMSEs = 40.13%	[124]
Rice	LNC	UAV (AIRPHEN multispectral camera)	VIs	Linear spectral mixture analysis Statistical analysis	R^2 = 0.78 RMSE = 0.26% RMSE = 10.4%	[116]
Wheat	LNC	UAV (hyperspectral camera)	VIs Texture	RR PLSR SVR RF	R^2 = 0.84 RMSE = 0.25	[123]
Wheat	LNC LNA	ASD FieldSpec Handheld 2 RealSense depth camera D435i	VIs Canopy structural	PLS RF	LNC: R^2 = 0.78 RMSE = 0.35% LNA: R^2 = 0.79 RMSE = 1.54 g/m^2	[86]
Rice	NNI	UAV (Parrot Sequoia camera)	Spectral bands VIs	LR, SMLR RF SVM ANN	RF accuracy is the highest: R^2 = 0.61 (stem elongation stage) R^2 = 0.79 (heading stage) RMSEs = 0.09	[37]
Rice	LNC PNC LNA PNA	UAV (Cubert S185 hyperspectral camera)	Spectral bands VIs	LR, MLR PLSR ANN RF SVM	At single growth stage, LR estimation N status based on VIs has the highest accuracy; at multiple growth stages, PLSR and ML are better.	[64]
Wheat	N content	HyMap sensor	Spectral bands	PROSAIL-PRO GP Heteroscedastic GP	RMSE = 2.1 g/m^2 The optimal N retrieval spectral bands are in the SWIR.	[31]
Wheat	LCC	Landsat8 ASD FieldSpec Pro	VIs	LR PROSPECT SAIL	Use hyperspectral leaf reflectance data to simulate Landsat-8 bands LR: R^2 = 0.59 PROSPECT: R^2 = 0.64	[75]
Wheat	LCC	Sentinel-2 RapidEye EnMAP	Spectral bands	PLSR	Sentinel-2: R^2 = 0.755 RapidEye: R^2 = 0.689 EnMAP: R^2 = 0.735	[74]
Wheat	LNC	ASD FieldSpec HandHeld	VIs	Statistical analysis	AIVI could overcomes the impact of VZAs: R^2 = 0.84 at −20° R^2 = 0.83 at −10° to −40°	[53]

Table 2. *Cont.*

Crop	N Status	Data Platforms	Indices	Retrieval Method	Results	References
Wheat	LNC	ASD FieldSpec	VIs Spectral bands	VIs BPNN XGBoost PLSR	$R^2 \geq 0.83$ at 0° to −30° VZAs range The accuracy of PLSR is better than VIs (16–17%), BPNN (15–16%) and XGBoost (29–58%) at VZAs ±60°	[166]
Rice	LNC_{Li}	ASD FieldSpec4	VIs	Vertical distribution model	LNC_{L1}: $R^2 = 0.768$ LNC_{L2}: $R^2 = 0.700$ LNC_{L3}: $R^2 = 0.623$ LNC_{L4}: $R^2 = 0.549$	[44]
Wheat	NNI	ASD FieldSpec Micro-Hyperspec and NIR-100 imager SC655 thermal camera	VIs Thermal indices	Statistical analysis	The combination of CCCI and DWI can overcome the influence of water to retrieve NNI, and the RMSE is reduced to 0.109.	[59]

Table 2 covers case studies from different regions.

In addition to physiological characteristics of vegetation, differences in soil background and canopy structure can cause difficulties in extracting whole leaf pixels, and noise from atmospheric transport processes also affects spectral accuracy. Existing studies mainly consider the effects of soil background and fractional vegetation cover (FVC). Before and after removing the background pixels, the accuracy of remote sensing inversion of crop biochemical parameters was significantly improved; for example, R^2 in LAI inversion could improve 0.27 [181], in N status inversion could improve 0.11 [82], and in Chl status inversion could improve 0.10 [182]. However, the applicable method of background elimination is also extremely important, as it requires high performance to adapt to the complex and changing field environment [182]. FVC correlates with background information and combining this information with spectra can also improve N estimates in different environmental contexts [183,184]. In addition, most imaging systems use top or side views to collect data, and the anisotropy of spectral information leads to different responses to crop N, and a suitable angle of spectral acquisition is important for accurate N monitoring. The blue band has atmospheric function, when the N content estimation model combined with blue band and other N-sensitive bands can improve the adaptability of the angle of data acquisition [53]. Under multiple conditions such as data acquisition, environmental stress and crop physiological stress, the spectral information is mixed with numerous non-target factors, which need to be decomposed to determine the precise response of crop N to the spectrum. Therefore, various influencing factors should be considered when estimating crop N by remote sensing to achieve high precision diagnosis.

(3) Improving the generalizability of models is key to crop N monitoring and assessment. To summarize the currently used models for inversion of N status and their effectiveness, a statistical model is currently the most commonly applied method in research experiments. When using statistical models, it is first necessary to determine the crop N indicator, the determination of which may result in experimental bias due to equipment or operational practices; subsequently, in the phase of selecting characteristic bands with high correlation to the crop N indicator, there is the possibility of wrong band selection. In essence, the reliability of statistical models to assess crop N status depends on the dataset used to train the algorithm and the model. When applied to separate datasets under different conditions, the models are less generalizable. To achieve regional scalability and explore the influence of environmental differences on modeling, datasets can be constructed by combining spectral information and ecological factors so that they contain a large amount of variability data to improve model generalizability. However, improving the predictive accuracy of the model by adjusting the input parameters still has limits. Considering the principles, there is a trade-off between model interpretation and model performance. The predictive principles of traditional statistical models are intuitive and easy to understand,

but at the expense of model performance; some machine learning models produce better predictive accuracy, while they are considered black box models because explaining how these models make decisions is a very difficult task (partial model prediction accuracy show in Table 2).

Physical models are highly advantageous in achieving model generalization, as they simulate the interaction between physical–chemical parameters and light from the physiological mechanisms of crops, thus providing explanations for the complex relationships between spectra and physical–chemical parameters at different fertility periods and under different growing conditions. However, the tedious and time-consuming inversion process limits its application. Existing studies usually analyze crop status at a small number of growth stages, so the statistical model has limited transferability across crops with different phenological status. The physical model overcomes this limitation and allows a wider range of crop canopy properties to be simulated. A hybrid model combining the mechanisms of statistical and physical models is not only efficient and flexible, but also explanatory for parameter inversion. In a hybrid model, the physical model is used to generate simulated spectra, which describes intra-canopy radiative transfer and interactions according to physics laws, thus providing information on spectral reflectance in relation to crop physicochemical variables [28]. Using simulated spectral data as input to train statistical models can provide physical constraints and explanations, and give a wider range of suitability [185]. Hybrid models have been applied successfully to estimate crop physical–chemical parameters (LAI, LCC, FVC, etc.) [185–187], but have rarely been used to invert N status. In recent years, physical models have moved from the previous indirect inversion of N through the physiological relationship between Chl and N to explore direct modeling of crop N status from spectral information, with more stable models and high response efficiency. Berger et al. [31] and Verrelst et al. [158] combined GP and PROSAIL-PRO models for inversion of crop N content, and confirmed the efficiency of hybrid models for direct N estimation. Therefore, how to achieve the complementary advantages between statistical and physical models, then construct a crop N estimation model with both mechanics and accuracy, will be the focus of future research.

6. Conclusions

Remote sensing technology is developing at a rapid pace and non-destructive monitoring and assessment of crop N status is gaining importance. This paper analyses the physiological mechanisms and spectral response characteristics of remote sensing monitoring for canopy N. Taking the remote sensing monitoring platform, the correlation between remotely sensed data and N status, and the remote sensing retrieval methods as the entry point, this paper provides an in-depth summary of the research techniques in the field of remote sensing monitoring for canopy N. The factors affecting the accuracy of remote sensing monitoring are also discussed. To date, the research at field scale has been well validated. The development of sensors and spectral carrying platforms facilitates high-precision remote sensing monitoring of crop N at farm scale. Due to the amount of information that can be extracted from remote sensing data, the efficiency of model use has become a key research concern. The efficiency and flexibility of machine learning models and the explanatory nature of physical models have their own advantages. The hybrid of the two models is beginning to show results in improving model stability. In addition, the effective use of multi-source data, and the removal of confounding factors in crop N monitoring need to be further explored. In-depth understanding of the limitations of current technology will be necessary to enhance the understanding of the link between canopy optical properties and crop N status, and to identify more appropriate N retrieval methods. In the context of the current rapid development of smart agriculture, the combination of sensors, remote sensing platforms and the Internet of Things results in the initial formation of a crop growth monitoring IoT platform. It provides the development direction for real-time monitoring and early forecasting of crop N, making it more widely application in the fields of growth monitoring, yield prediction and precision fertilization.

Author Contributions: Conceptualization, J.Z. and X.Y.; writing—original draft preparation, J.Z. and X.Y.; writing—review and editing, X.S., G.Y., X.D. and X.M. All authors have read and agreed to the published version of the manuscript.

Funding: This research was funded by Science and Technology Department of Guangdong Province (2019B020216001) and National Key Research and Development of China (2019YFE0125300).

Data Availability Statement: Not applicable.

Conflicts of Interest: The authors declare no conflict of interest.

References

1. Miao, Y.; Stewart, B.A.; Zhang, F. Long-term experiments for sustainable nutrient management in China. A review. *Agron. Sustain. Dev.* **2011**, *31*, 397–414. [CrossRef]
2. Liang, G.; Sun, P.; Waring, B.G. Nitrogen agronomic efficiency under nitrogen fertilization does not change over time in the long term: Evidence from 477 global studies. *Soil Tillage Res.* **2022**, *223*, 105468. [CrossRef]
3. Diacono, M.; Rubino, P.; Montemurro, F. Precision nitrogen management of wheat. A review. *Agron. Sustain. Dev.* **2013**, *33*, 219–241. [CrossRef]
4. Lu, J.J.; Wang, H.Y.; Miao, Y.X.; Zhao, L.Q.; Zhao, G.M.; Cao, Q.; Kusnierek, K. Developing an active canopy sensor-based integrated precision rice management system for improving grain yield and quality, nitrogen use efficiency, and lodging resistance. *Remote Sens.* **2022**, *14*, 2440. [CrossRef]
5. Chang, J.F.; Havlik, P.; Leclere, D.; de Vries, W.; Valin, H.; Deppermann, A.; Hasegawa, T.; Obersteiner, M. Reconciling regional nitrogen boundaries with global food security. *Nat. Food* **2021**, *2*, 700–711. [CrossRef]
6. Ren, K.; Xu, M.; Li, R.; Zheng, L.; Liu, S.; Reis, S.; Wang, H.; Lu, C.; Zhang, W.; Gao, H.; et al. Optimizing nitrogen fertilizer use for more grain and less pollution. *J. Clean. Prod.* **2022**, *360*, 132180. [CrossRef]
7. Yang, G.J.; Zhao, C.J.; Li, Z.H. *Quantitative Remote Sensing of Crop Nitrogen Nutrition and Its Application;* Science Press: Beijing, China, 2019.
8. Andrews, M.; Raven, J.A.; Lea, P.J. Do plants need nitrate? The mechanisms by which nitrogen form affects plants. *Ann. Appl. Biol.* **2013**, *163*, 174–199. [CrossRef]
9. Li, D.; Tian, M.; Cai, J.; Jiang, D.; Cao, W.; Dai, T. Effects of low nitrogen supply on relationships between photosynthesis and nitrogen status at different leaf position in wheat seedlings. *Plant Growth Regul.* **2013**, *70*, 257–263. [CrossRef]
10. Wang, L.; Xue, C.; Pan, X.; Chen, F.; Liu, Y. Application of controlled-release urea enhances grain yield and nitrogen use efficiency in irrigated rice in the Yangtze River Basin, China. *Front. Plant Sci.* **2018**, *9*, 999. [CrossRef]
11. Fu, Y.; Yang, G.; Li, Z.; Song, X.; Li, Z.; Xu, X.; Wang, P.; Zhao, C. Winter wheat nitrogen status estimation using UAV-based RGB imagery and gaussian processes regression. *Remote Sens.* **2020**, *12*, 3778. [CrossRef]
12. Moharana, S.; Dutta, S. Spatial variability of chlorophyll and nitrogen content of rice from hyperspectral imagery. *ISPRS J. Photogramm. Remote Sens.* **2016**, *122*, 17–29. [CrossRef]
13. Thorp, K.R.; Wang, G.; Bronson, K.F.; Badaruddin, M.; Mon, J. Hyperspectral data mining to identify relevant canopy spectral features for estimating durum wheat growth, nitrogen status, and grain yield. *Comput. Electron. Agric.* **2017**, *136*, 1–12. [CrossRef]
14. Fu, Y.; Yang, G.; Pu, R.; Li, Z.; Li, H.; Xu, X.; Song, X.; Yang, X.; Zhao, C. An overview of crop nitrogen status assessment using hyperspectral remote sensing: Current status and perspectives. *Eur. J. Agron.* **2021**, *124*, 126241. [CrossRef]
15. Sun, J.; Shi, S.; Wang, L.; Li, H.; Wang, S.; Gong, W.; Tagesson, T. Optimizing LUT-based inversion of leaf chlorophyll from hyperspectral lidar data: Role of cost functions and regulation strategies. *Int. J. Appl. Earth Obs. Geoinf.* **2021**, *105*, 102602. [CrossRef]
16. Vigneau, N.; Ecarnot, M.; Rabatel, G.; Roumet, P. Potential of field hyperspectral imaging as a non destructive method to assess leaf nitrogen content in Wheat. *Field Crops Res.* **2011**, *122*, 25–31. [CrossRef]
17. Shao, Y.; Zhao, C.; Bao, Y.; He, Y. Quantification of nitrogen status in rice by least squares support vector machines and reflectance spectroscopy. *Food Bioprocess Technol.* **2012**, *5*, 100–107. [CrossRef]
18. Li, S.; Ji, W.; Chen, S.; Peng, J.; Zhou, Y.; Shi, Z. Potential of VIS-NIR-SWIR spectroscopy from the Chinese soil spectral library for assessment of nitrogen fertilization rates in the paddy-rice region, China. *Remote Sens.* **2015**, *7*, 7029–7043. [CrossRef]
19. Maes, W.H.; Steppe, K. Perspectives for remote sensing with unmanned aerial vehicles in precision agriculture. *Trends Plant Sci.* **2019**, *24*, 152–164. [CrossRef]
20. Inoue, Y.; Sakaiya, E.; Zhu, Y.; Takahashi, W. Diagnostic mapping of canopy nitrogen content in rice based on hyperspectral measurements. *Remote Sens. Environ.* **2012**, *126*, 210–221. [CrossRef]
21. Kattenborn, T.; Schiefer, F.; Zarco-Tejada, P.; Schmidtlein, S. Advantages of retrieving pigment content [μg/cm2] versus concentration [%] from canopy reflectance. *Remote Sens. Environ.* **2019**, *230*, 111195. [CrossRef]
22. Cai, C.; Li, G.; Di, L.; Ding, Y.; Fu, L.; Guo, X.; Struik, P.C.; Pan, G.; Li, H.; Chen, W.; et al. The acclimation of leaf photosynthesis of wheat and rice to seasonal temperature changes in T-FACE environments. *Glob. Chang. Biol.* **2020**, *26*, 539–556. [CrossRef] [PubMed]

23. Boogaard, H.; Van Diepen, C.; Rotter, R.; Cabrera, J.; Van Laar, H. *WOFOST 7.1: User's Guide for the WOFOST 7.1 Crop Growth Simulation Model and WOFOST Control Center 1.5*; Wageningen University & Research: Wageningen, The Netherlands, 1998.
24. Li, D.; Chen, J.M.; Yan, Y.; Zheng, H.; Yao, X.; Zhu, Y.; Cao, W.; Cheng, T. Estimating leaf nitrogen content by coupling a nitrogen allocation model with canopy reflectance. *Remote Sens. Environ.* **2022**, *283*, 113314. [CrossRef]
25. Ohyama, T. Nitrogen as a major essential element of plants. *Nitrogen Assim. Plants* **2010**, *37*, 2–17.
26. Wang, Y.-P.; Chang, Y.-C.; Shen, Y. Estimation of nitrogen status of paddy rice at vegetative phase using unmanned aerial vehicle based multispectral imagery. *Precis. Agric.* **2022**, *23*, 1–17. [CrossRef]
27. Li, D.; Wang, X.; Zheng, H.; Zhou, K.; Yao, X.; Tian, Y.; Zhu, Y.; Cao, W.; Cheng, T. Estimation of area- and mass-based leaf nitrogen contents of wheat and rice crops from water-removed spectra using continuous wavelet analysis. *Plant Methods* **2018**, *14*, 76. [CrossRef]
28. Berger, K.; Verrelst, J.; Feret, J.-B.; Wang, Z.; Wocher, M.; Strathmann, M.; Danner, M.; Mauser, W.; Hank, T. Crop nitrogen monitoring: Recent progress and principal developments in the context of imaging spectroscopy missions. *Remote Sens. Environ.* **2020**, *242*, 111758. [CrossRef]
29. Botha, E.J.; Leblon, B.; Zebarth, B.J.; Watmough, J. Non-destructive estimation of wheat leaf chlorophyll content from hyperspectral measurements through analytical model inversion. *Int. J. Remote Sens.* **2010**, *31*, 1679–1697. [CrossRef]
30. Feret, J.-B.; Berger, K.; de Boissieu, F.; Malenovsky, Z. PROSPECT-PRO for estimating content of nitrogen-containing leaf proteins and other carbon-based constituents. *Remote Sens. Environ.* **2021**, *252*, 112173. [CrossRef]
31. Berger, K.; Verrelst, J.; Feret, J.-B.; Hank, T.; Wocher, M.; Mauser, W.; Camps-Valls, G. Retrieval of aboveground crop nitrogen content with a hybrid machine learning method. *Int. J. Appl. Earth Obs. Geoinf.* **2020**, *92*, 102174. [CrossRef]
32. Zhu, Y.; Yao, X.; Tian, Y.; Liu, X.; Cao, W. Analysis of common canopy vegetation indices for indicating leaf nitrogen accumulations in wheat and rice. *Int. J. Appl. Earth Obs. Geoinf.* **2008**, *10*, 1–10. [CrossRef]
33. Fitzgerald, G.J.; Rodriguez, D.; Christensen, L.K.; Belford, R.; Sadras, V.O.; Clarke, T.R. Spectral and thermal sensing for nitrogen and water status in rainfed and irrigated wheat environments. *Precis. Agric.* **2006**, *7*, 233–248. [CrossRef]
34. Zhu, Y.; Tian, Y.; Yao, X.; Liu, X.; Cao, W. Analysis of common canopy reflectance spectra for indicating leaf nitrogen concentrations in wheat and rice. *Plant Prod. Sci.* **2007**, *10*, 400–411. [CrossRef]
35. Hua, Q.; Yu, Y.; Dong, S.; Li, S.; Shen, H.; Han, Y.; Zhang, J.; Xiao, J.; Liu, S.; Dong, Q.; et al. Leaf spectral responses of Poa crymophila to nitrogen deposition and climate change on Qinghai-Tibetan Plateau. *Agric. Ecosyst. Environ.* **2019**, *284*, 106598. [CrossRef]
36. Tang, Y.L.; Wang, R.C.; Huang, J.F. Relations between red edge characteristics and agronomic parameters of crops. *Pedosphere* **2004**, *14*, 467–474.
37. Zha, H.; Miao, Y.; Wang, T.; Li, Y.; Zhang, J.; Sun, W.; Feng, Z.; Kusnierek, K. Improving unmanned aerial vehicle remote sensing-based rice nitrogen nutrition index prediction with machine learning. *Remote Sens.* **2020**, *12*, 215. [CrossRef]
38. Wang, W.; Yao, X.; Tian, Y.-C.; Liu, X.-J.; Ni, J.; Cao, W.-X.; Zhu, Y. Common spectral bands and optimum vegetation indices for monitoring leaf nitrogen accumulation in rice and wheat. *J. Integr. Agric.* **2012**, *11*, 2001–2012. [CrossRef]
39. Nguyen, H.T.; Lee, B.W. Assessment of rice leaf growth and nitrogen status by hyperspectral canopy reflectance and partial least square regression. *Eur. J. Agron.* **2006**, *24*, 349–356. [CrossRef]
40. He, J.; Ma, J.; Cao, Q.; Wang, X.; Yao, X.; Cheng, T.; Zhu, Y.; Cao, W.; Tian, Y. Development of critical nitrogen dilution curves for different leaf layers within the rice canopy. *Eur. J. Agron.* **2022**, *132*, 126414. [CrossRef]
41. Li, H.; Zhao, C.; Huang, W.; Yang, G. Non-uniform vertical nitrogen distribution within plant canopy and its estimation by remote sensing: A review. *Field Crops Res.* **2013**, *142*, 75–84. [CrossRef]
42. Wang, J.H.; Wang, Z.J.; Huang, W.J.; Ma, Z.H.; Liu, L.Y.; Zhao, C.J. The vertical dist ribution characteristic and spectral response of canopy nitrogen in different layer of winter wheat. *Natl. Remote Sens. Bull.* **2004**, *8*, 309–316.
43. Duan, D.-D.; Zhao, C.-J.; Li, Z.-H.; Yang, G.-J.; Zhao, Y.; Qiao, X.-J.; Zhang, Y.-H.; Zhang, L.-X.; Yang, W.-D. Estimating total leaf nitrogen concentration in winter wheat by canopy hyperspectral data and nitrogen vertical distribution. *J. Integr. Agric.* **2019**, *18*, 1562–1570. [CrossRef]
44. He, J.; Zhang, X.; Guo, W.; Pan, Y.; Yao, X.; Cheng, T.; Zhu, Y.; Cao, W.; Tian, Y. Estimation of vertical leaf nitrogen distribution within a rice canopy based on hyperspectral data. *Front. Plant Sci.* **2020**, *10*, 1802. [CrossRef] [PubMed]
45. Camino, C.; Gonzalez-Dugo, V.; Hernandez, P.; Sillero, J.C.; Zarco-Tejada, P.J. Improved nitrogen retrievals with airborne-derived fluorescence and plant traits quantified from VNIR-SWIR hyperspectral imagery in the context of precision agriculture. *Int. J. Appl. Earth Obs. Geoinf.* **2018**, *70*, 105–117. [CrossRef]
46. Aranguren, M.; Castellon, A.; Aizpurua, A. Crop sensor based non-destructive estimation of nitrogen nutritional status, yield, and grain protein content in wheat. *Agriculture* **2020**, *10*, 148. [CrossRef]
47. Staenz, K.; Secker, J.; Gao, B.C.; Davis, C.; Nadeau, C. Radiative transfer codes applied to hyperspectral data for the retrieval of surface reflectance. *ISPRS J. Photogramm. Remote Sens.* **2002**, *57*, 194–203. [CrossRef]
48. Peron-Danaher, R.; Russell, B.; Cotrozzi, L.; Mohammadi, M.; Couture, J. Incorporating multi-scale, spectrally detected nitrogen concentrations into assessing nitrogen use efficiency for winter wheat breeding populations. *Remote Sens.* **2021**, *13*, 3991. [CrossRef]

49. Jiang, X.; Zhen, J.; Miao, J.; Zhao, D.; Shen, Z.; Jiang, J.; Gao, C.; Wu, G.; Wang, J. Newly-developed three-band hyperspectral vegetation index for estimating leaf relative chlorophyll content of mangrove under different severities of pest and disease. *Ecol. Indic.* **2022**, *140*, 108978. [CrossRef]
50. Xiong, Y.; Liu, B.; Yue, Y. Inversion of nitrogen content of plant leaves based on ASD and FISS. *Ecol. Environ. Sci.* **2013**, *22*, 582–587.
51. Jiang, J.; Zhang, Z.; Cao, Q.; Liang, Y.; Krienke, B.; Tian, Y.; Zhu, Y.; Cao, W.; Liu, X. Use of an active canopy sensor mounted on an unmanned aerial vehicle to monitor the growth and nitrogen status of winter wheat. *Remote Sens.* **2020**, *12*, 3684. [CrossRef]
52. Wang, J.; Xu, B.; Wang, C.; Yang, G.; Yang, Z.; Mei, X.; Yang, X. Design and application of data acquisition and analysis system for CropSense. *Smart Agric.* **2019**, *1*, 91–104.
53. He, L.; Song, X.; Feng, W.; Guo, B.B.; Zhang, Y.S.; Wang, Y.H.; Wang, C.Y.; Guo, T.C. Improved remote sensing of leaf nitrogen concentration in winter wheat using multi-angular hyperspectral data. *Remote Sens. Environ.* **2016**, *174*, 122–133. [CrossRef]
54. Wu, B.; Huang, W.; Ye, H.; Luo, P.; Ren, Y.; Kong, W. Using multi-angular hyperspectral data to estimate the vertical distribution of leaf chlorophyll content in wheat. *Remote Sens.* **2021**, *13*, 1501. [CrossRef]
55. Liao, Q.; Zhang, D.; Wang, J.; Yang, G.; Yang, H.; Craig, C.; Wong, Z.; Wang, D. Assessment of chlorophyli content using a new vegetation index based on multi-angular hyperspectral image data. *Spectrosc. Spectr. Anal.* **2014**, *34*, 1599–1604.
56. Sun, J.; Shi, S.; Gong, W.; Yang, J.; Du, L.; Song, S.; Chen, B.; Zhang, Z. Evaluation of hyperspectral LiDAR for monitoring rice leaf nitrogen by comparison with multispectral LiDAR and passive spectrometer. *Sci. Rep.* **2017**, *7*, 40362. [CrossRef]
57. Huang, S.; Miao, Y.; Yuan, F.; Cao, Q.; Ye, H.; Lenz-Wiedemann, V.I.S.; Bareth, G. In-season diagnosis of rice nitrogen status using proximal fluorescence canopy sensor at different growth stages. *Remote Sens.* **2019**, *11*, 1847. [CrossRef]
58. Du, L.; Gong, W.; Yang, J. Application of spectral indices and reflectance spectrum on leaf nitrogen content analysis derived from hyperspectral LiDAR data. *Opt. Laser Technol.* **2018**, *107*, 372–379. [CrossRef]
59. Pancorbo, J.L.; Camino, C.; Alonso-Ayuso, M.; Raya-Sereno, M.D.; Gonzalez-Fernandez, I.; Gabriel, J.L.; Zarco-Tejada, P.J.; Quemada, M. Simultaneous assessment of nitrogen and water status in winter wheat using hyperspectral and thermal sensors. *Eur. J. Agron.* **2021**, *127*, 126287. [CrossRef]
60. Zhang, J.; Qiu, X.; Wu, Y.; Zhu, Y.; Cao, Q.; Liu, X.; Cao, W. Combining texture, color, and vegetation indices from fixed-wing UAS imagery to estimate wheat growth parameters using multivariate regression methods. *Comput. Electron. Agric.* **2021**, *185*, 106138. [CrossRef]
61. Oppelt, N.; Mauser, W. Hyperspectral monitoring of physiological parameters of wheat during a vegetation period using AVIS data. *Int. J. Remote Sens.* **2004**, *25*, 145–159. [CrossRef]
62. Lu, B.; Dao, P.D.; Liu, J.; He, Y.; Shang, J. Recent advances of hyperspectral imaging technology and applications in agriculture. *Remote Sens.* **2020**, *12*, 2659. [CrossRef]
63. Zhu, H.; Liu, H.; Xu, Y.; Yang, G. UAV-based hyperspectral analysis and spectral indices constructing for quantitatively monitoring leaf nitrogen content of winter wheat. *Appl. Opt.* **2018**, *57*, 7722–7732. [CrossRef] [PubMed]
64. Wang, L.; Chen, S.; Li, D.; Wang, C.; Jiang, H.; Zheng, Q.; Peng, Z. Estimation of paddy rice nitrogen content and accumulation both at leaf and plant levels from UAV hyperspectral imagery. *Remote Sens.* **2021**, *13*, 2956. [CrossRef]
65. Yi, X.; Lan, A.; Wen, X.; Zhang, Y.; Li, Y. Monitoring of heavy metals in farmland soils based on ASD and GaiaSky-mini. *Chin. J. Ecol.* **2018**, *37*, 1781–1788.
66. Wan, L.; Cen, H.; Zhu, J.; Zhang, J.; Du, X.; He, Y. Using fusion of texture features and vegetation indices from water concentration in rice crop to UAV remote sensing monitor. *Smart Agric.* **2020**, *2*, 58–67.
67. Zhu, J.; Cen, H.; He, L.; He, Y. Development and performance evaluation of a multi-rotor unmanned aircraft system for agricultural monitoring. *Smart Agric.* **2019**, *1*, 43–52.
68. Delloye, C.; Weiss, M.; Defourny, P. Retrieval of the canopy chlorophyll content from Sentinel-2 spectral bands to estimate nitrogen uptake in intensive winter wheat cropping systems. *Remote Sens. Environ.* **2018**, *216*, 245–261. [CrossRef]
69. Zhao, H.; Song, X.; Yang, G.; Li, Z.; Zhang, D.; Feng, H. Monitoring of nitrogen and grain protein content in winter wheat based on Sentinel-2A data. *Remote Sens.* **2019**, *11*, 1724. [CrossRef]
70. Houborg, R.; Cescatti, A.; Migliavacca, M.; Kustas, W.P. Satellite retrievals of leaf chlorophyll and photosynthetic capacity for improved modeling of GPP. *Agric. For. Meteorol.* **2013**, *177*, 10–23. [CrossRef]
71. Hunt, E.R., Jr.; Doraiswamy, P.C.; McMurtrey, J.E.; Daughtry, C.S.T.; Perry, E.M.; Akhmedov, B. A visible band index for remote sensing leaf chlorophyll content at the canopy scale. *Int. J. Appl. Earth Obs. Geoinf.* **2013**, *21*, 103–112. [CrossRef]
72. Huang, S.; Miao, Y.; Yuan, F.; Gnyp, M.L.; Yao, Y.; Cao, Q.; Wang, H.; Lenz-Wiedemann, V.I.S.; Bareth, G. Potential of RapidEye and WorldView-2 satellite data for improving rice nitrogen status monitoring at different growth stages. *Remote Sens.* **2017**, *9*, 227. [CrossRef]
73. Magney, T.S.; Eitel, J.U.H.; Vierling, L.A. Mapping wheat nitrogen uptake from RapidEye vegetation indices. *Precis. Agric.* **2017**, *18*, 429–451. [CrossRef]
74. Cui, B.; Zhao, Q.-J.; Huang, W.-J.; Song, X.-Y.; Ye, H.-C.; Zhou, X.-F. Leaf chlorophyll content retrieval of wheat by simulated RapidEye, Sentinel-2 and EnMAP data. *J. Integr. Agric.* **2019**, *18*, 1230–1245. [CrossRef]
75. Croft, H.; Arabian, J.; Chen, J.M.; Shang, J.; Liu, J. Mapping within-field leaf chlorophyll content in agricultural crops for nitrogen management using Landsat-8 imagery. *Precis. Agric.* **2020**, *21*, 856–880. [CrossRef]
76. Yao, X.; Liu, X.J.; Tian, Y.C.; Cao, W.X.; Zhu, Y.; Zhang, Y. Quantitative relationships between satellite channels-based spectral parameters and wheat canopy leaf nitrogen status. *Chin. J. Appl. Ecol.* **2013**, *24*, 431–437.

77. Tan, C.; Zhou, J.; Luo, M.; Du, Y.; Yang, X.; Ma, C. Using combined vegetation indices to monitor leaf chlorophyll content in winter wheat based on Hj-1a/1b images. *Int. J. Agric. Biol.* **2017**, *19*, 1576–1584.
78. Jiang, X.; Fang, S.; Huang, X.; Liu, Y.; Guo, L. Rice mapping and growth monitoring based on time series GF-6 images and red-edge bands. *Remote Sens.* **2021**, *13*, 579. [CrossRef]
79. Kang, Y.; Meng, Q.; Liu, M.; Zou, Y.; Wang, X. Crop classification based on red edge features analysis of GF-6 WFV data. *Sensors* **2021**, *21*, 4328. [CrossRef]
80. Xia, T.; He, Z.; Cai, Z.; Wang, C.; Wang, W.; Wang, J.; Hu, Q.; Song, Q. Exploring the potential of Chinese GF-6 images for crop mapping in regions with complex agricultural landscapes. *Int. J. Appl. Earth Obs. Geoinf.* **2022**, *107*, 102702. [CrossRef]
81. Wang, C.; Zhang, X.; Shi, T.; Zhang, C.; Li, M. Classification of medicinal plants astragalus mongholicus bunge and sophora flavescens aiton using GaoFen-6 and multitemporal Sentinel-2 data. *IEEE Geosci. Remote Sens. Lett.* **2022**, *19*, 2502805. [CrossRef]
82. Chen, P.; Liang, F. Cotton nitrogen nutrition diagnosis based on spectrum and texture feature of images from low altitude unmanned aerial vehicle. *Sci. Agric. Sin.* **2019**, *52*, 2220–2229, (In Chinese with English Abstract).
83. Yu, Z.; Wang, J.I.; Chen, L.I.; Fu, Y.U.; Zhu, H.O.; Feng, H.A.; Xu, X.I.; Li, Z.H. An entirely new approach based on remote sensing data to calculate the nitrogen nutrition index of winter wheat. *J. Integr. Agric.* **2021**, *20*, 2535–2551.
84. Justes, E.; Mary, B.; Meynard, J.M.; Machet, J.M.; Thelier-Huche, L. Determination of a critical nitrogen dilution curve for winter wheat crops. *Ann. Bot.* **1994**, *74*, 397–407. [CrossRef]
85. Zheng, H.; Ma, J.; Zhou, M.; Li, D.; Yao, X.; Cao, W.; Zhu, Y.; Cheng, T. Enhancing the nitrogen signals of rice canopies across critical growth stages through the integration of textural and spectral information from Unmanned Aerial Vehicle (UAV) multispectral imagery. *Remote Sens.* **2020**, *12*, 957. [CrossRef]
86. Li, H.; Li, D.; Xu, K.; Cao, W.; Jiang, X.; Ni, J. Monitoring of nitrogen indices in wheat leaves based on the integration of spectral and canopy structure information. *Agronomy* **2022**, *12*, 833. [CrossRef]
87. Zhou, K.; Cheng, T.; Zhu, Y.; Cao, W.; Ustin, S.L.; Zheng, H.; Yao, X.; Tian, Y. Assessing the impact of spatial resolution on the estimation of leaf nitrogen concentration over the full season of paddy rice using near-surface imaging spectroscopy data. *Front. Plant Sci.* **2018**, *9*, 964. [CrossRef]
88. Zheng, H.; Cheng, T.; Li, D.; Zhou, X.; Yao, X.; Tian, Y.; Cao, W.; Zhu, Y. Evaluation of RGB, color-infrared and multispectral images acquired from unmanned aerial systems for the estimation of nitrogen accumulation in rice. *Remote Sens.* **2018**, *10*, 824. [CrossRef]
89. Hansen, P.M.; Schjoerring, J.K. Reflectance measurement of canopy biomass and nitrogen status in wheat crops using normalized difference vegetation indices and partial least squares regression. *Remote Sens. Environ.* **2003**, *86*, 542–553. [CrossRef]
90. Dunn, B.W.; Dehaan, R.; Schmidtke, L.M.; Dunn, T.S.; Meder, R. Using field-derived hyperspectral reflectance measurement to identify the essential wavelengths for predicting nitrogen uptake of rice at panicle initiation. *J. Near Infrared Spectrosc.* **2016**, *24*, 473–483. [CrossRef]
91. Feng, M.-C.; Zhao, J.-J.; Yang, W.-D.; Wang, C.; Zhang, M.-J.; Xiao, L.-J.; Ding, G.-W. Evaluating winter wheat (*Triticum aestivum* L.) nitrogen status using canopy spectrum reflectance and multiple statistical analysis. *Spectrosc. Lett.* **2016**, *49*, 507–513. [CrossRef]
92. Tan, C.; Du, Y.; Zhou, J.; Wang, D.; Luo, M.; Zhang, Y.; Guo, W. Analysis of different hyperspectral variables for diagnosing leaf nitrogen accumulation in wheat. *Front. Plant Sci.* **2018**, *9*, 674. [CrossRef]
93. Wang, J.-J.; Song, X.-Y.; Mei, X.; Yang, G.-J.; Li, Z.-H.; Li, H.-L.; Meng, Y. Sensitive bands selection and nitrogen content monitoring of rice based on gaussian regression analysis. *Spectrosc. Spectr. Anal.* **2021**, *41*, 1722–1729.
94. Yu, F.; Feng, S.; Du, W.; Wang, D.; Guo, Z.; Xing, S.; Jin, Z.; Cao, Y.; Xu, T. A study of nitrogen deficiency inversion in rice leaves based on the hyperspectral reflectance differential. *Front. Plant Sci.* **2020**, *11*, 573272. [CrossRef] [PubMed]
95. Liu, T.; Xu, T.; Yu, F.; Yuan, Q.; Guo, Z.; Xu, B. A method combining ELM and PLSR (ELM-P) for estimating chlorophyll content in rice with feature bands extracted by an improved ant colony optimization algorithm. *Comput. Electron. Agric.* **2021**, *186*, 106177. [CrossRef]
96. Yang, B.; Chen, J.; Chen, L.; Cao, W.; Yao, X.; Zhu, Y. Estimation model of wheat canopy nitrogen content based on sensitive bands. *Trans. Chin. Soc. Agric. Eng.* **2015**, *31*, 176–182.
97. Cao, Y.L.; Xiao, W.; Liu, Y.D.; Jiang, K.L.; Guo, B.Y.; Yu, F.H. Dimension reduction of hyperspectral data and analysis of rice nitrogen content. *J. Shenyang Agric. Univ.* **2021**, *52*, 109–115.
98. Liu, H.; Zhu, H.; Wang, P. Quantitative modelling for leaf nitrogen content of winter wheat using UAV-based hyperspectral data. *Int. J. Remote Sens.* **2017**, *38*, 2117–2134. [CrossRef]
99. Lu, J.; Li, W.; Yu, M.; Zhang, X.; Ma, Y.; Su, X.; Yao, X.; Cheng, T.; Zhu, Y.; Cao, W.; et al. Estimation of rice plant potassium accumulation based on non-negative matrix factorization using hyperspectral reflectance. *Precis. Agric.* **2021**, *22*, 51–74. [CrossRef]
100. Xu, X.G.; Zhao, C.J.; Wang, J.H.; Li, C.J.; Yang, X.D. Associating new spectral features from visible and near infrared regions with optimal combination principle to monitor leaf nitrogen concentration in barley. *J. Infrared Millim. Waves* **2013**, *32*, 351–358+365. [CrossRef]
101. An, G.; Xing, M.; He, B.; Liao, C.; Huang, X.; Shang, J.; Kang, H. Using machine learning for estimating rice chlorophyll content from in situ hyperspectral data. *Remote Sens.* **2020**, *12*, 3104. [CrossRef]
102. Boochs, F.; Kupfer, G.; Dockter, K.; Kühbauch, W. Shape of the red edge as vitality indicator for plants. *Remote Sens.* **1990**, *11*, 1741–1753. [CrossRef]

103. Horler, D.N.H.; Barber, J.; Barringer, A.R. Effects of heavy metals on the absorbance and reflectance spectra of plants. *Int. J. Remote Sens.* **1980**, *1*, 121–136. [CrossRef]
104. Feng, W.; Yao, X.; Zhu, Y.; Tian, Y.C.; Cao, W. Monitoring leaf nitrogen status with hyperspectral reflectance in wheat. *Eur. J. Agron.* **2008**, *28*, 394–404. [CrossRef]
105. Tian, Y.; Yao, X.; Yang, J.; Cao, W.; Zhu, Y. Extracting red edge position parameters from ground- and space-based hyperspectral data for estimation of canopy leaf nitrogen concentration in rice. *Plant Prod. Sci.* **2011**, *14*, 270–281. [CrossRef]
106. Li, D.; Cheng, T.; Zhou, K.; Zheng, H.; Yao, X.; Tian, Y.; Zhu, Y.; Cao, W. WREP: A wavelet-based technique for extracting the red edge position from reflectance spectra for estimating leaf and canopy chlorophyll contents of cereal crops. *ISPRS J. Photogramm. Remote Sens.* **2017**, *129*, 103–117. [CrossRef]
107. Guo, B.-B.; Qi, S.-L.; Heng, Y.-R.; Duan, J.-Z.; Zhang, H.-Y.; Wu, Y.-P.; Feng, W.; Xie, Y.-X.; Zhu, Y.-J. Remotely assessing leaf N uptake in winter wheat based on canopy hyperspectral red-edge absorption. *Eur. J. Agron.* **2017**, *82*, 113–124. [CrossRef]
108. Blackburn, G.A. Wavelet decomposition of hyperspectral data: A novel approach to quantifying pigment concentrations in vegetation. *Int. J. Remote Sens.* **2007**, *28*, 2831–2855. [CrossRef]
109. Guo, J.; Zhang, J.; Xiong, S.; Zhang, Z.; Wei, Q.; Zhang, W.; Feng, W.; Ma, X. Hyperspectral assessment of leaf nitrogen accumulation for winter wheat using different regression modeling. *Precis. Agric.* **2021**, *22*, 1634–1658. [CrossRef]
110. Liang, T.; Duan, B.; Luo, X.; Ma, Y.; Yuan, Z.; Zhu, R.; Peng, Y.; Gong, Y.; Fang, S.; Wu, X. Identification of high nitrogen use efficiency phenotype in rice (*Oryza sativa* L.) through entire growth duration by unmanned aerial vehicle multispectral imagery. *Front. Plant Sci.* **2021**, *12*, 740414. [CrossRef]
111. Cui, B.; Zhao, Q.; Huang, W.; Song, X.; Ye, H.; Zhou, X. A new integrated vegetation index for the estimation of winter wheat leaf chlorophyll content. *Remote Sens.* **2019**, *11*, 974. [CrossRef]
112. Zhu, Y.; Zhou, D.; Yao, X.; Tian, Y.; Cao, W. Quantitative relationships of leaf nitrogen status to canopy spectral reflectance in rice. *Aust. J. Agric. Res.* **2007**, *58*, 1077–1085. [CrossRef]
113. Wang, W.; Yao, X.; Yao, X.; Tian, Y.; Liu, X.; Ni, J.; Cao, W.; Zhu, Y. Estimating leaf nitrogen concentration with three-band vegetation indices in rice and wheat. *Field Crops Res.* **2012**, *129*, 90–98. [CrossRef]
114. Tian, Y.C.; Yao, X.; Yang, J.; Cao, W.X.; Hannaway, D.B.; Zhu, Y. Assessing newly developed and published vegetation indices for estimating rice leaf nitrogen concentration with ground- and space-based hyperspectral reflectance. *Field Crops Res.* **2011**, *120*, 299–310. [CrossRef]
115. Yao, Y.; Miao, Y.; Cao, Q.; Wang, H.; Gnyp, M.L.; Bareth, G.; Khosla, R.; Yang, W.; Liu, F.; Liu, C. In-season estimation of rice nitrogen status with an active crop canopy sensor. *IEEE J. Sel. Top. Appl. Earth Obs. Remote Sens.* **2014**, *7*, 4403–4413. [CrossRef]
116. Wang, W.; Wu, Y.; Zhang, Q.; Zheng, H.; Yao, X.; Zhu, Y.; Cao, W.; Cheng, T. AAVI: A novel approach to estimating leaf nitrogen concentration in rice from unmanned aerial vehicle multispectral imagery at early and middle growth stages. *IEEE J. Sel. Top. Appl. Earth Obs. Remote Sens.* **2021**, *14*, 6716–6728. [CrossRef]
117. Patel, M.K.; Ryu, D.; Western, A.W.; Suter, H.; Young, I.M. Which multispectral indices robustly measure canopy nitrogen across seasons: Lessons from an irrigated pasture crop. *Comput. Electron. Agric.* **2021**, *182*, 106000. [CrossRef]
118. Li, F.; Mistele, B.; Hu, Y.; Yue, X.; Yue, S.; Miao, Y.; Chen, X.; Cui, Z.; Meng, Q.; Schmidhalter, U. Remotely estimating aerial N status of phenologically differing winter wheat cultivars grown in contrasting climatic and geographic zones in China and Germany. *Field Crops Res.* **2012**, *138*, 21–32. [CrossRef]
119. Palka, M.; Manschadi, A.M.; Koppensteiner, L.; Neubauer, T.; Fitzgerald, G.J. Evaluating the performance of the CCCI-CNI index for estimating N status of winter wheat. *Eur. J. Agron.* **2021**, *130*, 126346. [CrossRef]
120. Duan, B.; Fang, S.; Zhu, R.; Wu, X.; Wang, S.; Gong, Y.; Peng, Y. Remote estimation of rice yield with Unmanned Aerial Vehicle (UAV) data and spectral mixture analysis. *Front. Plant Sci.* **2019**, *10*, 204. [CrossRef]
121. Schirrmann, M.; Giebel, A.; Gleiniger, F.; Pflanz, M.; Lentschke, J.; Dammer, K.-H. Monitoring agronomic parameters of winter wheat crops with low-cost UAV imagery. *Remote Sens.* **2016**, *8*, 706. [CrossRef]
122. Wang, Y.; Wang, D.; Shi, P.; Omasa, K. Estimating rice chlorophyll content and leaf nitrogen concentration with a digital still color camera under natural light. *Plant Methods* **2014**, *10*, 36. [CrossRef]
123. Zhang, J.; Cheng, T.; Shi, L.; Wang, W.; Niu, Z.; Guo, W.; Ma, X. Combining spectral and texture features of UAV hyperspectral images for leaf nitrogen content monitoring in winter wheat. *Int. J. Remote Sens.* **2022**, *43*, 2335–2356. [CrossRef]
124. Xu, K.; Zhang, J.; Li, H.; Cao, W.; Zhu, Y.; Jiang, X.; Ni, J. Spectrum- and RGB-D-based image fusion for the prediction of nitrogen accumulation in wheat. *Remote Sens.* **2020**, *12*, 4040. [CrossRef]
125. Bausch, W.C.; Duke, H.R. Remote sensing of plant nitrogen status in corn. *Trans. ASAE* **1996**, *39*, 1869–1875. [CrossRef]
126. Cho, M.A.; Skidmore, A.K. A new technique for extracting the red edge position from hyperspectral data: The linear extrapolation method. *Remote Sens. Environ.* **2006**, *101*, 181–193. [CrossRef]
127. Chen, P.; Haboudane, D.; Tremblay, N.; Wang, J.; Vigneault, P.; Li, B. New spectral indicator assessing the efficiency of crop nitrogen treatment in corn and wheat. *Remote Sens. Environ.* **2010**, *114*, 1987–1997. [CrossRef]
128. Feng, W.; Zhang, H.-Y.; Zhang, Y.-S.; Qi, S.-L.; Heng, Y.-R.; Guo, B.-B.; Ma, D.-Y.; Guo, T.-C. Remote detection of canopy leaf nitrogen concentration in winter wheat by using water resistance vegetation indices from in-situ hyperspectral data. *Field Crops Res.* **2016**, *198*, 238–246. [CrossRef]

129. Barnes, E.; Clarke, T.; Richards, S.; Colaizzi, P.; Haberland, J.; Kostrzewski, M.; Waller, P.; Choi, C.; Riley, E.; Thompson, T. Coincident Detection of Crop Water Stress, Nitrogen Status and Canopy Density Using Ground Based Multispectral Data. In Proceedings of the Fifth International Conference on Precision Agriculture, Bloomington, MN, USA, 16–19 July 2000; p. 6.
130. Daughtry, C.S.; Walthall, C.; Kim, M.; De Colstoun, E.B.; McMurtrey Iii, J.E. Estimating corn leaf chlorophyll concentration from leaf and canopy reflectance. *Remote Sens. Environ.* **2000**, *74*, 229–239. [CrossRef]
131. Eitel, J.U.H.; Long, D.S.; Gessler, P.E.; Hunt, E.R. Combined spectral index to improve ground-based estimates of nitrogen status in dryland wheat. *Agron. J.* **2008**, *100*, 1694–1702. [CrossRef]
132. Lelong, C.C.D.; Burger, P.; Jubelin, G.; Roux, B.; Labbe, S.; Baret, F. Assessment of unmanned aerial vehicles imagery for quantitative monitoring of wheat crop in small plots. *Sensors* **2008**, *8*, 3557–3585. [CrossRef]
133. Chen, Z.; Miao, Y.; Lu, J.; Zhou, L.; Li, Y.; Zhang, H.; Lou, W.; Zhang, Z.; Kusnierek, K.; Liu, C. In-season diagnosis of winter wheat nitrogen status in smallholder farmer fields across a village using unmanned aerial vehicle-based remote sensing. *Agronomy* **2019**, *9*, 619. [CrossRef]
134. Brinkhoff, J.; Dunn, B.W.; Robson, A.J.; Dunn, T.S.; Dehaan, R.L. Modeling mid-season rice nitrogen uptake using multispectral satellite data. *Remote Sens.* **2019**, *11*, 1837. [CrossRef]
135. Yang, J.; Gong, W.; Shi, S.; Du, L.; Sun, J.; Song, S.L. Estimation of nitrogen content based on fluorescence spectrum and principal component analysis in paddy rice. *Plant Soil Environ.* **2016**, *62*, 178–183. [CrossRef]
136. YI, Q.-X.; Huang, J.-F.; Wang, F.-M.; Wang, X.-Z.; Liu, Z.-Y. Monitoring rice nitrogen status using hyperspectral reflectance and artificial neural network. *Environ. Sci. Technol.* **2007**, *41*, 6770–6775.
137. Du, L.; Yang, J.; Sun, J.; Shi, S.; Gong, W. Leaf biochemistry parameters estimation of vegetation using the appropriate inversion strategy. *Front. Plant Sci.* **2020**, *11*, 533. [CrossRef] [PubMed]
138. Yang, J.; Du, L.; Gong, W.; Shi, S.; Sun, J. Estimating leaf nitrogen concentration based on the combination with fluorescence spectrum and first-derivative. *R. Soc. Open Sci.* **2020**, *7*, 191941. [CrossRef]
139. Yang, J.; Song, S.; Du, L.; Shi, S.; Gong, W.; Sun, J.; Chen, B. Analyzing the effect of fluorescence characteristics on leaf nitrogen concentration estimation. *Remote Sens.* **2018**, *10*, 1402. [CrossRef]
140. Yao, X.; Huang, Y.; Shang, G.; Zhou, C.; Cheng, T.; Tian, Y.; Cao, W.; Zhu, Y. Evaluation of six algorithms to monitor wheat leaf nitrogen concentration. *Remote Sens.* **2015**, *7*, 14939–14966. [CrossRef]
141. Wang, L.; Zhou, X.; Zhu, X.; Guo, W. Estimation of leaf nitrogen concentration in wheat using the MK-SVR algorithm and satellite remote sensing data. *Comput. Electron. Agric.* **2017**, *140*, 327–337. [CrossRef]
142. Liang, L.; Di, L.; Huang, T.; Wang, J.; Lin, L.; Wang, L.; Yang, M. Estimation of leaf nitrogen content in wheat using new hyperspectral indices and a random forest regression algorithm. *Remote Sens.* **2018**, *10*, 1940. [CrossRef]
143. Van Wittenberghe, S.; Verrelst, J.; Rivera, J.P.; Alonso, L.; Moreno, J.; Samson, R. Gaussian processes retrieval of leaf parameters from a multi-species reflectance, absorbance and fluorescence dataset. *J. Photochem. Photobiol. B-Biol.* **2014**, *134*, 37–48. [CrossRef]
144. Chlingaryan, A.; Sukkarieh, S.; Whelan, B. Machine learning approaches for crop yield prediction and nitrogen status estimation in precision agriculture: A review. *Comput. Electron. Agric.* **2018**, *151*, 61–69. [CrossRef]
145. Shah, S.H.; Angel, Y.; Houborg, R.; Ali, S.; McCabe, M.F. A random forest machine learning approach for the retrieval of leaf chlorophyll content in wheat. *Remote Sens.* **2019**, *11*, 920. [CrossRef]
146. Jacquemoud, S.; Baret, F. PROSPECT: A model of leaf optical properties spectra. *Remote Sens. Environ.* **1990**, *34*, 75–91. [CrossRef]
147. Feret, J.-B.; Francois, C.; Asner, G.P.; Gitelson, A.A.; Martin, R.E.; Bidel, L.P.R.; Ustin, S.L.; Le Maire, G.; Jacquemoud, S. PROSPECT-4 and 5: Advances in the leaf optical properties model separating photosynthetic pigments. *Remote Sens. Environ.* **2008**, *112*, 3030–3043. [CrossRef]
148. Jacquemoud, S.; Verhoef, W.; Baret, F.; Bacour, C.; Zarco-Tejada, P.J.; Asner, G.P.; Francois, C.; Ustin, S.L. PROSPECT plus SAIL models: A review of use for vegetation characterization. *Remote Sens. Environ.* **2009**, *113*, S56–S66. [CrossRef]
149. Verhoef, W. Light scattering by leaf layers with application to canopy reflectance modeling: The SAIL model. *Remote Sens. Environ.* **1984**, *16*, 125–141. [CrossRef]
150. Danner, M.; Berger, K.; Wocher, M.; Mauser, W.; Hank, T. Retrieval of biophysical crop variables from multi-angular canopy spectroscopy. *Remote Sens.* **2017**, *9*, 726. [CrossRef]
151. Sun, J.; Shi, S.; Yang, J.; Chen, B.; Gong, W.; Du, L.; Mao, F.; Song, S. Estimating leaf chlorophyll status using hyperspectral lidar measurements by PROSPECT model inversion. *Remote Sens. Environ.* **2018**, *212*, 1–7. [CrossRef]
152. Luo, Y.; Weng, E.; Wu, X.; Gao, C.; Zhou, X.; Zhang, L. Parameter identifiability, constraint, and equifinality in data assimilation with ecosystem models. *Ecol. Appl.* **2009**, *19*, 571–574. [CrossRef]
153. Combal, B.; Baret, F.; Weiss, M.; Trubuil, A.; Mace, D.; Pragnere, A.; Myneni, R.; Knyazikhin, Y.; Wang, L. Retrieval of canopy biophysical variables from bidirectional reflectance—Using prior information to solve the ill-posed inverse problem. *Remote Sens. Environ.* **2003**, *84*, 1–15. [CrossRef]
154. Thorp, K.R.; Wang, G.; West, A.L.; Moran, M.S.; Bronson, K.F.; White, J.W.; Mon, J. Estimating crop biophysical properties from remote sensing data by inverting linked radiative transfer and ecophysiological models. *Remote Sens. Environ.* **2012**, *124*, 224–233. [CrossRef]
155. Yang, G.; Zhao, C.; Pu, R.; Feng, H.; Li, Z.; Li, H.; Sun, C. Leaf nitrogen spectral reflectance model of winter wheat (*Triticum aestivum*) based on PROSPECT: Simulation and inversion. *J. Appl. Remote Sens.* **2015**, *9*, 095976. [CrossRef]

156. Li, Z.; Jin, X.; Yang, G.; Drummond, J.; Yang, H.; Clark, B.; Li, Z.; Zhao, C. Remote sensing of leaf and canopy nitrogen status in winter wheat (*Triticum aestivum* L.) based on N-PROSAIL model. *Remote Sens.* **2018**, *10*, 1463. [CrossRef]
157. Li, Z.; Li, Z.; Fairbairn, D.; Li, N.; Xu, B.; Feng, H.; Yang, G. Multi-LUTs method for canopy nitrogen density estimation in winter wheat by field and UAV hyperspectral. *Comput. Electron. Agric.* **2019**, *162*, 174–182. [CrossRef]
158. Verrelst, J.; Berger, K.; Rivera-Caicedo, J.P. Intelligent sampling for vegetation nitrogen mapping based on hybrid machine learning algorithms. *IEEE Geosci. Remote Sens. Lett.* **2021**, *18*, 2038–2042. [CrossRef]
159. Stagakis, S.; Markos, N.; Sykioti, O.; Kyparissis, A. Monitoring canopy biophysical and biochemical parameters in ecosystem scale using satellite hyperspectral imagery: An application on a Phlomis fruticosa Mediterranean ecosystem using multiangular CHRIS/PROBA observations. *Remote Sens. Environ.* **2010**, *114*, 977–994. [CrossRef]
160. Nagol, J.R.; Sexton, J.; Kim, D.-H.; Anand, A.; Morton, D.; Vermote, E.; Townshend, J.R. Bidirectional effects in Landsat reflectance estimates: Is there a problem to solve? *ISPRS J. Photogramm. Remote Sens.* **2015**, *103*, 129–135. [CrossRef]
161. Song, X.; Feng, W.; He, L.; Xu, D.; Zhang, H.-Y.; Li, X.; Wang, Z.-J.; Coburn, C.A.; Wang, C.-Y.; Guo, T.-C. Examining view angle effects on leaf N estimation in wheat using field reflectance spectroscopy. *ISPRS J. Photogramm. Remote Sens.* **2016**, *122*, 57–67. [CrossRef]
162. Lu, N.; Wang, W.; Zhang, Q.; Li, D.; Yao, X.; Tian, Y.; Zhu, Y.; Cao, W.; Baret, F.; Liu, S.; et al. Estimation of nitrogen nutrition status in winter wheat from unmanned aerial vehicle based multi-angular multispectral imagery. *Front. Plant Sci.* **2019**, *10*, 1601. [CrossRef]
163. Sun, T.; Fang, H.; Liu, W.; Ye, Y. Impact of water background on canopy reflectance anisotropy of a paddy rice field from multi-angle measurements. *Agric. For. Meteorol.* **2017**, *233*, 143–152. [CrossRef]
164. Huang, W.; Yang, Q.; Pu, R.; Yang, S. Estimation of nitrogen vertical distribution by bi-directional canopy reflectance in winter wheat. *Sensors* **2014**, *14*, 20347–20359. [CrossRef]
165. Ma, C.; Zhai, L.; Li, C.; Wang, Y. Hyperspectral estimation of nitrogen content in different leaf positions of wheat using machine learning models. *Appl. Sci.* **2022**, *12*, 7427. [CrossRef]
166. He, L.; Liu, M.-R.; Guo, Y.-L.; Wei, Y.-K.; Zhang, H.-Y.; Song, X.; Feng, W.; Guo, T.-C. Angular effect of algorithms for monitoring leaf nitrogen concentration of wheat using multi-angle remote sensing data. *Comput. Electron. Agric.* **2022**, *195*, 106815. [CrossRef]
167. Zhang, J.; Miao, Y.; Batchelor, W.D.; Lu, J.; Wang, H.; Kang, S. Improving high-latitude rice nitrogen management with the CERES-rice crop model. *Agronomy* **2018**, *8*, 263. [CrossRef]
168. Zhao, P.; Zhou, Y.; Li, F.; Ling, X.; Deng, N.; Peng, S.; Man, J. The adaptability of APSIM-wheat model in the middle and lower reaches of the Yangtze River Plain of China: A case study of winter wheat in Hubei Province. *Agronomy* **2020**, *10*, 981. [CrossRef]
169. Si, Z.; Zain, M.; Li, S.; Liu, J.; Liang, Y.; Gao, Y.; Duan, A. Optimizing nitrogen application for drip-irrigated winter wheat using the DSSAT-CERES-wheat model. *Agric. Water Manag.* **2021**, *244*, 106592. [CrossRef]
170. Cao, J.; Liu, X.J.; Tang, L.; Cao, W.X.; Zhu, Y. Model designing for suitable nitrogen index dynamics of rice and wheat. *Chin. J. Appl. Ecol.* **2010**, *21*, 359–364.
171. Zhang, X.; Friedl, M.A.; Schaaf, C.B. Global vegetation phenology from Moderate Resolution Imaging Spectroradiometer (MODIS): Evaluation of global patterns and comparison with in situ measurements. *J. Geophys. Res.-Biogeosci.* **2006**, *111*. [CrossRef]
172. Zheng, H.; Cheng, T.; Yao, X.; Deng, X.; Tian, Y.; Cao, W.; Zhu, Y. Detection of rice phenology through time series analysis of ground-based spectral index data. *Field Crops Res.* **2016**, *198*, 131–139. [CrossRef]
173. Zhang, M.; Zhu, D.; Su, W.; Huang, J.; Zhang, X.; Liu, Z. Harmonizing multi-source remote sensing images for summer corn growth monitoring. *Remote Sens.* **2019**, *11*, 1266. [CrossRef]
174. Shu, M.; Gu, X.; Zhou, L.; Xu, B.; Yang, G. Establishing NDRE dynamic models of winter wheat under multi-nitrogen rates based on a field spectral sensor. *Appl. Opt.* **2021**, *60*, 993–1002. [CrossRef] [PubMed]
175. Schlemmer, M.R.; Francis, D.D.; Shanahan, J.F.; Schepers, J.S. Remotely measuring chlorophyll content in corn leaves with differing nitrogen levels and relative water content. *Agron. J.* **2005**, *97*, 106–112. [CrossRef]
176. Kusnierek, K.; Korsaeth, A. Simultaneous identification of spring wheat nitrogen and water status using visible and near infrared spectra and powered partial least squares regression. *Comput. Electron. Agric.* **2015**, *117*, 200–213. [CrossRef]
177. Klem, K.; Zahora, J.; Zemek, F.; Trunda, P.; Tuma, I.; Novotna, K.; Hodanova, P.; Rapantova, B.; Hanus, J.; Vavrikova, J.; et al. Interactive effects of water deficit and nitrogen nutrition on winter wheat. Remote sensing methods for their detection. *Agric. Water Manag.* **2018**, *210*, 171–184. [CrossRef]
178. Berger, K.; Machwitz, M.; Kycko, M.; Kefauver, S.C.; Van Wittenberghe, S.; Gerhards, M.; Verrelst, J.; Atzberger, C.; van der Tol, C.; Damm, A.; et al. Multi-sensor spectral synergies for crop stress detection and monitoring in the optical domain: A review. *Remote Sens. Environ.* **2022**, *280*, 113198. [CrossRef]
179. Devadas, R.; Lamb, D.W.; Backhouse, D.; Simpfendorfer, S. Sequential application of hyperspectral indices for delineation of stripe rust infection and nitrogen deficiency in wheat. *Precis. Agric.* **2015**, *16*, 477–491. [CrossRef]
180. Raj, R.; Walker, J.P.; Pingale, R.; Banoth, B.N.; Jagarlapudi, A. Leaf nitrogen content estimation using top-of-canopy airborne hyperspectral data. *Int. J. Appl. Earth Obs. Geoinf.* **2021**, *104*, 102584. [CrossRef]
181. Wu, S.; Deng, L.; Guo, L.; Wu, Y. Wheat leaf area index prediction using data fusion based on high-resolution unmanned aerial vehicle imagery. *Plant Methods* **2022**, *18*, 68. [CrossRef]
182. Qiao, L.; Gao, D.; Zhang, J.; Li, M.; Sun, H.; Ma, J. Dynamic influence elimination and chlorophyll content diagnosis of maize using UAV spectral imagery. *Remote Sens.* **2020**, *12*, 2650. [CrossRef]

183. Xu, X.; Fan, L.; Li, Z.; Meng, Y.; Feng, H.; Yang, H.; Xu, B. Estimating leaf nitrogen content in corn based on information fusion of multiple-sensor imagery from UAV. *Remote Sens.* **2021**, *13*, 340. [CrossRef]
184. Yao, X.; Ren, H.; Cao, Z.; Tian, Y.; Cao, W.; Zhu, Y.; Cheng, T. Detecting leaf nitrogen content in wheat with canopy hyperspectrum under different soil backgrounds. *Int. J. Appl. Earth Obs. Geoinf.* **2014**, *32*, 114–124. [CrossRef]
185. Upreti, D.; Huang, W.; Kong, W.; Pascucci, S.; Pignatti, S.; Zhou, X.; Ye, H.; Casa, R. A comparison of hybrid machine learning algorithms for the retrieval of wheat biophysical variables from Sentinel-2. *Remote Sens.* **2019**, *11*, 481. [CrossRef]
186. Verger, A.; Baret, F.; Camacho, F. Optimal modalities for radiative transfer-neural network estimation of canopy biophysical characteristics: Evaluation over an agricultural area with CHRIS/PROBA observations. *Remote Sens. Environ.* **2011**, *115*, 415–426. [CrossRef]
187. Doktor, D.; Lausch, A.; Spengler, D.; Thurner, M. Extraction of plant physiological status from hyperspectral signatures using machine learning methods. *Remote Sens.* **2014**, *6*, 12247–12274. [CrossRef]

MDPI

Article

Estimating Agricultural Cropping Intensity Using a New Temporal Mixture Analysis Method from Time Series MODIS

Jianbin Tao [1], Xinyue Zhang [1], Yiqing Liu [2], Qiyue Jiang [1] and Yang Zhou [1,*]

[1] Key Laboratory for Geographical Process Analysis & Simulation of Hubei Province/School of Urban and Environmental Sciences, Central China Normal University, Wuhan 430079, China; taojb@mail.ccnu.edu.cn (J.T.); xyzhang@mails.ccnu.edu.cn (X.Z.); qyjiang99@mails.ccnu.edu.cn (Q.J.)

[2] Institute of Disaster Risk Science, Faculty of Geographical Sciences, Beijing Normal University, Beijing 100875, China; liuyiqing@mail.bnu.edu.cn

* Correspondence: yzhou2017@ccnu.edu.cn

Abstract: Agricultural cropping intensity plays an important role in evaluating the food security and the sustainable development of agriculture. The existing indicators measuring cropping intensity include cropping frequency and multiple cropping index. As a nominal measurement, cropping frequency classifies crop patterns into single-cropping and/or double-cropping and leads to information loss. Multiple cropping index is calculated on the basis of statistical data, ignoring the spatial heterogeneity within the administrative region. Neither of these indicators can meet the requirements of precision agriculture, and new methods for fine cropping intensity mapping are still lacking. Time series remote sensing data provide vegetation phenology information and reveal temporal development of vegetation, which can be used to facilitate the fine cropping intensity mapping. In this study, a new temporal mixture analysis method is introduced to estimate the abundance level cropping intensity from time series remote sensing data. By analyzing phenological characteristics of major land-cover types in time series vegetatiosacan indices, a novel feature space was constructed by using the selected PCA components, and three unique endmembers (double-cropping, natural vegetations and water bodies) were found. Then, a linear spectral mixture analysis model was applied to decompose mixed pixels by replacing spectral data with multi-temporal data. The spatio-temporal continuous, fine resolution, abundance level cropping intensity maps were produced for the North China Plain and the middle and lower reaches of the Yangtze River Valley. The experiments indicate a good result at both county and pixel level validation. The method of manually delineating endmembers can well balance the accuracy and efficiency. We also found the size of the study area has little effect on the unmixing accuracy. The results demonstrated that the proposed method can model cropping intensity finely at large scale and long temporal span, at the same time with high efficiency and ease of implementation.

Keywords: cropping intensity; temporal mixture analysis; endmember; unmixing; time series images

Citation: Tao, J.; Zhang, X.; Liu, Y.; Jiang, Q.; Zhou, Y. Estimating Agricultural Cropping Intensity Using a New Temporal Mixture Analysis Method from Time Series MODIS. *Remote Sens.* **2023**, *15*, 4712. https://doi.org/10.3390/rs15194712

Academic Editor: Georgios Mallinis

Received: 9 June 2023
Revised: 10 September 2023
Accepted: 11 September 2023
Published: 26 September 2023

1. Introduction

Agricultural production is the cornerstone for the survival and development of human beings [1]. Due to the impact of urbanization, development of the market economy, the international food trade and climate change, dramatic changes have taken place in agricultural land systems. The most famous examples are the conversion of croplands into built-up areas in the Yangtze River Delta and Pearl River Delta of China [2,3], the phenomenon of double-cropping transitioning to single-cropping in the middle reaches of the Yangtze River Valley [4], the non-grain use of croplands in China [5], the collapse of soybean planting in Northeast China [6,7], and the abandoned croplands in mountainous area in South China [8]. Furthermore, locust disasters [9], armed conflict [10], COVID-19 [11,12], etc., aggravated the uncertainty of food production. The world has encountered the most serious food crisis in the past 50 years [13]. On the other hand, continuous agricultural intensification in some regions has caused serious ecological and environmental problems, such as excessive consumption of water resources [14], land degradation [15], agricultural non-point source pollution [16] and greenhouse gas emissions [17]. Therefore, agricultural cropping intensity is an important input for evaluating the sustainable development of agriculture.

The existing methods of cropping intensity mapping include cropping frequency (CF) and the multiple cropping index (MCI). The calculation of MCI is mostly based on statistical data, ignoring the spatial heterogeneity within the administrative regions [18]. CF is a nominal measurement that divides crop patterns into single-cropping and double-cropping, lacking quantitative measurement of cropping intensity. Most existing methods for mapping CF are based on counting the number of peaks in coarse resolution remote sensing images such as the Moderate Resolution Imaging Spectroradiometer (MODIS) vegetation index profiles [19–23]. When a coarse resolution image is used to monitor cropping frequency, the mixed pixel problem is particularly serious due to the fragmented landscapes, small patch size of croplands and complicated crop patterns in Central and South China.

There has been some progress recently on the construction of new indicators or new methods for cropping intensity mapping. Time series high-resolution images such as those from Sentinel-2 are used to monitor cropping intensity, alleviating the problem of mixed pixels to some extent [24–26]. However, these methods are highly dependent on data availability and constrained by complicated data pre-processing. The authors used Bayesian network to obtain cropping intensity with interval measurement using time series MODIS data [27]. This method is highly dependent on training samples, and the accuracy of the model output depends on the quality of the samples. In any case, the above existing methods of cropping intensity mapping cannot meet the requirements of precision agriculture, and new methods for fine cropping intensity mapping suitable for Central and South China are still lacking.

Although continuous accumulated high-resolution remote sensing images are helpful for precise cropping intensity mapping, this convenience has only existed since the launch of Landsat 8 in 2013 and the launch of Sentinel-2 in 2015. Medium-resolution data over long time series can bridge the gap and push back observations to the year 2000 or even earlier. Time series remote sensing data contain seasonal variation of vegetation and abundant vegetation phenology information [28]. Intra-annual dense time series images carry information on multiple cropping of crops. However, this information is usually used for crop mapping [29,30], phenology monitoring [31], etc. Vegetation phenology information contained in time series remote sensing data is not yet fully explored for cropping intensity mapping. Dense time-series remote sensing images have enough repeated observation (MODIS MOD13Q1 has 23 observations a year) and make this work possible [32].

Mixed pixels are more common in coarse resolution images and require mixture analysis technology to decompose pixels to the abundance level. There are two mixture analysis techniques: spectral mixture analysis (SMA) and temporal mixture analysis (TMA). SMA methods are generally used to decompose spectral remote sensing data, among which linear spectral mixture analysis (LSMA) is more often used. Based on the principle of SMA, TMA is a method to decompose mixed pixels by replacing spectral data with multi-temporal data.

The selection of endmembers (including their number and types) is another important issue for unmixing. The two-endmember model is suitable for the decomposition of natural land-cover types, while the three-endmember model is suitable for the decomposition of land-cover types under human disturbance [33]. For built-up areas, the most commonly used endmember selection method is the vegetation–impervious surface–soil endmember model proposed by Ridd et al. [34]. For non-built-up areas, vegetation–soil–shadow (or dry vegetation) endmember model is generally used [35]. The endmember selection is usually scene-dependent when the TMA method is used: for example, the vegetation–impervious surface–soil model [36], forests–multiple cropping–single cropping–non-vegetation model [32], and grass–corn–winter wheat–alfalfa model [37]. With proper endmember design, large scale, long temporal spans and fine cropping intensity are possible.

To aim for large-scale, long-temporal-span and fine-resolution agricultural cropping intensity estimation, this study presents a new TMA method for estimating cropping intensity from historical archived time series coarse resolution remote sensing data. The method includes: (1) constructing the feature space, finding the unique endmembers to estimate the abundance level cropping intensity; (2) exploring the feasibility of manually delineating endmembers and the effectiveness of the method on different regions with varied completeness of endmember land-cover types. This work will be of great significance for fine cropping intensity mapping at large scale and long-time series.

2. Materials and Methods

2.1. Study Area and Data

2.1.1. Study Area

The study area was the North China Plain and the middle and lower reaches of the Yangtze River Valley (Figure 1). Major land-cover types were natural vegetation, croplands, water bodies and built-up areas. Forests and scrublands were synthesized to natural vegetation since they have similar phenology. Double-cropping and single-cropping coexist in the research area. The dominant crop patterns are single-cropping paddy rice, winter wheat-soybean, winter rapeseed-paddy rice, winter wheat-maize and double-cropping paddy rice.

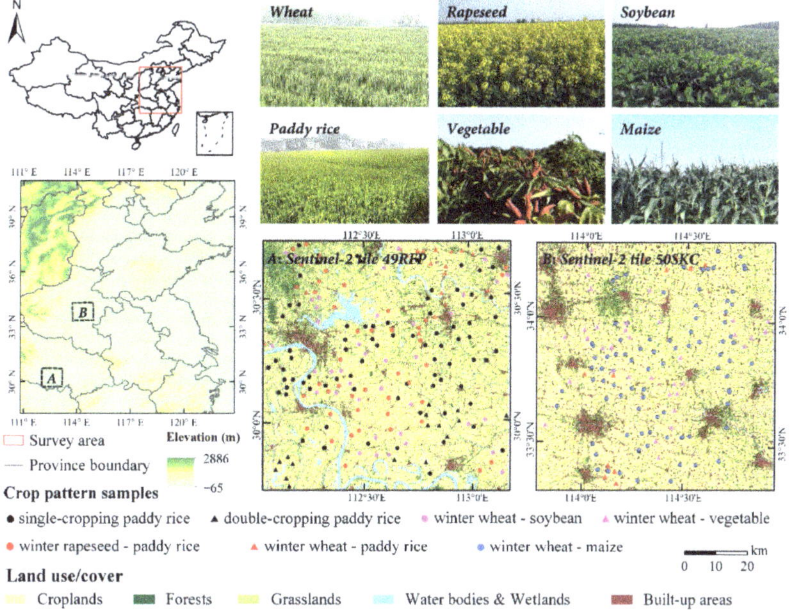

Figure 1. The location, major land-cover types and reference samples of the study area (A and B are two test regions, corresponding to Sentinel-2 tiles 49RFP and 50SKC). The main land cover types are from the FROM-GLC10 datasets (data source: http://data.ess.tsinghua.edu.cn/, accessed on 1 March 2023) in 2017.

2.1.2. Data

MODIS MOD13Q1 Enhanced Vegetation Index (EVI) products in 2018 with 250 m resolution were used and were synthesized over 16 days based on the maximum value composite principle. The image tiles including h27v05, h27v06, h28v05, and h28v06 (h: horizontal, v: vertical) were downloaded from the National Aeronautics and Space

Administration website (https://modis.gsfc.nasa.gov/, accessed on 1 March 2023). Image preprocessing, including mosaic, resampling and reprojection, was conducted using the MODIS Reprojection Tool. To eliminate the interference of clouds, snow, shadows and other factors, Savitzky–Golay filtering was adopted to reconstruct the original time series. Data quality information was extracted based on the pixel reliability layer of MOD13Q1 products.

Sentinel-2 data with a spatial resolution of 10 m were used to generate validation data. Sentinel-2A (launched 2015) and Sentinel-2B (launched 2017) sensors together offer 5-day revisit with global coverage [38]. NDVI was calculated for the Sentinel-2 data by using near infrared and red band.

Sentinel-2 NDVI composite from three key phenological phases was used to prepare the reference cropping intensity data. The first phase was in mid March, which represents the peak of the first growing season of double-cropping (termed GS1). The second phase was from late May to early June, which is the transition period between the two growing seasons (termed TGS). Both single-cropping and double-cropping had low EVI values at this stage. The third phase was in mid July, which is the peak of the second growing season of double-cropping (termed GS2). Double-cropping is shown in magenta, and single-cropping in blue in the false-color composite image (Figure 2a,b).

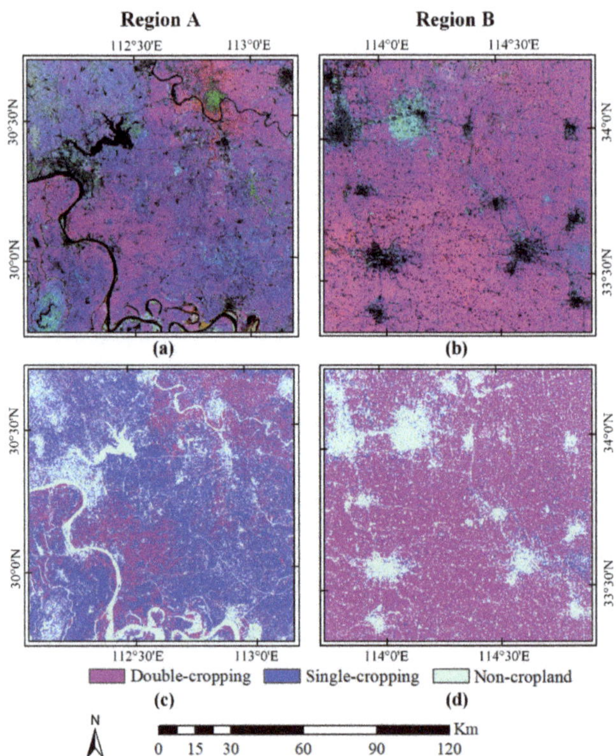

Figure 2. Unsupervised classification of Sentinel-2 images: (**a**,**b**) are false-color composites of Sentinel-2 NDVI from three phenological phases (GS1, TGS, GS2); (**c**,**d**) are ISODATA classification results for the two regions (A) and (B).

The classification was conducted based on these NDVI composites using ISODATA, in which the classification number was between 10 and 20, and the iteration time was set to 20. The classification results were merged into three categories: double-cropping, single-cropping and non-cropland (Figure 2c,d). Finally, they were aggregated to 250 m fractional images to match the spatial resolution of MODIS data.

The crop pattern sample data were used to evaluate the accuracy of the Sentinel-2 derived cropping intensity data. The producers' accuracies (PA) of single-cropping and double-cropping were 91.06% and 91.16%, respectively, their users' accuracies (UA) were 92.44% and 90.58%, respectively (Table 1), and the overall accuracy (OA) and Kappa coefficient were 92.6% and 0.888, respectively, suggesting the reliability of the Sentinel-2 derived cropping intensity data.

Table 1. Accuracies of Sentinel-2 derived cropping intensity data using crop pattern sample data.

Cropping Intensity	Single-Cropping	Double-Cropping	Other	Total	UA (%)
single-cropping	489	28	12	529	92.44
double-cropping	42	567	17	626	90.58
other	6	27	591	624	94.71
total	537	622	620	1779	
PA (%)	91.06	91.16	95.32		OA = 0.926 Kappa = 0.888

To evaluate the accuracy of the Sentinel-2 derived cropping intensity data, we used 1779 crop samples (537 single-cropping samples, 622 double-cropping samples and 620 other samples) as reference data. These crop patterns were transferred into cropping intensity: single cropping, double cropping and other (including non-crop cultivation and non-cropland). These samples were from a filed survey in 2018 as ground truth (Figure 1), and augmented by visual interpreting high-resolution images on the Google Earth Engine platform. The ground truth data were collected using a mobile application, GPSTool 4.0. The augmentation were delineated manually by overlaying ground truth data with high-resolution images. The sample data were roughly distributed evenly throughout the validation area and were independent and identically distributed.

In addition to the Sentinel-2 derived reference data, three cropping intensity products were further used for the validation. The first was the MCD12Q2 V6 Land Cover Dynamics product, which provides global estimates of the timing of vegetation phenology at 500 m resolution [39]. The NumCycles layer in MCD12Q2 provides the total number of valid vegetation cycles in a year. The annual average of the NumCycles was calculated and then used as reference data. The second product was the Global Cropping Intensity (GCI) dataset, which is an annual global multi-cropping index distribution map covering the period from 2001 to 2019 at 250 m resolution [40]. The third was a global cropping intensity map dataset at 30 m resolution (GCI30) from 2016 to 2018 [41].

2.2. Methods

The geometric method was used to implement the TMA model. The geometric method is an important method for mixture analysis, and from a geometric perspective, the multidimensional images can be viewed as a convex simplex. The convex geometry method [42] was introduced to construct the simplex structure of time series remote sensing data in the feature space, and the endmembers were found in this feature space.

The major steps of the methodology, including feature selection, feature space construction, endmember selection, cropping intensity estimation and validation, are presented in the flowchart (Figure 3).

2.2.1. Feature Selection

Intra-annual time series remote sensing images provide land surface phenological information and reveal spatio-temporal development of vegetation [43]. The EVI profiles of different land-cover types demonstrated the unique phenological characteristics of double-cropping, single-cropping, evergreen forests, deciduous forests, water bodies, built-up areas, etc. (Figure 4). The temporal profiles of crops are obviously different from those of other land-cover types. Moreover, there are great differences between single-cropping and double-cropping. There are two peaks for double-cropping, one is around DOY (day of year) 065 to 097, and the other is around DOY 193 to 225. However, there is only one peak for single-cropping, which is around DOY 177 to 209. Similarly, there is only one green cycle for forests, but the green cycle is much longer and wider compared to that of single-cropping. In addition, both water bodies and built-up areas have low EVI values throughout the year. These phenomena make recovering cropping intensity information from time series remote sensing data possible.

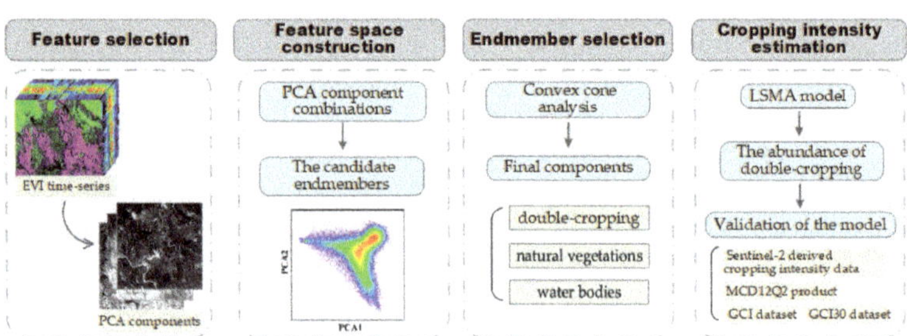

Figure 3. Flowchart of the TMA method.

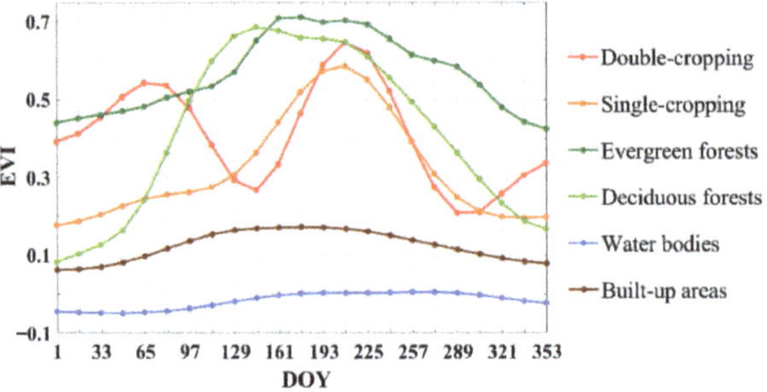

Figure 4. EVI profiles of major land-cover types in Hubei province. A total of 3000 samples were collected to generate EVI profiles, including 537 of double-cropping, 622 single-cropping, 486 evergreen forests, 432 deciduous forests, 517 water bodies and 406 built-up areas.

Principal component analysis (PCA) has been proven to be effective at detecting seasonal changes in vegetation when applied to the time series vegetation index [44]. PCA transformation projects original data into new k-dimensional components ordered by variance, with the majority of information provided by the first several components. PCA components have geographic meanings; for example, Henderson et al. [45] found the components corresponding closely to typical vegetation density or degree of seasonality, Wang et al. [46] found the components coinciding with the average normalized difference vegetation index (NDVI) (PCA component 1) and accounting for the most prominent man-induced vegetation alterations (PCA component 2).

PCA transformation was applied to the original time series MODIS EVI to obtain components with geographic meanings while reducing feature dimensions. In this study, the first three components reserved 93.18% of the original information. From PCA component 1 (PCA 1), vegetation and non-vegetation (water bodies and built-up areas) could be easily distinguished, and the difference between natural vegetation and croplands was also very significant (Figure 5a,b). PCA component 2 (PCA 2) could discriminate double-cropping from other land-cover types, and the difference between double-cropping and single-cropping was also very obvious (Figure 5c). Moreover, PCA component 3 (PCA 3) could discriminate single-cropping from other land-cover types (Figure 5d). Therefore, the combination of the first few components could be used to discriminate the major land-cover types.

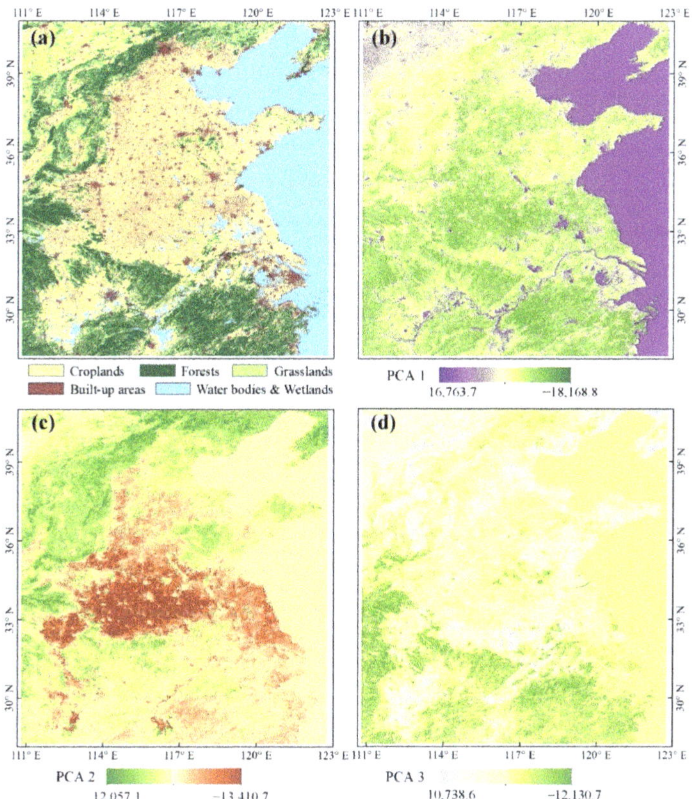

Figure 5. Land-cover types (**a**) and the first three PCA components of the time series EVI, (**b**) PCA 1, (**c**) PCA 2 and (**d**) PCA 3.

2.2.2. Feature Space Construction and Endmember Selection

The key for the unmixing is the selection of appropriate endmembers. The accuracy of unmixing was directly affected by the quality and quantity of the selected endmembers. A triangle is the simplest simplex when the convex geometry is introduced for the unmixing [47]. Convex cone analysis takes the boundary points of the convex cone constructed from the observed spectra as the endmembers [48].

Two PCA components were employed to implement a projection from image space to feature space and visually examine the distribution of the candidate endmembers in the feature space. The combinations were from any two of the first three components. Candidate endmember land-cover classes were sourced from the major land-cover types of the study area and include natural vegetation, water bodies, double-cropping, single-cropping and built-up areas. Since the purpose of the study was to map cropping intensity, the cropland was subdivided into single-cropping and double-cropping, while other land-cover types were used in broad categories.

Through observing these combinations, four candidate endmembers were obtained: double-cropping, single-cropping, natural vegetations and water bodies (Figure 6).

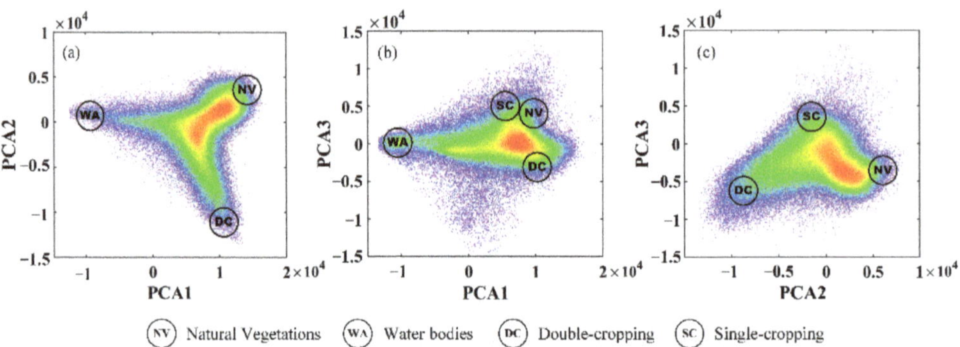

Figure 6. The feature space and the distributions of major land-cover types with different PCA component combinations, in which the gray dots are all pixels of the images. (**a**) PCA 1, PCA 2 combination, (**b**) PCA 1, PCA 3 combination, (**c**) PCA 2, PCA 3 combination.

Through visually examining the scatter points based on the N-dimensional visualization tool in ENVI, we found that the PCA 1 and PCA 3 combination and PCA 2 and PCA 3 combination should be excluded, and the PCA 1 and PCA 2 combination would be the best choice (Figure 6). The mixture of single-cropping and natural vegetation could explain the exclusion of the PCA 1 and PCA 3 combination. As for the exclusion of the PCA 2 and PCA 3 combination, the single-cropping could not be used as an endmember because its cropping intensity was about half that of double-cropping theoretically and could only be inside the boundary of the feature space. With the PCA 1 and PCA 2 combination, all pixels clustered closest to a triangle in the feature space, with three vertexes representing double-cropping, natural vegetation and water bodies, which could be the candidate endmembers (Figure 7).

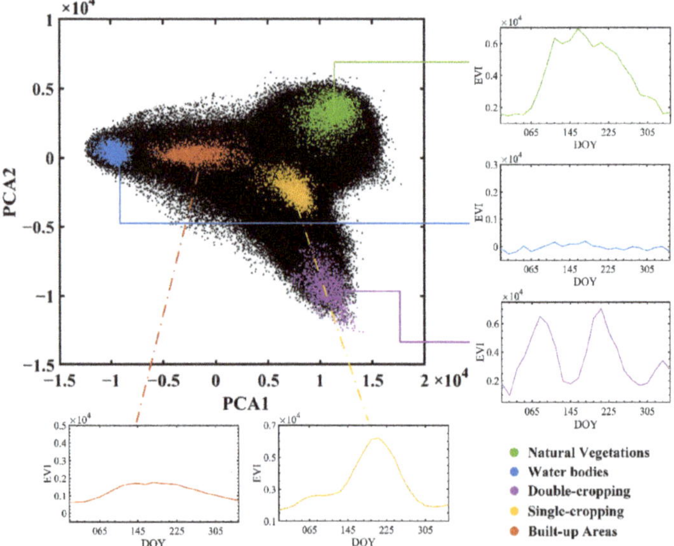

Figure 7. The triangular feature space, in which the black dots are all pixels of the images, and the colored dots are candidate endmember land-cover types. The EVI profiles of relevant land-cover types are also presented.

We also found that single-cropping was located in the transition zone from double-cropping to the center of the feature space, and built-up areas were located in the transition zone from water bodies to the center of the feature space. Double-cropping occupied one of the corners and had the highest cropping intensity; natural vegetation and water bodies were located in the other two corners and had zero cropping intensity; single-cropping was on the transition zone from double-cropping to the center of the feature space and had decreased cropping intensity. From these analyses, the unmixing based on this triangle feature space met the requirements of our research. Therefore, three endmembers were selected finally: double-cropping, natural vegetation and water bodies.

To guarantee the purity of endmembers, the Sentinel-2 images and historical images on the Google Earth platform were used to identify large fields to assist the manual endmember collection. High accuracy can be obtained by manually selecting endmember pixels through the visual interpretation method [49]. The pure pixels corresponding to the endmembers were widely distributed across the study area, and the amount reached about 0.5% of the total pixels.

2.2.3. Cropping Intensity Estimation

The LSMA model was used to implement the TMA method. The fully constrained LSMA model was used to estimate the abundance of each endmember. The unmixing was conducted in the time dimension by replacing the original spectral information with multi-temporal vegetation indices. The linear unmixing method assumed that the EVI temporal spectra of a pixel are a linear combination of each endmember. The formulas of linear spectral unmixing (1) and constraints (2) and (3) are as follows:

$$EVI_i = \Sigma_{j=1}^{n}\left(f_j EVI_{i,j}\right) + \varepsilon_i \quad (1)$$

$$\Sigma_{j=1}^{n} f_j = 1 \quad (2)$$

$$0 \le f_i \le 1 \quad (3)$$

where EVI_i is the EVI value for each phase i in the temporal EVI image, n is the number of end members, f_i is the fraction for each end member j, $EVI_{i,j}$ is the EVI value of endmember j in phase i (also the abundance of each endmember), and ε_i is the residual.

The abundances of all endmembers were under the constraint of being non-negative and added up to 1. The abundance of double-cropping was extracted and was regarded as cropping intensity, in which pixel value 0 represented abandoned cropland, and pixel value 1 represented homogeneous double-cropping areas.

2.2.4. Validation of the Model

In addition to visual examination, the estimated cropping intensity was validated by using the Sentinel-2 derived cropping intensity data, MCD12Q2 product, GCI dataset and GCI30 dataset as references.

The accuracy of the model output was evaluated using the coefficient of determination R^2, root mean square error (RMSE) of the linear fitting with respect to the reference data.

The validation was conducted at grid level and county level, respectively. On grid level, both the modeled result and the reference data were aggregated to 2 km fractural images by summarizing the values within each grid. On county level, zonal statistics were used to summarize the data to county level for those complete counties in the research area.

The software packages used in this research for constructing the feature space and the model validation were ENVI (ENVI version 4.8, Exelis Visual Information Solutions, Boulder, CO, USA) and MATLAB (MATLAB 2018a, The MathWorks, Inc., Natick, MA, USA).

3. Results

3.1. Cropping Intensity Map

The developed TMA method was used to estimate cropping intensity in the research area in 2018 (Figure 8). Relatively high cropping intensity could be found in the North China Plain and the center and north of Hubei Province. Furthermore, natural vegetation, water bodies, and built-up areas featured very low intensity values.

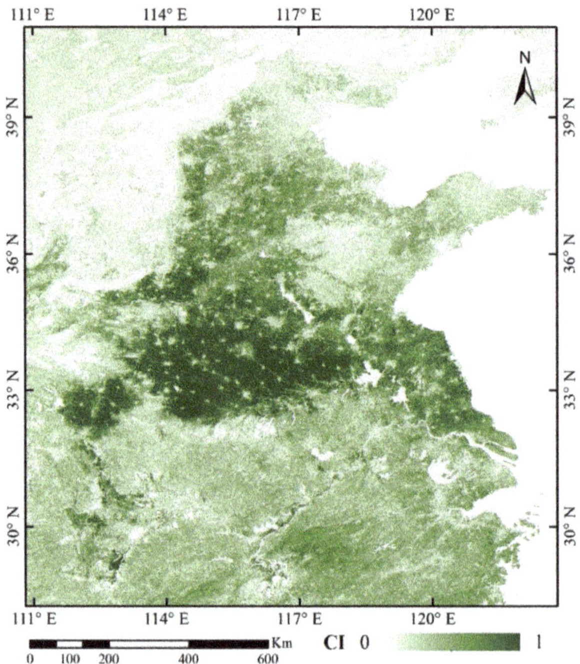

Figure 8. Cropping intensity of the study area in 2018.

A comparison of the estimated cropping intensity with MODIS false-color composite images and cropping frequency is presented in Figure 9. Since double-cropping areas had high EVI values on DOY 065 and DOY 209 and low EVI values on DOY 145 (Figure 4), they have a magenta tone in the false-color composite images. It can be observed that the high cropping intensity pixels in the estimated cropping intensity correlate well with those magenta pixels in the MODIS image. Additionally, the estimated cropping intensity avoids dividing the cropping intensity into fixed categories like cropping frequency, and thus contains detailed information.

3.2. Validation Results

When using linear fitting at the grid level, 1000 cases were randomly selected to analyze the agreement between the Sentinel-2 derived cropping intensity data and the estimated cropping intensity in the two test regions. The coefficient of determination (R^2) of the linear fitting reached 0.85 and 0.94 in the two test regions, respectively (Figure 10), suggesting that our cropping intensity result well captures the variations in the reference samples.

The agreement between the three cropping frequency datasets (MCD12Q2, GCI and GCI30) and estimated cropping intensity at the county level is presented in Figure 11. Their R^2 all achieved 0.93, demonstrating that the performance of our method is satisfactory (Figure 11).

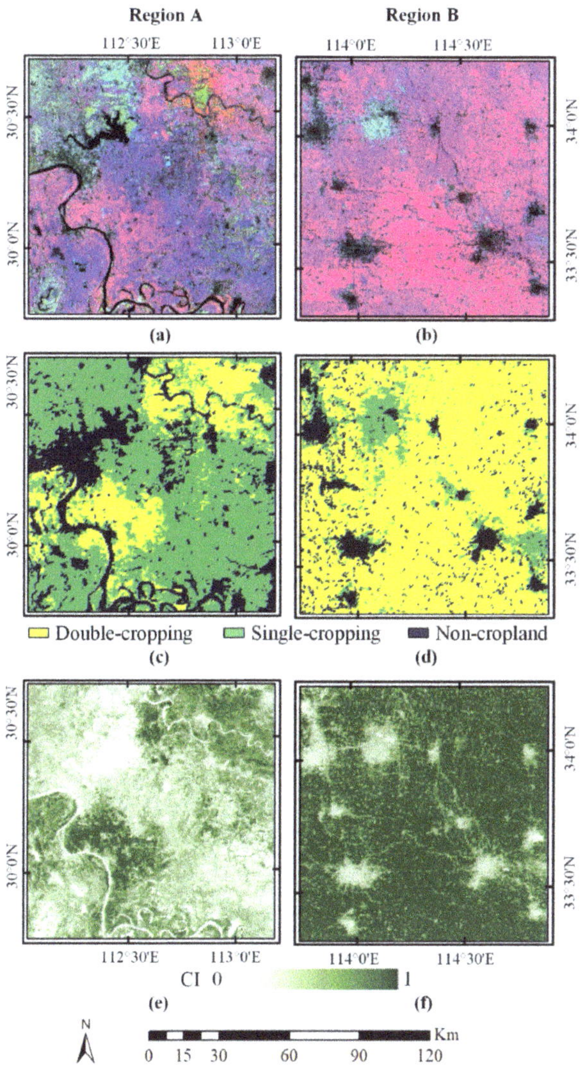

Figure 9. Visual comparison of cropping intensity in 2018; (**a**,**b**) are MODIS false-color composites (DOY 065, 145, 225); (**c**,**d**) are cropping frequency from MCD12Q2; (**e**,**f**) are the estimated cropping intensity for the test regions A and B.

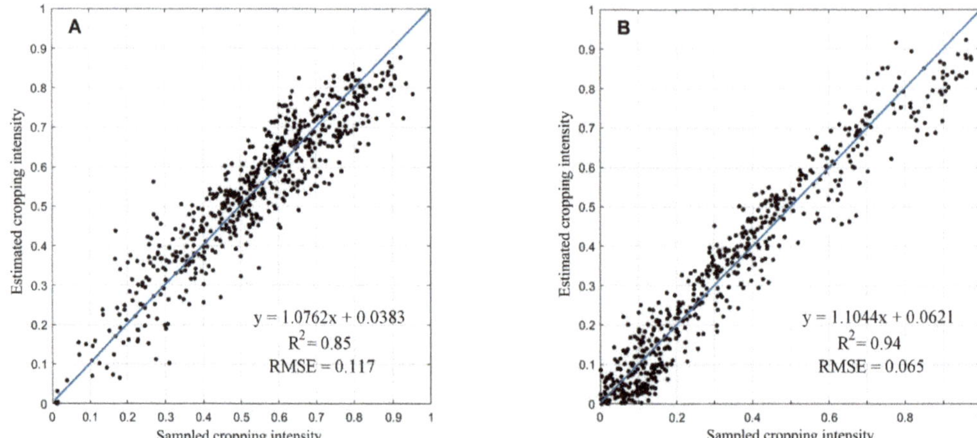

Figure 10. Linear fitting of the Sentinel-2 derived cropping intensity and estimated cropping intensity. (**A**,**B**) are two test regions.

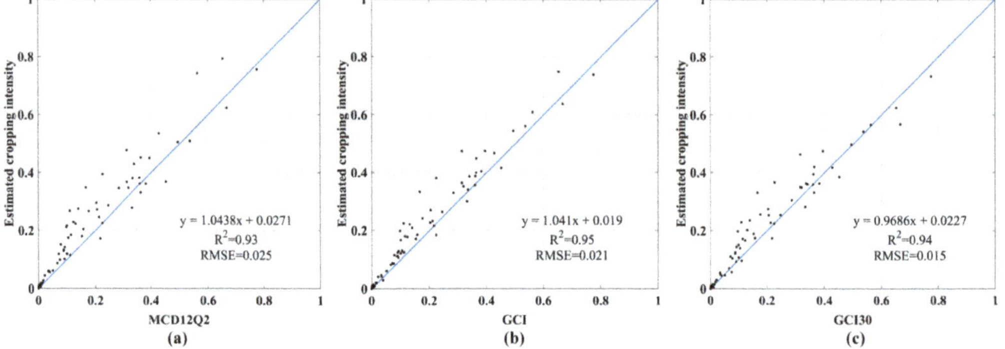

Figure 11. Linear fitting of the estimated cropping intensity and three cropping frequency datasets at the county level: (**a**) MCD12Q2, (**b**) GCI and (**c**) GCI30.

4. Discussions

4.1. Determining the Final Endmembers

To demonstrate the validity of the final endmembers, three PCA components were used to visually examine the distribution of the candidate endmembers in the feature space (Figure 12). Through rotating the feature space, different PCA component combinations could be obtained and be represented in this three-dimensional space. For the two-dimensional case, it was viewed by rotating the feature space to compress one of the dimensions. It was difficult to visualize the one-dimensional case, so it was carried out in the two-dimensional space and assumed to be projected into one of the features. PCA 2 was found to be the most ideal feature when one feature was used. Table 2 lists all possible combinations using the three PCA components.

Through the visual examination, we found that three models might meet our requirements (Table 2, in bold font). With the first model, it was difficult to meet the purity requirements for the endmembers because it lacked major land-cover types. The four-endmember model (the fifth model) did not fit the definition of cropping intensity in our study because single-cropping was also taken as one of the endmembers. Therefore, the three-endmember model using PCA 1 and PCA 2 was finally selected.

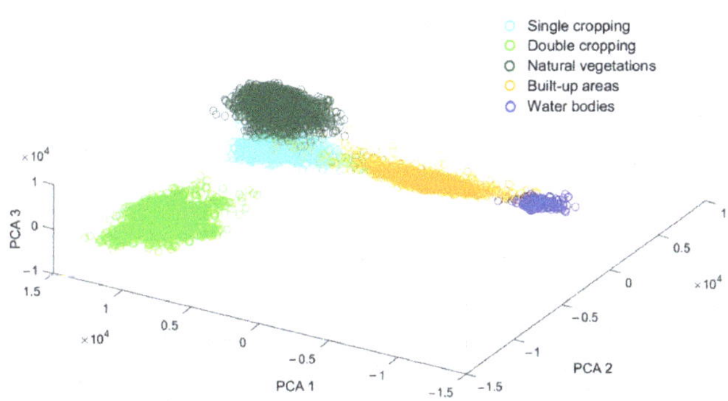

Figure 12. The three-dimensional feature space.

Table 2. Multiple endmember models.

Number of Features	Features	Types of Models	Endmembers
1	PCA 2	two-endmember model	**double-cropping, natural vegetation,**
2	PCA 1, PCA 3	two-endmember model	double-cropping, water bodies
2	PCA 1, PCA 2	three-endmember model	**double-cropping, natural vegetation, water bodies**
2	PCA 2, PCA 3	three-endmember model	double-cropping, single-cropping, natural vegetation
3	PCA 1, PCA 2, PCA 3	four-endmember model	**double-cropping, single-cropping, natural vegetation, water bodies**

4.2. Delineating the Endmembers Manually

Accurate and fast endmember selection is the key to the successful application of the method. Although the endmember selection based on high-resolution remote sensing images (named ESRS) could ensure the purity and representativeness of the endmembers, it suffered from the heavy load of work for pure pixel selection and was highly dependent on the availability of high-resolution images.

Therefore, we explored the feasibility of manually delineating endmembers (named MDE). This was achieved by drafting the vertex of the convex simplex through human–computer interaction in the feature space. We first delineated those pixels around the vertices as endmembers, as close to the vertices as possible. As this work was susceptible to the operator's knowledge of what a "vertex" is, the operation was repeated three times by covering different area sizes (Figure 13).

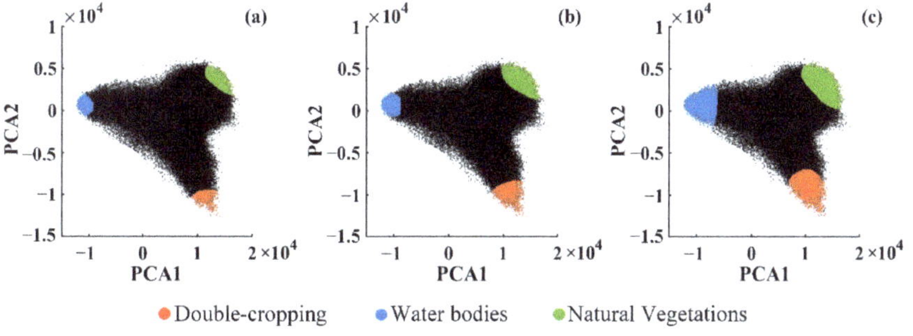

Figure 13. Delineating endmembers with different area sizes: size 1 (**a**), size 2 (**b**), size 3 (**c**). The sizes of the area gradually expand.

The estimated cropping intensity with different vertex sizes was compared with MCD12Q2 at the county level. The R^2 of the correlation between the estimation and MCD12Q2 were all above 0.87 (Figure 14), which confirmed that the proposed endmember selection method is applicable and robust. Compared with ESRS, MDE has the advantages of high efficiency and ease of implementation, and the accuracy and efficiency can be well-balanced.

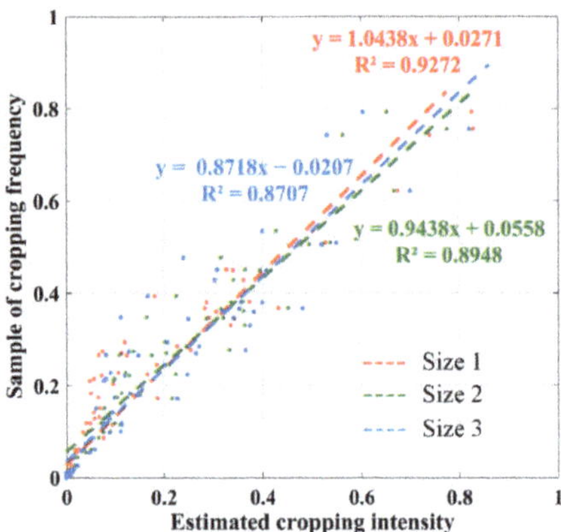

Figure 14. Correlation between MCD12Q2 and MDE estimated at the county level.

The accuracy of the unmixing decreased slightly when the sizes of endmember areas expanded (Figure 14). This is explainable since the closer the endmember is to the vertex of the convex simplex, the purer the endmember will be. Therefore, to ensure the accuracy, the delineating area of endmembers should be as small as possible and as close to the vertex as possible.

4.3. Unmixing in Regions with Different Sizes and Varied Endmember Land-Cover Types

The shape of the feature space varied when the research area covered different regions with different sizes. The shape of the feature space affected the unmixing accuracy since the approximate triangle was the basis of the method. Therefore, the relationships of the size of the research area and the completeness of the land-cover types with the unmixing accuracies were explored.

The estimated cropping intensity and MCD12Q2 were compared for test areas with different sizes (Figure 15). The mean cropping intensity values were compared at 10 km $\times$ 10 km block level. The area sizes, the corresponding feature space and the unmixing accuracies for each test area are given in Table 3.

The completeness of the three endmembers in the study area was the precondition for the successful application of this method. The three vertices in the feature spaces were obvious in all test areas with varied sizes. R^2 values were above 0.87 in all test areas, demonstrating that the size of the study area had little effect on the unmixing accuracy as long as the study area had all necessary land-cover types.

The effectiveness of the proposed method also depends on the completeness of endmember land-cover types. Our method can be applied directedly to the North China Plain and the middle and lower reaches of the Yangtze River Valley, which are double-cropping or double-single-mixed cropping areas with relatively large patches of croplands. The successful application of the method requires the concurrence of three land-cover types (double-cropping, natural vegetation and water bodies). The method will need further work (for example, to construct a new feature space, to find the new optimal endmembers again) when lacking any of the three necessary endmember land-cover types, such as those areas where crops have only one growing season (Table 4). However, expanding the test area does aid the inclusion of all three necessary endmember land-cover types.

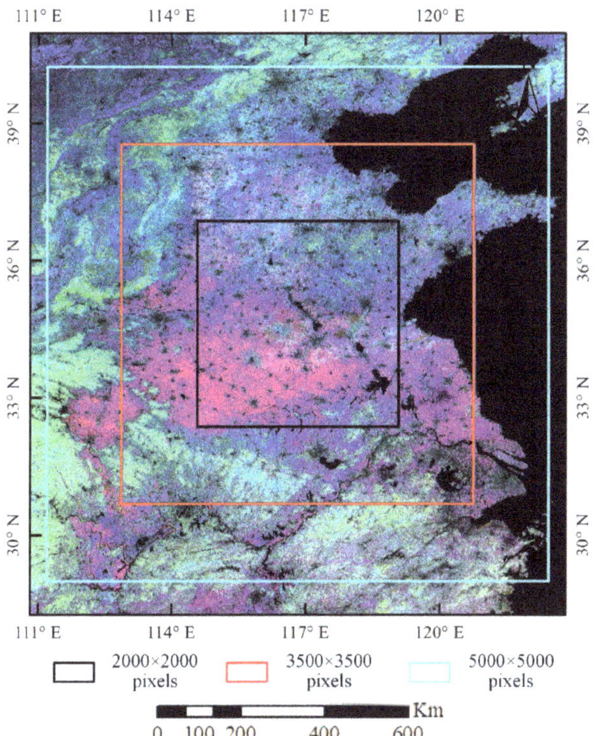

Figure 15. Test areas with different sizes.

Table 3. The unmixing accuracies for areas with different sizes.

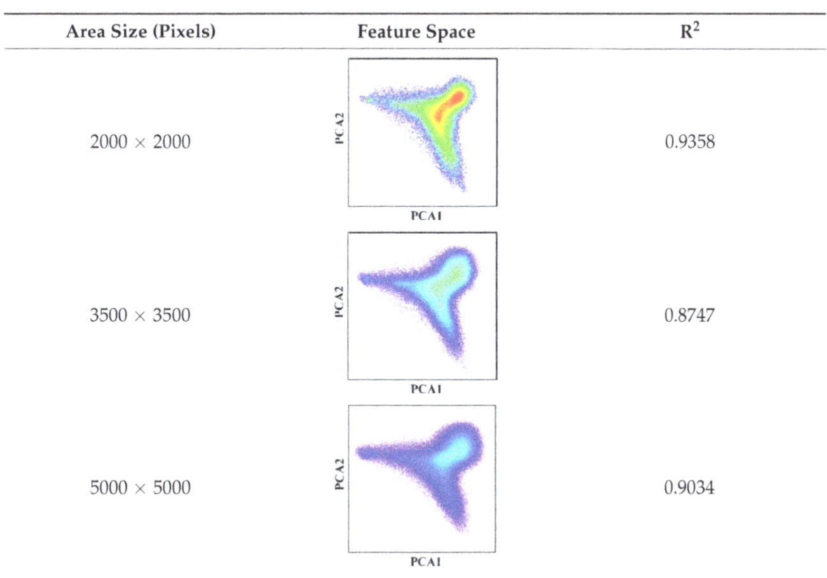

Area Size (Pixels)	Feature Space	R^2
2000 × 2000	PCA2 / PCA1	0.9358
3500 × 3500	PCA2 / PCA1	0.8747
5000 × 5000	PCA2 / PCA1	0.9034

Table 4. Completeness of endmembers when test areas are located in regions with varied land-cover types.

Endmembers	MODIS False-Color Composite (DOY 065, 145, 065)	Feature Space	Completeness of Endmembers
double-cropping, water bodies	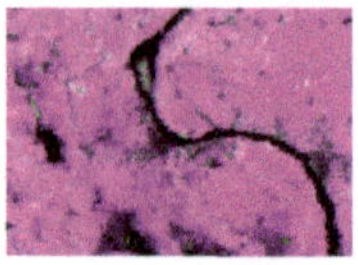	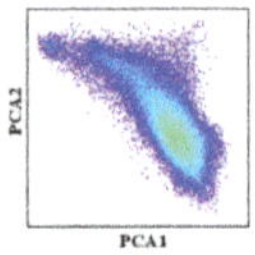 PCA2 PCA1	No
double-cropping, natural vegetation	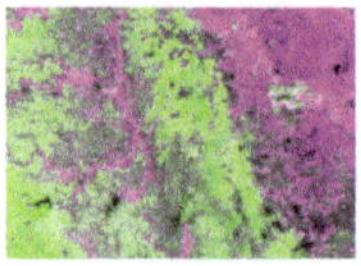	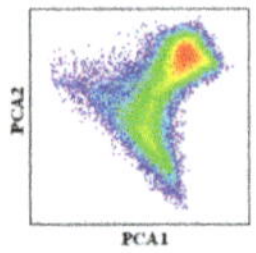 PCA2 PCA1	No
natural vegetation, water bodies	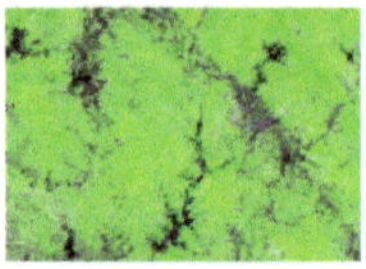	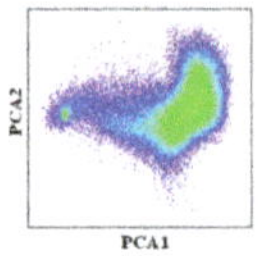 PCA2 PCA1	No
double-cropping, natural vegetation, water bodies	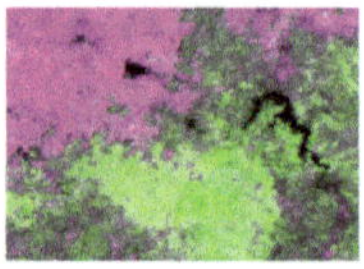	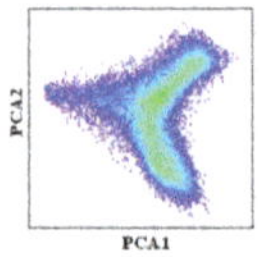 PCA2 PCA1	Yes

More effort is needed to find the proper endmembers when the method is extended to areas other than China. Since the interpretation for PCA images is scene-dependent and there has been no well outlined procedure for it, the interpretation needs more elaboration, and that is where innovation is possible.

The abundance of natural vegetation and water bodies can also be estimated as a "by-product" of this research. MODIS MOD13Q1 products within a year were used and the focus was cropping intensity estimation in this study. If the method is transplanted to time series vegetation indices with different temporal spans, other land-cover types can also be unmixed.

5. Conclusions

Fine cropping intensity mapping is essential for agricultural production and the sustainable development of agriculture. This study reports our work on developing a new method to estimate cropping intensity from time series remote sensing data for a specific region of China. A novel feature space was constructed, and three unique endmembers (double-cropping, natural vegetation and water bodies) were found. A new TMA method was developed to map cropping intensity at the abundance level. The estimated results were compared with sample data and cropping intensity product data at the pixel and county levels, respectively. The experiments demonstrated that the method is a highly accurate, semi-automatic, and easily implemented approach suitable for large-scale and long time-series cropping intensity mapping.

The study provided a novel method for cropping intensity estimation from historical archived time-series coarse-resolution remote sensing data. Firstly, a new TMA method was developed to conduct spatio-temporal continuous fine-resolution cropping intensity mapping from coarse-resolution remote sensing data. The phenology information was fully mined considering the seasonal variation in vegetation, including the phenological difference between crops and other land-cover types. Secondly, a unique feature space was constructed, along with three endmembers: double-cropping, natural vegetation and water bodies. The estimated results expressed crop extent and cropping intensity at the abundance level, improving the precision of cropping intensity estimation and avoiding dividing crop patterns rigidly into double-cropping or single-cropping. Thirdly, the MDE method has the advantage of high efficiency and ease of implementation, facilitating the endmember selection and unmixing process. The research provided insights into TMA-based cropping intensity mapping.

Future work will involve extending the method to a wider area and discussing the impact of regional differentiation on the unmixing accuracy.

Author Contributions: Conceptualization, J.T.; Methodology, J.T.; Validation, X.Z. and Q.J.; Data curation, X.Z. and Y.L.; Writing—original draft, Y.L.; Writing—review & editing, Y.Z. All authors have read and agreed to the published version of the manuscript.

Funding: This work was supported by the National Natural Science Foundation of China (Grant No. 41971371 and 32001368), the National Key Technologies Research and Development Program (Grant No. 2022YFB3903504) and the Fundamental Research Funds for the Central Universities (CCNU22JC022).

Data Availability Statement: No new data were created or analyzed in this study. Data sharing is not applicable to this article.

Acknowledgments: The authors appreciate the comments and suggestions from anonymous reviewers.

Conflicts of Interest: The authors declare no conflict of interest.

References

1. Zhu, H.; Sun, M. Main progress in the research on land use intensification. *Acta Geogr. Sin.* **2014**, *69*, 1346–1357.
2. Liu, G.; Zhang, L.; Zhang, Q.; Musyimi, Z. The response of grain production to changes in quantity and quality of cropland in yangtze river delta, China. *J. Sci. Food Agric.* **2015**, *95*, 480–489. [CrossRef]
3. Liu, J.; Kuang, W.; Zhang, Z.; Xu, X.; Qin, Y.; Ning, J.; Zhou, W.; Zhang, S.; Li, R.; Yan, C.; et al. Spatiotemporal characteristics, patterns, and causes of land-use changes in China since the late 1980s. *J. Geogr. Sci.* **2014**, *24*, 195–210. [CrossRef]
4. Tao, J.; Wang, Y.; Qiu, B.; Wu, W. Exploring cropping intensity dynamics by integrating crop phenology information using bayesian networks. *Comput. Electron. Agric.* **2022**, *193*, 106667. [CrossRef]
5. Su, Y.; Qian, K.; Lin, L.; Wang, K.; Guan, T.; Gan, M. Identifying the driving forces of non-grain production expansion in rural China and its implications for policies on cultivated land protection. *Land Use Policy* **2020**, *92*, 104435. [CrossRef]
6. Liu, X.; Herbert, S.J. Fifteen years of research examining cultivation of continuous soybean in northeast China: A review. *Field Crops Res.* **2002**, *79*, 1–7. [CrossRef]
7. Sun, J.; Mooney, H.; Wu, W.; Tang, H.; Tong, Y.; Xu, Z.; Huang, B.; Cheng, Y.; Yang, X.; Wei, D.; et al. Importing food damages domestic environment: Evidence from global soybean trade. *Proc. Natl. Acad. Sci. USA* **2018**, *115*, 5415–5419. [CrossRef]
8. Li, S.; Li, X.; Sun, L.; Cao, G.; Fischer, G.; Tramberend, S. An estimation of the extent of cropland abandonment in mountainous regions of China. *Land Degrad. Dev.* **2018**, *29*, 1327–1342. [CrossRef]
9. Peng, W.; Ma, N.L.; Zhang, D.; Zhou, Q.; Yue, X.; Khoo, S.C.; Yang, H.; Guan, R.; Chen, H.; Zhang, X. A review of historical and recent locust outbreaks: Links to global warming, food security and mitigation strategies. *Environ. Res.* **2020**, *191*, 110046. [CrossRef]
10. Mottaleb, K.A.; Kruseman, G.; Snapp, S. Potential impacts of ukraine-russia armed conflict on global wheat food security: A quantitative exploration. *Glob. Food Secur.* **2022**, *35*, 100659. [CrossRef]
11. Rahimi, P.; Islam, M.S.; Duarte, P.M.; Tazerji, S.S.; Sobur, M.A.; El Zowalaty, M.E.; Ashour, H.M.; Rahman, M.T. Impact of the COVID-19 pandemic on food production and animal health. *Trends Food Sci. Technol.* **2022**, *121*, 105–113. [CrossRef] [PubMed]
12. Yu, X.; Liu, C.; Wang, H.; Feil, J.-H. The impact of COVID-19 on food prices in China: Evidence of four major food products from beijing, shandong and hubei provinces. *China Agric. Econ. Rev.* **2020**, *12*, 445–458. [CrossRef]
13. Dongyu, Q. *Role and Potential of Potato in Global Food Security*; FAO: Rome, Italy, 2022.
14. Liu, W.; Yang, H.; Liu, Y.; Kummu, M.; Hoekstra, A.Y.; Liu, J.; Schulin, R. Water resources conservation and nitrogen pollution reduction under global food trade and agricultural intensification. *Sci. Total Environ.* **2018**, *633*, 1591–1601. [CrossRef] [PubMed]
15. Rachele, R. *Briefing-Desertification and Agriculture*; European Parliament: Strasbourg, France, 2020.
16. Xu, B.; Niu, Y.; Zhang, Y.; Chen, Z.; Zhang, L. China's agricultural non-point source pollution and green growth: Interaction and spatial spillover. *Environ. Sci. Pollut. Res.* **2022**, *29*, 60278–60288. [CrossRef] [PubMed]
17. Liu, Y.; Tang, H.; Muhammad, A.; Huang, G. Emission mechanism and reduction countermeasures of agricultural greenhouse gases—A review. *Greenh. Gases Sci. Technol.* **2019**, *9*, 160–174. [CrossRef]

18. Xie, H.; Liu, G. Spatiotemporal difference and determinants of multiple cropping index in China during 1998–2012. *Acta Geogr. Sin.* **2015**, *70*, 604–614.
19. Guo, Y.; Xia, H.; Pan, L.; Zhao, X.; Li, R. Mapping the northern limit of double cropping using a phenology-based algorithm and google earth engine. *Remote Sens.* **2022**, *14*, 1004. [CrossRef]
20. Jiang, M.; Xin, L.; Li, X.; Tan, M.; Wang, R. Decreasing rice cropping intensity in southern China from 1990 to 2015. *Remote Sens.* **2019**, *11*, 35. [CrossRef]
21. Löw, F.; Biradar, C.; Dubovyk, O.; Fliemann, E.; Akramkhanov, A.; Narvaez Vallejo, A.; Waldner, F. Regional-scale monitoring of cropland intensity and productivity with multi-source satellite image time series. *Gisci. Remote Sens.* **2018**, *55*, 539–567. [CrossRef]
22. Qiu, B.; Lu, D.; Tang, Z.; Song, D.; Zeng, Y.; Wang, Z.; Chen, C.; Chen, N.; Huang, H.; Xu, W. Mapping cropping intensity trends in China during 1982–2013. *Appl. Geogr.* **2017**, *79*, 212–222. [CrossRef]
23. Yan, H.; Liu, F.; Qin, Y.; Doughty, R.; Xiao, X. Tracking the spatio-temporal change of cropping intensity in China during 2000–2015. *Environ. Res. Lett.* **2019**, *14*, 035008. [CrossRef]
24. Pan, L.; Xia, H.; Yang, J.; Niu, W.; Qin, Y. Mapping cropping intensity in huaihe basin using phenology algorithm, all sentinel-2 and landsat images in google earth engine. *Int. J. Appl. Earth Obs. Geoinf.* **2021**, *102*, 102376. [CrossRef]
25. Liu, L.; Xiao, X.; Qin, Y.; Wang, J.; Xu, X.; Hu, Y.; Qiao, Z. Mapping cropping intensity in China using time series landsat and sentinel-2 images and google earth engine. *Remote Sens. Environ.* **2020**, *239*, 111624. [CrossRef]
26. Liu, C.; Zhang, Q.; Tao, S.; Qi, J.; Ding, M.; Guan, Q.; Wu, B.; Zhang, M.; Nabil, M.; Tian, F. A new framework to map fine resolution cropping intensity across the globe: Algorithm, validation, and implication. *Remote Sens. Environ.* **2020**, *251*, 112095. [CrossRef]
27. Tao, J.; Wu, W.; Xu, M. Using the bayesian network to map large-scale cropping intensity by fusing multi-source data. *Remote Sens.* **2019**, *11*, 168. [CrossRef]
28. Tao, J.; Wu, W.; Liu, W. Spatial-temporal dynamics of cropping frequency in hubei province over 2001–2015. *Sensors* **2017**, *17*, 2622. [CrossRef]
29. Qiu, B.; Li, W.; Tang, Z.; Chen, C.; Qi, W. Mapping paddy rice areas based on vegetation phenology and surface moisture conditions. *Ecol. Indic.* **2015**, *56*, 79–86. [CrossRef]
30. Tian, H.F.; Huang, N.; Niu, Z.; Qin, Y.C.; Pei, J.; Wang, J. Mapping winter crops in China with multi-source satellite imagery and phenology-based algorithm. *Remote Sens.* **2019**, *11*, 820. [CrossRef]
31. d'Andrimont, R.; Taymans, M.; Lemoine, G.; Ceglar, A.; Yordanov, M.; van der Velde, M. Detecting flowering phenology in oil seed rape parcels with sentinel-1 and-2 time series. *Remote Sens. Environ.* **2020**, *239*, 111660. [CrossRef]
32. Pok, S.; Matsushita, B.; Fukushima, T. An easily implemented method to estimate impervious surface area on a large scale from modis time-series and improved dmsp-ols nighttime light data. *ISPRS J. Photogramm. Remote Sens.* **2017**, *133*, 104–115. [CrossRef]
33. Chen, Y.; Yun, W.; Zhou, X.; Peng, J.; Li, S.; Zhou, Y. Classification and extraction of land use information in hilly area based on mesma and rf classifier. *Trans. Chin. Soc. Agric. Mach* **2017**, *48*, 136–144.
34. Ridd, M.K. Exploring a vis (vegetation-impervious surface-soil) model for urban ecosystem analysis through remote sensing: Comparative anatomy for cities. *Int. J. Remote Sens.* **1995**, *16*, 2165–2185. [CrossRef]
35. Ji, C.; Li, X.; Wei, H.; Li, S. Comparison of different multispectral sensors for photosynthetic and non-photosynthetic vegetation-fraction retrieval. *Remote Sens.* **2020**, *12*, 115. [CrossRef]
36. Pi, X.; Zeng, Y.; He, C. High-resolution urban vegetation coverage estimation based on multi-source remote sensing data fusion. *Natl. Remote Sens. Bull.* **2021**, *25*, 1216–1226. [CrossRef]
37. Wang, Y.; Zhuo, R.; Xu, L.; Fang, Y. A spatial-temporal bayesian deep image prior model for moderate resolution imaging spectroradiometer temporal mixture analysis. *Remote Sens.* **2023**, *15*, 3782. [CrossRef]
38. Caballero, I.; Stumpf, R.P. Towards routine mapping of shallow bathymetry in environments with variable turbidity: Contribution of sentinel-2a/b satellites mission. *Remote Sens.* **2020**, *12*, 451. [CrossRef]
39. Gray, J.; Sulla-Menashe, D.; Friedl, M.A. User Guide to Collection 6 Modis Land Cover Dynamics (mcd12q2) Product. In *NASA EOSDIS Land Process*; DAAC: Missoula, MT, USA, 2019; Volume 6, pp. 1–8.
40. Liu, X.; Zheng, J.; Yu, L.; Hao, P.; Chen, B.; Xin, Q.; Fu, H.; Gong, P. Annual dynamic dataset of global cropping intensity from 2001 to 2019. *Sci. Data* **2021**, *8*, 283. [CrossRef]
41. Zhang, M.; Wu, B.; Zeng, H.; He, G.; Liu, C.; Tao, S.; Zhang, Q.; Nabil, M.; Tian, F.; Bofana, J. Gci30: A global dataset of 30 m cropping intensity using multisource remote sensing imagery. *Earth Syst. Sci. Data* **2021**, *13*, 4799–4817. [CrossRef]
42. Boardman, J.W. Automated Spectral Unmixing of Aviris Data Using Convex Geometry Concepts. In Proceedings of the Jplairborne Geoscience Workshop, Washington, DC, USA, 25–29 October 1993.
43. Melaas, E.K.; Friedl, M.A.; Zhu, Z. Detecting interannual variation in deciduous broadleaf forest phenology using landsat tm/etm+ data. *Remote Sens. Environ.* **2013**, *132*, 176–185. [CrossRef]
44. Bellón, B.; Bégué, A.; Seen, D.L.; Almeida, C.A.d.; Simões, M. A remote sensing approach for regional-scale mapping of agricultural land-use systems based on ndvi time series. *Remote Sens.* **2017**, *9*, 600. [CrossRef]
45. Henderson, C.; Petersen, K.; Redak, R. Spatial and temporal patterns in the seed bank and vegetation of a desert grassland community. *J. Ecol.* **1988**, *76*, 717–728. [CrossRef]
46. Wang, T.; Kou, X.; Xiong, Y.; Mou, P.; Wu, J.; Ge, J. Temporal and spatial patterns of ndvi and their relationship to precipitation in the loess plateau of China. *Int. J. Remote Sens.* **2010**, *31*, 1943–1958. [CrossRef]

47. Yang, C.; Wu, G.; Ding, K.; Shi, T.; Li, Q.; Wang, J. Improving land use/land cover classification by integrating pixel unmixing and decision tree methods. *Remote Sens.* **2017**, *9*, 1222. [CrossRef]
48. Singh, R.; Kumar, V. Evaluating automated endmember extraction for classifying hyperspectral data and deriving spectral parameters for monitoring forest vegetation health. *Environ. Monit. Assess.* **2022**, *195*, 72. [CrossRef]
49. Wang, H.; Wu, B.; Li, X.; Lu, S. Extraction of impervious surface in hai basin using remote sensing. *J. Remote Sens.* **2011**, *15*, 388–400. (In Chinese)

MDPI
St. Alban-Anlage 66
4052 Basel
Switzerland
www.mdpi.com

Remote Sensing Editorial Office
E-mail: remotesensing@mdpi.com
www.mdpi.com/journal/remotesensing

www.ingramcontent.com/pod-product-compliance
Lightning Source LLC
LaVergne TN
LVHW070201170726
843515LV00007B/1975

* 9 7 8 3 0 3 6 5 9 4 4 6 0 *